Dietary Supplements with Antioxidant Activity
Understanding Mechanisms and Potential Health Benefits

Food Chemistry, Function and Analysis

Series editors:
Gary Williamson, *Monash University, Australia*
Alejandro G. Marangoni, *University of Guelph, Canada*
Graham A. Bonwick, *AgriFoodX Limited, UK*
Catherine S. Birch, *AgriFoodX Limited, UK*

Titles in the series:
1: Food Biosensors
2: Sensing Techniques for Food Safety and Quality Control
3: Edible Oil Structuring: Concepts, Methods and Applications
4: Food Irradiation Technologies: Concepts, Applications and Outcomes
5: Non-extractable Polyphenols and Carotenoids: Importance in Human Nutrition and Health
6: Cereal Grain-based Functional Foods: Carbohydrate and Phytochemical Components
7: Steviol Glycosides: Cultivation, Processing, Analysis and Applications in Food
8: Legumes: Nutritional Quality, Processing and Potential Health Benefits
9: Tomato Chemistry, Industrial Processing and Product Development
10: Food Contact Materials Analysis: Mass Spectrometry Techniques
11: Vitamin E: Chemistry and Nutritional Benefits
12: Anthocyanins from Natural Sources: Exploiting Targeted Delivery for Improved Health
13: Carotenoid Esters in Foods: Physical, Chemical and Biological Properties
14: Eggs as Functional Foods and Nutraceuticals for Human Health
15: Rapid Antibody-based Technologies in Food Analysis
16: DNA Techniques to Verify Food Authenticity: Applications in Food Fraud
17: Advanced Gas Chromatography in Food Analysis
18: Handbook of Food Structure Development
19: Mitigating Contamination from Food Processing
20: Biogenic Amines in Food: Analysis, Occurrence and Toxicity
21: Nutrition and Cancer Prevention: From Molecular Mechanisms to Dietary Recommendations
22: Health Claims and Food Labelling
23: Nutraceuticals and Human Health: The Food-to-supplement Paradigm
24: Nutritional Signalling Pathway Activities in Obesity and Diabetes
25: The Chemistry and Bioactive Components of Turmeric
26: Foodomics
27: Food Proteins and Peptides: Emerging Biofunctions, Food and Biomaterial Applications
28: Handbook of Antioxidant Methodology: Approaches to Activity Determination
29: Fats and Associated Compounds: Consumption and Human Health

30: Oral Processing and Consumer Perception: Biophysics, Food Microstructures and Health
31: Development of Trans-free Lipid Systems and their Use in Food Products
32: Advanced Spectroscopic Techniques for Food Quality
33: Berries and Berry Bioactive Compounds in Promoting Health
34: Metabolism of Nutrients by Gut Microbiota
35: Bio- and Nano-sensing Technologies for Food Processing and Packaging
36: Edible Fungi: Chemical Composition, Nutrition and Health Effects
37: Chemistry and Nutritional Effects of Capsicum
38: Pigmented Cereals and Millets: Bioactive Profile and Food Applications
39: Dietary Supplements with Antioxidant Activity: Understanding Mechanisms and Potential Health Benefits

How to obtain future titles on publication:
A standing order plan is available for this series. A standing order will bring delivery of each new volume immediately on publication.

For further information please contact:
Book Sales Department, Royal Society of Chemistry, Thomas Graham House, Science Park, Milton Road, Cambridge, CB4 0WF, UK
Telephone: +44 (0)1223 420066, Fax: +44 (0)1223 420247,
Email: booksales@rsc.org
Visit our website at www.rsc.org/books

Dietary Supplements with Antioxidant Activity
Understanding Mechanisms and Potential Health Benefits

Edited by

Cesarettin Alasalvar
TÜBİTAK Marmara Research Center, Turkey
Email: cesarettin.alasalvar@tubitak.gov.tr

Fereidoon Shahidi
Memorial University of Newfoundland, Canada
Email: fshahidi@mun.ca

and

Chi-Tang Ho
Rutgers University, USA
Email: ctho@sebs.rutgers.edu

ROYAL SOCIETY OF CHEMISTRY

Food Chemistry, Function and Analysis No. 39

Print ISBN: 978-1-83916-262-6
PDF ISBN: 978-1-83916-611-2
EPUB ISBN: 978-1-83916-612-9
Print ISSN: 2398-0656
Electronic ISSN: 2398-0664

A catalogue record for this book is available from the British Library

The Royal Society of Chemistry is a charity, registered in England and Wales, Number 207890, and a company incorporated in England by Royal Charter (Registered No. RC000524), registered office: Burlington House, Piccadilly, London W1J 0BA, UK, Telephone: +44 (0) 20 7437 8656.

Visit our website at www.rsc.org/books

Printed in the United Kingdom by CPI Group (UK) Ltd, Croydon, CR0 4YY, UK

Preface

Antioxidant supplementation, either in the form of tablets/capsules and potions or in foods, is based on the notion that reactive oxygen species and associated free radicals contribute to many human diseases by causing oxidative stress. Thus, decreasing oxidative damage may help prevent their occurrence. Therefore, the assumption that antioxidants are good for health and that consuming antioxidants will improve health is commonly accepted by the general public.

The overall antioxidant defence in the human body depends on endogenously synthesized antioxidant enzymes, which serve as the primary defence, whereas diet-derived antioxidants (food and supplements) are categorized as secondary defence lines. Many diet-derived substances (such as vitamins, minerals, cofactors, certain fatty acids, and bioactive phytochemicals, among others) may act as antioxidants. Consumption of natural foods rich in antioxidants such as fruits, vegetables, nuts, grains, and medicinal plants may provide a better option for health promotion and disease risk reduction compared to taking high doses of dietary antioxidant supplements.

This book consists of 16 chapters including an introductory chapter (Chapter 1). The global regulations and health claims on natural dietary antioxidant supplements are comprehensively screened in Chapter 2. The multifunctional health benefits (cardiovascular health, vascular function, adiposity/weight control, bone health, gastrointestinal and gut health/microbiota, antiviral, antiaging, brain health, non-alcoholic liver disease, various types of cancers, immune health, diabetes, and lipid supplements, among others) of the dietary antioxidant supplements are reviewed thoroughly (Chapters 3–16).

Food Chemistry, Function and Analysis No. 39
Dietary Supplements with Antioxidant Activity: Understanding Mechanisms and Potential Health Benefits
Edited by Cesarettin Alasalvar, Fereidoon Shahidi and Chi-Tang Ho

Published by the Royal Society of Chemistry, www.rsc.org

We are most grateful to the contributors to this book, who are internationally renowned researchers, for their comprehensive account of the global perspective on the issues of concern related to dietary supplements with antioxidant activity. Their contributions will serve as a major resource for those interested in dietary supplements with antioxidant activity. Biochemists, food scientists, dietitians, nutritionists, and health professionals including medical doctors from academia, government, and nutrition clinics as well as industry will find the contents of this book of much interest. Although this book serves primarily as a reference manual, it also summarizes the current state of knowledge in a key research area and contains novel ideas for future research and development. In addition, it provides an easy-to-read text suitable for teaching senior undergraduate and postgraduate students in the relevant areas. Dietary supplement manufacturers, decision makers, regulatory bodies, and exporters/importers will obtain optimum benefit from this publication. Finally, we trust that this book paves the way for better appreciation of the concepts, products, and opportunities in the field for professionals, regulators, manufacturers, and consumers.

Cesarettin Alasalvar
Fereidoon Shahidi
Chi-Tang Ho

Contents

Food Chemistry, Function and Analysis No. 39
Dietary Supplements with Antioxidant Activity: Understanding Mechanisms and Potential Health Benefits
Edited by Cesarettin Alasalvar, Fereidoon Shahidi and Chi-Tang Ho

Published by the Royal Society of Chemistry, www.rsc.org

CHAPTER 1

Dietary Supplements with Antioxidant Activity – Understanding Mechanisms and Potential Health Benefits: An Overview

CESARETTIN ALASALVAR,*[a] SUI KIAT CHANG,[b] CHI-TANG HO[c] AND FEREIDOON SHAHIDI[d]

[a] TÜBİTAK Marmara Research Center, Life Sciences, Gebze-Kocaeli, Turkey; [b] Department of Allied Health Sciences, Faculty of Science, Universiti Tunku Abdul Rahman, Kampar, Perak, Malaysia; [c] Department of Food Science, Rutgers University, New Brunswick, NJ, USA; [d] Department of Biochemistry, Memorial University of Newfoundland, St. John's, NL, Canada
*Email: cesarettin.alasalvar@tubitak.gov.tr

1.1 Introduction

Antioxidant supplementation, either in the form of tablets/capsules and potions or in foods, is based on the notion that reactive oxygen species (ROS) and other free radicals contribute to many human diseases by causing oxidative stress and that decreasing oxidative damage may help to prevent their occurrence. Therefore, the assumption that antioxidants are good for health

Food Chemistry, Function and Analysis No. 39
Dietary Supplements with Antioxidant Activity: Understanding Mechanisms and Potential Health Benefits
Edited by Cesarettin Alasalvar, Fereidoon Shahidi and Chi-Tang Ho

Published by the Royal Society of Chemistry, www.rsc.org

and that consuming antioxidants will improve health is commonly accepted by the general public.[1]

Many diet-derived substances can act as antioxidants. The most familiar ones are some vitamins (such as vitamins C, E, K, β-carotene, and folic acid) and minerals (such as selenium and zinc), along with cofactors (such as coenzyme Q10). There are also many different classes of bioactive phytochemicals (such as alkaloids, betalains, carotenoids, flavonoids, phenolic acids, coumarins, phytoestrogens, non-protein nitrogen compounds, organosulfur compounds, stilbenes, and tannins, among others) which have been reported to possess strong antioxidant activities.[2–7] Moreover, long-chain omega-3 fatty acids such as eicosapentaenoic acid (EPA) and docosahexaenoic acid (DHA) from marine sources[8] and essential oils from medicinal herbs[9] either exert an antioxidant effect or are involved in the oxidation process that benefits from the use of antioxidants. Figure 1.1 shows the representative diet-derived antioxidant substances and their health outcomes.

In addition to diet-derived antioxidants, there are also endogenously-synthesized primary (such as superoxide dismutase, catalase, and glutathione peroxidase) and secondary (such as glutathione reductase enzymes) antioxidant enzymes. The overall antioxidant defence in the human body depends on endogenously-synthesized antioxidant enzymes, which serve as the primary defence, whereas diet-derived antioxidants (food and supplements) are categorized as secondary defence.[1,10,11] This contribution provides some important highlights about antioxidants from dietary supplements and natural sources by discussing their potential mechanisms and health benefits.

1.2 Antioxidants from Dietary Supplements

Excessive free radicals generated *in vivo* increase cellular oxidative stress. Chronic oxidative stress is a common pathophysiological condition that causes chronic oxidative damage and finally inflammation. Chronic inflammation is the major determinant of various non-communicable diseases (NCDs) later in human life, contributing to ageing, cardiovascular disease (CVD), diabetes, cancers, obesity, and neurodegenerative diseases (NDD) such as dementia and Alzheimer's disease.[10,11] Hence, taking dietary antioxidants in the form of food or supplements decreases oxidative damage *in vivo* and may help to prevent the severity of NCDs.[1,11]

However, the dosage of administered dietary antioxidants may not always be accurate as this often exceeds the recommended dietary allowances (RDA) when used as a supplement. Thus, lower doses and/or mixtures of antioxidants are superior to high doses of single agents.[10,12] For example, consumption of foods rich in vitamin C was shown to decrease oxidative damage physiologically, whereas the intake of vitamin C alone at high dosage did not. In this relation, a study demonstrated an increased plasma vitamin C level by 60% after daily supplementation with 500 mg of

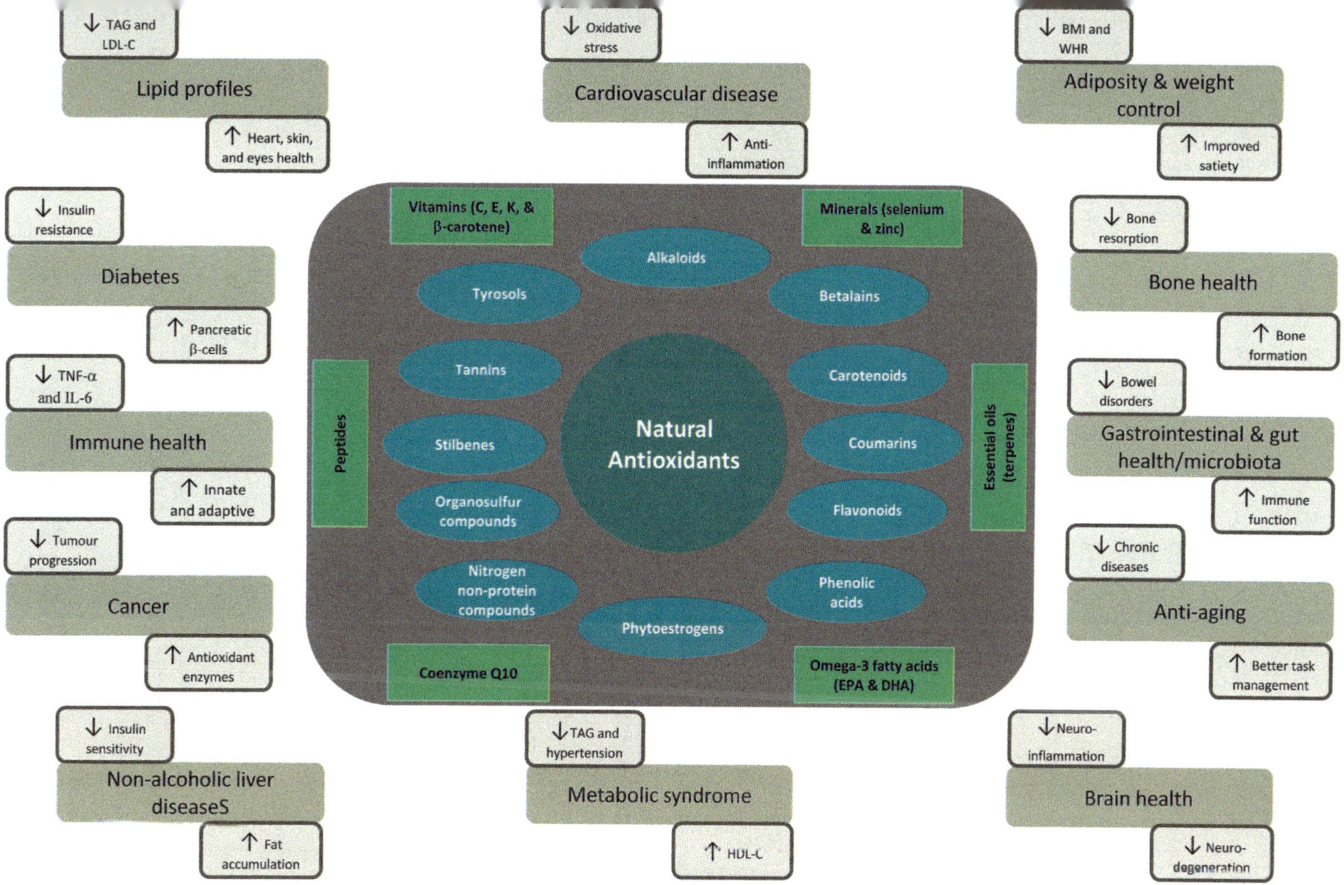

Figure 1.1 Representative diet-derived antioxidant substances and their health outcomes. Abbreviations: BMI, body mass index; HDL-C, high-density lipoprotein cholesterol; IL-6, interleukin-6; LDL-C, low-density lipoprotein cholesterol; TAG, triacylglycerols; TNF-α, tumour necrosis factor-alpha; WHR, waist–hip ratio.

vitamin C for 6 weeks. However, the pro-oxidant activity of vitamin C was also noticed as significantly higher levels of 8-oxoadenine in DNA were present in blood lymphocytes.[13] In fact, the use of vitamin C at a dosage of 1.000 mg for optimal health far exceeded the RDA of 75–90 mg per day.[14] Vitamin E, in the form of α-tocopherol, becomes a pro-oxidant at high concentrations (>700 ppm), both in food and in the body.[15]

Polyphenols are the most active naturally occurring antioxidants.[16] Most dietary polyphenols have both antioxidant and prooxidant properties depending on the dose and environment.[17] (−)-Epigallocatechin-3-gallate (EGCG) has been used as a dietary supplement for the control of obesity. However, it triggers adverse reactions, such as hepatotoxicity, among some consumers.[18] The prooxidant capacity of a polyphenol largely relies on its autoxidation, its ability to produce ROS, including hydrogen peroxide. Oxidation of phenolic compounds, such as gallic acid, (−)-epigallocatechin (EGC), catechin, quercetin, hydroxytyrosol, rosmarinic acid, and delphinidin, as well as thiol compounds, generates hydrogen peroxide which contributes to cellular oxidative stress.[19,20] Transition metal ions, such as copper, promote EGCG oxidation and produce highly harmful hydroxyl radicals.[21] Resveratrol, a stilbene with high antioxidant capacity, has a pro-oxidant effect in the presence of copper.[22] It is well known that every compound has negative effects if overdosed. This is the same for antioxidants. The consumption of high doses of antioxidants could be the main reason for the increased toxicity reported, leading to prooxidant effects or "antioxidative stress".[23]

There are numerous meta-analyses, systematic reviews, observational studies, and human intervention trials that demonstrate no beneficial effect of multi-vitamin or multi-mineral antioxidant supplements on all-cause mortality, cancer, CVD, and cognitive functions, among others.[24–29] Moreover, it was found that β-carotene, vitamin E, and possibly high doses of vitamin A supplements may be associated with higher all-cause mortality.[29]

In reality, certain antioxidant supplements demonstrate positive effects and good efficacy. Many supplements, including vitamins, minerals, and other functional ingredients such as probiotics, are important components of modern health care.[30] Dietary supplements are important in preventing vitamin and mineral deficiencies, and some studies have demonstrated that a specific combination of vitamins and minerals can delay progression of early age-related macular degeneration.[24] However, supplements that are useful in treating certain conditions are frequently overused among the general population to improve or maintain health.[27,30] Therefore, full attention needs to be paid to the dosages of all supplements consumed and moderate rather than mega-dose consumption of antioxidants (in the form of tablets or capsules) is recommended.[1]

1.3 Antioxidants from Natural Sources

Natural antioxidants are widely distributed in foods, beverages, and medicinal plants. The notion of "Let food be thy medicine and medicine be thy food" advocated by Hippocrates some 2500 years ago is still relevant today. It is a known fact that all foods are functional since they provide nutrients, flavour (taste and aroma), and bioactive phytochemicals.[16]

Consumption of fruits, vegetables, beverages, cereals, grains/seeds, nuts, mushrooms, spices, and medicinal herbs, among others, helps maintain good health and delay disease occurrence. In addition, industrial processing of agricultural by-products also provides potentially important natural sources of antioxidants.[1,7,31,32] The intake of 3–5 daily servings of fruits and vegetables would protect against NCDs such as coronary heart disease (CHD), cancer, NDD, diabetes, and obesity, among others.[33] This is because fruits and vegetables are chemically complex, where their health benefits could arise from the overall effects of several compounds or mixtures of components present, including vitamins, minerals, fibres, and bioactive phytochemicals. This mechanism is known as a synergistic action where the observed effects exceed the sum of individual effects in complex biological systems.[1,16]

These natural antioxidants, especially polyphenols and carotenoids, exhibit a wide range of biological effects, including anti-inflammatory, anti-aging, anti-atherosclerosis, and anti-cancer.[3,6] The effective extraction and proper assessment of antioxidants from food and medicinal plants are crucial to explore the potential antioxidant sources and promote the application in functional foods, pharmaceuticals, and food additives.[32]

1.4 Mechanism of Action of Antioxidant Compounds/Supplements

The potential mechanism of protective action of antioxidant compounds is thought to be due to their direct scavenging of free radicals that include ROS, reactive nitrogen species (RNS), and lipid peroxides.[16] Besides, antioxidant compounds inhibit the initiation step or interrupt the propagation step of lipid oxidation, thus reducing the formation of volatile decomposition products (*e.g.*, aldehydes and ketones) that cause rancidity. This action prolongs the oxidative stability of food.[16] Antioxidant compounds alleviate oxidative damage by activating endogenous antioxidant enzymes (superoxide dismutase, glutathione peroxidase, and catalase).[1] Mitochondria are not only the energy providers to maintain cell life activities, but also the regulators of cellular oxidative stress. Thus, changes in mitochondrial function are closely related to oxidative damage to cells.[34] Antioxidant compounds have been shown to alleviate oxidative damage by regulating mitochondrial function by adjusting the protein and gene expression levels of Bax, Bcl-2, caspase-3, and cleaved caspase-3.[34] Moreover, antioxidant compounds regulate the phosphatidylinositol 3-kinase (PI3K)/Akt and mitogen-activated protein kinase (MAPK) signaling pathways to upregulate genes [MAPK kinases, extracellular-signal related kinase (ERK)1/2, and c-Jun N-terminal kinase (JNK)1/2] related to oxidative stress, thus reducing the production of pro-inflammatory cytokines, which in turn decrease endogenous inflammation.[8,35] This, in turn, enhances the phosphorylation level of PI3K/Akt and its activity, which activates NF-E2-related factor 2 (Nrf2). Then, Nrf2 further promotes the antioxidant enzymes [quinone oxidoreductase 1 (NQO1) and heme oxygenase-1 (HO-1)] to remove ROS.[36] The available evidence indicates that the induction of HO-1 is in conjunction with neuroprotection against various stimuli, including glutamate and erastin.[37] In this relation, antioxidant compounds also enhance the phosphorylation level of MAPK to remove ROS. Hence, antioxidants recover the mitochondrial function to make cells metabolize and function normally.[34] Figure 1.2 summarizes the complex mechanisms of action of antioxidant compounds from food/supplements from previous studies.

There are several misconceptions about antioxidants and free radicals in human physiology. Humans have a well-balanced pool of free radicals and antioxidants that allow some free radicals to perform useful functions while alleviating oxidative damage *in vivo*. ROS detoxify pathogens as part of the human innate immune system.[38] In addition, ROS also detoxify some infectious microorganisms. Phagocytes use ROS to kill pathogens in innate immunity.[11,38] In fact, free radicals are needed by stem cells for proper functioning. However, excessive free radicals may affect their function.[39]

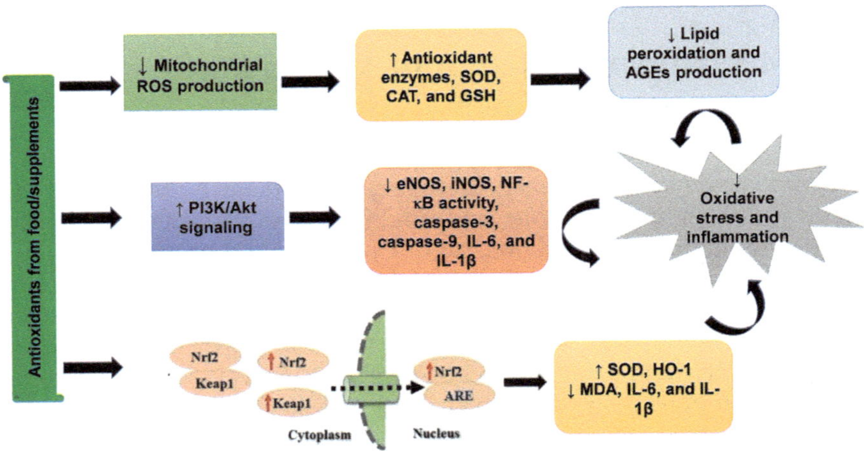

Figure 1.2 The complex mechanisms of action of antioxidant compounds from food/supplements. Abbreviations: AGEs, advanced glycation end-products; ARE, antioxidant response element; CAT, catalase; eNOS, endothelial nitric oxide synthase; GSH, glutathione; HO-1, heme oxygenase 1; IL-6, interleukin-6; IL-1β, interleukin-1β; iNOS, inducible nitric oxide synthase; Keap1, Kelch-like ECH-associated protein 1; MDA, malondialdehyde; NF-κB, nuclear factor kappa-light-chain-enhancer of activated B cells; Nrf2, nuclear factor erythroid 2-related factor 2; PI3K, phosphatidylinositol 3-kinase; ROS, reactive oxygen species; SOD, superoxide dismutase.

The overall antioxidant defence in the human body depends mainly on the endogenous antioxidant enzymes, such as reduced glutathione, catalase, peroxiredoxins, and superoxide dismutase. Meanwhile, diet-derived antioxidants (*e.g.*, vitamins A, C, and E, and phenolic compounds) are classified as secondary defence. The antioxidant defence system functions to minimize the levels of most harmful ROS on one hand, while allowing adequate levels of ROS for their useful functions, such as cell signaling and redox regulation. Therefore, the antioxidant defence system will be more effective if it is stimulated with some weak or mild prooxidants or free radicals. This, in turn, stimulates the level of endogenous antioxidants in the body, such as glutathione and superoxide dismutase to combat infections.[11,12] On the contrary, over-consumption of antioxidants could down-regulate important endogenous antioxidant enzymes, depressing the immune system partially.[11,12]

1.5 Conclusion

Antioxidant compounds from food/supplements exert their effects by directly scavenging free radicals, activating endogenous antioxidant enzymes, regulating mitochondrial function, and reducing cellular inflammation. Consumption of natural foods rich in antioxidants such as fruits, vegetables, nuts, grains, and medicinal plants may provide a better option for health promotion and disease risk reduction compared to taking high dosages of dietary antioxidant supplements. Since natural foods are chemically complex and their health benefits could arise from the overall effects of several bioactive compounds of mixtures of components present, including vitamins, minerals, fibres, and bioactive phytochemicals. High-dose antioxidant supplements may not always benefit human health, while moderate presence of prooxidants in the body may sometimes be helpful for stimulating the overall immune response to external hazardous compounds.

References

1. S. K. Chang, C. Alasalvar and F. Shahidi, Weighing the benefits of dietary antioxidant supplements, *Food Technol.*, 2018, 72, 44–53.
2. I. Kostova, S. Bhatia, P. Grigorov, S. Balkansky, V. S. Parmar and A. K. Prasad, *et al.*, Coumarins as antioxidants, *Curr. Med. Chem.*, 2011, **18**, 3929–3951.
3. A. N. Li, S. Li, Y. J. Zhang, X. R. Xu, Y. M. Chen and H. B. Li, Resources and biological activities of natural polyphenols, *Nutrients*, 2014, **6**, 6020–6047.
4. A. Dasgupta and K. Klein, Fruits, vegetables, and nuts: Good sources of antioxidants, in *Antioxidants in Food, Vitamins and Supplements – Prevention and Treatment of Disease*, ed. A. Dasgupta and K. Klein, Elsevier, Oxford, UK, 2014, pp. 209–235.
5. A. Dasgupta and K. Klein, Antioxidant vitamins and minerals, in *Antioxidants in Food, Vitamins and Supplements – Prevention and Treatment*

of Disease, ed. A. Dasgupta and K. Klein, Elsevier, Oxford, UK, 2014, pp. 277–294.

6. B. P. Arathi, P. Raghavendra-Rao Sowmya, K. Vijay, V. Baskaran and R. Lakshminarayana, Metabolomics of carotenoids: The challenges and prospects – A review, *Trends Food Sci. Technol.*, 2015, **45**, 105–117.
7. C. Alasalvar, S. K. Chang, B. Bolling, W. Y. Oh and F. Shahidi, Specialty seeds: Nutrients, bioactives, bioavailability, and health benefits: A comprehensive review, *Compr. Rev. Food Sci. Food Saf.*, 2021, **20**, 2382–2427.
8. E. J. Anderson, K. A. Thayne, M. Harris, S. R. Shaikh, T. M. Darden and D. S. Lark, *et al.*, Do fish oil omega-3 fatty acids enhance antioxidant capacity and mitochondrial fatty acid oxidation in human atrial myocardium *via* PPARγ activation?, *Antioxid. Redox Signaling*, 2014, **21**, 1156–1163.
9. S. M. B. Hashemi, S. B. Khorram and M. Sohrabi, Antioxidant activity of essential oils in foods, in *Essential Oils in Food Processing: Chemistry, Safety and Applications*, ed. S. M. B. Hashemi, A. M. Khaneghah and A. de Souza Sant'Ana, John Wiley & Sons Ltd., Oxford, UK, 2018, pp. 247–265.
10. A. Bast and G. R. Haenen, Ten misconceptions about antioxidants, *Trends Pharmacol. Sci.*, 2013, **34**, 430–436.
11. B. Halliwell, Free radicals and antioxidants – quo vadis?, *Trends Pharmacol. Sci.*, 2011, **32**, 125–130.
12. B. Halliwell, Reflections of an aging free radical, *Free Radicals Biol. Med.*, 2020, **161**, 234–245.
13. I. D. Podmore, H. R. Griffiths, K. E. Herbert, N. Mistry, P. Mistry and J. Lunec, Vitamin C exhibits pro-oxidant properties, *Nature*, 1998, **392**, 559.
14. Institute of Medicine, Ascorbic acid, in *Dietary Reference Intakes for Ascorbic Acid, Vitamin E, Selenium, and Carotenoids*, The National Academy Press, Washington, DC, 2000, pp. 95–185.
15. F. Shahidi and A. C. de Camargo, Tocopherols and tocotrienols in common and emerging dietary sources: Occurrence, applications, and health benefits, *Int. J. Mol. Sci.*, 2016, **17**, 1745–1773.
16. F. Shahidi and P. Ambigaipalan, Phenolics and polyphenolics in foods, beverages and spices: Antioxidant activity and health effects – A review, *J. Funct. Foods*, 2015, **18**, 820–897.
17. J. D. Lambert and R. J. Elias, The antioxidant and pro-oxidant activities of green tea polyphenols: a role in cancer prevention, *Arch. Biochem. Biophys.*, 2010, **501**, 65–72.
18. V. J. Navarro, H. L. Bonkovsky, S.-I. Hwang, M. Vega, H. Barnhart and J. Serrano, Catechins in dietary supplements and hepatotoxicity, *Dig. Dis. Sci.*, 2013, **58**, 2682–2690.
19. M. Grzesik, G. Bartosz, I. Stefaniuk, M. Pichla, J. Namieśnik and I. Sadowska-Bartosz, Dietary antioxidants as a source of hydrogen peroxide, *Food Chem.*, 2019, **278**, 692–699.

20. L. H. Long, A. Hoi and B. Halliwell, Instability of, and generation of hydrogen peroxide by phenolic compounds in cell culture media, *Arch. Biochem. Biophys.*, 2010, **501**, 162–169.
21. J. Wang, X. Wang, Y. He, L. Jia, C. S. Yang and R. J. Reiter, *et al.*, Antioxidant and pro-oxidant activities of melatonin in the presence of copper and polyphenols *in vitro* and *in vivo*, *Cells*, 2019, **8**, 903.
22. C. A. de la Lastra and I. Villegas, Resveratrol as an antioxidant and pro-oxidant agent: mechanisms and clinical implications, *Biochem. Soc. Trans.*, 2007, **35**, 1156–1160.
23. B. Poljsak and I. Milisav, The neglected significance of "antioxidative stress", *Oxid. Med. Cell. Longevity*, 2012, 480895.
24. G. Angelo, V. J. Drake and B. Frei, Efficacy of multivitamin/mineral supplementation to reduce chronic disease risk: a critical review of the evidence from observational studies and randomized controlled trials, *Crit. Rev. Food Sci. Nutr.*, 2015, **55**, 1968–1991.
25. G. Bjelakovic, D. Nikolova and C. Gluud, Antioxidant supplements to prevent mortality, *J. Am. Med. Assoc.*, 2013, **310**, 1178–1179.
26. S. P. Fortmann, B. U. Burda, C. A. Senger, J. S. Lin and E. P. Whitlock, Vitamin and mineral supplements in the primary prevention of cardiovascular disease and cancer: an updated systematic evidence review for the US Preventive Services Task Force, *Ann. Intern. Med.*, 2013, **159**, 824–834.
27. A. J. Kuszak, D. C. Hopp, J. S. Williamson, J. M. Betz and B. C. Sorkin, Approaches by the US National Institutes of Health to support rigorous scientific research on dietary supplements and natural products, *Drug Test. Anal.*, 2016, **8**, 413–447.
28. S. K. Myung, W. Ju, B. Cho, S. W. Oh, S. M. Park and B. K. Koo, *et al.*, Efficacy of vitamin and antioxidant supplements in prevention of cardiovascular disease: systematic review and meta-analysis of randomised controlled trials, *Br. Med. J.*, 2013, **346**, f10.
29. Y. Ye, J. Li and Z. Yuan, Effect of antioxidant vitamin supplementation on cardiovascular outcomes: a meta-analysis of randomized controlled trials, *PLoS One*, 2013, **8**, e56803.
30. P. A. Cohen, The supplement paradox: negligible benefits, robust consumption, *J. Am. Med. Assoc.*, 2016, **316**, 1453–1454.
31. G. F. Deng, C. Shen, X. R. Xu, R. D. Kuang, Y. J. Guo and L. S. Zeng, *et al.*, Potential of fruit wastes as natural resources of bioactive compounds, *Int. J. Mol. Sci.*, 2012, **13**, 8308–8323.
32. D. P. Xu, Y. Li, X. Meng, T. Zhou, Y. Zhou and J. Zhang, *et al.*, Natural antioxidants in foods and medicinal plants: extraction, assessment and resources, *Int. J. Mol. Sci.*, 2017, **18**, 96.
33. J. L. Slavin and B. Lloyd, Health benefits of fruits and vegetables, *Adv. Nutr.*, 2012, **3**, 506–516.
34. F. J. Bock and S. W. G. Tait, Mitochondria as multifaceted regulators of cell death, *Nat. Rev. Mol. Cell Biol.*, 2020, **21**, 85–100.

35. Y. Yu, S. Mo, M. Shen, Y. Chen, Q. Yu and Z. Li, *et al.*, Sulfated modification enhances the immunomodulatory effect of *Cyclocarya paliurus* polysaccharide on cyclophosphamide-induced immunosuppressed mice through MyD88-dependent MAPK/NF-κB and PI3K-Akt signaling pathways, *Food Res. Int.*, 2021, **150**, 110756.
36. M. Abdalkader, R. Lampinen, K. M. Kanninen, T. M. Malim and J. R. Liddell, Targeting Nrf2 to suppress ferroptosis and mitochondrial dysfunction in neurodegeneration, *Front. Neurosci.*, 2018, **12**, 1–9.
37. M. Takashima, K. Ichihara and Y. Hirata, Neuroprotective effects of Brazilian green propolis on oxytosis/ferroptosis in mouse hippocampal HT22 cells, *Food Chem. Toxicol.*, 2019, **132**, 110669.
38. J. A. Imlay, Cellular defences against superoxide and hydrogen peroxide, *Annu. Rev. Biochem.*, 2008, **77**, 755–776.
39. M. M. Juntilla, V. D. Patil, M. Calamito, R. P. Joshi, M. J. Birnbaum and G. A. Koretzky, AKT1 and AKT2 maintain hematopoietic stem cell function by regulating reactive oxygen species, *Blood*, 2010, **115**, 4030–4038.

CHAPTER 2

The Global Regulations and Health Claims on Natural Dietary Antioxidant Supplements

JERZY ZAWISTOWSKI

Food, Nutrition, and Health, University of British Columbia, Vancouver, BC, Canada
Email: jerzy.zawistowski@ubc.ca

2.1 Introduction

There is no doubt that supplemental products with health benefits are sustainable on the global market. In 2020, the market of the dietary supplement of major shareholders in the US, Canada, Europe, Japan, and China was estimated at US$41.1 billion with a growth of 5.9% in the compound annual growth rate. It has been predicted that the size of the global market of dietary supplements will reach US$74.8 billion in 2026.[1] The importance of the health benefits of dietary supplements is well documented and supported by scientific research. The important group of dietary supplement consumers is the elderly population who experience health problems and want to be healthy by supplementation of diet rather than rely entirely on medication. There is a link between the ingestion of supplements and the prevention of certain chronic diseases. Antioxidant supplements may offer

Food Chemistry, Function and Analysis No. 39
Dietary Supplements with Antioxidant Activity: Understanding Mechanisms and Potential Health Benefits
Edited by Cesarettin Alasalvar, Fereidoon Shahidi and Chi-Tang Ho

Published by the Royal Society of Chemistry, www.rsc.org

protection of cells from oxidative stress as well as protection of DNA, proteins, and lipids from oxidative damage. Some clinical studies support the view that the consumption of antioxidants may reduce the risk of chronic diseases such as coronary heart disease (CHD) and cancer. An antioxidant-rich diet associated with a healthy lifestyle may promote longevity.[2] Antioxidants such as catechins, anthocyanins, carotenoids, isoflavones, and various other phytochemicals have been the subject of extensive research, particularly with respect to their ability for scavenging free radicals and reducing oxidative stress. This may, in turn, contribute to numerous health effects including the risk reduction of CHD, macular degeneration, and some types of cancer. However, despite a plethora of publications on antioxidants and their modes of action, there is still a lack of plausibility related to antioxidant health benefits.[3] Although consuming fruits and vegetables in a daily diet may provide various antioxidants, supplements may offer higher amounts of these compounds in the form of capsules and pills. Therefore, many consumers prefer supplementing their diet with a higher concentration of antioxidants. Yang *et al.*[4] estimated that the major source of total antioxidant intake in US adults comes from dietary supplements accounting for 25% of the total dietary values. Dietary supplements are common on the market and well recognized by consumers as well as the industry for their health benefits. The US Congressional findings have revealed that about 50% of the 260 million American adults regularly consume dietary supplements of vitamins, minerals, or herbs as a means of improving their nutrition.[5]

Recognizing the importance of this type of product, many world jurisdictions have implemented regulations to govern the approval and to some extent grant health claims. Three leading regulatory jurisdictions in this respect are the USA, EU, and Canada. The regulators in the USA and EU categorize supplemental products as dietary supplements, while in Canada these are under natural health products. This chapter focuses on reviewing the regulations and health claims on antioxidant supplements in selected countries which also includes any updates introduced over the recent years.

2.2 Regulations and Health Claims on Antioxidant Supplements in Selected Countries

2.2.1 Canada

Dietary supplements with antioxidant activity are regulated under the Natural Health Product Regulations by the Natural and Non-prescription Health Product (NNPHP) Directorate as a part of the Health Canada structure.[6] The directorate was established in 2000 and came into effect in 2004. Natural Health Products regulations underwent important changes that are no longer allowing products consumed as foods. However, natural health products (NHP) with bioactive compounds (nutraceuticals) may be approved if in medicinal forms, such as capsules, tablets, and potions. Unlike the regulations in many other countries, NHP regulations allow for a full range of health claims, including treatment, prevention, and diagnostic claims that are prohibited to use on functional foods with natural bioactive compounds in many countries including Canada. In addition, NHP regulations allow disease risk reduction and general health maintenance claims. Treatment/prevention claims are related to permanent or partial elimination or prevention of serious diseases/conditions such as treatment of prostate cancer and prevention of rheumatoid arthritis. Risk reduction claims are like claims on functional foods and may include lowering a major factor of diseases such as blood lipids and plasma cholesterol levels with reducing the risk of CHD. On the other hand, antioxidant claims are in the category of maintenance claims and may include "source of antioxidants", "provides antioxidants", or "providing antioxidants for the maintenance of good health".[7] To make the claim, products must contain at least one ingredient that has antioxidant properties. Most of the antioxidant claims are in low-risk category requirements and can be supported by a wide range of evidence. If an antioxidant claim is more specific, for example, related to eye health, the claim is evaluated according to the conditions/diseases specified within the claim.[7]

Currently, Health Canada has introduced a review process for approval of natural health products.[8] The process is divided into three classes based on the level of certainty in the evidence (Table 2.1). The fastest route to receive approval for the NHP is to submit a dossier under Class 1.[8] In addition, if the industry contemplates a license application for a product with more than one ingredient, they may consider information under the product monographs (Class 2).[8] There are numerous monographs for products with antioxidants. Examples of single ingredient monographs with antioxidant properties are

Table 2.1 Classes of applications for approval of natural health products in Canada.[8]

Classes	Level of certainty	Evidence	Timelines of review	Examples
Class 1	High	Based on a single monograph, pre-cleared by NNHPD	60 days	Vitamins, minerals, and various herbal products
Class 2	Medium	Based on two or more monographs, pre-cleared by NNHPD but additional published literature is required	180 days	Current authorized product with a new claim
Class 3	Low	Requiring full pre-market assessment	180 days	A new product with ingredient combination risk factors (*e.g.*, weight management and diuretics)

Abbreviation: NNHPD, Natural and Non-prescription Health Product Directorate.

Table 2.2 Selected examples of single-ingredient monographs of NHP with antioxidant properties.[9]

Name	Form	Uses/claims
Acai (*Euterpe oleracea*)	Dry/fluid extract	Source of/provides antioxidants: helps fight/protect (cell) against/reduce (the oxidative effect of/the oxidative damage caused by/cell damage caused by) free radicals
Bilberry (*Vaccinium myrtillus*)	Standardized dry extract with 37.6% anthocyanins	Source of/provides antioxidants
Citrus bioflavonoids	Dry/fluid extract	Source of/provides antioxidants
Coenzyme Q10	—	Antioxidant: helps to maintain and/or support cardiovascular health
Curcumin (*Curcuma longa*)	—	Antioxidant: used in herbal medicine to help relieve joint inflammation
Grape seed extract	Dry/fluid extract	Source of/provides antioxidants
Green coffee bean extract	Dry/fluid extract	Source of/provides antioxidants: helps support cardiovascular health
Green tea extracts	Standardized extract with up to 690 mg total catechins and 150 mg of caffeine	Source of/provides antioxidants
Lycopene	Gummies, tablets, caplets, capsules, strips, lozenges, powders, or liquids. The product must contain more than 95% lycopene	Provides antioxidants for the maintenance of good health
Olive leaf (*Olea europaea*)	Dry/fluid extract	Source of/provides antioxidants
Quercetin	—	Source of/provides antioxidants
Red wine extract	Dry/fluid extract	Source of/provides antioxidants
Resveratrol	—	An antioxidant
Rutin	—	An antioxidant
Tomato extract	Standardized extract with lycopene	Source of/provides antioxidants
Turmeric (*Curcuma longa*)	—	Source of/provides antioxidants: traditionally used in herbal medicine as an anti-inflammatory agent to help relieve joint pain and in Ayurveda to relieve pain and inflammation

Abbreviation: NHP, natural health products.

listed in Table 2.2.[9] Most of the approved antioxidant products can carry a general claim label such as "source of antioxidants" or "provide antioxidants". However, some health claims are more specific under the

Table 2.3 Examples of product monographs of vitamins and minerals with antioxidant properties.[10]

Name	Uses/claims
Vitamin C	Source of/an antioxidant for the maintenance of good health
	Antioxidant for good health
	Source of/an antioxidant that helps fight/protect (cell) against/reduce (the oxidative effect of/the oxidative damage caused by/cell damage caused by) free radicals
Vitamin E	Source of/an antioxidant for the maintenance of good health
	Antioxidant for good health
	Source of/an antioxidant that protects the fat in body tissues from oxidation
	Source of/an antioxidant that helps (to) (a) fight, (b) protect (cells) against, or (c) reduce (the oxidative effect of/the oxidative damage caused by/cell damage caused by) free radicals
Selenium	Source of/an antioxidant for the maintenance of good health
	Antioxidant for good health
	Source of/an antioxidant that helps protect against oxidative stress
	Source of/an antioxidant that helps (to) (a) fight, (b) protect (cells) against, or (c) reduce (the oxidative effect of/the oxidative damage caused by/cell damage caused by) free radicals
	Helps to maintain the normal function of the thyroid gland

product monographs. Product monographs, which include more than one ingredient, are published in the NHP ingredients database. A monograph on antioxidants is a part of this list and contains more specific health claims.[9] This includes marigold extract and isolates, lutein and zeaxanthin, and their esters. These are important antioxidants for the maintenance of eye health. Therefore, specific claims are "helps to maintain eyesight/support eye health in conditions (associated with sunlight damage), such as cataracts and age-related macular degeneration", "helps to reduce the risk of developing cataracts", and "helps to improve macular pigment optical density".

NHP categories also include vitamins and vitamin precursors, minerals, and trace elements, some of them with well-established antioxidant characteristics. Table 2.3 shows examples of vitamins and minerals with uses/claims in this category.[10]

It is important to know that approval of any NHP undergoes a thorough and detailed review process by the NNPHP Directorate. Once a product is approved, the label on the package will carry a unique identification number, natural product number (NPN), in much the same way as pharmaceuticals. This indicates that the product has undergone and passed a review process including product formulation, labeling, and instruction for use. The NPN unique indicator also allows for easy and fast recall of products in case of safety issues.[8]

2.2.2 USA

In 1994, the US Senate passed and enacted the Dietary Supplement Health and Education Act (DSHEA), which amended the Federal Food, Drug, and

Cosmetic Act (FD&C).[5] This law is designed to regulate dietary supplements for the US market. The goal of this law is to provide consumers broad access to dietary supplements as well as information about the intended uses of dietary supplements. DSHEA defines a dietary supplement as a product to supplement the diet, such as vitamins, minerals, botanicals, amino acids, and any dietary substance for increasing the total dietary intake. This may include a concentrate, metabolite, extract, or combination of ingredients described in the law. There are many forms of dietary supplements including capsules, tablets, powders, liquids, and energy bars.[5] Although dietary supplements are regulated by the Food and Drug Administration (FDA) as foods for oral ingestion, it is worth noting that dietary supplements are not intended to be used as a conventional food or a sole item of a meal or the diet. It must also be labeled as a dietary supplement.[5] However, there is no legal requirement for approval by the FDA. Companies that are planning to introduce dietary supplements must notify the FDA 75 days prior to marketing products and must comply with the regulations.[11] Products must be safe and not present a significant or unreasonable risk of illness or injury under the conditions of use, although federal law does not require confirmation of safety by regulatory authorities.[11] Once a dietary supplement is on a market, the onus is on a company to monitor the safety of the product and report any serious adverse effects described by consumers to the FDA. In addition, this is also the responsibility of the FDA to monitor the safety of products. Dietary supplements may bear structure or function health claims, but the law does not require obtaining pre-approval for the claim whether the claim is accurate or truthful. The FDA must be informed within 30 days after marketing a product to use a claim, which may be reviewed for substantiation by the authorities.[11] Claims to mitigate, prevent, or cure disease are prohibited. Dietary supplement products must be properly labeled including supplement facts and direction of usage. The law also requires a disclaimer to be included. "This statement has not been evaluated by the FDA. This product is not intended to diagnose, treat, cure, or prevent any disease", as a part of the label of the dietary supplement (§6. Sec. 403(r)21 U.S.C. 343(r).[5]

The major contributors of antioxidants in the American diet are dietary supplements, which account for 25% of the total antioxidant capacity (TAC), about 120 mg of vitamin C equivalents per day of TAC in adults.[4] Dietary supplements may bear a nutrient content claim using the term antioxidant or antioxidants if a product contains dietary antioxidant ingredients with a reference dietary intake (RDI) [21 CFR §101.54 (g)].[12] This type of claim shows the level of one or more antioxidants in a dietary supplement. It is mainly related to antioxidant vitamins such as vitamins C and E. The level of the vitamin must qualify for such a claim. A product that contains nutrients at 20% or a higher level of the RDI may bear a "high" claim, while a product with 10–19% of nutrients is eligible for a "good source" claim. A claim such as "more" is a relative claim and could be made with respect to the level of vitamins if the products contain at least 10% more of the RDI for the vitamins (expressed as a percent of the daily value) per reference amount.[12] The names of the nutrients that are subject of the claim must be included in the claim "high in antioxidant vitamins C and E" or alternatively the claim "high in antioxidants" must be linked with a symbol (*e.g.*, an asterisk) to the name of antioxidant vitamin (or vitamins) that appears on the product label. Some of the nutrients with antioxidant activities but without RDI may also be the subject of antioxidant claims. For example, β-carotene is recognized as a precursor of vitamin A. In such a case, β-carotene which is a subject of the claim "good source of antioxidant β-carotene" must be in sufficient concentration to produce at least 10% of vitamin A (retinol) RDI upon conversion in the human body.[12] Structure/function claims regarding antioxidants (*e.g.*, X is involved in antioxidant processes) are allowed as long as such claims are truthful, not misleading, and meet the requirements described in section 403(r)(6)(b) of the Federal Food, Drug, and Cosmetic Act.[13] Claims must be substantiated by competent and reliable scientific evidence. A claim may not refer to an effect of a specific disease or symptoms of a disease. For example, "helps maintain joint health" is permissible but "lowering the arthritis pain" is not. Some permissible claims may be linked to a bibliography. Citation to a publication is allowed if it does not imply a disease reduction, prevention, or treatment. For example, reference to "K. Bailey, Antioxidants may lower blood cholesterol levels, *Food Sci.*, 2007, **12**, 34–56 or B. Garth, Antioxidant properties of green tea, *Am. J. Med.*, 2009, **14**, 92–109" is permissible.[14]

It is worth mentioning that health claims on dietary supplements may be considered by the FDA based on industry petitions under the qualified health claim (QHC) regulations.[15] The claims are contemplated when the strength of scientific evidence falls below more rigorous "significant scientific agreement" FDA requirements, which are necessary to support disease risk reduction claims under the Nutrition Labeling and Education Act (NLEA). There are three categories (levels B–D) of QHCs depending on the strength of scientific evidence. Products with QHCs must also display a label disclaimer (Table 2.4). The agency does not issue the affirmative approval for QHCs; instead, it issues a Letter of Enforcement Discretion based on scientific evidence. If evidence is satisfactory, the FDA will not object to using the QHC. Currently, as of 2020, there are 38 QHCs. Amongst these claims, there are QHCs on products with antioxidant vitamins C and E and antioxidant selenium. Antioxidant QHCs have linked the substance of the claim (antioxidant) with some type of cancer (Table 2.5). The wording of the claims such as "may reduce the risk of specific cancer" and "based on limited studies" must be appropriate in order not to mislead consumers about the efficacy of the substance.[15] Any petition for QHCs submitted to the FDA by manufacturers that meets requirements specified in 21 CFR 101.70[16] will be acknowledged within 15 days of receipt by the agency. Within 45 days of receipt, the FDA will file the petition and a docket number will be assigned. On or before 270 days after receipt of the petition, a final decision

Table 2.4 Qualified health claims.[15]

Scientific ranking	FDA category	Appropriate qualifying language[a]
First level	A (unqualified claims)	Claim is supported by significant scientific agreement and evidence is derived from well-designed studies – Nutrition Labeling and Education Act claim.
Second level	B	Although there is scientific evidence supporting the claim, the evidence is not conclusive.
Third level	C	Some scientific evidence suggests... however, the FDA has determined that this evidence is limited and not conclusive.
Fourth level	D	Very limited and preliminary scientific research suggests... the FDA concludes that there is little scientific evidence supporting this claim.

[a]The language reflects wording used in qualified health claims as to which the agency has previously exercised enforcement discretion for certain dietary supplements. During this interim period, the precise language as to which the agency considers exercising enforcement discretion may vary depending on the specific circumstances of each case.

Table 2.5 Qualified health claims for antioxidants.[15]

Name of antioxidant	Qualified health claims considered under the enforcement discretion	FDA category[a]
Vitamin C	One weak study and one study with inconsistent results suggest that vitamin C supplements may reduce the risk of gastric cancer. Based on these studies, the FDA concludes that it is highly uncertain that vitamin C supplements reduce the risk of gastric cancer.	D
Vitamin E	One small study suggests that vitamin E supplements may reduce the risk of bladder cancer. However, two small studies showed no reduction of risk. Based on these studies, the FDA concludes that it is highly unlikely that vitamin E supplements reduce the risk of bladder cancer.	D
	Two weak studies and one study with inconsistent results suggest that vitamin E supplements may reduce the risk of colorectal cancer. However, another limited study showed no reduction of risk. Based on these studies, the FDA concludes that it is highly unlikely that vitamin E supplements reduce the risk of colorectal cancer.	D
	One weak and limited study suggests that vitamin E supplements may reduce the risk of renal cell cancer. The FDA concludes that it is highly uncertain that vitamin E supplements reduce the risk of renal cell cancer.	D
Selenium	Selenium may reduce the risk of certain cancers. Some scientific evidence suggests that consumption of selenium may reduce the risk of certain forms of cancer. However, the FDA has determined that this evidence is limited and not conclusive.	C
	Selenium may produce anti-carcinogenic effects in the body. Some scientific evidence suggests that consumption of selenium may produce anti-carcinogenic effects in the body. However, the FDA has determined that this evidence is limited and not conclusive.	C

[a]Refer to Table 2.4.

(letter of enforcement discretion) will be sent to the petitioner. Extensions beyond 270 days can be granted upon mutual agreement between the petitioner and the agency.[16]

2.2.3 European Union

Dietary supplements or more precisely food supplements, are defined by the European Food Safety Authority (EFSA) as concentrated sources of nutrients, which include vitamins, minerals, as well as other ingredients with "a nutritional or physiological effect" in dose forms such as pills, capsules, tablets, powders, and measurable liquids. Food supplements may also derive from various plants and herbs. Like the US regulation, in the EU supplements are regulated as food.[17] Directive 2002/46/EC (Annex I and Annex II), which is the main law to regulate supplements, provides a list of vitamins and minerals that can be used for the formulation of supplements. Annex I of the directive lists all approved vitamins and minerals, while Annex II lists the specific chemical forms that may be used in the manufacturing of food supplements. Among all approved vitamins and minerals, there are substances with antioxidant activities, including vitamins E, C, and A (beta-carotene), as well as minerals such as selenium and zinc.[17]

Table 2.6 shows antioxidant vitamins and minerals and the form of their use as food supplements. This directive also describes labeling requirements for supplements such as daily dose, net content, and warnings as listed below:[17]

1. The name under which products are sold shall be "food supplement".
2. The labeling and advertising of the product must not be related to preventing, treating, or curing human disease.
3. The labeling shall bear the following specifics:
   - the names of the categories of nutrients or substances that characterize the product or an indication of the nature of those nutrients or substances,
   - the portion of the product recommended for daily consumption,
   - a warning not to exceed the stated recommended daily dose,
   - a statement to the effect that food supplements should not be used as a substitute for a varied diet, and
   - a statement to the effect that the products should be stored out of the reach of young children.

The amount of nutrients or substances present in the product shall be declared in the numerical form per portion of the product as recommended per daily consumption. Vitamins and minerals shall also be expressed as a percentage of the reference values. Health claims on foods are governed and approved under EC 1924/2006 regulations on Nutrition and Health Claims made of foods.[18] A list of approved health claims has been in effect since December of 2012.[19,20]

Table 2.6 Selected vitamins and minerals as well as their forms with potential antioxidant properties that can be used for food supplements in the EU.[17]

Nutrients	Approved chemical forms	Note
Vitamin A	Beta-carotene	—
Vitamin E	D-Alpha-tocopherol, DL-Alpha-tocopherol D-Alpha-tocopheryl acetate DL-Alpha-tocopheryl acetate D-Alpha-tocopheryl acid succinate Mixed tocopherols* Tocotrienol tocopherol**	*Alpha-tocopherol <20%, beta-tocopherol <10%, gamma-tocopherol 50–70%, and delta-tocopherol 10–30% **Typical levels of individual tocopherols and tocotrienols: - 115 $mg\,g^{-1}$ alpha-tocopherol (101 $mg\,g^{-1}$ minimum) - 5 $mg\,g^{-1}$ beta-tocopherol (<1 $mg\,g^{-1}$ minimum) - 45 $mg\,g^{-1}$ gamma-tocopherol (25 $mg\,g^{-1}$ minimum) - 12 $mg\,g^{-1}$ delta-tocopherol (3 $mg\,g^{-1}$ minimum) - 67 $mg\,g^{-1}$ alpha-tocotrienol (30 $mg\,g^{-1}$ minimum) - <1 $mg\,g^{-1}$ beta-tocotrienol (<1 $mg\,g^{-1}$ minimum) - 82 $mg\,g^{-1}$ gamma-tocotrienol (45 $mg\,g^{-1}$ minimum) - 5 $mg\,g^{-1}$ delta-tocotrienol (<1 $mg\,g^{-1}$ minimum)
Vitamin C	L-Ascorbic acid Sodium-L-ascorbate Calcium-L-ascorbate* Potassium-L-ascorbate L-Ascorbyl 6-palmitate Magnesium L-ascorbate Zinc L-ascorbate	*May contain up to 2% of threonate
Selenium	L-Selenomethionine Selenium enriched yeast* Selenious acid Sodium selenate Sodium hydrogen selenite Sodium selenite	*Selenium-enriched yeasts produced by culture in the presence of sodium selenite as the selenium source and containing, in the dried form as marketed, not more than 2.5 mg Se per g. The predominant organic selenium species present in the yeast is selenomethionine (between 60 and 85% of the total extracted selenium in the product). The content of other organic selenium compounds including selenocysteine shall not exceed 10% of the total extracted selenium. Levels of inorganic selenium normally shall not exceed 1% of the total extracted selenium.

Table 2.6 (*Continued*)

Nutrients	Approved chemical forms	Note
Zinc	Zinc acetate Zinc L-ascorbate Zinc L-aspartate Zinc bisglycinate Zinc chloride Zinc citrate Zinc gluconate Zinc lactate Zinc L-lysinate Zinc malate Zinc mono-L-methionine sulphate Zinc oxide Zinc carbonate Zinc L-pidolate Zinc picolinate Zinc sulphate	

Claims are divided into three categories:[18]

1. general health claims (Article 13.1 of the regulations)
   - the role of a nutrient or other substances in growth, development, and the functions of the body,
   - referring to psychological and behavioral functions, and
   - slimming or weight-control
2. disease risk reduction claims (Article 14(1)(a) of the regulations)
3. children's development and health (Article 14(1)(b) of the regulations)

In accordance with the above regulations, the EU Commission published a register on nutrition and health claims with status.[21] Many antioxidants or substances containing antioxidants are subject to submission for health claims, for example "due to its antioxidant properties, lutein/zeaxanthin contributes to the hindrance of harmful oxidative processes in the body, in this way to the maintenance of our health/to the defense of the tissues" or "lutein/zeaxanthin are eye protection factors, maintain normal vision".[20] Requirements for the scientific substantiation of health claims related to antioxidants are published with updates in 2018.[22] Over 200 submissions for health claims related to antioxidants are not authorized under the above regulations due to non-compliance with the regulation, because claims are not substantiated by scientific evidence.[21] However, there are permitted health claims for few antioxidant nutrients and other substances:[20,21]

1. Riboflavin (vitamin B2), vitamin C, vitamin E, manganese, selenium, and zinc contribute to the protection of cells from oxidative stress.
   - The claim may be used only for food that is at least a source of vitamins or minerals as referred to in the claimed source of

[name of vitamin/s] and/or [name of mineral/s] as listed in the Annex to Regulation (EC) No. 1924/2006.[18]

2. Olive oil polyphenols contribute to the protection of blood lipids from oxidative stress.
   - The claim may be used only for olive oil which contains at least 5 mg of hydroxytyrosol and its derivatives (*e.g.*, oleuropein complex and tyrosol) per 20 g of olive oil. To bear the claim information shall be provided to the consumer that the beneficial effect is obtained with a daily intake of 20 g of olive oil.

Food supplements can also be approved under novel food regulations.[23] This regulation describes rules that the industry must follow to place novel products on the EU market. Novel foods are defined as foods and ingredients that have not been "previously used to a significant degree for human consumption" within the EU before May 15, 1997. The definition also includes food that has been manufactured, prepared, preserved, or packaged by a process that has not previously been applied to that food or/and causes the food to undergo a major change. Novel food categories include food supplements, insects, new sources of vitamins and minerals, nanomaterials, food obtained from new technologies (*e.g.*, ultraviolet), and products outside the EU (noni juice, chia seeds, *etc.*). EU Commission established a list of all approved novel foods,[24] including novel supplements with antioxidant properties (Table 2.7).

Substances with antioxidant properties which resemble the format of dietary supplements such as pills, capsules, or tinctures can also be approved under EU legislation as traditional herbal medicine (2008/911/EC).[25] This type of product is defined as medicine containing one or more herbal ingredients as bioactive compounds. European Commission has established a list of herbal medicinal products.[25] An example of an approved herbal preparation is a tincture containing extract from Calendula flowers for treatment of minor inflammations of the skin, wounds, mouth, and throat. Similarly, a distillate from fresh leaves and bark of common witch hazel has been approved for relief of minor skin inflammation.[25]

2.2.4 Japan

The modern term "Functional Foods" was created in Japan in the late 1980s as a result of 9 years of cooperation among the government, academia, and industry associations.[26] Based on the extensive research that focused on the functions of food for adjusting the physical conditions of the human body, the Japanese Ministry of Health, Labor, and Welfare established a regulatory system and policy for approving Food for Specified Health Use (FOSHU) in 1991.[27,28] Although the approval of FOSHU was restricted to food for about ten years, in 2001, the Japanese Government ameliorated the regulations to allow the use of dietary supplements under FOSHU governance.[29] Like dietary supplements regulated under DSHEA in the US, supplements in

Table 2.7 Examples of approved novel food supplements with antioxidant properties under the novel food regulations.[24]

Specified food supplement as defined in directive 202/46/EC[17]	Maximum levels	Additional labeling requirements
Astaxanthin-rich oleoresin from *Haematococcus pluvialis* algae	40–80 mg per day of oleoresin, resulting in ≤8 mg astaxanthin per day	The designation of the novel food on the labeling of the foodstuffs containing it shall be "Astaxanthin".
Extract of defatted cocoa powder	1 g per day and 300 mg polyphenols corresponding to not more than 550 mg of extract of defatted cocoa powder in one portion of food (or food supplement)	Consumers shall be instructed not to consume more than 600 mg polyphenols corresponding to 1.1 g of extract of defatted cocoa powder per day.
Epigallocatechin gallate as a purified extract from green tea leaves (*Camellia sinensis*)	150 mg of extract in one portion of food or food supplement	The labeling shall bear a statement that consumers should not consume more than 300 mg of extract per day
Flavonoids from *Glycyrrhiza glabra*	120 mg per day	The designation of the novel food on the labeling of the foodstuffs containing it shall be "Flavonoids from *Glycyrrhiza glabra* L."
Lycopene	15 mg per day	The designation of the novel food on the labeling of the foodstuffs containing it shall be "Lycopene".
Noni fruit juice powder (*Morinda citrifolia*)	6.6 g per day (equivalent to 30 mL of noni juice)	The designation of the novel food on the labeling of the foodstuffs containing it shall be "Noni juice powder" or "Juice powder of *Morinda citrifolia*".
Noni leaves (*Morinda citrifolia*)	2.4 g per day	—
trans-Resveratrol	150 mg per day for adult population (capsule or tablet form)	The designation of the novel food on the labeling of the food supplements containing it shall be "*trans*-resveratrol". The labeling of food supplements containing *trans*-resveratrol shall bear a statement that people using medicines should only consume the product under medical supervision.

Table 2.7 (*Continued*)

Specified food supplement as defined in directive 202/46/EC[17]	Maximum levels	Additional labeling requirements
Zeaxanthin	2 mg per day	The designation of the novel food on the labeling of the foodstuffs containing it shall be "synthetic zeaxanthin".
Zinc L-pidolate	3 g per day	The designation of the novel food on the labeling of the foodstuffs containing it shall be "Zinc L-pidolate".

medicinal formats such as pills, tablets, capsules, and other forms are allowed under FOSHU regulations. Like other world regulatory jurisdictions, the Japanese Government defines dietary supplements as products to supplement the diet with dietary ingredients such as vitamins, minerals, and botanical substances, among others. This type of product with standardized vitamins and minerals allows bearing nutrient function claims. Among the approved claims for vitamin supplements, there are also claims for antioxidants such as:[30,31]

1. Vitamin E "is a nutrient which helps to protect fat in the body from being oxidized and to maintain the cell healthy".
2. Vitamin C "is a nutrient which helps to maintain skin and mucosa's health and has antioxidant effect".

Products with nutrient function claims do not need to be approved by the governmental authorities prior to marketing; however, they need to bear the warning "Taking large amounts of this product will not cure disease or promote health. The indicated daily intake should be followed".[31]

Further reform of Japanese regulations came in 2015 with the creation of the Foods with Function Claims (FFC) regulatory venue.[32] This was rationalized by the high cost associated with the FOSHU approval. Under these regulations, marketing products with health claims is easier, faster, and low cost. In contrast to FOSHU, FFC is based on the notification system and is the responsibility of the industry. The industry is responsible to manufacture FFC under well-established systems and evaluate the safety and effectiveness of products following governmental regulations. All needed information (safety, efficacy, and manufacturing) must be submitted to the Secretary-General of the Consumer Affairs Agency within 60 days before marketing the product.[32,33] Like the US dietary supplements (DSHEA), FFC may bear structure/function claims based on clinical trials but also on the systematic literature review of a product or functional ingredients.

Table 2.8 Japanese antioxidant supplements as Food with Functional Claims. Reproduced from ref. 35 with permission from Elsevier, Copyright 2022.

Product name	Producer	Functional ingredient	Function
Rohto V5 capsule	Rohto Pharmaceutical Co. Ltd	Lutein and zeaxanthin	Maintenance of good eyesight
Mebae	Yawata Corporation	Lutein	Maintenance of good eye condition
Blueberry & acai mix	ITO EN, Ltd	Astaxanthin	Maintenance of good eye condition
Hitomi-no-megumi-lutein 40	Fine Japan Co. Ltd	Lutein esters	Protection of eyes by stimulation of light
Nature made lutein	Otsuka Pharmaceutical Co. Ltd	Lutein	Support of eye sensitivity of contrast
Nature made astaxanthin	Otsuka Pharmaceutical Co. Ltd	Astaxanthin	Support of eyes to focus adjustment
RUTEIN	Kobayashi Pharmaceutical Co. Ltd	Lutein	Maintenance of good ophthalmic yellow spot
Sunphenon EGCG capsule	Taiyo Kagaku Co. Ltd	EGCG	Decrease of body fat
Dear-Natura gold lutein	Asahi Food & Healthcare Co. Ltd	Lutein	Maintenance of good eyesight
Blueberry lutein	Kabushikigaisya Refre	Bilberry extract, anthocyanin, and lutein	1. Improving ophthalmic tiredness 2. Maintenance of good ophthalmic yellow spot
ASUTAAIKEA	Fuji Chemical Industries Co. Ltd	Astaxanthin	Decrease of eyestrain
Egao-no-lutein	Egao Co. Ltd	Lutein	Maintenance of good eye condition
Mainichino-daizu-isofurabon	Fujicco Co. Ltd	Soy isoflavone	Maintenance of good bone

Abbreviation: EGCG, epigallocatechin gallate.

In addition, the final product must disclose a label disclaimer: "This product has not been individually evaluated by the Secretary-General of the Consumer Affairs Agency".[32,33] Only one year after the launching of the regulations, there were 395 FFC products on the Japanese market, including antioxidant supplements such as lycopene, lutein, astaxanthin, *etc.* It is worth noting that 149 products were supplements including capsules, tablets, liquids, and powders.[32]

In 2019, there were 1785 FFC products on the Japanese market manufactured by 300 companies.[34] Among FFC with approved functions, there are antioxidant supplements such as lutein, zeaxanthin, astaxanthin, and lycopene, among others.[35] Dietary supplements and their functions are listed in Table 2.8. It is worth noting that the primary function of the listed antioxidant supplements is the maintenance of good eyesight.

2.2.5 Australia and New Zealand

Dietary supplements are regulated by the Australian Therapeutic Goods Act of 1989,[36] and New Zealand Dietary Supplements Regulations of 1985.[37] However, New Zealand supplements with bioactives at higher pharmacological doses are regulated by the Medicines Act 1981.[38] The latter supplements may be considered complementary medicine in Australia. As of 9 December 2021, following permitted indications are accepted in relation to the inclusion of the Medicine in the ARTG:[39]

1. antioxidant/reduce free radicals formed in the body,
2. helps reduce/decrease free radical damage to body cells,
3. maintain/support healthy eye function,
4. maintain/support eye health, and
5. maintain/support healthy eyesight/vision.

New Zealand Dietary Supplements Regulation defines an antioxidant supplement as any substance that has the ability to arrest or retard oxidative rancidity. Furthermore, the regulations are very specific regarding which antioxidants and at what levels are allowed in dietary supplements:[37]

1. propyl gallate, dodecyl gallate, octyl gallate, butylated hydroxyanisole, butylated hydroxytoluene, and tertiary butylhydroquinone, where the proportion of those antioxidants, singly or in combination, does not exceed 100 ppm,
2. ascorbyl palmitate and ascorbyl stearate, where the proportion of those antioxidants, singly or in combination, does not exceed 500 ppm,
3. natural tocopherols, synthetic tocopherols, citric acid, and sodium citrate, and
4. isopropyl citrate mixture, monoglyceride citrate, and phosphoric acid, where the proportion of those antioxidants, whether singly or in combination, does not exceed 100 ppm.

2.2.6 China

The health benefits of natural products such as herbs, nuts, and tea were recognized for centuries in China and were used in ancient times for the protection and treatment of various diseases. Today, health foods such as supplements are popular among Chinese consumers. In fact, China is a very big market for dietary supplements. In 2013, the sale of health foods reached US$22.2 billion, and it was estimated to double in 2019.[40] Food Hygiene Law of 1995[41] with amendments defines health foods as a supplement with a specific effect on the biological functions of the body. This definition includes herbs, botanicals, vitamins, minerals, dietary supplements, and some Traditional Chinese Medicine (TCM).[42] Chinese health foods can be either foods or dietary supplements in the form of pills, capsules, tablets, powders, and tinctures. State Food and Drug Administration (SFDA) is the main regulatory body governing the approval of domestic and imported health foods. The application process for product registration must be within the Ministry of Health or SFDA.[42] Administrative measures on the registration and record filing of health foods took effect on July 1, 2016 and now require a notification procedure for approval.[43] Once the product is approved, a health food approval certificate is issued along with a seal of approval. Furthermore, imported health foods and supplements can be registered in China but only if products were on the market of exporting country for at least one year.

The products may allow bearing function health claims if supported by scientific substantiation including human clinical studies, animal studies, or both. Twenty-seven claims are allowed on health products.[42] New health function claims may be allowed after submission is evaluated and verified.[40] Antioxidant ingredients as the most popular supplements and approved health foods are soy isoflavones, lycopene, proanthocyanidins, and flavonoids. These ingredients are present in ginseng, goji berry, and astragalus roots which are used as raw materials for manufacturing supplements.[44] Some of the claims are directly or indirectly associated with the presence of antioxidants in the formulated products. Function claims such as antioxidant functions, vision improvement, removing skin chloasma, and removing acne have a more direct association.[42] SFDA has issued a notice for nine health function evaluation methods including the assessment of antioxidant function.[45] The notice protocol includes proper selection of animal models, biomarkers, duration of experiments, and criteria used for the assessment.[44,45]

2.2.7 India

Dietary supplements that include vitamins, minerals, herbal products, protein supplements, and Chyawanprash account for 40% of the Indian nutraceutical market.[46] The Food Safety Standards Authority of India is the regulatory agency responsible for the approval of food products as well as dietary supplements. Unfortunately, the regulations are not clearly defined and dietary supplements in the form of pills, tablets, and capsules may be categorized as drugs.[46] The most exceptional and powerful antioxidant formulation is Chyawanprash, a mix of 50 herbs and spices, in the form of a paste.[47] This mixture is considered a traditional Indian Ayurvedic health supplement. Indian traditional medicine and Ayurveda products are governed by the Ministry of Ayurveda, Yoga, Unani, Siddha, and Homeopathy. According to Indian law, the Chyawanprash product can be sold either as an Ayurvedic medicine following the exact composition and manufacturing that are approved by the governmental authorities or as a proprietary Ayurvedic medicine with some modifications but using approved ingredients.[47] Another ayurvedic supplement, ashwagandha extract, is gaining popularity in North America. Ashwagandha, which has very potent antioxidant and anti-inflammatory properties, has recently been approved by Health Canada as a natural health product (NPN 80102381) with recognized health benefits relating to stress, sleep, and energy, among others.[48] The Food Safety and Standards Act 2006[49] defines health supplements as "products that may be formulated in the form of powders, granules, tablets, capsules, liquids, jelly, and other dosage forms but not parenteral, and are meant for oral administration". It may contain one or more ingredients obtained from plants or botanicals or their parts, minerals, or vitamins and intended for supplementation.

2.2.8 South Korea

According to a survey conducted between 2005 and 2006, over 60% of Korean adults had used various types of dietary supplements.[50] The latest survey data showed the frequency of the use of dietary supplements increased from 26% in 2005 to 34% in 2009.[51,52] Antioxidant vitamins E and C are the top vitamin supplements among users.[52] In addition, anti-inflammatory supplements such as glucosamine/chondroitin are among the top ten dietary supplements used by elderly adults.[51]

Dietary/nutritional supplements are governed by the Ministry of Food and Drug Safety under the Functional Health Food Act,[53] Regulation on Approval of Functional Ingredient for Health Functional Food,[54] and Regulations Concerning Recognition of Functional Ingredients and Standards and Specifications for Health Functional Foods.[55] The act was introduced in 2002 and amended several times. The act provides regulatory bases for oversight of functional foods, which are health foods and supplements.[56] Health functional food is defined as "processed food used with the intention to enhance and maintain human health by (consuming) physiologically functional ingredients and/or components in forms of tablets, capsules, powder granules, liquid, and pills".[57,58] According to Korea Health Supplements Association, there are two types of health foods:[59]

1. Foods manufactured or processed using functional ingredients are notified in the Code of Health Functional Food[60] in accordance with the standards and specifications for health functional foods.
2. Foods manufactured or processed using functional ingredients are individually recognized by the government pursuant to the regulations.

The code provides standards and specifications for functional ingredients and also instructions for manufacturing, distribution, storage, sale, and import of products with functional compounds.[60] The code refers to vitamins and minerals, and functional ingredients including antioxidant ingredients such as lutein, lycopene, soybean isoflavone, and green tea extract, among others.

The Korean Food and Drug Administration (KFDA) established a positive list of products that may be sold in Korea. The government recognizes 37 functional ingredients/food categories including nutritional supplements (vitamins and minerals), essential amino acids, proteins, dietary fibers, essential fatty acids, and other nutrient supplements such as ginseng products, aloe, squalene, and chlorella.[61,62] Dietary supplements that are produced in other countries but not on a list may not be considered health functional foods in Korea.[62] Therefore, foreign products are subject to an import notification. The foreign producer must work with a Korean company since KFDA only issues import notification certificates to Korea-based firms.[62]

Nutrient function claims can be made on nutritional supplements containing vitamins and minerals only if they have their own recommended dietary allowances (RDA). Products must contain at least 30% of the RDA of eligible nutrients to qualify for claims.[60] Furthermore, function claims are allowed on the products containing functional ingredients. Permitted nutrient function claims on supplements containing nutrients and functional ingredients with antioxidant properties are listed in Table 2.9.

2.2.9 Association of Southeast Asian Nations

Dietary supplements and traditional medicine including herbal and botanical supplements as health foods are part of regulatory frameworks in other Asian countries. Depending on countries, the term "health foods" has a different meaning. For example, in the Philippines supplements are included in this term, while in Malaysia it may be an interface between food and drug. Malaysian regulations include ingredients such as vitamins, minerals, amino acids, natural substances of plant/animal origin, enzymes, and substances with nutritional/physiological functions within a dietary supplement category in the form of pills, capsules, and other medicinal formats.[63] However, classifications and regulations are based on product composition. If a product contains at least 20% or less bioactives with pharmacological and/or therapeutic properties, such a product is regulated by the Food Safety and Quality Division, while a product with more than 20% bioactives and less than 80% food-based ingredients is regulated by the National Pharmaceutical Control Bureau.[63] The guide to assist in determining food–drug interphase products has recently been updated by the Ministry of Health Malaysia.[64] In the Philippines, a dietary supplement is defined as "a processed food product intended to supplement the diet bears or contains one or more of the following dietary ingredients: vitamin,

Table 2.9 List of nutrients and other functional ingredients with antioxidant properties and permitted health claims in South Korea.[60]

Ingredients	Health claims	RDA
Vitamins and minerals		
Vitamin A	Necessary for vision adaptation in dark places	210–1000 μg retinol
β-Carotene	Necessary for vision adaptation in dark places	0.42–7.0 mg
Vitamin E	Necessary for the protection of cells from free radicals	3.3–400 mg α-tocopherol
Vitamin C	Necessary for the protection of cells from free radicals by antioxidant activity	30–1000 mg
Copper	Necessary for the protection of cells from free radicals	0.45–7.0 mg
Selenium	Necessary for the protection of cells from free radicals	16.5–135 μg
Manganese	Necessary for the protection of cells from free radicals	0.9–3.5 mg
Functional ingredients		
Chlorella	May help to maintain healthy skin and antioxidant activity	8–150 mg as total chlorophyll
Plants containing chlorophyll	May help to maintain healthy skin and antioxidant activity	8–150 mg as total chlorophyll
Spirulina	May help to maintain healthy skin and antioxidant activity	8–150 mg as total chlorophyll
Green tea extract	May help to support antioxidant activity and reduce body fat	0.3–1 g as catechin
Propolis extract	May help to support antioxidant activity and antimicrobial activity in the oral cavity	16–17 mg as total flavonoids
Coenzyme Q10	May help to support antioxidant activity and maintain healthy blood pressure	90–100 mg as coenzyme Q10
Soybean isoflavone	May help to maintain bone health	24–27 mg as aglycone soybean isoflavone
Squalene	May help to support antioxidant activity	10 g as squalene
Marigold flower extract	May help the eye health by maintaining the density of macular pigments which can be decreased by ageing	10–20 mg as lutein
Haematococcus pluvialis extract	May help to improve eye fatigue	4–12 mg as astaxanthin
Glucosamine	May help to maintain healthy joint and cartilage	1.5 g as glucosamine hydrochloride or sulphate
Bilberry extract	May help to improve eye fatigue	72–108 mg as anthocyanosides
Tomato extract	May help to support antioxidant activity	5.7–15 mg as all-*trans*-lycopene

Abbreviation: RDA, recommended dietary allowances.

mineral, amino acid, herb, or other dietary substance of botanical, animal, artificial, or natural origin to increase the total daily in amounts conforming to the latest Philippine Recommended Energy and Nutrient Intake or international daily requirements. It is usually in the form of capsule, tablet, liquid, gel, powder, or pill and is not represented for use as a conventional food or as the sole item of a meal or diet or a replacement for drugs and medicines".[65] It is worth noting that, under the Philippine Food Fortification Act of 2000,[66] there is a mandatory requirement for the fortification of some stable foods to help reduce the problem of malnutrition and vitamin deficiency. Antioxidant supplementation plays an important role in the fortification of staple and processed foods. Since vitamin A deficiency is endemic to the Philippines, fortification with β-carotene/vitamin A is required.

Like Canadian natural health products, in Brunei, a dietary supplement (named health supplement) is defined as "any product that is used to supplement a diet and to maintain, enhance and improve the healthy function of the human body". It must be in medicinal forms such as capsules, tablets, powders, and liquids and may contain one or more or a combination of the following:[67]

1. vitamins, minerals, amino acids, fatty acids, enzymes, probiotics, and other bioactive substances; substances derived from natural sources, including animal, mineral, and botanical materials in the form of extracts, isolates, concentrates, and metabolites, and
2. synthetic sources of ingredients mentioned above but only where the safety of these has been proven.

In 2014, the Vietnamese Ministry of Health introduced regulations on the management of health foods.[68] These regulations govern health supplements, food supplements, and dietary supplements under the name of health protection foods, which are products in the form of capsules, pills, tablets, powders, and liquids. Like other Asian countries, in Vietnam, dietary supplements are vitamins, minerals, amino acids, fatty acids, enzymes, probiotics, and other bioactives originating from animals, minerals, and vegetables in the form of extracts, isolates, and concentrates including compounds with antioxidant properties. Although there is no positive list of permitted claims, supplements may bear nutrient function claims associated with bioactives, if scientifically justified.[69]

In Singapore, the health supplement "is a product that is used to supplement a diet and to support or maintain, enhance, and improve the healthy functions of the human body. It cannot be injectable or a preparation that needs to be sterile, such as injections and eyedrops. It cannot be an item of a meal or diet" and it must be administered in small doses (capsules, softgels, tablets, and liquids).[70] Health claims are allowed on health supplements, except for disease risk reduction, and must be substantiated by evidence that is relevant to the claims. The updated list of health claims for traditional medicine and health supplements was recently published and contains over 800 entries in various categories.[71] Antioxidant claims on health supplements are included in the general health or body part category, *e.g.*, "an antioxidant that protects the fat in body tissues from oxidation, antioxidants that help fight/protect (cell) against/reduce (the oxidative effects of/the oxidative damage caused by/cell damage caused by) free radicals, and beta-carotene, a provitamin A, helps to maintain eyesight, skin, membranes, and immune function".[71] Health supplements and other health-related products are regulated by the Health Sciences Authority.[71]

In Thailand, a health supplement is regulated by the Thai Food and Drug Administration,[72] which defines this kind of product as a supplement if it "contains common herbal ingredients and other bioactive ingredients at a daily dosage recognized by the Thai FDA as safe for food". Vitamin and mineral supplements must have the Thai RDI,[73] and the levels must be within the range of 15–100% of the Thai RDI. Antioxidant claims on supplements such as "vitamin A for night vision" are allowed if it is properly substantiated. Health supplements containing ingredients not approved as food ingredients or containing vitamins and minerals that exceed the Thai RDI value must be registered as traditional medicines, drugs, or any other related sub-categories.[74] Echinacea extracts that function as antioxidants, help defend cells against oxidative stress, reduce inflammation, and have other antioxidant effects are considered traditional medicine. Products classified as food supplements require a notification process.[74]

In Indonesia, the National Agency of Drug and Food Control of the Republic of Indonesia (Indonesian Badan Pengawas Obat dan Makanan) or Badan POM, a government agency of Indonesia) is responsible for governing the approval and registration of health and food supplements.[75] Thus, antioxidant components including vitamins A, C, and E, carotenoids, soy isoflavones, and tea polyphenols are approved for use in health and food supplements. Disease risk reduction claims "as an antioxidant, vitamin E may help reduce the possible risk of CHD" or soya isoflavones "may reduce the risk of atherosclerosis and CHD" are acceptable.[63]

Recently, the Association of Southeast Asian Nations (ASEAN) member states (Malaysia, Thailand, Vietnam, Laos, Myanmar, Singapore, Indonesia, the Philippines, and Brunei) implemented harmonization measures regarding traditional medicines and health supplements. Consequently, many ASEAN countries adopted terminology and guidelines on claims and claims substantiation for health supplements, and traditional medicine. Health supplements refer to "any products that are used to supplement a diet and to maintain, enhance and improve the healthy function of the human body and contain one or more vitamins, minerals, isolates, and other substances in the form of natural materials, as well as ingredients derived from synthetic sources", while traditional medicines refer to any medicinal products for human use consisting of active ingredients derived from natural sources (plants, animals, and/or minerals) used in the system of traditional Chinese or Indian medicinal practice or Jamu theory.[70] These products may bear traditional health maintenance claims, *e.g.*, "tonic traditionally used to strengthen the body by nourishing blood and invigorating vital energy" or "traditionally used to relieve cold and sore throat". In addition, functional claims, *e.g.*, "promoting healthy skin", "maintain/support alertness", "helps to relieve post-menopausal discomforts", and "support health in ageing", on health supplements are allowed. The use of the above claims is to be determined by Health Sciences Association.[70,71] Many of these claims are associated with antioxidant ingredients, for example, isoflavones are known to reduce vasomotor symptoms in postmenopausal women.

2.3 Conclusion

This chapter provided a comprehensive review of the regulations and health claims on antioxidant supplements. The importance of regulation is to provide consumers with safe and efficacious products. Antioxidant supplements may offer protection of cells from oxidative stress as well as protection of DNA, proteins, lipids from oxidative damage, and maintenance of good eye condition. Therefore, the information is shown as a health claim on the supplement label after the regulatory review process is complete and the claim is approved. Antioxidants such as catechins, anthocyanins, carotenoids, isoflavones, and various other phytochemicals, as well as antioxidant vitamins and minerals, are popular dietary supplements and approved by regulators in many countries. Although dietary supplements may be called by different names such as natural health products, health foods, herbal supplements, or traditional medicine, the form is common and includes capsules, tablets, tinctures, powders, or pills at a daily dosage level. In addition, a dietary supplement is not represented for use as a conventional food or as the sole item of a meal or diet or as a replacement for drugs and medicines.

References

1. Dietary Supplements Segment, Nutraceuticals – Global Market Trajectory and Analytics: Research and Markets, Published online at: https://www.researchandmarkets.com/reports/338402/nutraceuticals_global_market_trajectory_and (accessed March 20, 2022).
2. F. Fogacci and A. F. G. Cicero, Metabolic and vascular effects of an apple polyphenols extract. The results of a clinical trial on a highly standardized extract, *Agro Food Ind. Hi-Tech*, 2021, **32**, 28–30.
3. A. Das and C. K. Sen, Nutritional supplements and functional foods: functional significance and global regulations, in *Nutraceuticals and Functional Foods Regulations in the United States and Around the World*, ed. D. Bagchi, Elsevier, Oxford, UK, 2nd edn, 2014, pp. 13–39.
4. M. Yang, S.-J. Chung, C. E. Chung, D.-O Kim, W. O. Song and S. I. Koo, *et al.*, Estimation of total antioxidant capacity from diet and supplements in US adults, *Br. J. Nutr.*, 2011, **106**, 254–263.

5. Dietary Supplement Health and Education Act of 1994, *Public Law 103-417, 103rd Congress, 108 STAT, Congressional Record, Vol. 140*, National Institutes of Health, Bethesda, MD, 1994.
6. Natural Health Products Regulations, SOR/2003-196, June 5, 2003, Last Amended on February 11, 2022, Published online at https://laws-lois.justice.gc.ca/ (accessed March 15, 2022).
7. *Pathway for Licensing Natural Health Products Making Modern Health Claims, Version 1.0*, Health Canada, Ottawa, ON, Canada, 2006.
8. *Natural Health Products Management of Applications Policy*, Health Canada, Ottawa, ON, Canada, 2019.
9. Listing of Monographs, Health Canada, Modified December 7, 2021, Published online at: http://webprod.hc-sc.gc.ca/nhpid-bdipsn/monosReq.do?lang=eng (accessed March 15, 2022).
10. Multi-Vitamin/Mineral Supplements Monograph, Health Canada, Modified December 7, 2021, Published online at: http://webprod.hc-sc.gc.ca/nhpid-bdipsn/atReq.do?atid=multi_vitmin_suppl&lang=eng (accessed March 15, 2022).
11. *Electronic Code of Federal Regulations (e-CFR), 21 CFR 190.6, Requirement for Premarket Notification, Subchapter B, Food for Human Consumption, Part 190, Dietary Supplements, Subpart B, New Dietary Ingredient Notification*, The US National Archives and Records Administration, College Park, MD.
12. *Electronic Code of Federal Regulations (e-CFR), 21 CFR 101.54 (g), Specific Requirements for Nutrient Content Claims § 101.54 Nutrient Content Claims for "Good Source", "High", "More", and "High Potency." (g) Nutrient Content Claims Using the Term "Antioxidant", Part 101. Food Labeling, Subchapter D Specific Requirements for Nutrients Content Claims*, The US National Archives and Records Administration, College Park, MD.
13. *Guidance for Industry: Substantiation for Dietary Supplement Claims Made Under Section 403(r) (6) of the Federal Food, Drug, and Cosmetic Act. Docket No. FDA-2004-D-0303*, Center for Food Safety and Applied Nutrition, Washington, DC.
14. Permissible *vs.* Impermissible Structure/Function Claims for Dietary Supplements, Venable LLP, 2014, Published online at: https://www.venable.com/files/upload/FDLI-Dietary_Supplements.pdf (accessed February 28, 2022).
15. Qualified Health Claims, Guidance for Industry: Evidence-Based Review System for the Scientific Evaluation of Health Claims, Docket No. FDA-2007-D-0371, Center for Food Safety and Applied Nutrition, Washington, DC.
16. *Electronic Code of Federal Regulations (e-CFR), 21 CFR 101.70, Petitions for Health Claims, Part 101 Food Labeling, Subchapter E – Specific Requirements for Health Claims*, The US National Archives and Records Administration, College Park, MD.
17. Directive 2002/46/EC of the European Parliament and of the Council of 10 June 2002 on the approximation of the laws of the member states

relating to food supplements, *Official Journal of the European Union*, 2002, 183, pp. 51–57.
18. Regulations (EC) No. 1924/2006 of the European Parliament and of the Council of 20 December 2006 on nutrition and health claims made on foods, *Official Journal of the European Union*, 2006, 404/9, pp. 9–25.
19. I. Wohlfahrt, The regulatory situation of food supplements in Europe, *Agro Food Ind. Hi-Tech*, 2018, **29**, 22–24.
20. Commission Regulations (EU) No. 432/2012 of 16 May 2012 establishing a list of permitted health claims made on foods, other than those referring to the reduction of disease risk and to children's development and health, *Official Journal of the European Union*, 2012, 136, pp. 1–40.
21. EU Register of Health Claims, European Commission, Updated March 11, 2022, Published online at: https://ec.europa.eu/food/safety/labelling_nutrition/claims/register/public/?event=register.home (accessed March 28, 2022).
22. EFSA, Guidance for the scientific requirements for health claims related to antioxidants, oxidative damage and cardiovascular health, *EFSA J.*, 2018, **16**, 1–21.
23. Regulation (EU) 2015/2283 of the European Parliament and of the Council of 25 November 2015 on novel foods, amending Regulation (EU) No. 1169/2011 of the European Parliament and of the Council and repealing Regulation (EC) No. 258/97 of the European Parliament and of the Council and Commission Regulation (EC) No. 1852/2001, *Official Journal of the European Union*, 2015, 327/1, pp. 1–22.
24. Commission Implementing Regulation (EU) 2017/2470 of 20 December 2017 establishing the Union list of novel foods in accordance with Regulation (EU) 2015/2283 of the European Parliament and of the Council on novel foods, *Official Journal of the European Union*, 2017, 351/1, pp. 72–141.
25. Commission Decision of 21 November 2008 establishing of a list of herbal substances, preparations and combinations thereof for use in traditional herbal medicinal products, 2008/911/EC, *Official Journal of the European Union*, 2008, 328, pp. 42–75.
26. S. Arai, Studies of functional foods in Japan – State of the Art, *Biosci., Biotechnol., Biochem.*, 1996, **60**, 9–15.
27. *Ministerial Ordinance of 1991, The Nutrition Improvement Law Enforcement Regulations, Ministerial Ordinance No. 41, July 1991, Amendment to Ministerial Ordinance No. 33*, Tokyo, Japan, 1996.
28. S. Arai, Global view on functional foods: Asian perspectives, *Br. J. Nutr.*, 2002, **88**(Suppl.), S139–S143.
29. H. Ohama, H. Ikeda and H. Mariyama, Health foods and foods with health claims in Japan, in *Nutraceuticals and Functional Foods Regulations in the United States and Around the World*, ed. D. Bagchi, Elsevier, Oxford, UK, 2nd edn, 2014, pp. 265–299.
30. T. Shimizu, Health claims and scientific substantiation of functional foods – Japanese System Aiming the Global Standard, *Curr. Top. Nutraceutical Res.*, 2003, **1**, 1–12.

31. H. Tanaka, F. Kaneda, R. Suguro and H. Baba, Current system for regulation of health foods in Japan, *J. Jpn. Med. Associ.*, 2004, **47**, 436–450.
32. H. Ikeda, Food for specified health uses and Japanese regulations: The inventor of functional foods in the world, in *IFT Annual Meeting and Food Expo*, 16–19 July 2016, Chicago.
33. What are Food with Functional Claims? The Consumer Affairs Agency, Food Labeling Division, Published online at: https://www.caa.go.jp/policies/policy/food_labeling/informatio/pamphlets/pdf/151224_1.pdf (accessed March 22, 2022).
34. T. Koe, Foods with function claims in Japan: four years on, which firms have benefited most from the regulations? *Nutra Ingredients-Asia*, May 23, 2019.
35. H. Kamioka, K. Tsutani, H. Origasa, T. Yoshizaki, J. Kitayuguchi and M. Shimada, *et al.*, Quality of systematic reviews of the foods with function claims registered at the consumer affairs agency web site in Japan: a prospective systematic review, *Nutr. Res.*, 2017, **40**, 21–31.
36. *Therapeutic Goods Act 1989, No. 21, 1990, Compilation No. 68 dated July 1, 2017*, Australian Government, The Office of Parliamentary Counsel, Canberra, Australia.
37. *Dietary Supplements Regulations 1985, SR 1985/208, Amended under the Legislation Act 2012*, The Ministry of Health, Wellington, New Zealand.
38. *Medicines Act 1981, Public Act No. 118, Version as of December 7, 2021*, The Ministry of Health, Wellington, New Zealand.
39. Notice Preventing the Making of Representations under Section 42DKB of the Therapeutic Goods Act 1989, Attachment A, Australian Government, Department of Health, Therapeutic Goods Administration, March 4, 2022, Published online at: https://www.tga.gov.au/prevention-notice-contract-manufacturing-packaging-services-pty-ltd (accessed March 28, 2022).
40. Chinese New Regulation of Health Food Products, Natural Products Insider, October 20, 2015, Published online at: https://www.naturalproductsinsider.com/archive/chinese-new-regulation-health-food-products (accessed March 20, 2022).
41. *Food Hygiene Law of the People's Republic of China 1995, Order of the President of the People's Republic of China No. 59, adopted at the 16th Meeting of the Standing Committee of the Eighth National People's Congress*, Beijing, the People's Republic of China, 1995.
42. J. Zawistowski, Legislation of functional foods in Asia, in *Functional Foods: Concept to Profit*, ed. M. Saarela, Woodhead Publishing Ltd., Oxford, Cambridge, UK, 2nd edn, 2011, pp. 75–108.
43. *Decree of the State Food and Drug Administration No. 22, The Administrative Measures for the Registration and Filing of Health Foods*, Beijing, the People's Republic of China, 2016.
44. C. Hu, Historical changes of raw materials and claims of health food regulations in China, in *Nutraceuticals and Functional Foods Regulations*

in the United States and Around the World, ed. D. Bagchi, Elsevier, Oxford, UK, 2nd edn, 2014, pp. 363–388.

45. Notice on the Issuance of Nine Health Function Evaluation Methods, Including Oxidation Function Evaluation Methods, State Food and Drug Administration, April 23, 2012, Published online at: http://www.law-lib.com/law/law_view.asp?id=382899 (accessed March 20, 2022).
46. G. N. K. Ganesh, A. Ramachandran, K. R. Suresh, V. Senthil and R. B. Priyadharshini, Nutraceuticals – a regulatory review, *Int. J. Drug Regul. Aff.*, 2015, **3**, 22–29.
47. R. Sharma, N. Natália Martins, K. Kuca, A. Chaudhary, A. Kabra and M. M. Rao, *et al.*, A traditional Indian bioactive health supplement, *Biomolecules*, 2019, **9**, 1–24.
48. *Ashwagandha – Withania Somnifera, Natural Health Product Monograph*, Health Canada, Ottawa, ON, Canada, 2019.
49. *The Food Safety and Standards Act, 2006, No. 34, Ministry of Law and Justice*, Legislative Department, New Delhi, India, 2006.
50. S.-M. Ock, S.-S. Hwang, J.-S. Lee, C.-H. Song and C.-M. Ock, Dietary supplement use by South Korean adults: Data from the national complementary and alternative medicine use survey (NCAMUS) in 2006, *Nutr. Res. Pract.*, 2010, **4**, 69–74.
51. H. A. Park, Top 10 dietary supplements of Korean adults from the 4th Korea National Health and Nutrition Examination Survey, *Korean J. Fam. Med.*, 2011, **32**, 263–266.
52. M. Kang, D. W. Kim, Y. J. Baek, S.-H. Moon, H. J. Jung and Y. J. Song, *et al.*, Dietary supplement use and its effect on nutrient intake in Korean adult population in the Korea National Health and Nutrition Examination Survey IV (2007–2009) data, *Eur. J. Clin. Nutr.*, 2014, **68**, 804–810.
53. Health Functional Food Act, *Act No. 6727, Aug. 26, 2002, Amended by Acts No. 7211, March 22, 2004; No. 7428, March 31, 2005; No. 8033, Oct. 4, 2005; No. 8365, Apr. 11, 2007; No. 8852, Feb. 29, 2008; No. 8941, March 21, 2008; No. 9932, Jan. 18, 2010; No. 10128, March 17, 2010; No. 10219, March 31, 2010, No. 11690, Mar. 23, 2013, No. 11985, Jul. 30, 2013, No. 12106, Aug. 13, 2013, No. 12389, Jan. 28, 2014, No. 12669, May 21, 2014*, Korea Ministry of Food and Drug Safety, Seoul, Republic of Korea.
54. Regulation on Approval of Functional Ingredient for Health Functional Food, *Note No. 2004-12, Jan. 31, 2004. Amended by Notes No. 2006-36, Aug. 29, 2006, No. 2007-51, Jul. 11, 2007*, Korea Food and Drug Administration, Seoul, Republic of Korea.
55. Regulations Concerning Recognition of Functional Ingredients and Standards and Specifications for Health Functional Foods, *Note No. 201426, Feb. 12, 2014, Amended by Note No. 2016141, Dec. 21, 2016*, Korea Food and Drug Administration, Seoul, Republic of Korea.
56. GAIN Report – KS2022-0008, *Food and Agricultural Import Regulations and Standards Country Report, Republic of Korea, Global Agricultural Information Network, USDA Foreign Agriculture Service*, Washington, DC, 2022.

57. J. Zawistowski, Regulation of functional foods in selected Asian countries in the Pacific Rim, in *Nutraceuticals and Functional Foods Regulations in the United States and Around the World*, ed. D. Bagchi, Elsevier, Oxford, UK, 2nd edn, 2014, pp. 419–464.
58. South Korea Regulatory Overview, New Hope Network, October 19, 2019, Published online at: https://www.newhope.com/supply-news-amp-analysis/south-korea-regulatory-overview (accessed April 9, 2022).
59. Korea Health Supplements Association, Introduction to the Health Functional Foods in Korea, May 2017, Published online at: https://www.khsa.or.kr/assets/images/user/contents/KHSA_english.pdf (accessed April 9, 2022).
60. Health Functional Food Code 2021, English Version No. 2020-92, September 23, 2020, Korea Ministry of Food and Drug Safety, Published online at: https://www.scribd.com/document/539736704/Health-Fuctional-Food-Code-No-2020-92 (accessed April 9, 2022).
61. J. Y. Kim, S. J. Kim and H. J. Lee, Regulations on health functional foods in Korea, in *Nutraceuticals and Functional Foods Regulations in the United States and Around the World*, ed. D. Bagchi, Elsevier, Oxford, UK, 2nd edn, 2014, pp. 389–400.
62. GAIN Report – KS1239, *Functional Food Market Brief Update, Republic of Korea, Global Agricultural Information Network*, USDA Foreign Agriculture Service, Washington, DC, 2012.
63. J. Zawistowski, Health foods and foods with health claims in the Asian subcontinent – countries of the Association of Southeast Asian Nations, in *Developing New Functional Food and Nutraceutical Products*, ed. D. Bagchi and S. Nair, Elsevier/Academic Press, Amsterdam, The Netherlands, 2017, pp. 323–367.
64. Guide to the Classification of Food-Drug Interface Products, Food-Drug Interphase Products, The National Pharmaceutical Regulatory Division or by the Food Safety And Quality Division of the Ministry Of Health Malaysia, Published online at: https://www.npra.gov.my/index.php/en/classification-guideline/product-classification-guideline.html (accessed April 1, 2022).
65. *Food Safety Act of 2013, The Implementing Rules and Regulations of Republic Act No. 10611, "An Act to Strengthen the Food Safety Regulatory System in the Country to Protect Consumer Health and Facilitate Market Access of Local Foods and Food Products, and for Other Purposes" Otherwise Known as the "Food Safety Act of 2013"*, Republic of the Philippines Department of Agriculture and Department of Health, Manila, Philippines, 2015.
66. *Republic Act No. 8976, An Act Establishing the Philippine Food Fortification Program and for Other Purposes, The Implementing Rules and Regulations of Republic Act No. 8976*, Department of Health, Office of the Secretary, Republic of the Philippines, Manila, Philippines, 2000.
67. *Brunei Darussalam Guidelines for Manufacturing and Handling of Halal Medicinal Products, Traditional Medicines, and Health Supplements,*

Guidance Document 2010, The Religious Council Negara Brunei Darussalam, Bandar Seri Begavan, Brunei, 2010.
68. *Regulations on Management of Functional Foods, Circular No. 43/2014/TT-BYT dated November 24, 2014*, The Ministry of Health, Hanoi, Vietnam, 2014.
69. E. S. Tee, Health claims and scientific substantiation in SEA – process and key learnings, in *9th Seminar on Nutrition Labeling, Claims and Communication Strategies*, 2015 August 4–5, ILSI SEA Region, Manila, Philippines.
70. *Guidelines for Claims and Claims Substantiation of Traditional Medicines and Health Supplements, Regulatory Guidance*, Health Science Authority, Singapore, 2022.
71. List of Health Claims for Traditional Medicines and Health Supplements, Health Science Authorities, March 1, 2022, Published online at: https://www.hsa.gov.sg/docs/default-source/hprg-tmhs/chpb-tmhs/tmhs_claims_list.pdf (accessed March 15, 2022).
72. Thai FDA, Food and Drug Administration Thailand, Ministry of Public Health, Published online at: https://www.fda.moph.go.th/sites/FDA_EN/Pages/Main.aspx (accessed March 22, 2022).
73. *Nutrition Labeling, Notification of Ministry of Public Health No. 182 B.E. 2541, The Government Gazette Volume 115, Part 47 (Ngor)*, Bangkok, Thailand, 1998.
74. *Healthcare Resource Guide: Healthcare Procurement*, Bangkok, Thailand, 2016.
75. *Regulation of the Head of National Agency of Drug and Food Control of the Republic of Indonesia, No. HK.00.05.41.1381 on Procedure of Food Supplement Registration*, Head of National Agency of Drug and Food Control of the Republic of Indonesia, Jakarta, Indonesia, 2005.

CHAPTER 3

Reduction of Cardiovascular Disease Risks with Dietary Antioxidant Supplements

LIANLIANG LIU,[a] YUEQIN LI,[a] LEZHEN DONG,[a] YUNZHEN ZHANG,[a] ZUFANG WU[a] AND JIANBO XIAO*[b,c]

[a] College of Food and Pharmaceutical Sciences, Deep Processing Technology Key Laboratory of Zhejiang Province Animal Protein Food, Ningbo University, Ningbo, PR China; [b] Institute of Food Safety and Nutrition, Jinan University, Guangzhou 510632, China; [c] Nutrition and Bromatology Group, Department of Analytical and Food Chemistry, Faculty of Sciences, Universidade de Vigo, 32004 Ourense, Spain
*Email: jianboxiao@yahoo.com

3.1 Introduction

Heart and cerebrovascular diseases are collectively referred to as cardiovascular disease (CVD). It has also the highest incidence and mortality rate in the world, mainly including coronary heart disease (CHD), cerebrovascular disease (stroke), ischemic heart disease (IHR), peripheral arterial disease (PAD), heart failure, and congenital heart disease.[1] Atherosclerosis (AS) is the basis of many CVDs, which is induced by oxidative stress and inflammation. Formation of AS lesions arises from lipid metabolism disorder, lipid pathological oxidation, and platelet migration and aggregation.

Food Chemistry, Function and Analysis No. 39
Dietary Supplements with Antioxidant Activity: Understanding Mechanisms and Potential Health Benefits
Edited by Cesarettin Alasalvar, Fereidoon Shahidi and Chi-Tang Ho

Published by the Royal Society of Chemistry, www.rsc.org

Hyperlipidaemia and dyslipidaemia are also essential risk factors for AS lesions. Hyperlipidaemia, which is characterized by increased circulating lipid levels [*e.g.*, total cholesterol (TC), triacylglycerols (TAG), and low-density lipoprotein cholesterol (LDL-C)], doubles the chance of developing CVD.[2] It changes the composition of lipoproteins and promotes the accumulation of inflammatory mediators and reactive oxygen species (ROS), resulting in systemic immunity activation and inflammation response.[3] When ROS are overproduced in the body, LDL-C easily enters the vascular endothelium and oxidative modification is enhanced, and thus LDL-C is converted into oxidized LDL (ox-LDL) and then swallowed by macrophages and vascular smooth muscle cells (VSMC) to form foam cells which increase the risk of IHR and stroke.[4]

Hyperlipidaemia and hypertension can cause vascular endothelial cell (VEC) dysfunction.[4] VEC dysfunction is closely related to the reduction of nitric oxide (NO) or bioavailability. Under physiological conditions, the vasodilator substances synthesized and released by endothelial cells (such as NO) maintain a dynamic balance with vasoconstrictor substances [*e.g.*, angiotensin-II (Ang-II) and endothelin-1 (ET-1)].[5] In addition, sustained hypertension stimulation (such as ROS and Ang-II) can inhibit endothelial nitric oxide synthase (eNOS) expression and decrease the bioavailability of NO, leading to a decrease in endothelium-dependent vasodilation in the vascular system.[6] The blood pressure increase promotes the infiltration of plasma lipids into vascular intimal cells, causing platelet accumulation and VSMC proliferation, leading to AS occurrence.[7]

Hyperlipidaemia and hypertension inevitably cause an increase in ROS production, which can induce serious dysfunction of cardiomyocytes and endothelial cells, leading to enzyme and protein denaturation.[6] Continuous oxidative stress can cause inflammation, which mainly manifested in changing the synthesis and expression of various cell inflammatory factors and adhesion molecules, such as interleukin-6 (IL-6) and tumour necrosis factor-α (TNF-α).[6] Three major risk factors for CVD, namely, hyperlipidaemia, hypertension, and oxidative stress, are illustrated in Figure 3.1.

Hence, reducing oxidative stress and inflammation occurrence and improving VEC function is an effective strategy to reduce CVD risk. The effects of dietary supplements on oxidative stress and inflammation in the CVD process reported in recent epidemiological studies, *in vitro* and *in vivo* animal studies, and human intervention studies are reviewed in detail. The main mechanisms of dietary supplements in preventing CVD are clarified. This provides a new direction to prevent CVD and lay a theoretical foundation for comprehensively exploring the potential health benefits of biologically active ingredients in foods that help reduce CVD risks.

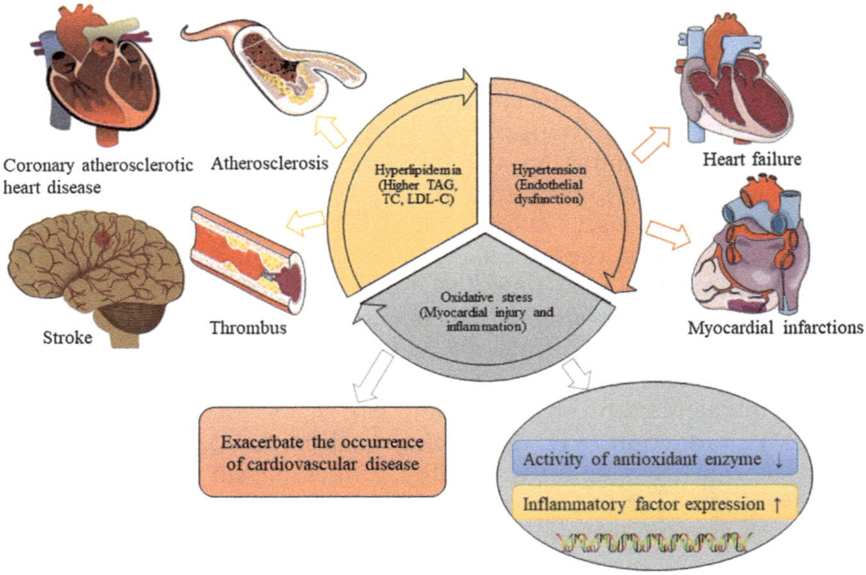

Figure 3.1 Three major risk factors for CVD: hyperlipidaemia, hypertension, and oxidative stress. Oxidative stress can aggravate hyperlipidaemia and hypertension, and then accelerate the occurrence of CVD. Abbreviations: CVD, cardiovascular disease; LDL-C, low-density lipoprotein cholesterol; TAG, triacylglycerols; TC, total cholesterol.

3.2 Health Benefits of Dietary Supplements with Antioxidant Activity for CVD

Dietary supplements are a widely studied intervention that can effectively reduce CVD risk in high-risk groups. Evidence from an increasing number of epidemiological and clinical trials shows that adequate intake of foods with antioxidant bioactive substances rich in phenolic acids (gallic acid, sinapic acid, and ferulic acid), stilbenes (resveratrol), flavonoids [quercetin, naringenin, hesperidin, fisetin, isorhamnetin, anthocyanins, curcumin, dihydromyricetin, epigallocatechin gallate (EGCG), and epicatechin gallate (ECG)], alkaloids (berberine), saponins (notoginsenoside R1 and astragaloside), vitamins (lycopene, vitamin E, vitamin D, and coenzyme Q10), and monounsaturated and polyunsaturated fatty acids (MUFA and PUFA, respectively), such as oleic acid, alpha-linolenic acid (ALA), eicosapentaenoic acid (EPA), and docosahexaenoic acid (DHA), can effectively lower cholesterol and blood pressure levels as well as ameliorate endothelial function, thereby decreasing CVD risk. Thus, foods rich in antioxidant bioactive substances can serve as possible candidates for CVD prevention. They are abundant in Mediterranean diet (MedDiet) related foods, which exhibit potential antioxidant and anti-inflammatory properties by blocking oxidative stress and inhibiting proinflammatory mediators.

For a long time, people have been recommended to consume adequate amounts of plant foods with strong antioxidant properties, which can effectively reduce CVD risk. These recommendations are mainly in view of epidemiological study results. Dietary supplements are beneficial to prevent and slow down CVD occurrence. Increasing intake of foods with antioxidant bioactive substances has been strongly correlated with reduced risk of CVD, stroke, and all-cause mortality.[8,9]

3.3 Epidemiological Studies

In subsequent sections, the intake of phenolic acids, flavonoids, alkaloids, saponins, vitamins, and MUFA and PUFA is discussed with respect to CVD risk reduction.

3.3.1 Phenolic Acid and Flavonoids

In a prospective cohort study conducted in Denmark, 55 169 Danish residents aged 52–60 years without history of ischemic stroke were followed up for 21 years. It was found that the intake of total flavonoids (719 mg per day), flavonols (696 mg per day), and flavanol oligos + polymers (717 mg per day) led to a reduction in the morbidity of ischemic stroke of 12% [hazard ratio (HR) (95% confidence interval (CI)): 0.88 (0.81, 0.96)], 10% [0.90 (0.82, 0.98)], and 18% [0.82 (0.75, 0.89)].[10] Similarly, a prospective study aimed to reveal the relationship between the intake of flavonoid-rich fruits and the risk of stroke and subtype stroke in the Japanese population, in which 87 177 Japanese people aged 44–76 years were treated for an average of 13.1 years tracking survey. The total consumption of flavonoid-rich fruits, especially citrus fruits, strawberries, and grapes, was related to lower stroke risk in women [HR (95% CI): 0.70 (0.58, 0.84)].[11] Of note, a dose–response meta-analysis pointed out that increasing dietary total flavonoid consumption (500 mg per day) had a linear relationship with reduced CVD risk. Increasing the intake of anthocyanins and flavan-3-ols had the best effect on reducing CVD risk, while increasing flavonol and flavone consumption was negatively correlated with CHD. There is an inverse correlation between flavanone intake and stroke risk. Kaempferol and quercetin intake was linearly correlated with lower risks of CVD and CHD, respectively.[12] This provides strong evidence for the potential cardiovascular benefits of dietary supplementation with foods rich in flavonoids. A daily consumption of 800 g of fruit can effectively reduce the relative risk of CVD. Dietary supplementation of foods rich in flavonoids can also effectively lower blood pressure. Recently, a systematic review and meta-analysis of randomized controlled trials on frequent consumption of berries rich in anthocyanins and proanthocyanidins on CVD risk factors showed that the higher intake of berries decreased systolic blood pressure (SBP) and diastolic blood pressure (DBP) by 3.68 mmHg [(95% CI): (6.79, 0.58), $P = 0.02$] and 1.52 mmHg [(95% CI): (2.87, 0.18), $P = 0.04$], respectively.[8] Therefore, dietary supplementation of fruits rich in anthocyanins and proanthocyanidins can reduce CVD risk.

It is noteworthy that the intake of grain brans, abundant in polyphenols and dietary fibre, has great potential in reducing CVD risk. A systematic review and meta-analysis of prospective cohort studies pointed out that there was an inverse relationship between the consumption of more whole grains and various causes of mortality and CVD. For every supply, three servings of whole grains per day reduced CVD mortality by 25%.[9] Similarly, a 9.5 year follow-up survey was conducted on 137 130 subjects aged 30–75 years. The highest intake of refined grains (≥350 g per day) was related to a higher risk of overall mortality [HR (95% CI): 1.27 (1.11, 1.46), $P = 0.004$] and major cardiovascular events [HR (95% CI): 1.33 (1.16, 1.52), $P < 0.001$].[13] There were no significant correlations ($P > 0.05$) between whole grain or white rice consumption and cardiovascular events. There was no consistent conclusion that the intake of whole grains was positively correlated with reduced CVD risk. However, since whole grains contain abundant polyphenols and dietary fibre, it is still strongly recommended to consume more whole grains and reduce consumption of refined grains.[13]

Moreover, EGCG, ECG, epicatechin (EC), and epigallocatechin (EGC) also have beneficial effects on patient health after a CVD event. A meta-analysis based on population studies found that consuming a cup (237 mL) of tea every day (338 mg per day and 280 mg per day of total flavonoids for green and black teas, respectively) reduced the rate of mortality from CVD by an average of 4%. The risk of all-cause mortality and stroke was reduced by 1.5% and 4%, respectively.[14] Similarly, a large long-term cohort study of Japanese men and women found that stroke survivors who consumed more than 7 cups per day of green tea [HR (95% CI): 0.38 (0.20, 0.71)] had about 60% lower all-cause mortality than those who rarely consumed green tea, while this connection was not observed in people who did without a myocardial infarction or stroke history.[15] Moreover, a study using a two-sample Mendelian randomization method found that genes predict that drinking an extra cup of tea a day will reduce small-vessel stroke risk [odds ratio (OR) (95% CI): 0.79 (0.69, 0.91); $P = 0.506$].[16]

3.3.2 Alkaloids and Saponins

A recent meta-analysis of indole and isoquinoline alkaloids clarified that attenuating aortic AS in apolipoprotein E (Apo E) −/− mice using alkaloid intervention was significantly related to the reduction of the aortic AS plaque area [standardized mean difference (SMD): −3.19, 95% CI: (−3.88, −2.51)]. Their subgroup analysis showed that indole and isoquinoline alkaloids significantly reduced the burden of AS, respectively [SMD: −2.73, −4.19; 95% CI: (−3.56, −1.90), (−5.18, −3.20)].[17] This indicates that indole and isoquinoline alkaloids have a direct anti-atherosclerotic effect in Apo E −/− mice. Of note, a preclinical systematic review of the effects of notoginsenoside R1 on organ ischemia/reperfusion (I/R) injury also pointed out that notoginsenoside R1 therapeutic intervention can significantly reduce the area of myocardial infarction and cerebral infarction caused by I/R and reduce serum creatinine concentration in the kidneys. The findings indicated that the multiple-organ protection of notoginsenoside R1 after I/R injury is mainly through antioxidant, anti-inflammatory, and anti-apoptosis mechanisms, accelerating angiogenesis and ameliorating energy metabolism.[18]

3.3.3 Vitamins

It is obvious that lycopene may have a great potential in reducing mortality from CVD. A recent meta-analysis of epidemiological focus on lycopene and tomatoes and CVD risk showed that the risk of stroke [HR (95% CI): 0.74 (0.62–0.89), $P=0.02$; $I^2=32$] and CVD [HR (95% CI): 0.86 (0.77–0.95), $P=0.003$; $I^2=0$] in individuals with high lycopene intake or high serum lycopene metabolite was significantly reduced as well as individual mortality [HR (95% CI): 0.63 (0.49–0.81), $P<0.001$; $I^2=46$].[19] Moreover, a recent epidemiological study found that high vitamin D consumption was negatively correlated with the risk of CVD in the Mexican population. Around 6294 Mexican people were followed for a ten-year study. It was found that the subjects with the highest vitamin D intake [516.21 international unit (IU)] had lower TAG levels (14.6 mg dL^{-1}), their waist circumference (WC) was decreased by 2.0 cm, and hypertriglyceridemia and obesity were also reduced. Subjects with intake of the highest amount of vitamin D (516.21 IU) had a lower risk of CVD within 10 years compared with those with the lowest intake (70.4 IU) [OR (95% CI): 0.51 (0.33, 0.77); $P=0.007$].[20]

Similarly, vitamin E inhibits lipid peroxidation by scavenging ROS. A meta-analysis of vitamin E consumption and stroke risk in view of prospective study results indicated that a higher intake of vitamin E in the diet was associated with a lower risk of stroke. People with high dietary vitamin E consumption had a 17% lower risk of stroke.[21] Higher vitamin E consumption was negatively correlated with the overall risk of stroke [relative risk (RR) = 0.83, 95% CI: (0.73, 0.94)]. In a follow-up of <10 years, dietary vitamin E consumption was related to decreased stroke risk [RR = 0.84, 95% CI: (0.72, 0.91)]. Dietary vitamin E intake in the range of 15–39.45 mg per day seems to be conducive to the prevention of stroke.[21] Noteworthily, coenzyme Q10 supplement can enhance the prognosis of critically ill patients in the intensive care unit. A 3.2 year follow-up study was carried out on 242 patients with CVD (mean age: 67 ± 16 years, mean coenzyme Q10 level: 0.55 ± 0.35 mg L^{-1}), in which the patients were classified based on serum coenzyme Q10 levels into high (>0.46 mg L^{-1}) and low (≤0.46 mg L^{-1}) coenzyme Q10 groups.[22] The average serum coenzyme Q10 levels of those who died from CVD were significantly lower than those of the survivors (0.48 *versus* 0.58 mg L^{-1}; $P=0.035$). Meanwhile, creatinine and high-sensitivity C-reactive protein (hs-CRP) levels were higher than those of the survivors [HR (95% CI): 0.14 (0.02, 0.81); $P=0.027$]. Patients with low coenzyme Q10 levels also had a significantly higher incidence of all-cause mortality.[22]

This indicates that low serum coenzyme Q10 levels can be used as a predictor and potential treatment target.

3.3.4 MUFA and PUFA

MUFA + PUFA exhibit the potential of protecting the cardiovascular system and the heart. A meta-analysis of current prospective cohort studies showed that there is an obvious reverse dose relationship between total nut intake and morbidity and mortality of CVD, CHD, and atrial fibrillation risk.[23] As far as specific types of nuts were concerned, the intake of tree nuts was related to lower CVD incidence and mortality, respectively [risk ratio: 0.85, 0.77; 95% CI: (0.80–0.91), (0.72–0.82); $I^2 = 0\%$, $I^2 = 3\%$]. Intake of peanuts and walnuts was related to lower CHD incidence and mortality, respectively [risk ratio: 0.82, 0.76; 95% CI: (0.69–0.96), (0.67–0.86); $I^2 = 74\%$, $I^2 = 46\%$], and lower stroke mortality [risk ratio: 0.83; 95% CI: (0.75–0.93); $I^2 = 0\%$].[23] This conclusion was similar to almond consumption and CVD risk factors: a meta-analysis of randomized controlled trials. Almond diet intervention led to the greatest reduction in apolipoprotein B (Apo B), TC, and LDL-C. Almond intervention with dietary intake >42.5 g per day significantly reduced TC, body weight, DBP, and fasting blood glucose (FBG). Moreover, almond intervention with dietary intake ≤42.5 g per day significantly reduced TC/high density lipoprotein cholesterol (HDL-C), LDL-C/HDL-C, and Apo B compared to subjects with no intake of nuts.[24] Consuming almonds decreased CVD risk *via* reducing blood lipids, body weight, and Apo B, which was mainly attributed to the richness of MUFA + PUFA and phenolic compounds.

Recent epidemiological data have shown that there is a positive correlation between consumption of foods with strong antioxidant activity (fresh fruits, vegetables, whole grains, nuts, and insisting on drinking tea) and reduced CVD risk. Although some dietary supplements are not significantly related to reduced CVD risk, it is still recommended to increase the intake of food with antioxidant bioactive substances rich in phenolic acids, stilbenes, flavonoids, alkaloids, saponins, vitamins, and MUFA and PUFA. Some epidemiological studies have also shown limitations related to confounding factors, which may reduce the accuracy of the research results. For example, it does not consider socioeconomic status, education, gender, age, diet or lifestyle information, and disease history. Hence, it is recommended to consider all potential confounding factors and more randomized controlled trials with enough participants to improve the validity of the research results.

3.4 *In Vitro* and *In Vivo* Animal Studies

There is growing evidence that hyperlipidaemia is closely related to AS and plays an important role in CVD development. Hyperlipidaemia is a result of abnormal lipid metabolism, which can cause plasma lipid levels to become higher than normal.[2] A lower concentration of HDL-C is negatively correlated with CVD development.[25] The circulating LDL particles penetrate the endothelium of the artery wall and are subsequently oxidized. These ox-LDL particles induce oxidative stress and inflammation in the overlying endothelium cells and surrounding smooth muscle cells (SMC).[3]

ROS play a crucial role in the pathophysiological process of hypertension. Endothelial oxidative stress is an outcome of increased uncoupled eNOS synthase production and intracellular enzyme nicotinamide adenine dinucleotide phosphate hydrogenase (NADPH) activity and the damage of mitochondrial activity, which leads to a reduction in the synthesis and bioavailability of NO, causing endothelial dysfunction.[7] Inhibiting the activity of angiotensin converting enzyme (ACE) is the main target for lowering blood pressure, and thus promoting the production of NO is also an effective way to lower blood pressure. The increase of ROS and oxidative stress during MI/R can cause significant heart damage. Excessive production of ROS in the ischemic area of the myocardium can directly trigger cardiomyocyte apoptosis, inflammation, and energy metabolism disorders.[26]

The cardioprotective effects of dietary antioxidants are directly correlated with the relatively low morbidity of CVD. Regarding the cardioprotective effects of dietary supplements, many *in vitro* and *in vivo* animal studies have focused on evaluating the capacity of dietary supplements effectively to reduce CVD risk by resisting hyperlipidaemia and hypertension, improving endothelial cell (EC) dysfunction, and ameliorating oxidative stress and proinflammatory marker expression (Table 3.1).

In the subsequent sections, bioactive substances (phenolic acids, stilbenes, flavonoids, alkaloids, saponins, vitamins, and MUFA + PUFA) responsible for reducing the risk of CVD are discussed in detail.

3.4.1 Phenolic Acids

Phenolic acids can effectively lower blood pressure and improve endothelial dysfunction. What cannot be ignored is that phenolic acid alleviates hypertension *via* improving myocardial hypertrophy, fibrosis, and reducing oxidative stress. The administration of gallic acid alleviates myocardial hypertrophy and dysfunction through regulation of oxidative stress, inflammatory response, and myocardial fibrosis. Gallic acid can also attenuate the vascular oxidative stress of Ang-II perfusion of the aorta. During gallic acid treatment, proteasome trypsin-like and chymotrypsin-like activities were reduced and the degradation of eNOS and the reduction of NO levels were inhibited.[7] Additionally, it is also worth noting that the *N*′-nitro-L-arginine methyl ester hydrochloride (L-NAME) was used to induce hypertension in rats. After L-NAME induced hypertension in rats, hydrophobicity changes of free radicals in the myofibril structure, extensive vacuoles in the cytoplasm, myocardial hypertrophy and nuclear pyknosis were observed. After treatment with sinapic acid, the above changes were alleviated. Combined administration of sinapic acid and captopril can effectively reduce hypertension, enhance cardiovascular function, and reduce myocardial hypertrophy in hypertensive rats.[27] This shows that sinapic

Table 3.1 Main results of *in vitro* and *in vivo* animal studies to evaluate the relationship between dietary supplementation and CVD.

Molecule	Type of compound	Model of study	Treatment protocol	Underlying mechanism of the observed results	Ref.
Gallic acid	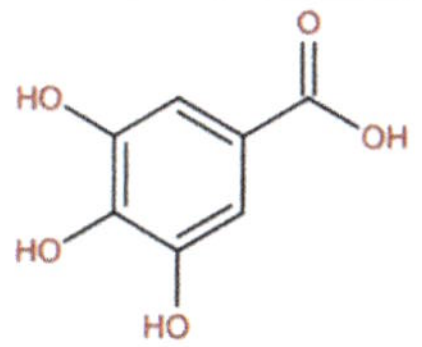	Wild-type C57BL/6J male mice	5–20 mg kg^{-1} (for 2 weeks)	- Ameliorated hypertension, vascular inflammation, and fibrosis - Abolished vascular endothelial dysfunction and oxidative stress	7
Sinapic acid	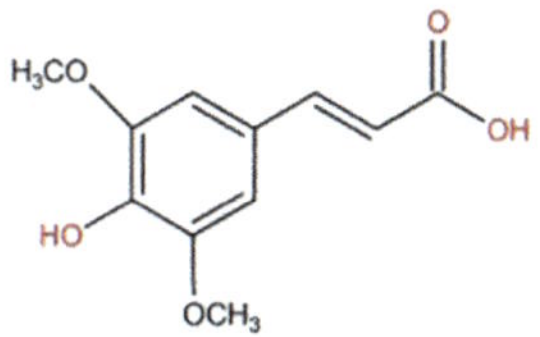	Albino Wistar male rats induced by L-NAME	40 mg kg^{-1} (for 4 weeks)	- TC, TAG, VLDL-C, LDL-C, Ang II, AT2R, ET-1, and Ang-II↓ - Serum HDL-C, SOD activity, NOS3, and GSH↑ - Hyperlipidaemia, oxidative stress, and myocardial hypertrophy↓	27
Quercetin	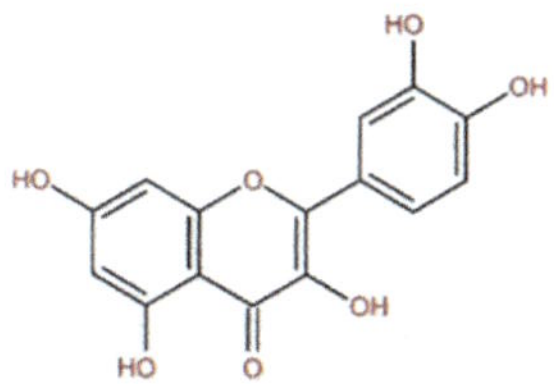	Male Wistar rats induced by HFD with streptozotocin	30 mg kg^{-1} (for 2 weeks)	- TC, TAG, NF-κB, and IL-1β levels↓ - HDL-C and IL-10 levels, the activity of SOD, CAT, and GPX↑ - AS lesions, atherogenic index, and MDA levels↓	30
Naringenin	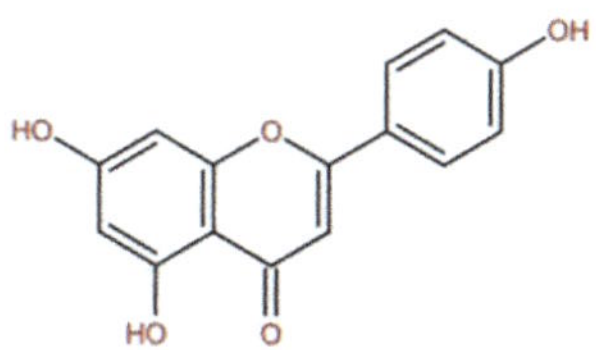	Male Sprague-Dawley rats H9c2 cardiomyoblasts	50 mg kg^{-1}; oral gavage (for 5 days) 80 µmol L^{-1} (for 6 hours)	- Myocardial apoptosis and infarction↓ - MI/R-induced oxidative stress and ER stress↓ - Superoxide generation and myocardial MDA level↓ - ATF6, p-PERK, and EIF2α↓	31

Fisetin	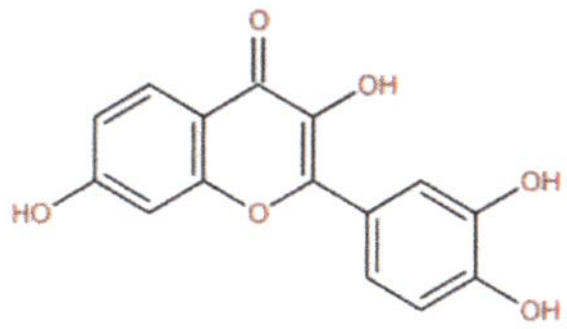	Male Wistar rats with MI/R injury	20 mg kg^{-1}	– Myocardial infarct size, apoptosis, LDH, and creatine kinase content↓ – The mRNA expression of PI3k, AKT, p-PI3K, and p-AKT↑	26
Dihydromyricetin	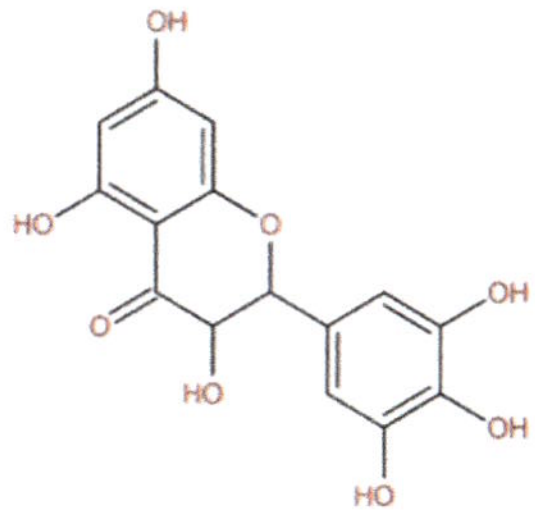	Male Apo E −/− mice and C57BL/6J mice	500 mg kg^{-1} (for 12 weeks)	– White blood cells, AS plaque formation, and microRNA-21 expression↓ – NO production and eNOS phosphorylation↑ – Improves EC function, glucose, and lipid metabolism	32
Resveratrol	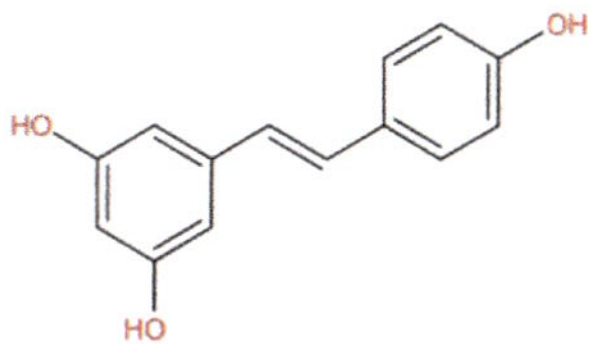	Human umbilical vein endothelial cells induced ox-LDL	5, 10, 20, 40, 80, and 120 μg mL^{-1} (for 24 hours)	– Bcl-2/Bax ratio, mitochondrial membrane potential, and SOD activity↑ – The release of mitochondrial cytochrome *c* into the cytoplasm and lipid peroxidation level↓	28
		Spontaneously and malignantly hypertensive rats	10 mg kg^{-1} (for 4 weeks)	– Blood pressure, lipid peroxidation, and MPO levels – Hypertrophic and apoptotic↓ – NO release and bioavailability, the activity of SOD, CAT, and GPX↑ – Ameliorated morphological changes in the heart tissue	29

Table 3.1 (*Continued*)

Molecule	Type of compound	Model of study	Treatment protocol	Underlying mechanism of the observed results	Ref.
Isorhapontigenin	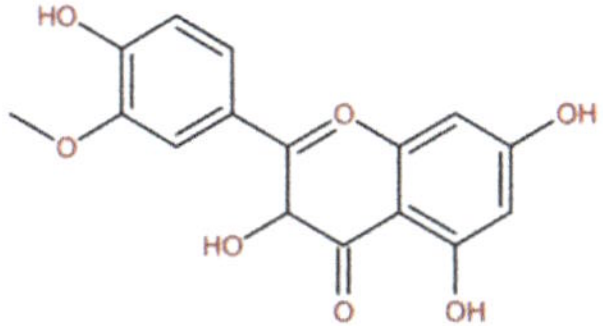	Rats subjected to transient middle cerebral artery occlusion/reperfusion injuries	6.25–100 mg kg^{-1} (for 24 hours)	– Infarct volumes, brain water contents, neurological deficit, MDA, and Bcl-2 level↓ – The activity of SOD, GSH, and GSH-PX, levels of caspase-3 and Bax, and expression of PI3K and AKT↑	33
Anthocyanins	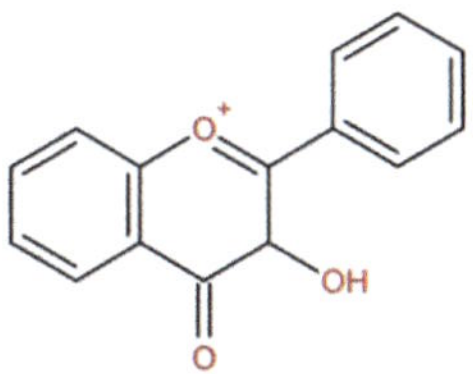	Mice homozygous for the absence of the LDL-C receptor gene	High-fat diet + 2 g per kg per day grape juice or a high-fat diet + simvastatin (20 mg per kg per day) (for 60 days)	– Dyslipidaemia↓ – HDL-C level↑ – Strong potential for lowering blood lipids and preventing CVD	35
		D-Gal induced senescence rats	100–300 mg kg^{-1} (for 8 weeks)	– Endothelial senescence, oxidative stress and nitrotyrosine in aortas, and eNOS uncoupling↓ – NO levels and bioavailability, eNOS phosphorylation, and SIRT1 expression↑	36
Curcumin	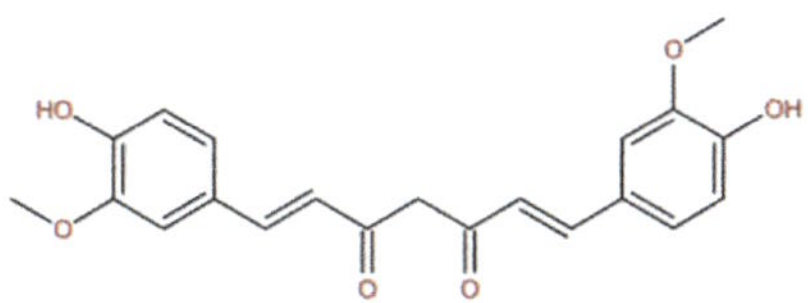	Mouse neuroblastoma N2a cells; male C57BL/6 mice	5, 15, 25, and 35 μmol L^{-1} (for 24 hours) 100, 200, 300, and 400 mg kg^{-1}	– Infarct volume, morphological damage of neurons, Bax, and caspase-3 expression↓ – Bcl-2 protein levels and neuronal survival rate after I/R injury *in vivo*↑ – Improved neurological function and cell viability – Maintained mitochondrial membrane integrity	37

EGCG	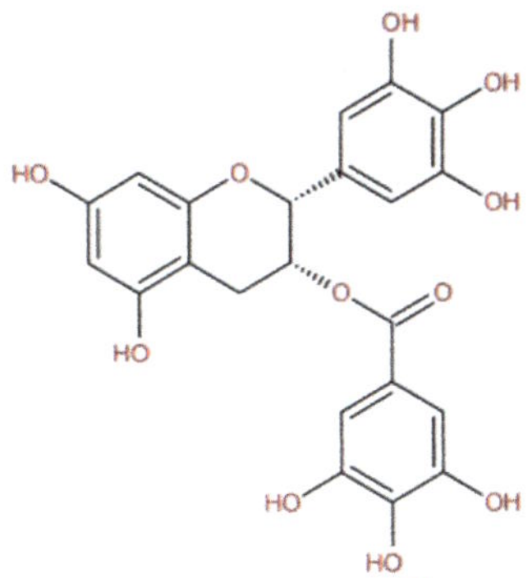	SVZ NPCs stimulated with LPS	20–40 μmol L^{-1} (for 7 days)	– Proliferation and neuronal differentiation of SVZ NPCs↑ – Neuroprotective against acute ischemic stroke – The migration of SVZ neuroblasts and functional recovery	39
	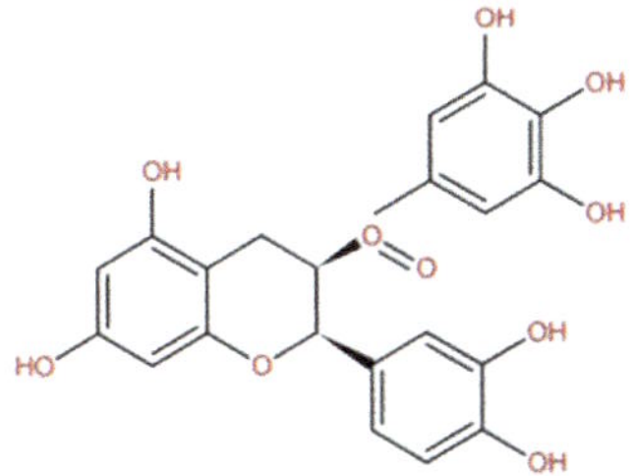	Vascular smooth muscle cells	0, 1, 2.5, 5, 10, 20, 50, and 100 μmol L^{-1} (for 6, 12, 24, and 48 hours)	– The expression of proliferating cell nuclear antigen and cyclin D1↓ – TC content, ROS activity, and inflammatory response↓ – The expression of p-PMAPK 38, p-JNK, p-ERK1/2, p-IκBα, p-NF-κBp65, and TLR4↓ – Regulates the cell cycle and improves cell foaming	40
Berberine	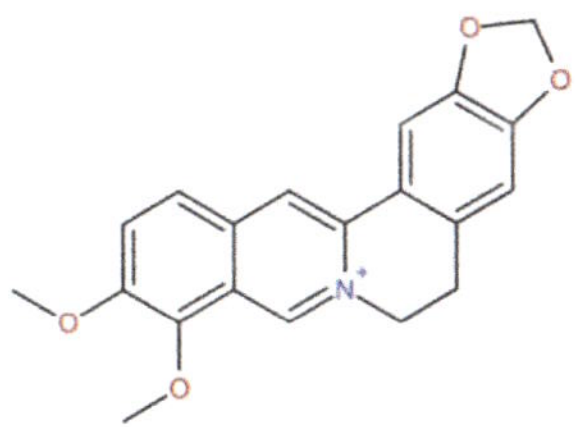	Human coronary artery endothelial cells	20 μmol L^{-1} (for 24 hours)	– ROS production and the protein expression of ATF4, p-EIF2α, p-PERK, XBP1, and p-IRE1↓ – Improving vascular health and protecting coronary artery EC	41
		C57BL/6 mice with I/R injury	10 mg kg^{-1} (for 30 days)	– The expression levels of NF-κBp65, caspase-3, and caspase-9↓ – The expression of Bad, Bcl-xl, cellular tumour antigen p53, PI3K, and AKT↑	42

Table 3.1 (*Continued*)

Molecule	Type of compound	Model of study	Treatment protocol	Underlying mechanism of the observed results	Ref.
Astragaloside	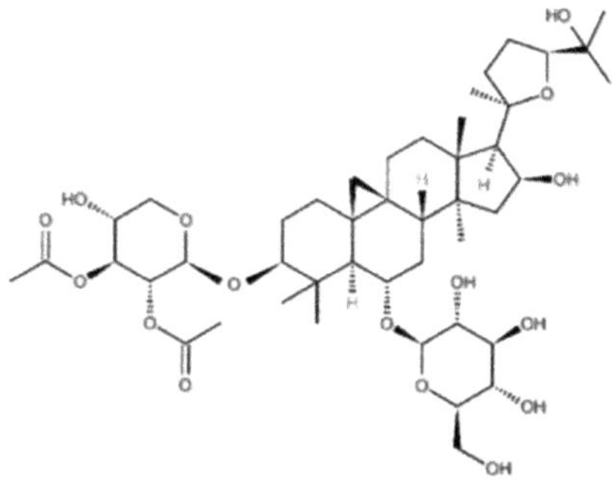	Spontaneously hypertensive rats	200 mg kg^{-1}	– Inhibits ACE activity in a dose-dependent manner, and the IC_{50} was 1.85 ± 0.01 μg mL^{-1} – SBP level↓	43
Ilex pilosa total saponins	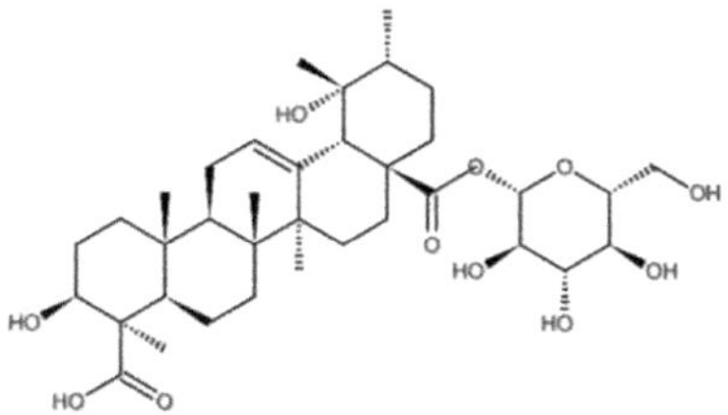	Oxidative damage of H9c2 cells induced by H_2O_2	800 μg mL^{-1} (for 24 hours)	– Cell viability, the activity of SOD, CAT, and Bcl-2 protein expression↑ – The activity of LDH, MDA, and creatine kinase↓ – The expression of caspase-3 and Bax protein↓	44
Lycopene	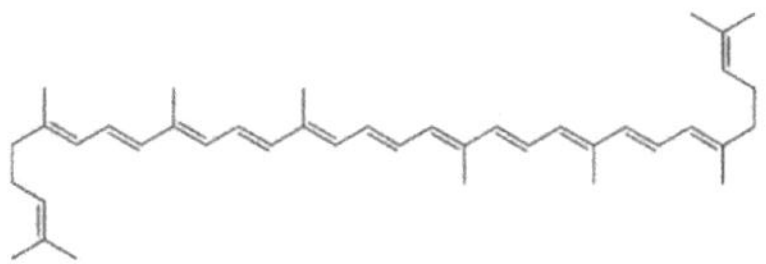	Wistar rats that received intravenous infusion of Ang-II	10 mg per kg per day (for 14 days)	– SBP level, left ventricular weight, left ventricular hypertrophy index, and cardiomyocyte hypertrophy↓ – Collagen deposition and the subsequent fibrosis area, O_2^- production, and MDA activity↓	45

Calcitriol or paricalcitol	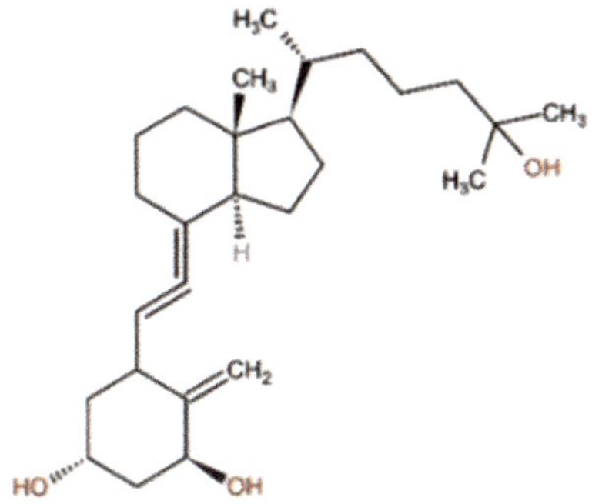 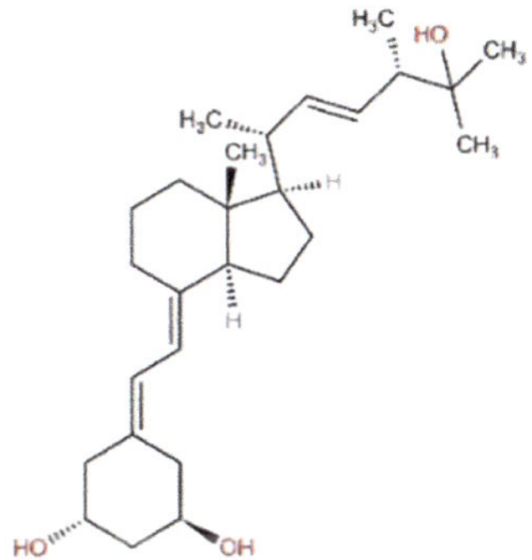	Vascular smooth muscle cells	Calcitriol (10^{-8} M) and paricalcitol (3×10^{-8} M) (for 24 hours)	– The expression of PAR-2, miR-346, TF, and nuclear translocation of NF-kBp65↓ – The expression of TFPI and vitamin D receptor↑	46
Vitamin E	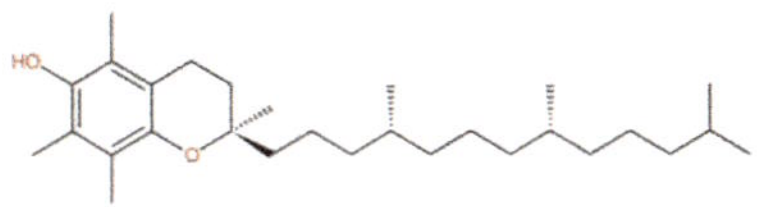	Apo E knockout mice	100 mg kg^{-1} (for 8 weeks)	– AS lesion area, the content of ox-LDL and TBARS↓ – The mRNA expression of CD36↓ – The mRNA expression of PPARγ, LXRα, and ABCA1↑	47

Table 3.1 (*Continued*)

Molecule	Type of compound	Model of study	Treatment protocol	Underlying mechanism of the observed results	Ref.
Coenzyme Q10	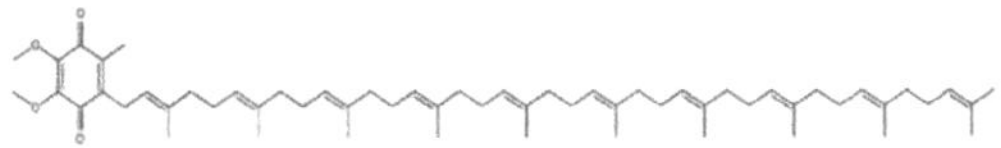	Male Wistar rat stroke model	30 mg kg^{-1}	– Mortality rate, brain necrosis area in acute↓ – Improvement in neurological status – Delayed period after cerebral ischemia	48

Abbreviations: ABCA1, ATP-binding cassette transporter, class A1; ACE, angiotensin converting enzyme; AKT, p-protein kinase B; Ang-II, angiotensin-II; Apo E, apolipoprotein E; AS, atherosclerosis; AT2R, Ang II type 2 receptor; ATF4, activating transcription factor 4; ATF6, activating transcription factor 6; Bax, Bcl-2-associated X; Bcl-2, B-cell lymphoma-2; CAT, catalase; CHOP, CCAAT/enhancer-binding protein homologous protein; CVD, cardiovascular disease; EC, endothelial cells; EGCG, epigallocatechin gallate; EIF 2α, eukaryotic initiation factor 2 alpha; eNOS, endothelial nitric oxide synthase; ER, endoplasmic reticulum; ET-1, endothelin-1; GPX, glutathione peroxidase; GSH, glutathione; GSH-Px, glutathione peroxidase; HDL-C, high density lipoprotein cholesterol; HFD, high-fat diet; IC_{50}, half maximal inhibitory concentration; IL-1β, interleukin-1 beta; IL-6, interleukin-6; IL-10, interleukin-10; LDH, lactate dehydrogenase; LDL-C, low-density lipoprotein cholesterol; L-NAME, L-arginine methyl ester hydrochloride; LPS, lipopolysaccharide; LXRα, liver-x-receptor alpha; MDA, malondialdehyde; MI/R, myocardial ischemia–reperfusion; MPO, myeloperoxidase; miR-346, microRNA-346; NF-κB, nuclear factor-kappa B; NF-κBp65, nuclear translocation of nuclear factor κ-light-chain-enhancer of activated B cells protein 65; NO, nitric oxide; NOS3, nitric oxide synthase 3; ox-LDL, oxidized low-density lipoprotein; p-AKT, phosphorylation of p-protein kinase B; PAR-2, protease-activated receptor 2; p-EIF 2α, phosphorylation of eukaryotic initiation factor 2 alpha; p-ERK 1/2, phosphorylation of extracellular regulated protein kinase 1/2; p-PMAPK 38, phosphorylation of phosphorylated mitogen activated protein kinase 38; p-JNK, phosphorylation of Jun N-terminal kinase; p-PERK, phosphorylation of protein kinase RNA-like endoplasmic reticulum kinase; PPARγ, peroxisome proliferator-activated receptor gamma; PI3K, phosphoinositide-3-kinase; p-PI3K, phosphorylation of phosphoinositide-3-kinase; p-IRE1, phosphorylation of inositol-requiring ER-to-nucleus signalling protein 1; p-IκBα, p-inhibitor of NF-κB alpha; p-NF-κBp65, phosphorylation of nuclear translocation of nuclear factor κ-light-chain-enhancer of activated B cell protein 65; SBP, systolic blood pressure; SOD, superoxide dismutase; SIRT1, silent information regulator protein 1; SVZ NPCs, subventricular zone neural progenitor cells; TAG, triacylglycerols; TBARS, thiobarbituric acid reactive substances; TC, total cholesterol; TF, tissue factor; TFPI, tissue factor pathway inhibitor; TLR4, Toll-like receptor 4; VLDL-C, very low density lipoprotein cholesterol; XBP1, X-box binding protein 1.

acid can be used as an auxiliary drug to lower blood pressure and protect cardiovascular health.

3.4.2 Stilbenes

It is worth mentioning that resveratrol also has a protective effect on endothelial cell apoptosis and oxidative damage. It prevents apoptosis and damage of human umbilical vein endothelial cells (HUVECs) through inhibiting oxidative stress caused by mitochondria. After resveratrol treatment, the proportion of B-cell lymphoma-2/Bcl-2-associated X (Bcl-2/Bax) in HUVECs, mitochondrial membrane potential and superoxide dismutase (SOD) activity increased, and the release of mitochondrial cytochrome *c* to the cytoplasm and the lipid peroxidation level were reduced.[28] Resveratrol also shows a strong ability to lower blood pressure. After long-term treatment with resveratrol, the increase in blood pressure of spontaneously and malignant hypertensive rats decreased by 18.2% and 16.7%, respectively, indicating that resveratrol can effectively reduce blood pressure.[29] Resveratrol improves NO bioavailability in spontaneously and malignant hypertensive rats by inhibiting the increase of myeloperoxidase (MPO) levels. Moreover, it also enhances the antioxidant defence by increasing antioxidant enzymes activities [such as SOD, catalase (CAT), and peroxidase (GPX)] or by reducing blood pressure increase, which presented a positive effect on the heart tissue.[29]

3.4.3 Flavonoids

Treatment of diabetic rats on a high-fat diet (HFD) with quercetin, a representative flavonoid, reduced the consistency of TC and TAG in plasma and decreased the number of AS lesions while increasing HDL-C concentration. The antioxidant enzyme activities (CAT, SOD, and GPX) were enhanced in the carotid artery.[30] Quercetin also inhibited the inflammatory response, which is manifested by the decrease of the interleukin-1 beta (IL-1β) level and nuclear factor-kappa B (NF-κB), and the increase of the interleukin-10 (IL-10) level. 5′-Adenosine monophosphate-activated protein kinase (AMPK) inhibits NF-κB expression by increasing silent information regulator protein 1 (SIRT1) expression, thus reducing the inflammatory response. These protective effects of quercetin were achieved *via* regulating the AMPK/SIRT1/NF-κB signalling pathway to inhibit oxidative stress and inflammation.[30] Figure 3.2 shows that biologically active substances (phenolic acids, stilbenes, flavonoids, alkaloids, saponins, and vitamins) exert their vasorelaxant effects primarily through increasing NO levels *via* multiple mechanisms.

In addition, naringenin can also reduce cardiomyocyte apoptosis and myocardial infarction, and effectively improve heart function in male Sprague-Dawley rats. Treatment with naringenin was found to protect cardiomyocytes from MI/R injury by activating the cyclic guanosine monophosphate/protein kinase G type I alpha (cGMP/PKGIα) signal to reduce endoplasmic reticulum stress (Figure 3.2). The superoxide production and

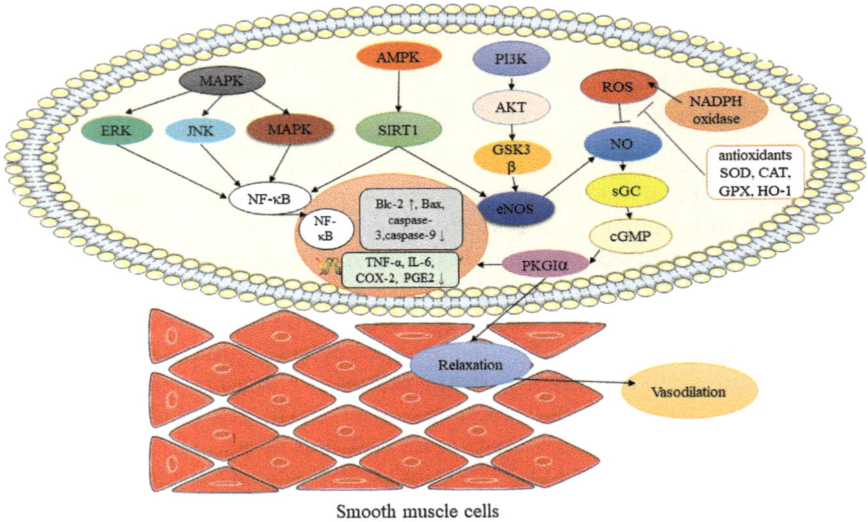

Figure 3.2 Biologically active substances (phenolic acids, stilbenes, flavonoids, alkaloids, saponins, and vitamins) exert their vasorelaxant effects primarily through increasing NO levels *via* multiple mechanisms: (i) increase of NO bioavailability through reducing oxidative stress, (ii) decrease of ROS production or *via* increase of antioxidant capacity, and (iii) increase of activation of eNOS or inhibition of inflammatory factor expression through the PI3K/AKT/GSK3β/eNOS cascade, or AMPK/SIRT1/eNOS cascade, MAPK/NF-κB cascade. Abbreviations: AKT, p-protein kinase B; AMPK, 5′-adenosine monophosphate-activated protein kinase; Bax, Bcl-2-associated X; Bcl-2, B-cell lymphoma-2; CAT, catalase; cGMP, cyclic guanosine monophosphate; COX-2, cyclooxygenase-2; eNOS, endothelial nitric oxide synthase; ERK, extracellular regulated protein kinases; GPX, glutathione peroxidase; GSK3β, glycogen synthase kinase 3 beta; HO-1, heme oxygenase-1; IL-6, interleukin-6; JNK, Jun N-terminal kinase; MAPK, mitogen-activated protein kinase; NADPH, nicotinamide adenine dinucleotide phosphate; NF-κB, nuclear factor-kappa B; NO, nitric oxide; PGE2, prostaglandin E2; PI3K, phosphoinositide-3-kinase; PKGIα, protein kinase G type I alpha; ROS, reactive oxygen species; sGC, soluble guanylate cyclase; SIRT1, silent information regulator protein 1; SOD, superoxide dismutase; TNF-α, tumour necrosis factor-alpha.

myocardial malondialdehyde (MDA) levels were decreased. It also inhibited the expression of activating transcription factor 6 (ATF 6) and phosphorylation of protein kinase RNA-like endoplasmic reticulum kinase (p-PERK), PKGIα, and eukaryotic initiation factor 2 alpha (EIF 2α).[31]

Furthermore, the natural flavonoid fisetin was found to have the potential to protect the heart from MI/R injury, mainly by suppressing the level of oxidative stress and inflammation and improving mitochondrial dysfunction. Intraperitoneal administration of fisetin in MI/R model rats reduced the myocardial infarction area and lowered the content of lactate dehydrogenase (LDH) and creatine kinase. Notably, fisetin protected the heart from I/R injury and preserved cardiac mitochondria by activating the phosphoinositide-3-kinase/protein kinase B/glycogen synthase kinase 3 beta (PI3K/AKT/GSK3β) signalling pathway (Figure 3.2).[26] Similarly, in Apo E −/− mice, dihydromyricetin treatment alleviated cardiomyocyte hypertrophy induced by inflammatory factors or Ang-II by increasing phosphorylation of eNOS and NO production. The accumulation of white blood cells in the liver and aortic vessel walls was reduced, which improved lipid metabolism and reduced AS plaque formation.[32]

In addition, isorhamnetin has also antioxidant and cardioprotective capabilities. Intraperitoneal injection of isorhamnetin in rats with transient middle cerebral artery occlusion/reperfusion injury exhibited a certain neuroprotective effect. It significantly reduced the infarct volume, neurological deficits, and MDA activity, and increased the antioxidant enzyme activities [such as SOD, CAT, and glutathione (GSH)]. The use of wortmannin to block PI3K activation eliminated the reduction in infarct volume and neurological deficit score caused by isorhamnetin-induced ischemic stroke, and inhibited phosphorylation of AKT (p-AKT) expression up-regulation.[33] Isorhamnetin provides neuroprotection counteracting I/R injury *via* activating the PI3K/AKT signal (Figure 3.2).

Anthocyanins show great potential in lowering blood lipids and protecting the heart, which can bind to cholesterol and prevent its absorption, thereby increasing bile acid excretion in faeces. It can also increase lipoprotein lipase (LPL) activity and promote TAG hydrolysis.[34] For instance, 30 male mice homozygous for the absence of the LDL receptor gene (LDLr −/−) were divided into groups and received a 60 day HFD + 2 g per kg per day grape juice or a HFD + simvastatin (20 mg per kg per day) diet. The group of mice that consumed grape juice (rich in anthocyanins) showed significantly reduced dyslipidaemia and increased HDL-C, which was similar to the results of the group treated with simvastatin.[35] This is a good proof that anthocyanins have a strong potential for lowering blood lipids and preventing CVD. Moreover, the intake of food rich in anthocyanins is related to the significant improvement of blood flow-mediated dilation (FMD) and blood pressure, which can effectively protect the cardiovascular system. Anthocyanins also increased NO bioavailability by reducing eNOS uncoupling. It restored the NO level in endothelial cells by blocking ROS formation and intracellular enzyme nicotinamide adenine dinucleotide phosphate hydrogenase (NADPH) activity, and enhanced eNOS enzyme activity by mediating SIRT1 deacetylation (Figure 3.2).[36]

Curcumin exerts a neuroprotective effect in mice with ischemic stroke. Curcumin treatment in mice with I/R injury reduced the infarct volume and neuronal morphological damage, and improved the nerve function and neuron viability rate *in vivo.* It is worth noting that curcumin suppresses ischemia which causes mitochondrial apoptosis by inhibiting Bax activation. It maintained mitochondrial membrane integrity, increased Bcl-2 protein levels, and reduced the expression of Bax and caspase-3 in mouse N2a cells after oxygen–glucose deprivation/reoxygenation injury, exerting a neuroprotective effect.[37]

Finally, catechins can also compete with cholesterol and suppress the activity of 3-hydroxy-3-methylglutaryl-coenzyme A (HMG-CoA) reductase, thereby reducing cholesterol production.[25] In a recent study, 40 male LDL receptor knockout mice on a HFD were fed with 25 mg per kg per day *Camellia sinensis* tea (green tea, white tea, and black tea) for up to 60 days. It was found that *Camellia sinensis* tea partially prevented hyperlipidaemia, reduced HDL-C in serum levels, and increased hs-CRP levels.[38]

These beneficial effects are mainly due to the abundant catechins in *Camellia sinensis* tea. This shows that catechins can be used as an adjuvant therapy related to controlling lipid intake to combat hyperlipidaemia and protect the heart. It is worth noting that the PI3K/AKT signalling pathway plays an essential role in EGCG-mediated proliferation of neural progenitor cells (NPCs) and neuronal differentiation. Supplementation with EGCG can boost the proliferation and functional recovery of NPCs in the late recovery period of stroke in mice.[39] M1 phenotype markers [*e.g.*, TNF-α and inducible nitric oxide synthase (iNOS)] of activated microglia were significantly reduced. The main M2 phenotype of microglia after EGCG treatment can eliminate inflammation through secreting anti-inflammatory cytokines [*e.g.*, IL-10, Arg-1, and transforming growth factor-beta (TGF-β)] after ischemic stroke, and p-AKT levels increased significantly, providing a great environment for NPC proliferation.[39] In addition, mitogen-activated protein kinase (MAPK) and NF-κB signalling pathways also play an essential role in influencing cell proliferation and migration and inflammatory factor release. Treating VSMC caused by ox-LDL with ECG improved cell foaming and reduced TC content, ROS activity, and inflammatory factor expression [such as phosphorylation of phosphorylated mitogen activated protein kinase 38 (p-PMAPK 38), phosphorylation of extracellular regulated protein kinase 1/2 (p-ERK 1/2), phosphorylation of Jun N-terminal kinase (p-JNK), phosphorylation of NF-κBp65, and phosphorylation of inhibitor of NF-κB alpha (p-IκBα) and Toll-like receptor 4 (TLR4)].[40] In general, ECG inhibits MAPK/NF-κB pathway activation in VSMC by suppressing ROS overproduction, thereby reducing cellular oxidative stress and inhibiting the proliferation and migration of VSMC (Figure 3.2).

3.4.4 Alkaloids and Saponins

Alkaloids show a great potential in protecting the cardiovascular system. Berberine treatment of human coronary artery endothelial cells (HCAECs) attenuated the increase in ROS generation and the decrease of endoplasmic reticulum stress marker expression [ATF 4, p-PERK, p-EIF 2α, X-box-binding protein 1 (XBP 1), and phosphorylation of inositol-requiring enzyme 1 (p-IRE 1)] and reduced endothelial inflammation protecting HCAECs.[41] Berberine also protected cardiomyocytes as confirmed by *in vivo* animal experiments. One hundred male C57BL/6 mice with I/R injury were treated with 10 $mg\,kg^{-1}$ berberine for 30 days. Berberine decreased the expression levels of NF-κBp65, caspase-3, and caspase-9 in mouse cardiomyocytes, suppressed cardiomyocyte apoptosis, and upregulated the expression of Bad, Bcl-xl, PI3K, and AKT, thus alleviating the inflammatory response (Figure 3.2). This indicates that NF-κB and PI3K/AKT signalling pathways participate in the anti-apoptotic and anti-inflammatory effects of berberine on cardiomyocytes.[42] These results demonstrate that berberine is a promising candidate for I/R injury treatment.

Saponins also have a strong antioxidant and blood pressure lowering ability. Recent studies have found that *Astragalus membranaceus* (rich in saponins) extract can reduce the SBP in spontaneously hypertensive rats by inhibiting angiotensin converting enzyme (ACE) activity. It was found to inhibit ACE activity in a dose-dependent manner.[43] When the rats were administered 200 $mg\,kg^{-1}$ of astragalus extract, the SBP level decreased by 22.33 mmHg. Noteworthily, a small peptide [named *Astragalus membranaceus* root-1 (AM-1)] was also found, which also inhibited ACE activity with an IC_{50} (indicating the fraction concentration that scavenged 50%) value of 415 μM. When 10 $\mu mol\,kg^{-1}$ AM-1 was administered to rats, the SBP level decreased by 42 mmHg.[43] Similarly, the oxidative damage to H9c2 cells induced by H_2O_2 was treated with 800 $\mu g\,mL^{-1}$ total saponins of *Ilex pilosa*. The cell viability was significantly enhanced and the activities of LDH, MDA, and creatine kinase were reduced. Cellular antioxidant capacity was significantly improved, mainly manifested in the increase of SOD and CAT activities. The expression of caspase-3 and Bax protein was inhibited, while Bcl-2 protein expression was up-regulated. These indicate that the total saponins of *Ilex pilosa* can protect cardiomyocytes from apoptosis by inhibiting oxidative stress and mitochondrial-induced intrinsic apoptosis.[44]

3.4.5 Vitamins

Vitamins also have a strong ability to regulate oxidative stress and protect cardiovascular health. For instance, Wistar rats received intravenous infusion of Ang-II (0.3 mg per kg per day) for 14 days, and half of them were given 10 mg per kg per day lycopene at the same time. It was found that intravenous infusion of Ang-II gradually improved SBP in rats (from 117 mmHg to 196 mmHg), while lycopene intervention therapy significantly reduced the increase in SBP (145 mmHg at the end of treatment, $P < 0.001$). It also reduced the increase in left ventricular weight (582 mg) inhibiting the increase in cardiomyocyte hypertrophy. The collagen sedimentation and the subsequent fibrosis area in the heart caused by Ang-II infusion in rats were significantly reduced when lycopene was used for intervention.[45] Moreover, VSMC-derived tissue factor (TF) is supposed to be the key to arterial thrombosis. Calcitriol or paricalcitol intervention prevented protease-activated receptor-2 (PAR-2) expression, increased the expression of tissue factor pathway inhibitor (TFPI) and vitamin D receptor, and inhibited the upregulation of microRNA-346 (miR-346) and TF. The upregulation of vitamin D receptor expression improved the anti-thrombotic effect of vitamin D.[46] This proved that vitamin D can suppress inflammation-induced thrombosis.

Interestingly, one study found that Apo E knockout mice spontaneously develop AS in an age-dependent (14–46 weeks old) manner under a normal diet. Supplementation with vitamin E (100 mg kg^{-1}) for 8 weeks at 6–30 weeks of age can reduce the content of serum ox-LDL and thiobarbituric acid reactive substances (TBARS). Vitamin E supplementation at 6, 14, and 22 weeks of age significantly reduced the AS lesion area, and the expression of peroxisome proliferator-activated receptor gamma (PPARγ), liver-x-receptor alpha (LXRα), and ATP-binding cassette transporter, class A1 (ABCA1) involved in cholesterol reverse transport was up-regulated. However, vitamin E supplementation at ≥30 weeks of age hardly affected the AS lesion area and the expression of these genes.[47] This suggests that vitamin E had an inhibitory effect on early AS, but not on late AS. The molecular mechanism of vitamin E inhibiting early atherosclerosis may partially be ascribed to PPARγ-LXRα-ABCA1 transduction pathway activation, which arrests the formation of foam cells due to cholesterol efflux induced by macrophages.

It is worth noting that increasing the activity of endogenous antioxidants can also effectively protect the cardiovascular system from oxidative stress. The stroke model of rats caused by middle cerebral artery occlusion and intravenous injection of coenzyme Q10 (ubiquinol, reduced form) after cerebral ischemia reduced mortality, improved neurological function, and reduced the area of brain death in acute and delayed phases. When administered intravenously immediately before thrombectomy or thrombolytic therapy, ubiquinol was found to exert a neuroprotective effect.[48]

3.4.6 MUFA and PUFA

Dietary intake of MUFA + PUFA also has a strong effect in reducing hyperlipidaemia. C57BL/6J mice were fed tomato seed oil (containing abundant MUFA + PUFA) instead of two-thirds lard for 12 weeks. The weight of the mice, TC, TAG, LDL-C, LDL-C/HDL-C, and total fatty acids were significantly decreased, while the HDL-C level in plasma was significantly increased. Supplementing tomato seed oil in the diet of mice improved fatty acid metabolism, reduced cholesterol absorption, and promoted cholesterol efflux, effectively alleviating hyperlipidemia.[49] The great results of these *in vitro* and *in vivo* experiments have laid a solid foundation for further human intervention clinical trials.

3.5 Human Intervention Studies

Several human intervention clinical trials have demonstrated the benefits of dietary supplements of antioxidant foods and their active ingredients in reducing CVD risk. Here, the research results of the protective effects of some dietary supplements on CVD are summarized (Table 3.2).

For a long time, many epidemiological and human intervention studies have shown that adhering to a healthy diet is correlated with a lower

Table 3.2 Main results of human intervention studies to evaluate the relationship between dietary supplementation and CVD.

Compound	Sample and design	Treatment protocol	Underlying mechanism of the observed results	Ref.
Mediterranean diet and low-fat vegan diet	– $n = 62$ – Overweight adults (mean age: 56.6 years)	50 g per day (for 32 weeks)	– A low-fat vegan diet improved body weight, lipid concentrations, and insulin sensitivity – SBP and DBP were lowered more in the MedDiet group	50
Citrus flavones and flavanones and olive polyphenols	– $n = 96$ – Healthy adults (40–75 years)	250–300 mg per day (for 8 weeks)	– TC, LDL-C, ox-LDL, SBP, and DBP ↓ – The expression of protein carbonyl, hs-CRP, IL-6, and TNF-α ↓ – The ratio of oxidized/reduced glutathione↑	51
Fruits, vegetables, and whole grains	– $n = 75$ – Overweight and obese women (18–50 years)	Whole grains or fruits and vegetables or whole grains + fruits and vegetables (35 g per day) (for 10 weeks)	– A weight loss diet rich in whole grains (HDL-C↑, LDL-C, and TC) – Diets rich in whole grains have a beneficial effect on CVD risk factors compared to diets rich in fruits and vegetables or whole grains + fruits and vegetables	52
Ferulic acid	– $n = 48$ – Subjects with hyperlipidaemia (20–60 years)	1000 mg per day (for 6 weeks)	– TC, LDL-C, TAG, MDA, and ox-LDL ↓ – HDL-C↑ – The expression of NF-κB, TNF-α, and hs-CRP ↓	54
Quercetin	– $n = 30$ – Subjects with elevated BP or moderately elevated blood sugar concentrations (20–60 years)	Two apples (once in the morning and once in the evening) (for 4 weeks)	– LDL-C levels ↓ – Plasma quercetin metabolites↑ – Lower risk of CVD with higher apple consumption	56
Anthocyanins and proanthocyanidins	– $n = 60$ – Subjects with chest pain suggesting myocardial ischemia at least 30 minutes before admission	40 g per day (for 8 weeks)	– LDL-C levels ↓ – Plasma anthocyanin-derived metabolites↑ – Significant correlation between LDL-C and ox-LDL levels and acute coronary syndrome	58

Table 3.2 (*Continued*)

Compound	Sample and design	Treatment protocol	Underlying mechanism of the observed results	Ref.
Curcumin	– $n = 39$ – Healthy men and postmenopausal women (45–74 years)	2000 mg per day (for 12 weeks)	– Vascular NO bioavailability↑ – Oxidative stress↓ – Improving resistance artery endothelial function	59
EGCG	– $n = 102$ – Women with central obesity (18–65 years)	857 mg per day (for 12 weeks)	– Body weight and waist circumference↓ – Plasma levels of TC and LDL↓	60
Berberine	– $n = 144$ – Healthy Caucasian adults	500 mg twice a day (for 3 months)	– Body weight and BMI↓ – TC, LDL-C, and TAG ↓ – HDL-C↑	61
Ginsenosides	– $n = 24$ – Healthy Caucasian adults (45–75 years)	3 g per day (for 8 weeks)	– FBG and SBP, LDL-C, and LDL-C/HDL ↓ – Serum nitrate/nitrite content and NO production↑	62
Tomato polyphenols	– $n = 26$ – Subjects with moderate overweight and hyperlipidaemia (45–60 years)	200 g per day (for 4 weeks)	– TC and LDL-C ↓ – Have favourable effects on lipid metabolism	63
Lycopene	– $n = 54$ – Subjects with moderate overweight (40–65 years)	224–350 mg per week (for 12 weeks)	– SAA, HDL levels, and cholesteryl ester transfer protein activity ↓ – PON-1 and lecithin cholesterol acyltransferase activity↑	64
Vitamin D	– $n = 362$ – Subjects with hypertriglyceridemia and vitamin D deficiency (18–60 years)	14 000 IU and 28 000 IU per week (for 12 months)	– Hypertriglyceridemia, TAG, and CVD risk ↓	65

DHA or EPA	– $n = 121$ – Healthy, normal-lipid subjects	Olive oil placebo (6 g per day) or EPA (600 mg per day or 1800 mg per day) and DHA (600 mg per day)	– DHA (600 mg per day) lower TC ability > EPA (1800 mg per day) – Lp-PLA2 concentration, hypertriglyceridemia, and LDL-C (EPA, 1800 mg per day) ↓	68
Sesame oil + vitamin E	– $n = 75$ – Subjects with suboptimal cardiovascular health (30–70 years)	Sesame oil (30 mL per day) enriched with vitamin E (400 mg per day) (for 8 weeks)	– Serum TC, TAG, FBG, HOMA-IR, MDA, hs-CRP, HDL-C, SBP, and DBP↓	71
Cavacurcumin, EPA, astaxanthin, and GLA	– $n = 80$ – Healthy adults (mean age: 48.8 ± 16.0 years)	Cavacurcumin (500 mg) + ω3 (675 mg) + astaxanthin (3 mg) + GLA (9.5 mg), mixed tocotrienols (12.5 mg) + hydroxytyrosol (6.25 mg) + vitamin D3 (1000 IU) + potassium (12.75 mg) (for 4 weeks)	– Inflammatory markers and SBP ↓ – EPA levels and the fatty acid index↑ – Endothelial function improvement	72

Abbreviations: BP, blood pressure; BMI, body mass index; CVD, cardiovascular disease; DBP, diastolic blood pressure; DHA docosahexaenoic acid; EGCG, epigallocatechin gallate; EPA, eicosapentaenoic acid; FBG, fasting blood glucose; GLA, gamma-linolenic acid; HDL-C, high density lipoprotein cholesterol; hs-CRP, high-sensitivity C-reactive protein; HOMA-IR, homeostatic model assessment; IL-6, interleukin-6; IU, international unit; LDL-C, low-density lipoprotein cholesterol; Lp-PLA2, lipoprotein-associated phospholipase A2; MedDiet, Mediterranean diet; MDA, malondialdehyde; NF-κB, nuclear factor-kappa B; NO, nitric oxide; ox-LDL, oxidized low-density lipoprotein; PON-1, paraoxonase; SAA, serum amyloid A; SBP, systolic blood pressure; TAG, triacylglycerols; TC, total cholesterol; TNF-α, tumour necrosis factor-alpha; ω3, omega-3.

incidence and risk of CVD, especially in those who adhere to MedDiet and low-fat vegan diet. These foods are often rich in phenolic acids, flavonoids, stilbenes, alkaloids, saponins, vitamins, and MUFA + PUFA, among others. Evidence from the clinic-adequate intake of foods with bioactive substances (phenolic acids, flavonoids, alkaloids, saponins, and MUFA + PUFA) showed reduced CVD risk. Detailed information about these bioactive substances is given in detail below.

3.5.1 Phenolic Acids and Flavonoids

In a randomized clinical trial (RCT), 62 overweight adults were randomly assigned to a MedDiet or low-fat vegan diet for 32 weeks of dietary intervention. The low-fat vegan diet resulted in a greater reduction in body weight, TC, and LDL-C compared to the MedDiet in the first 16 weeks of the study, while SBP and DBP were lowered more in the MedDiet group. The SBP and DBP of the MedDiet group were reduced by 9.3 and 7.3 mmHg, while in the low-fat vegan group the SBP and DBP were reduced by 3.4 and 4.1 mmHg, respectively.[50]

Similarly, nutritional supplements with 250–300 mg per day flavanone glycosides (neohesperidin and naringin), 175–200 mg per day flavanones (luteolin, apigenin, and podocarpus), or 85–90 mg per day olive polyphenols (secoiriodoids and phenols) for subjects up to 8 weeks were effective in reducing the risk of CVD. This is mainly manifested in a flow-mediated vasodilation increase, a reduction in SBP and DBP, lower TC, LDL-C, and ox-LDL, an increase in the ratio of oxidized/reduced glutathione, and significantly reduced expression of protein carbonyl, IL-6, hs-CRP, and TNF-α.[51] In addition, in a RCT, 75 overweight and obese women aged 18–50 were randomly assigned to consume diets rich in whole grains (35 g per day) or fruits and vegetables (35 g per day) or whole grains + fruits and vegetables (35 g per day) for a period of 10 weeks. Within 10 weeks, the subjects who consumed only whole grains had a greater reduction in body weight, SBP and TC, and a greater increase in HDL-C levels. Only the intake of whole grains was found to be more effective in reducing CVD risk factors than consumption of fruits and vegetables or whole grains + fruits and vegetables.[52] This may be because whole grains contain more phenolic acids and flavonoids, thus showing a stronger beneficial effect in preventing CVD.

It is worth mentioning that ferulic acid exerts an antioxidant effect by scavenging excessive ROS, protecting cells from oxidative stress damage. Ferulic acid can improve lipid levels by reducing the activity of 3-hydroxy-3-methylglutaryl-coenzyme A (HMG-CoA) reductase, which is an important enzyme for cholesterol synthesis. Ferulic acid exerts anti-inflammatory effects through inhibiting the expression of cyclooxygenase-2 (COX-2), prostaglandin E2 (PGE2), and TNF-α.[53] In a RCT, 48 subjects with hyperlipidemia were randomly given ferulic acid (1000 mg per day) intervention treatment for 6 weeks. It was found that the TC, TAG, and LDL-C levels were lower, while HDL-C levels increased after 6 weeks of supplementation with ferulic acid. It also reduced lipid peroxidation, which was manifested in the reduction of MDA levels and ox-LDL. The expression of NF-κB, TNF-α, and hs-CRP was suppressed.[54] This is a good proof that ferulic acid has a strong potential to lower CVD risk factors.

Furthermore, hesperidin was found to lower blood pressure and improve endothelial dysfunction by increasing the secretion of NO in endothelial cells, cyclic adenosine monophosphate (AMP) and cGMP, and suppression of cyclic nucleotide phosphodiesterase and ACE activity. After 8.5 weeks of oral administration of hesperidin (42 mg per day) in hypercholesterolemic subjects, TC and LDL-C levels were decreased by 0.47 mmol L^{-1} and 0.49 mmol L^{-1}, respectively.[55] Consuming orange juice every day also reduced SBP and DBP and increased total plasma antioxidant capacity or reduced lipid peroxidation. Of note, a randomized controlled crossover intervention study was conducted on 30 subjects aged 20–60 years with CVD risk factors who were given quercetin-rich apples for 4 weeks of intervention. After 4 weeks of dietary supplementation with peeled apples, the LDL-C levels in subjects became lower. The plasma quercetin metabolites significantly increased and improved the cardiovascular function.[56]

The intake of flavanones was found to depend on the type and amount of berries (rich in anthocyanins and proanthocyanidins) (250–300 g for raw fruits, 250–1000 mL per day for beverages/smooths, or 600 mg for capsules) which can improve blood vessel function and reduce all-cause mortality.[57] Moreover, in an 8 week RCT, patients with chest pain suggesting myocardial ischemia ≥30 min before admission, and the onset of symptoms for <24 hours were given bilberry powder (40 g per day, equal to 480 g of fresh bilberry) within 24 hours after percutaneous coronary intervention. During the intervention treatment, the anthocyanin-derived metabolites in the subjects' blood significantly increased and ox-LDL decreased significantly. There was a significant correlation between LDL-C and ox-LDL levels and acute coronary syndrome, indicating a dose–response relationship between the cholesterol levels and acute coronary syndrome.[58] Hence, foods rich in anthocyanins and proanthocyanidins may have clinically relevant beneficial effects after acute myocardial infarction.

Moreover, 39 healthy men and postmenopausal women were randomly assigned to supplement with curcumin (2000 mg per day) for 12 weeks, which showed good safety and tolerability. Curcumin treatment enhanced acute reduction of forearm blood flow to acetylcholine infusion induced by the NO synthase inhibitor NG-monomethyl-L-arginine and reduced the acute increase in forearm blood flow response to acetylcholine infusion of the antioxidant vitamin C. This indicates that the improvement of resistance artery endothelial function is mediated by the increase of NO bioavailability and oxidative stress reduction. After curcumin intervention treatment, the brachial artery blood FMD increased by 36%. While it is affected by the stiffness of large elastic arteries or oxidative stress, circulating markers of inflammation resulted in only a low impact.[59]

Catechin has also received widespread attention in protecting cardiovascular health. It can interfere with cholesterol absorption and increase its bile excretion, inhibiting the regulatory enzyme HMG-CoA reductase activity in cholesterol synthesis. Among them, EGCG is the most biologically active weight loss and lipid-lowering component in green tea, which can significantly reduce the proliferation and differentiation of adipocytes.[38] In a randomized, double-blind clinical trial, 102 women with central obesity [body mass index (BMI) ≥ 27 kg m^{-2} and a waist circumference (WC) ≥ 80 cm] were administrated with a high dose of green tea extract. It was found that after 12 weeks of treatment (EGCG: 857 mg per day), the body weight of the subject significantly decreased from 76.8 to 75.7 kg and WC decreased. The levels of TC and LDL-C in plasma showed a decreasing trend.[60]

3.5.2 Alkaloids and Saponins

After a 6 month break-in period of diet and exercise, the subjects were randomly assigned to a placebo or berberine (500 mg twice a day) supplement for 3 months. A significant decrease in body weight and BMI (−4.5%, compared to baseline $P < 0.05$) was observed during the 6 month running-in period of diet and exercise. Compared with the placebo group, TC (−11.6%), TAG (−21.2%), and LDL-C (−16.4%) were observed to decrease significantly, while the HDL-C value increased significantly (+9.1%) after taking berberine for 3 months.[61] This showed that proper exercise and intake of berberine have a better effect on mildly improving the blood lipid levels of subjects with low risk of CVD.

Ginsenosides also have potential cardiovascular benefits. Recently, in a randomized crossover clinical trial 24 subjects with type 2 diabetes (T2D) were randomly assigned to receive 1 g per meal (3 g per day) American ginseng extract (rich in ginsenosides) or placebo for up to 8 weeks of intervention. After receiving the intervention treatment of American ginseng extract, compared with the placebo group, glycosylated haemoglobin, FBG, and SBP were significantly reduced by 0.29%, 0.71 mmol L^{-1}, and 5.6 mmHg, respectively, while the serum nitrate/nitrite content increased (1.85 μmol L^{-1}). The average end difference of LDL-C and LDL-C/HDL decreased by 12.3% and 13.9%, respectively.[62] The improvement of blood sugar is along with a decrease in SBP and blood lipids and an increase in NO production, which improves endothelial function. This shows that American ginseng extract may help to better reduce the risk factors of CVD.

3.5.3 Vitamins

Noteworthily, lycopene may bind to LDL particles and act as a protective barrier to prevent LDL from being oxidized, reduce cholesterol synthesis by suppressing the activities of HMG-CoA reductase and acyl-coenzyme A cholesterol acyltransferase (ACAT) and regulate LDL receptor expression.

Recently, 26 subjects (45–60 years) who were moderately overweight and hyperlipidaemic were randomly assigned to receive tomato juice (200 g per day) for 4 weeks in a controlled study. Tomato juice may help decrease certain CVD risk factors, mainly manifested in lowering of the contents of TC and LDL-C.[63] Similarly, 54 subjects were randomly assigned to a control diet (lycopene < 10 mg per week), a diet rich in lycopene (224–350 mg per week), and lycopene supplements (70 mg per week) for 12 weeks of intervention. Lycopene intervention reduced inflammation, which was manifested as a decrease in serum amyloid A (SAA) levels. Lycopene restores paraoxonase (PON-1) activity in HDL-C by reducing the connection between SAA and HDL. In addition, after supplementing with lycopene, the activity of cholesteryl ester transfer protein in serum was reduced, while lecithin cholesterol acyltransferase activity in serum and HDL was increased.[64] This shows that lycopene can reduce systemic and HDL-related inflammation, and enhance the ability of HDL to resist AS.

Similarly, a randomized, double-blind clinical trial was conducted over a 12 month supplementation with vitamin D (14 000 IU and 28 000 IU per week) intervention using 362 (18–60 years) shift workers to reduce their hypertriglyceridemia and CVD risk.[65] For people suffering from hypertriglyceridemia and vitamin D deficiency, vitamin D supplement was found to effectively reduce TAG content and CVD risk.

3.5.4 MUFA and PUFA

MUFA and PUFA have also received much attention in human intervention and treatment of CVD. They may cause PPARγ and LXR upregulation, thereby increasing LDL uptake and exerting anti-inflammatory and antioxidant effects.[66] Lipoprotein-associated phospholipase A2 (Lp-PLA2) is a known marker of inflammation and its increase is related to CHD severity. Subjects with higher plasma levels of DHA and EPA had lower Lp-PLA2 quality and activity.[67] In a randomized double-blind trial involving 121 healthy, normal-lipid subjects who were supplemented with olive oil placebo (6 g per day) and DHA (600 mg per day), or EPA (600 mg per day, 1800 mg per day), DHA showed a stronger ability to lower TC than the same dose (600 mg per day) or even a higher dose of EPA (1800 mg per day). In the DHA treatment group, postprandial TC concentration was significantly reduced, while there was no significant change in inflammation biomarkers. In the EPA treatment group (1800 mg per day), Lp-PLA2 concentration was significantly reduced. The unique cardiovascular protective effect of EPA may be partly attributed to its ability to reduce the inflammatory response of the body.[68] It is worth mentioning that the ratio of omega-6 (ω6) to omega-3 (ω3) fatty acids affects the tissue fatty acid composition. When the ratio of ω6 to ω3 fatty acids exceeds 5 : 1, CVD risk is expected to increase. A lower proportion of ω6 to ω3 fatty acids is more ideal for decreasing CVD risk in western societies and developing countries. In the secondary prevention of CVD, when the proportion of ω6 to ω3 fatty acids was 4 : 1, the overall mortality rate was reduced by 70%.[69]

Of note, sunflower oil contains about 67% ω6 fatty acids and a high amount of vitamin E, which can effectively inhibit lipid peroxidation. The mechanism of action of sunflower oil in lowering cholesterol may be due to the high content of MUFA, which is achieved by inhibiting HMG-CoA reductase activity. Tocopherol can also inhibit cholesterol synthesis, effectively reducing ROS production, decreasing platelet aggregation, ox-LDL, and inflammation levels, and delaying intra-arterial thrombosis formation.[70] Dietary supplementation of sunflower oil rich in vitamin E can effectively reduce the cardiometabolic indices of patients with metabolic syndrome (MetS). Seventy-five subjects (30–70 years) with MetS were randomly assigned to consume sunflower oil rich in vitamin E. Compared with subjects who only consumed sunflower oil, serum TC, TAG, FBG, homeostatic model assessment, MDA, hs-CRP, SBP, and DBP were found to be significantly reduced and HDL-C levels were enhanced in the group consuming sunflower oil rich in vitamin E.[71]

Furthermore, dietary supplements containing cavacurcumin (500 mg), ω3 fatty acids [675 mg, including EPA (450 mg) + DHA (100 mg) + other ω3 fatty acids (125 mg)], astaxanthin (3 mg), gamma-linolenic acid (GLA, 9.5 mg), mixed tocotrienols (12.5 mg), hydroxytyrosol (6.25 mg), vitamin D3 (1000 IU), and potassium (12.75 mg) were given to 80 subjects over a 4 week period in a randomized, double-blind clinical trial. After taking the above supplements, mean SBP and inflammatory markers (hs-CRP and IL-6) were significantly reduced, the EPA levels and fatty acid index were significantly increased, and the endothelial function was also improved. This may be due to their own antioxidant and anti-inflammatory properties.[72] The above supplements improved NO bioavailability and endothelial function by lowering blood pressure. Unlike the long-term administration of NO donor anti-hypertensive drugs, dietary supplements did not seem to cause the development of tolerance and hypotension. It is necessary to further study the effects of natural antioxidants and derivative products on the absorption, distribution, and metabolism of fatty acids to reveal their effects on CVD.

There is an increasing amount of strong evidence that diets rich in phenolic acids, stilbenes, flavonoids, alkaloids, saponins, vitamins, and MUFA + PUFA can effectively prevent CVD. It supports the consumption of large amounts of fruits, juices, vegetables, whole grains, and nuts constituting MedDiet or a low-fat vegan diet, and adherence to drinking tea can reduce CVD risk factors. RCTs are considered to be the most effective way to determine the causal relationship between interventions and their effects. However, there are still some dietary intervention studies that are not strongly linked to reducing CVD risk. Different population characteristics, research designs, different dosages and types of supplements, and intervention time may all cause great differences in research results. Therefore, more and larger RCTs in humans and a longer follow-up time are needed to better verify the impact of dietary interventions on CVD risk factors, to assess certain types of dietary supplements that can reduce or prevent the incidence of CVD. It is strongly recommended that further RCTs be conducted in different well-defined target populations to strengthen the evidence of the effectiveness of dietary interventions on cardiovascular health and function, and to further clarify the underlying mechanisms of its impact. It is worthwhile to further explore the dose and time response of bioactive ingredients in different dietary supplements, and the bioavailability and pharmacokinetics in the human body.

3.6 Understanding the Mechanisms of Dietary Supplements with Antioxidant Activity

3.6.1 Hypolipidemic Activity

Polyphenols can reduce cholesterol synthesis *via* suppressing the activities of HMG-CoA and ACAT. They can also inhibit the expression of fatty acid synthase (FAS) and glycerol triphosphate dehydrogenase by regulating the transcription and expression of nuclear transcription factors (PPAR) and sterol regulatory element binding protein (SREBP) genes to prevent SREBP mediated lipogenesis, and reduce the synthesis of cholesterol, cholesterol esters, and endogenous fatty acids.[73] The downstream target gene of LXRs-ABCA1 can transfer free cholesterol in peripheral tissues and cells to HDL and regulate cholesterol reverse transport. Polyphenols through synergizing ABCA1 and Apo A enhance the HDL effect, and induce the gene expression of fatty acid transporter and acyl-CoA synthase, and LPL enhances fatty acid β oxidation, thus accelerating lipid metabolism. The activated LXR can also upregulate the rate-limiting enzyme cholesterol 7α hydroxylase expression in the bile acid synthesis pathway and accelerate cholesterol conversion to bile acids.[34] Polyphenols can also form complexes with cholesterol or generate esters in the intestine to lower cholesterol levels. They also promote the conversion of cholesterol to bile acids, accelerate the excretion of bile acids, block the hepato-intestinal circulation of bile acids, and reduce the content of cholesterol in the blood vessel wall, thereby accelerating cholesterol metabolism and reducing blood lipid absorption.[74] Furthermore, polyphenols have strong reducibility. They release electrons through self-oxidation and directly scavenge various free radicals, thus suppressing the oxidation of LDL. Polyphenols can also inhibit the oxidation of LDL by chelating divalent metal ions, effectively improving endothelial function, and reducing the morbidity and mortality of CVD.

3.6.2 Hypotensive Activity

NO is a crucial blood pressure regulator in the body. Endothelial cells are the main source of NO in blood vessels. Endothelial cells participate in regulating blood vessel tension and blood pressure through the synthesis and secretion of NO. It can relax vascular smooth muscle, resist platelet aggregation, and inhibit the proliferation of VEC through the cGMP pathway.[75]

When the receptors on VEC receive acetylcholine, the intracellular Ca^{2+} concentration is increased. Subsequently, it activates eNOS, releases a large amount of NO, diffuses directly into smooth muscle cells (SMC), and combines with guanylate cyclase in the cytoplasm, leading to an increase in cGMP synthesis, which can reduce the Ca^{2+} concentration in VSMC, causing relaxation and vasodilation. In addition, polyphenols can also reduce the production of ROS by neutralizing oxygen radicals, which improves NO bioavailability, enhances the antioxidant system or inhibits phosphodiesterase (PDE) activity, thereby increasing the expression of cGMP and eNOS.[76] Polyphenols also lower blood pressure by regulating the renin–angiotensin–aldosterone system (RAAS). Among them, Ang II and ACE are the key parts of the RAAS. ACE can catalyse Ang I into Ang II, which can inactivate bradykinin. Thus, inhibiting ACE activity can effectively lower blood pressure.[77] Certain flavonoids can inhibit ACE activity, and the flavonoid skeleton has an elemental composition that increases angiotensin-converting enzyme inhibitory (ACEI) activity, for example, the catechol group in the B ring, the double bond between C2 and C3 of the C ring, and the ketone group of C4 on the C ring.[78]

3.6.3 Free Radical Scavenging Activity

Many oxidases in organisms such as lipoxygenase (LPO), COX, and NO synthase are related to the generation of free radicals. Oxidation–reduction of peroxides and certain metal ions can cause the body to produce free radicals. Among them, Fe^{2+} catalyses H_2O_2 to produce $^{\bullet}OH$ (Fenton reaction) and transition metal ions catalyse the homolysis of LOOH to produce lipid oxygen radicals, which are most common.[79] Polyphenols can suppress the generation of free radicals by inhibiting the oxidative enzyme system and the metal ion pathway that induce oxidation by complexation.

For the inherent free radicals in the body, polyphenols have a direct scavenging effect. Polyphenols are used not only as preventive antioxidants to scavenge free radicals, but also as chain-blocking antioxidants to scavenge lipid free radicals. Lipids produced free radicals in the presence of ROS, which initiates the chain reaction of lipid free radicals. They can react with lipid chain oxidation intermediates (lipid free radicals or lipid oxygen free radicals) to suspend the chain reaction and suppress lipid oxidation.[4]

3.6.4 Antioxidant Activity

Polyphenols can also enhance the antioxidant effect by activating the body's own free radical scavenging mechanism. Under normal circumstances, the free radicals of the body are maintained in a balanced state below the damage threshold. The maintenance of this balanced state depends on the antioxidant system of the body, including the antioxidant enzyme system and non-enzymatic antioxidant defence system. Antioxidant enzymes in organisms mainly include SOD, glutathione peroxidase (GSH-Px), and CAT. The non-enzymatic antioxidant defence system contains fat/water-soluble antioxidants, such as vitamin E and reduced GSH and trace elements (selenium and zinc).[30] Polyphenols can effectively enhance the activity of these antioxidant enzymes (Figure 3.3).

It is worth mentioning that polyphenols also induce apoptosis of VSMC by regulating Bax and Bcl-x pathways and inhibiting the proliferation and migration of VSMC induced by inflammation factors [*e.g.*, TNF-α, NO, IL-6, vascular cell adhesion molecule 1 (VCAM-1), intercellular adhesion molecule 1 (ICAM-1), and prostaglandins (PGs)], and then relax the arteries.

Inflammation and oxidative stress interact and participate in the pathogenesis of CVD. The common key position may be NF-κB.[80] The excessive ROS produced by oxidative stress can increase the activity of COX and lipoxygenase (LPO), and down-regulate the activity of eNOS, thereby activating NF-κB (Figure 3.3). Polyphenols can inhibit the phosphorylation, ubiquitination and proteasome degradation of IκB, resulting in the failure of the heterodimer composed of p65 and p50 to be released.[30] The failure of the activation of the heterodimer composed of p65 and p50 by post-translational modification (phosphorylation, acetylation, and glycosylation) effectively prevents the heterodimer from translocating to the nucleus, and gene transcription cannot be achieved, which in turn reduces the expression of inflammation cytokines and the occurrence of CVD.[40,42]

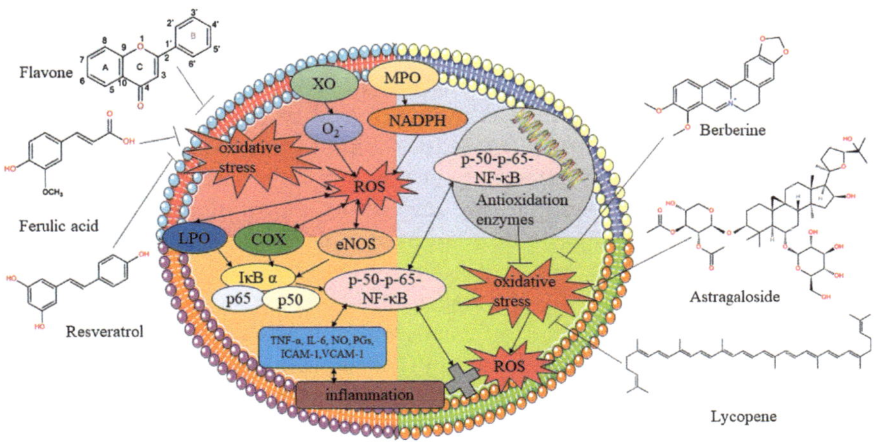

Figure 3.3 Mechanism of biologically active substances (flavonoids, phenolic acids, stilbenes, alkaloids, saponins, and vitamins) on inhibiting oxidative stress. Abbreviations: COX, cyclooxygenase; eNOS, endothelial nitric oxide synthase; ICAM-1, intercellular adhesion molecule-1; IL-6, interleukin-6; IκBα, I kappa B alpha; LPO, lipoxygenase; MPO, myeloperoxidase; NADPH, nicotinamide adenine dinucleotide phosphate; NF-κB, nuclear factor-kappa B; NO, nitric oxide; PGs, prostaglandins; ROS, reactive oxygen species; TNF-α, tumour necrosis factor-alpha; VCAM-1, vascular cell adhesion molecule 1; XO, xanthine oxidase.

3.7 Conclusion

Natural compounds in dietary supplements have received extensive attention for their low toxicity nature and a wide range of pharmacological activities. They have great potential as natural cardiovascular protective agents, which provide a theoretical basis for the clinical development of new drugs based on natural compounds. Dietary supplement foods rich in bioactive compounds such as phenolic acids, stilbenes, flavonoids, alkaloids, saponins, vitamins, and MUFA + PUFA, among others, can control the occurrence and development of CVD through multiple mechanisms and different ways, mainly including improving lipid metabolism, lowering blood pressure, reducing oxidative stress, and inflammation. While there are still many controversies about the role and mechanism of dietary supplements in the prevention and treatment of CVD, further research is needed to shed light on the outstanding issues. Moreover, the bioavailability of different bioactive substances in dietary supplements needs to be further studied. The presence of different structures and structure–activity relationships of various biologically active substances, measures such as molecular modification, compounding multiple active substances, changing delivery methods (such as microencapsulation, nano-delivery systems, and microemulsions) and targeting positioning must be considered to improve the bioavailability of the compounds of interest.

In epidemiological studies, it was found that adherence to drinking tea and intake of plenty of vegetables, fruits, whole grains, and nuts are related to lower all-cause mortality, CVD morbidity, and mortality. In *in vitro* and *in vivo* animal studies and RCT investigations, dietary supplementation of foods rich in phenolic acids, stilbenes, flavonoids, alkaloids, saponins, vitamins, and MUFA + PUFA was mostly found to lower cholesterol levels, improve vascular function, and reduce blood pressure, oxidative stress, and inflammation, thus effectively protecting the cardiovascular system. Considering the small sample size and short duration of intervention in most clinical trial investigations, a larger sample size and longer duration of research are needed to better measure the role of dietary supplements in reducing the risk of CVD and determine the best safe dose required. In general, dietary supplementation with various foods rich in phenolic acids, stilbenes, flavonoids, alkaloids, saponins, vitamins, and MUFA + PUFA as well as adherence to healthy dietary habits and lifestyles can effectivity promote cardiovascular health and ultimately prevent CVD.

References

1. Y. Deng, Y. Tu, S. Lao, M. Wu, H. Yin and L. Wang, *et al.*, The role and mechanism of citrus flavonoids in cardiovascular diseases prevention and treatment, *Crit. Rev. Food Sci. Nutr.*, 2022, **62**, 7591–7614.
2. L. L. Liu, R. Jin, J. Q. Hao, J. Zeng, D. Yin and Y. M. Yi, *et al.*, Consumption of the fish oil high-fat diet uncouples obesity and mammary

tumor growth through induction of reactive oxygen species in protumor macrophages, *Cancer Res.*, 2020, **80**, 2564–2574.
3. Y. S. Yao, T. D. Li and Z. H. Zeng, Mechanisms underlying direct actions of hyperlipidemia on myocardium: an updated review, *Lipids Health Dis.*, 2020, **19**, 23.
4. Y. C. Cheng, J. M. Sheen, W. L. Hu and Y. C. Hung, Polyphenols and oxidative stress in atherosclerosis-related ischemic heart disease and stroke, *Oxid. Med. Cell. Longevity*, 2017, **2017**, 8526438.
5. M. A. Incalza, R. D'Oria, A. Natalicchio, S. Perrini, L. Laviola and F. Giorgino, Oxidative stress and reactive oxygen species in endothelial dysfunction associated with cardiovascular and metabolic diseases, *Vascul. Pharmacol.*, 2018, **100**, 1–19.
6. X. Chang, T. Zhang, W. Zhang, Z. Zhao and J. Sun, Natural drugs as a treatment strategy for cardiovascular disease through the regulation of oxidative stress, *Oxid. Med. Cell. Longevity*, 2020, **2020**, 5430407.
7. X. Yan, Q. Y. Zhang, Y. L. Zhang, X. Han, S. B. Guo and H. H. Li, Gallic acid attenuates angiotensin II-induced hypertension and vascular dysfunction by inhibiting the degradation of endothelial nitric oxide synthase, *Front. Pharmacol.*, 2020, **11**, 1121.
8. Y. Y. Wang, J. L. Gallegos, C. Haskell-Ramsay and J. K. Lodge, Effects of chronic consumption of specific fruit (berries, citrus and cherries) on CVD risk factors: a systematic review and meta-analysis of randomised controlled trials, *Eur. J. Nutr.*, 2021, **60**, 615–639.
9. S. Benisi-Kohansal, P. Saneei, M. Salehi-Marzijarani, B. Larijani and A. Esmaillzadeh, Whole-grain intake and mortality from all causes, cardiovascular disease, and cancer: a systematic review and dose–response meta-analysis of prospective cohort studies, *Adv. Nutr.*, 2016, 7, 1052–1065.
10. B. H. Parmenter, F. Dalgaard, K. Murray, A. Cassidy, C. P. Bondonno and J. R. Lewis, *et al.*, Habitual flavonoid intake and ischemic stroke incidence in the Danish diet, cancer, and health cohort, *Am. J. Clin. Nutr.*, 2021, **114**, 348–357.
11. Q. Gao, J. Y. Dong, R. Z. Cui, I. Muraki, K. Yamagishi and N. Sawada, *et al.*, Consumption of flavonoid-rich fruits, flavonoids from fruits and stroke risk: a prospective cohort study, *Br. J. Nutr.*, 2021, 1–8.
12. A. Micek, J. Godos, D. Del Rio, F. Galvano and G. Grosso, Dietary flavonoids and cardiovascular disease: a comprehensive dose–response meta-analysis, *Mol. Nutr. Food Res.*, 2021, **65**, 2001019.
13. S. Swaminathan, M. Dehghan, J. M. Raj, T. Thomas, S. Rangarajan and D. Jenkins, *et al.*, Associations of cereal grains intake with cardiovascular disease and mortality across 21 countries in Prospective Urban and Rural Epidemiology study: prospective cohort study, *Br. Med. J.*, 2021, **372**, m4948.
14. M. Chung, N. Zhao, D. Wang, M. Shams-White, M. Karlsen and A. Cassidy, *et al.*, Dose–response relation between tea consumption and risk of cardiovascular disease and all-cause mortality: a systematic

review and meta-analysis of population-based studies, *Adv. Nutr.*, 2020, **11**, 790–814.
15. M. Teramoto, I. Muraki, K. Yamagishi, A. Tamakoshi and H. Iso, Green tea and coffee consumption and all-cause mortality among persons with and without stroke or myocardial infarction, *Stroke*, 2021, **52**, 957–965.
16. M. M. Wang, Y. J. Bai, Z. Wang, Z. Z. Zhang, D. D. Liu and X. G. Lian, Higher tea consumption is associated with decreased risk of small vessel stroke, *Clin. Nutr.*, 2021, **40**, 1430–1435.
17. Y. B. Zhang, M. Li, X. J. Li, T. Zhang, M. Qin and L. Q. Ren, Isoquinoline alkaloids and indole alkaloids attenuate aortic atherosclerosis in apolipoprotein e deficient mice: A systematic review and meta-analysis, *Front. Pharmacol.*, 2018, **9**, 602.
18. Q. Tong, P. C. Zhu, Z. Zhuang, L. H. Deng, Z. H. Wang and H. Zeng, *et al.*, Notoginsenoside R1 for organs ischemia/reperfusion injury: A preclinical systematic review, *Front. Pharmacol.*, 2019, **10**, 1204.
19. H. M. Cheng, G. Koutsidis, J. K. Lodge, A. W. Ashor, M. Siervo and J. Lara, Lycopene and tomato and risk of cardiovascular diseases: A systematic review and meta-analysis of epidemiological evidence, *Crit. Rev. Food Sci. Nutr.*, 2019, **59**, 141–158.
20. P. Munoz-Aguirrel, E. Denova-Gutierrez, M. Flores, E. Salazar-Martinez and J. Salmeron, High vitamin D consumption is inversely associated with cardiovascular disease risk in an urban mexican population, *PLoS One*, 2016, **11**, e0166869.
21. P. F. Cheng, L. H. Wang, S. J. Ning, Z. C. Liu, H. Lin and S. L. Chen, *et al.*, Vitamin E intake and risk of stroke: a meta-analysis, *Br. J. Nutr.*, 2018, **120**, 1181–1188.
22. M. Shimizu, T. Miyazaki, A. Takagi, Y. Sugita, S. Ouchi and T. Aikawa, *et al.*, Low coenzyme Q10 levels in patients with acute cardiovascular disease are associated with long-term mortality, *Heart Vessels*, 2021, **36**, 401–407.
23. N. Becerra-Tomas, I. Paz-Graniel, C. Kendall, H. Kahleova, D. Rahelic and J. L. Sievenpiper, *et al.*, Nut consumption and incidence of cardiovascular diseases and cardiovascular disease mortality: a meta-analysis of prospective cohort studies, *Nutr. Rev.*, 2019, 77, 691–709.
24. M. A. Lee-Bravatti, J. F. Wang, E. E. Avendano, L. King, E. J. Johnson and G. Raman, Almond consumption and risk factors for cardiovascular disease: a systematic review and meta-analysis of randomized controlled trials, *Adv. Nutr.*, 2019, **10**, 1076–1088.
25. M. Jabarpour, N. Rashtchizadeh, H. Argani, A. Ghorbanihaghjo, M. Ranjbarzadhag and D. Sanajou, *et al.*, The impact of dyslipidemia and oxidative stress on vasoactive mediators in patients with renal dysfunction, *Int. Urol. Nephrol.*, 2019, **51**, 2235–2242.
26. K. Shanmugam, S. R. Boovarahan, P. Prem, B. Sivakumar and G. A. Kurian, Fisetin attenuates myocardial ischemia-reperfusion injury by activating the reperfusion injury salvage kinase (RISK) signaling pathway, *Front. Pharmacol.*, 2021, **12**, 566470.

27. M. A. Aldubayan, A. S. Ahmed, A. M. Emara, A. A. Ahmed and R. M. Elgharabawy, Sinapic acid attenuates cardiovascular disorders in rats by modulating reactive oxygen species and angiotensin receptor expression, *Oxid. Med. Cell. Longevity*, 2020, **2020**, 1436858.
28. Y. J. Liu, X. Z. Chen and J. Li, Resveratrol protects against oxidized low-density lipoprotein-induced human umbilical vein endothelial cell apoptosis *via* inhibition of mitochondrial-derived oxidative stress, *Mol. Med. Rep.*, 2017, **15**, 2457–2464.
29. J. Grujic-Milanovic, V. Jacevic, Z. Miloradovic, D. Jovovic, I. Milosavljevic and S. D. Milanovic, *et al.*, Resveratrol protects cardiac tissue in experimental malignant hypertension due to antioxidant, anti-inflammatory, and anti-apoptotic properties, *Int. J. Mol. Sci.*, 2021, **22**, 5006.
30. F. W. Zhang, J. Feng, J. Y. Zhang, X. Kang and D. Qian, Quercetin modulates AMPK/SIRT1/NF-kappa B signaling to inhibit inflammatory/oxidative stress responses in diabetic high fat diet-induced atherosclerosis in the rat carotid artery, *Exp. Ther. Med.*, 2020, **20**, 280.
31. L. M. Yu, X. Dong, J. Zhang, Z. Li, X. D. Xue and H. J. Wu, *et al.*, Naringenin attenuates myocardial ischemia-reperfusion injury *via* cGMP-PKGI alpha signaling and *in vivo* and *in vitro* studies, *Oxid. Med. Cell. Longevity*, 2019, **2019**, 7670854.
32. D. F. Yang, Z. S. Yang, L. Chen, D. B. Kuang, Y. Zou and J. Li, *et al.*, Dihydromyricetin increases endothelial nitric oxide production and inhibits atherosclerosis through microRNA-21 in apolipoprotein E-deficient mice, *J. Cell. Mol. Med.*, 2020, **24**, 5911–5925.
33. X. Sun and X. Cui, Isorhapontigenin alleviates cerebral ischemia/reperfusion injuries in rats and modulated the PI3K/Akt signaling pathway, *Naunyn-Schmiedebergs Arch. Pharmacol.*, 2020, **393**, 1753–1760.
34. J. F. Reis, V. V. S. Monteiro, R. D. Gomes, M. M. do Carmo, G. V. da Costa and P. C. Ribera, *et al.*, Action mechanism and cardiovascular effect of anthocyanins: a systematic review of animal and human studies, *J. Transl. Med.*, 2016, **14**, 315.
35. A. M. Martins, D. Sarto, K. D. Caproni, J. Silva, J. Silva and P. S. Souza, *et al.*, Grape juice attenuates left ventricular hypertrophy in dyslipidemic mice, *PLoS One*, 2020, **15**, e0238163.
36. G. H. Lee, T. H. Hoang, E. S. Jung, S. J. Jung, S. K. Han and M. J. Chung, *et al.*, Anthocyanins attenuate endothelial dysfunction through regulation of uncoupling of nitric oxide synthase in aged rats, *Aging Cell*, 2020, **19**, e13279.
37. C. J. Xie, A. P. Gu, J. Cai, Y. Wu and R. C. Chen, Curcumin protects neural cells against ischemic injury in N2a cells and mouse brain with ischemic stroke, *Brain Behav.*, 2018, **8**, e00921.
38. M. C. L. Ferreira, L. N. Lima, L. H. T. Cota, M. B. Costa, P. M. E. Orsi and R. P. Espindola, *et al.*, Effect of *Camellia sinensis* teas on left ventricular hypertrophy and insulin resistance in dyslipidemic mice, *Braz. J. Med. Biol. Res.*, 2020, **53**, e9303.

39. J. C. Zhang, H. Xu, Y. Yuan, J. Y. Chen, Y. J. Zhang and Y. Lin, *et al.*, Delayed treatment with green tea polyphenol EGCG promotes neurogenesis after ischemic stroke in adult mice, *Mol. Neurobiol.*, 2017, **54**, 3652–3664.
40. W. F. Li, J. J. Yu, X. Xiao, W. Q. Li, L. L. Zang and T. F. Han, *et al.*, The inhibitory effect of (−)-epicatechin gallate on the proliferation and migration of vascular smooth muscle cells weakens and stabilizes atherosclerosis, *Eur. J. Pharmacol.*, 2021, **891**, 173761.
41. M. G. Xu, Q. Qi, L. N. Men, S. S. Wang, M. Li and M. Xiao, *et al.*, Berberine protects Kawasaki disease-induced human coronary artery endothelial cells dysfunction by inhibiting of oxidative and endoplasmic reticulum stress, *Vasc. Pharmacol.*, 2020, **127**, 106660.
42. L. Wang, H. Ma, Y. Xue, H. Shi, T. Ma and X. Cui, Berberine inhibits the ischemia-reperfusion injury induced inflammatory response and apoptosis of myocardial cells through the phosphoinositide 3-kinase/RAC-α serine/threonine-protein kinase and nuclear factor-κB signaling pathways, *Exp. Ther. Med.*, 2018, **15**, 1225–1232.
43. J. S. Wu, J. M. Li, H. Y. Lo, C. Y. Hsiang and T. Y. Ho, Anti-hypertensive and angiotensin-converting enzyme inhibitory effects of Radix Astragali and its bioactive peptide AM-1, *J. Ethnopharmacol.*, 2020, **254**, 8.
44. M. Z. Ying, J. J. Wu, J. M. Yin, X. Q. Shen and F. Li, Inhibitory effects of total saponins from *Ilex pubescens* Hook against hydrogen peroxide-induced cardiomyocyte apoptosis, *Trop. J. Pharm. Res.*, 2017, **16**, 2681–2688.
45. P. Ferreira-Santos, R. Aparicio, R. Carron, M. A. Sevilla, J. Monroy-Ruiz and M. J. Montero, Lycopene-supplemented diet ameliorates cardiovascular remodeling and oxidative stress in rats with hypertension induced by Angiotensin II, *J. Funct. Foods*, 2018, **47**, 279–287.
46. J. M. Martinez-Moreno, C. Herencia, A. M. de Oca, J. R. Munoz-Castaneda, M. E. Rodriguez-Ortiz and J. M. Diaz-Tocados, *et al.*, Vitamin D modulates tissue factor and protease-activated receptor 2 expression in vascular smooth muscle cells, *FASEB J.*, 2016, **30**, 1367–1376.
47. F. T. Tang, M. L. Lu, S. P. Zhang, M. Mei, T. Q. Wang and P. Q. Liu, *et al.*, Vitamin E conditionally inhibits atherosclerosis in apoe knockout mice by anti-oxidation and regulation of vasculature gene expressions, *Lipids*, 2014, **49**, 1215–1223.
48. O. O. Nikolaevna, G. E. Aronovna, K. E. Igorevna, B. M. Alekseevna, G. M. Vladimirovich and M. V. Gennadievich, *et al.*, Intravenous administration of coenzyme Q10 in acute period of cerebral ischemia decreases mortality by reducing brain necrosis and limiting its increase within 4 days in rat stroke model, *Antioxidants*, 2020, **9**, 1240.
49. W. S. He, L. L. Li, J. X. Rui, J. J. Li, Y. Y. Sun and D. D. Cui, *et al.*, Tomato seed oil attenuates hyperlipidemia and modulates gut microbiota in C57BL/6J mice, *Food Funct.*, 2020, **11**, 4275–4290.
50. N. D. Barnard, J. Alwarith, E. Rembert, L. Brandon, M. Nguyen and A. Goergen, *et al.*, A Mediterranean diet and low-fat vegan diet to

improve body weight and cardiometabolic risk factors: a randomized, cross-over trial, *J. Am. Coll. Nutr.*, 2021, **5**, 1–13.
51. M. S. Macarro, J. P. M. Rodriguez, E. B. Morell, S. Perez-Pinero, D. Victoria-Montesinos and A. M. Garcia-Munoz, *et al.*, Effect of a combination of citrus flavones and flavanones and olive polyphenols for the reduction of cardiovascular disease risk: an exploratory randomized, double-blind, placebo-controlled study in healthy subjects, *Nutrients*, 2020, **12**, 1475.
52. S. Fatahi, E. Daneshzad, H. Kord-Varkaneh, N. Bellissimo, N. R. Brett and L. Azadbakht, Impact of diets rich in whole grains and fruits and vegetables on cardiovascular risk factors in overweight and obese women: a randomized clinical feeding trial, *J. Am. Coll. Nutr.*, 2018, **37**, 568–577.
53. S. Ghosh, P. Basak, S. Dutta, S. Chowdhury and P. C. Sil, New insights into the ameliorative effects of ferulic acid in pathophysiological conditions, *Food Chem. Toxicol.*, 2017, **103**, 41–55.
54. A. Bumrungpert, S. Lilitchan, S. Tuntipopipat, N. Tirawanchai and S. Komindr, Ferulic acid supplementation improves lipid profiles, oxidative stress, and inflammatory status in hyperlipidemic subjects: a randomized, double-blind, placebo-controlled clinical trial, *Nutrients*, 2018, **10**, 713.
55. L. Pla-Paga, J. Companys, L. Calderon-Perez, E. Llaurado, R. Sola and R. M. Valls, *et al.*, Effects of hesperidin consumption on cardiovascular risk biomarkers: a systematic review of animal studies and human randomized clinical trials, *Nutr. Rev.*, 2019, **77**, 845–864.
56. N. P. Bondonno, C. P. Bondonno, L. C. Blekkenhorst, M. J. Considine, G. Maghzal and R. Stocker, *et al.*, Flavonoid-rich apple improves endothelial function in individuals at risk for cardiovascular disease: a randomized controlled clinical trial, *Mol. Nutr. Food Res.*, 2018, **62**, 1700674.
57. D. Martini, M. Marino, D. Angelino, C. Del Bo, D. Del Rio and P. Riso, *et al.*, Role of berries in vascular function: a systematic review of human intervention studies, *Nutr. Rev.*, 2020, **78**, 189–206.
58. L. Arevstrom, C. Bergh, R. Landberg, H. X. Wu, A. Rodriguez-Mateos and M. Waldenborg, *et al.*, Freeze-dried bilberry (*Vaccinium myrtillus*) dietary supplement improves walking distance and lipids after myocardial infarction: an open-label randomized clinical trial, *Nutr. Res.*, 2019, **62**, 13–22.
59. J. R. Santos-Parker, T. R. Strahler, C. J. Bassett, N. Z. Bispham, M. B. Chonchol and D. R. Seals, Curcumin older adults by increasing nitric oxide bioavailability and reducing oxidative stress, *Aging*, 2017, **9**, 187–208.
60. I. J. Chen, C. Y. Liu, J. P. Chiu and C. H. Hsu, Therapeutic effect of high-dose green tea extract on weight reduction: A randomized, double-blind, placebo-controlled clinical trial, *Clin. Nutr.*, 2016, **35**, 592–599.
61. G. Derosa, A. D'Angelo, A. Bonaventura, L. Bianchi, D. Romano and P. Maffioli, Effects of berberine on lipid profile in subjects with low cardiovascular risk, *Expert Opin. Biol. Ther.*, 2013, **13**, 475–482.

62. V. Vuksan, Z. Z. Xu, E. Jovanovski, A. L. Jenkins, U. Beljan-Zdravkovic and J. L. Sievenpiper, *et al.*, Efficacy and safety of American ginseng (*Panax quinquefolius* L.) extract on glycemic control and cardiovascular risk factors in individuals with type 2 diabetes: a double-blind, randomized, cross-over clinical trial, *Eur. J. Nutr.*, 2019, **58**, 1237–1245.
63. D. Michalickova, M. Belovic, N. Ilic, J. Kotur-Stevuljevic, O. Slanar and S. Sobajic, Comparison of polyphenol-enriched tomato juice and standard tomato juice for cardiovascular benefits in subjects with stage 1 hypertension: a randomized controlled study, *Plant Food Hum. Nutr.*, 2019, **74**, 122–127.
64. J. McEneny, L. Wade, I. S. Young, L. Masson, G. Duthie and A. McGinty, *et al.*, Lycopene intervention reduces inflammation and improves HDL functionality in moderately overweight middle-aged individuals, *J. Nutr. Biochem.*, 2013, **24**, 163–168.
65. V. C. Fajardo, F. L. P. de Oliveira, G. L. L. Machado-Coelho, F. A. P. Pimenta, S. N. de Freitas and A. L. P. Ribeiro, *et al.*, Effects of vitamin D supplementation on cardiovascular risk factors in shift workers study protocol for randomized, double-blind, placebo-controlled clinical trial, *Medicine*, 2019, **98**, e15417.
66. G. Tang, L. Y. Zhang, J. Tao and Z. Q. Wei, Effect of *Nigella sativa* in the treatment of nonalcoholic fatty liver disease: A systematic review and meta-analysis of randomized controlled trials, *Phytother. Res.*, 2021, **35**, 4183–4193.
67. P. Zacek, M. Bukowski, L. Johnson, S. K. Raatz and M. Picklo, Selective enrichment of n-3 fatty acids in human plasma lipid motifs following intake of marine fish, *J. Nutr. Biochem.*, 2018, **54**, 57–65.
68. I. B. Asztalos, J. A. Gleason, S. Sever, R. Gedik, B. F. Asztalos and K. V. Horvath, *et al.*, Effects of eicosapentaenoic acid and docosahexaenoic acid on cardiovascular disease risk factors: a randomized clinical trial, *Metabolism*, 2016, **65**, 1636–1645.
69. A. Sut, K. Chizynski, M. Rozalski and J. Golanski, Dietary intake of omega fatty acids and polyphenols and its relationship with the levels of inflammatory markers in men with chronic coronary syndrome after percutaneous coronary intervention, *Kardiol. Pol.*, 2020, **78**, 117–123.
70. H. H. Huang, G. H. Zhou, R. Pu, Y. J. Cui and D. Liao, Clinical evidence of dietary supplementation with sesame on cardiovascular risk factors: An updated meta-analysis of randomized controlled trials, *Crit. Rev. Food Sci. Nutr.*, 2022, **62**, 5592–5602.
71. A. Farajbakhsh, S. M. Mazloomi, M. Mazidi, P. Rezaie, M. Akbarzadeh and S. P. Ahmad, *et al.*, Sesame oil and vitamin E co-administration may improve cardiometabolic risk factors in patients with metabolic syndrome: a randomized clinical trial, *Eur. J. Clin. Nutr.*, 2019, **73**, 1403–1411.
72. D. Birudaraju, L. Cherukuri, A. Kinninger, B. T. Chaganti, K. Shaikh and S. Hamal, *et al.*, A combined effect of cavacurcumin, eicosapentaenoic acid (omega-3s), astaxanthin and gamma -linoleic acid (omega-6)

(CEAG) in healthy volunteers- a randomized, double-blind, placebo-controlled study, *Clin. Nutr. ESPEN*, 2020, **35**, 174–179.

73. S. Ul Islam, M. B. Ahmed, H. Ahsan and Y. S. Lee, Recent molecular mechanisms and beneficial effects of phytochemicals and plant-based whole foods in reducing LDL-C and preventing cardiovascular disease, *Antioxidants*, 2021, **10**, 784.
74. X. Gong, X. Li, Y. Xia, J. F. Xu, Q. Y. Li and C. H. Zhang, *et al.*, Effects of phytochemicals from plant-based functional foods on hyperlipidemia and their underpinning mechanisms, *Trends Food Sci. Technol.*, 2020, **103**, 304–320.
75. Y. S. Gao, T. J. Chen and J. U. Raj, Endothelial and smooth muscle cell interactions in the pathobiology of pulmonary hypertension, *Am. J. Respir. Cell Mol. Biol.*, 2016, **54**, 451–460.
76. D. Maaliki, A. A. Shaito, G. Pintus, A. El-Yazbi and A. H. Eid, Flavonoids in hypertension: a brief review of the underlying mechanisms, *Curr. Opin. Pharmacol.*, 2019, **45**, 57–65.
77. A. Daiber and S. Chlopicki, Revisiting pharmacology of oxidative stress and endothelial dysfunction in cardiovascular disease: Evidence for redox-based therapies, *Free Radic. Biol. Med.*, 2020, **157**, 15–37.
78. L. Guerrero, J. Castillo, M. Quinones, S. Garcia-Vallve, L. Arola and G. Pujadas, *et al.*, Inhibition of angiotensin-converting enzyme activity by flavonoids: structure–activity relationship studies, *PLoS One*, 2012, 7, e49493.
79. M. Quinones, M. Miguel and A. Aleixandre, Beneficial effects of polyphenols on cardiovascular disease, *Pharmacol. Res.*, 2013, **68**, 125–131.
80. H. Yi, R. Peng, L. Y. Zhang, Y. Sun, H. M. Peng and H. D. Liu, *et al.*, LincRNA-Gm4419 knockdown ameliorates NF-kappa B/NLRP3 inflammasome-mediated inflammation in diabetic nephropathy, *Cell Death Dis.*, 2017, **8**, 14.

CHAPTER 4

Green Tea Catechins and Nonalcoholic Fatty Liver Disease

PRIYANKAR DEY*[a] AND RICHARD S. BRUNO[b]

[a] Department of Biotechnology, Thapar Institute of Engineering and Technology, Patiala, Punjab, India; [b] Human Nutrition Program, The Ohio State University, Columbus, OH, USA
*Email: priyankar.dey@thapar.edu

4.1 Introduction

Tea is the most frequently consumed prepared beverage worldwide with an estimated 2 billion servings consumed daily. Consistent with its popularity, tea is harvested globally in 60 countries to yield >5 million tons annually.[1] It is consumed widely on a recreational basis, and its consumption has increased markedly due to growing scientific recognition that its polyphenolic catechins help to promote human health. Tea has been suggested to alleviate the risk of several chronic disorders and protect against premature mortality.[2] Chronic liver disease in particular has become a global public health burden of the 21st century which has manifested due to poor diet and decreased physical activity which has created a substantial imbalance between energy storage and expenditure. This drives not only the risk of obesity but also the rising prevalence of obesity-associated nonalcoholic fatty liver disease (NAFLD), a disorder that encompasses simple steatosis or nonalcoholic fatty liver, and its progressive advancement to

Food Chemistry, Function and Analysis No. 39
Dietary Supplements with Antioxidant Activity: Understanding Mechanisms and Potential Health Benefits
Edited by Cesarettin Alasalvar, Fereidoon Shahidi and Chi-Tang Ho

Published by the Royal Society of Chemistry, www.rsc.org

nonalcoholic steatohepatitis, fibrosis, and cirrhosis.[3] At present, there is no pharmacological treatment for NAFLD. This has resulted in a heightened emphasis on dietary interventions to prevent and treat this disorder that substantially increases the risk of liver-related mortality.

Numerous dietary agents have been investigated to ameliorate NAFLD. Various tea products are recognized for their bioactive constituents that can alleviate pro-inflammatory insults implicated in NAFLD including black, fermented, Pu-erh, kombucha, and Fuzhuan brick teas.[4] However, green tea is especially rich in polyphenolic catechins (Figure 4.1),[5] which is widely consumed, and may provide superior benefits on liver health. This chapter discusses the hepatoprotective activities of green tea and its catechins on NAFLD risk. Specifically, the antioxidant, anti-inflammatory, lipid-lowering, anti-obesity, and gut health-promoting activities of green tea are discussed in relation to improving liver health. Although most reported evidence in this area is derived from preclinical models and epidemiological studies, recent but limited efforts for research translation to human health are discussed. In addition, this chapter provides an important discussion concerning the safe

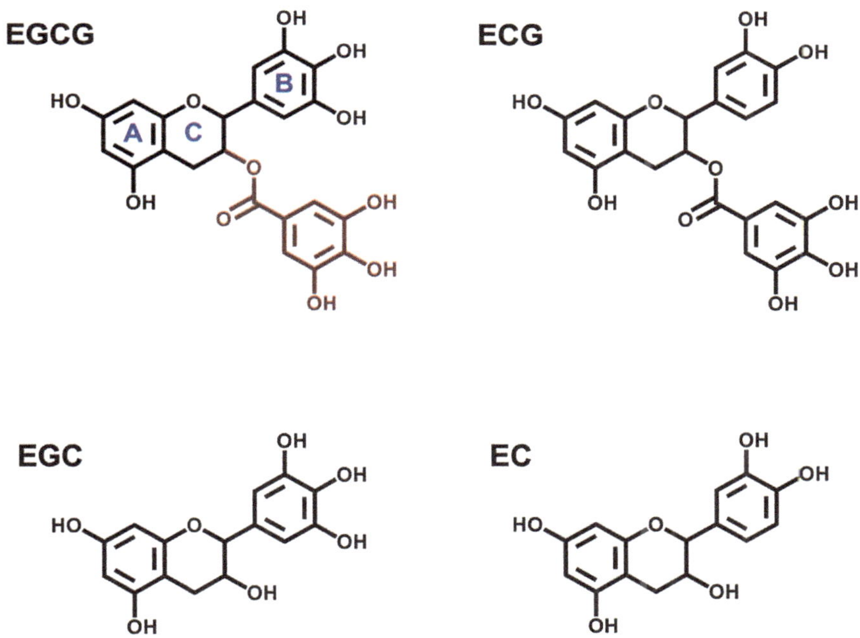

Figure 4.1 Structures of major green tea catechins. The general backbone of all catechins has aromatic A- and B-rings that are condensed to a 3-carbon oxygenated heterocyclic C-ring. EGCG and ECG also contain a galloyl group (highlighted in red) that is esterified to the 3-carbon of the C-ring. The A-ring of catechins has dihydroxy substitutions, whereas the B-rings of EC and ECG are di-hydroxylated compared with tri-hydroxylated EGC and EGCG. Abbreviations: EC, epicatechin; ECG, epicatechin gallate; EGC, epigallocatechin; EGCG, epigallocatechin gallate.

consumption of green tea and its catechins. Therefore, it provides an overview of the putative mechanisms by which green tea catechins alleviate NAFLD risk relative to an important public health goal of establishing health-evidence-based dietary recommendations in humans.

4.2 Green Tea

4.2.1 Phytochemical and Nutrient Composition

Green tea is well-recognized for being rich in polyphenolic catechins (Figure 4.1). However, other phytochemicals are present, albeit in much lower levels, including linoleic and α-linolenic acids, vitamins B_2, B_3, C, and E, and certain pro-vitamin A carotenoids.[6,7] The major catechins in green tea are epicatechin (EC), epicatechin gallate (ECG), epigallocatechin (EGC), and epigallocatechin gallate (EGCG). The total content of catechins comprises ~40% of the dry weight of the tea leaf,[8,9] but EGCG accounts for 50–80% of the total catechin content.[10]

The compositional profile of freshly brewed tea depends on several factors; these include tea plant variety, dry weight of the tea leaves, brewing duration, and brewing temperature.[11] For example, soaking green tea leaves (2.5 g) for 3 minutes in 250 mL water at 80 °C extracts approximately 30% of its water-soluble phytochemicals.[12] Brewing green tea for 30 minutes at 80 °C gives the highest extraction of total catechins (127 $mg\,g^{-1}$) and EGCG (43% w/w of total catechins).[13] The catechin content of commercially available tea can, therefore, differ based on tea brewing practices but also how tea leaves are processed post-harvest. Indeed, the tea plant (*Camellia sinensis*) is used to produce green tea, oolong tea, and black tea. These tea products have similar protein, fibre, carbohydrate, lipid, mineral, and caffeine contents. However, the polyphenol content of green tea is superior to those of oolong and black teas.[5] This is because leaves intended for green tea production are subject to rapid heat inactivation of enzymes (*e.g.*, polyphenol oxidase) that would otherwise degrade catechins, such as that occurring during partial and full fermentation of leaves intended for oolong tea and black tea, respectively. L-Theanine is a non-protein amino acid and the main taste compound found almost exclusively in tea leaves. It represents 1–2% of the dry tea leaf weight to provide 20–60 mg per serving of freshly brewed tea.[14] Brewed tea also contains 0.4–17% caffeine,[15] which does not differ appreciably between green, black, and oolong teas.

4.2.2 Antioxidant Activities

In vitro antioxidant activities are generally quantified using biochemical assays that are based on the reduction potential of the test compound to participate in proton or electron transfer. With reduction potentials of 430–570 mV, antioxidant activities of green tea catechins fall between those of α-tocopherol (480 mV) and ascorbic acid (280 mV).[16] Polyphenol-rich green tea extract (GTE) is reported to have superior activity compared with ascorbic acid, vitamin E, and β-carotene to scavenge superoxide, hydrogen peroxide, and hypochlorous acid.[17] The galloyl group of EGCG is also demonstrated to provide the functionality to scavenge the highly reactive hydroxyl and superoxide radicals.[18] Due to its greater polyphenol content, green tea has higher antioxidant activity than oolong and black teas.[19] Separate biochemical studies have also demonstrated that the antioxidant activity of green tea is attributed to its polyphenols,[20] which have been quantified at 300 $mg\,g^{-1}$ dried leaves.[21]

The antioxidant activities of green tea catechins have been documented in cellular and animal models and in some studies in humans. For example, EGCG-rich GTE protected against oxidative injury otherwise induced by hydroxyl radicals in human T lymphocytes.[22] Green tea also inhibits nitric oxide (NO) release in a macrophage cell line, lipid peroxidation in erythrocytes, and hydrogen peroxide generation in keratinocytes, and mitigates bleomycin-induced DNA damage in leukocytes.[23] In rats, supplementation of green tea polyphenols lowered oxidative stress-induced DNA damage in primary lymphocyte, colonocyte, and hepatocyte cells.[24] GTE supplementation in aged mice also decreased lipid peroxidation in association with increased circulating glutathione and total thiol concentrations.[25] In these same-aged mice, oxidative injury of the brain was suggested to be decreased due to GTE-mediated increases in cytoprotective defences (superoxide dismutase, catalase, glucose-6-phosphate dehydrogenase, glutathione peroxidase, and glutathione reductase) that are involved in the detoxification of reactive oxygen species (ROS).

Humans consuming green tea are reported to have increased total antioxidant status. For example, ingestion of freshly brewed green tea significantly increased total plasma antioxidant capacity at 60 min post-consumption.[26] Another study demonstrated a significant reduction of the lipid peroxidation biomarkers, malondialdehyde and 4-hydroxynonenal, in the serum of individuals with metabolic syndrome (MetS) following the daily consumption of 4 cups of green tea.[27] It is clear that green tea or its catechins are capable of providing antioxidant benefits to reduce oxidative injury. However, it remains unclear whether catechins are capable of directly scavenging reactive species *in vivo* consistent with their documented activities *in vitro*. Indeed, it is often reported that catechins have poor oral bioavailability which results in circulating concentrations that are substantially lower than other dietary antioxidants (*e.g.*, vitamin C and vitamin E). It is likely that catechins exert indirect antioxidant activities *in vivo* by upregulating endogenous antioxidant defences that detoxify reactive species or that their anti-inflammatory activities downregulate the generation of reactive species. Regardless of the mechanism, the antioxidant function of green tea catechins supports the notion that they may protect against oxidative stress-mediated hepatocellular injury that drives the progression of NAFLD (Figure 4.2).

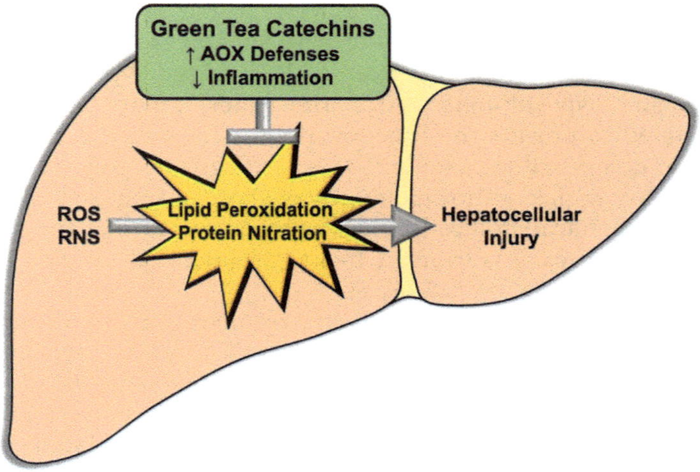

Figure 4.2 Antioxidant and anti-inflammatory activities of green tea catechins in NAFLD. ROS and RNS derived from cellular metabolism and activated inflammatory cells cause hepatocellular injury in NAFLD downstream of hepatic lipid peroxidation and protein nitration. Dietary supplementation with green tea catechins has been shown to upregulate cytoprotective antioxidant defenses (*e.g.*, superoxide dismutase, catalase, glutathione peroxidase, and glutathione biosynthesis) that help to detoxify ROS. Green tea catechins also attenuate NADPH oxidase and RNS generation from inflammatory cells. These antioxidant and anti-inflammatory activities protect against lipid peroxidation and protein nitration in association with attenuated NAFLD pathology. Abbreviations: AOX, antioxidant; NAFLD, nonalcoholic fatty liver disease; RNS, reactive nitrogen species; ROS, reactive oxygen species.

4.3 Hepatoprotective Activities of Green Tea in NAFLD

NAFLD is the leading liver disease worldwide due to its close association with obesity. Its pathogenesis has been described as a "two-hit" mechanism such that a "first-hit", mediated by excess energy intake and limited physical activity, causes insulin resistance and excess hepatocellular triacylglycerol accumulation (*e.g.*, steatosis). A "second-hit" driven by pro-inflammatory cytokines, adipokines, mitochondrial dysfunction, and oxidative stress then provokes hepatocellular injury of the liver with steatosis.[28,29] This historic view that NAFLD occurs sequentially has been revised in recent years to a "multiple-hit" or "parallel-hit" model to better demonstrate that hepatocellular lipid accumulation, insulin resistance, impaired β-oxidation, lipotoxicity, endotoxemia, and inflammation occur concurrently to induce NAFLD-associated liver injury.[30] Observational evidence reported during the past decades supports that higher intake of green tea is associated with liver health.[31–33] This has been complemented by studies *in vitro*, in rodent models, and in humans to help define the mechanism of action and efficacy of green tea catechins to NAFLD risk.[34] The subsequent sections discuss the preclinical and clinical evidence in support of green tea and its catechins to manage NAFLD risk while identifying knowledge that precludes research translation towards establishing evidence-based dietary recommendations in humans.

4.3.1 Hepatocellular Studies

Cellular studies have provided important mechanistic insight confirming that green tea catechins protect against hepatocellular injury implicated in NAFLD development. For example, GTE treatment in HepG2 liver cells protected against fenofibrate-induced oxidative injury.[35] EGCG treatment up to 5 μM in hepatocytes has also been shown to increase the accumulation of nuclear factor erythroid 2-related factor 2 (Nrf2), a transcription factor that upregulates the expression of cytoprotective antioxidant defences.[36] Likewise, green tea treatment in HepG2 cells prevented hydrogen peroxide-induced apoptosis by limiting oxidative stress and improving intracellular antioxidant enzyme levels.[37] Apart from antioxidant activities, others have reported important metabolic effects of green tea or its catechins. Sirtuin-1, which has reduced expression in patients[38] and rodent models,[39] is thought to contribute to dysregulated energy homeostasis, poor insulin sensitivity, and inflammatory and oxidative stress in NAFLD.[40] Studies in HepG2 cells have shown that GTE treatment increases sirtuin-1 expression, promotes deacetylation of liver kinase B1, and decreases hepatocellular triacylglycerol (TAG) accumulation.[41] EGCG treatment at high doses in hepatocytes also promotes cholesterol efflux, whereas lower doses inhibit cholesterol biosynthesis.[42] Others have shown that EGCG increases low-density lipoprotein (LDL) receptor expression and binding capacity in cultured hepatocytes while also upregulating the expression of the lipogenic transcription factor sterol regulatory element-binding protein-1 (SREBP-1).[42] GTE or EGCG also exerts favourable effects in increasing glycogen synthesis and inhibiting lipogenesis by activating adenosine monophosphate-activated protein kinase (AMPK) alpha and acetyl-CoA carboxylase in HepG2 cells.[43] Apart from limiting lipogenesis, EGCG stimulates hepatocellular lipid clearance through a mechanism involving the increased formation of autophagosomes, lysosomal acidification, and autophagic flux in HepG2 cells.[44] Overall, cellular studies in "healthy" cells demonstrate strong potential for green tea and its catechins to prevent oxidative injury through antioxidant activities and favourably influence energy metabolism to limit lipid biosynthesis and storage while enhancing lipid clearance.

Cellular models of lipid-loaded hepatocytes have also been utilized to study whether green tea catechins can ameliorate steatosis, which is regarded as the earliest stage of NAFLD severity. Treatment with catechins in fatty acid-loaded HepG2 cells reduced intracellular lipid accumulation by activating AMPK, a key regulator of lipid metabolism.[45] Proteomic analysis in this lipid-loaded cellular model also revealed that EGCG regulates proteins involved in lipid metabolism, glycometabolism, antioxidant defence, DNA repair, cellular respiration, and iron storage.[46] Protein expression studies in the same fatty acid-loaded HepG2 cells show that EGCG treatment lowers peroxiredoxin-6 and galactokinase while upregulating the succinate dehydrogenase flavoprotein subunit. Others have reported that EGCG treatment in lipid-loaded hepatocytes decreases the expression of certain microRNAs including miR-33a and miR-122.[47] While microRNAs are an active area of investigation relative to NAFLD risk, accumulating evidence indicates that they are important factors to modulate cellular oxidative stress, inflammation, and metabolic pathways in NAFLD,[48] and that they may have prognostic or predictive value for NAFLD.[49] Indeed, studies in HepG2 cells show that the direct binding of EGCG to miR-33a and miR-122 represses their expression and helps to explain the increased expression levels of the adenosine triphosphate (ATP) binding cassette transporter ABCA1 protein, a cholesterol efflux transporter, which occurs independent of changes in the expression of the lipogenic transcription factor sterol regulatory element-binding protein-2.[47] Taken together, studies on steatotic hepatocytes and "healthy" cells suggest that green tea catechins may play a role in not only reversing NAFLD but also functioning in a prophylactic manner to thwart the development of NAFLD.

4.3.2 Diet-induced Obese Models of NAFLD

Obesogenic dietary components contribute to NAFLD. This recognition has driven the use of diet-induced obese rodents to examine the hepatoprotective benefits of GTE and its catechins. Indeed, GTE or its catechins have been reported to limit NAFLD development in obese rodents by attenuating nuclear factor-kappa B (NFκB)-dependent inflammation. When mice were fed a high-fat diet (60% of energy from fat), dietary supplementation of powdered GTE at 2% (w/w) attenuated the DNA binding capacity of the p65 subunit of NFκB and expression of several genes associated with NFκB-dependent inflammatory responses.[50] Separate studies have shown that GTE supplementation in rats fed a high-fat diet attenuates cyclooxygenase-2 and prostaglandin E_2 while lowering hepatic lipid peroxidation and serum alanine aminotransferase.[51] Others have reported that dietary GTE supplementation can reverse the histologic evidence of NAFLD (*e.g.*, steatosis and hepatocellular ballooning) in mice fed a high-fat diet, likely by downregulating hepatic tumour necrosis factor receptor-1 (TNFR1) and Toll-like receptor-4 (TLR4), both of which induce NFκB activation and its downstream pro-inflammatory responses.[52] Separate studies also indicate that EGCG itself protects against hepatic inflammation by inhibition of NFκB-dependent signalling.[53,54] These findings support that targeting NFκB activation by dietary GTE or its catechins mitigates NAFLD by downregulating inflammatory responses that are known to cause cellular injury and dysregulate metabolic responses that favour hepatic lipid accumulation.

Nrf2 is a transcription factor that regulates cytoprotective defences and is regarded as a therapeutic target in liver diseases including NAFLD.[55] Consistent with EGCG inducing the expression of glutathione *S*-transferase, glutathione peroxidase, glutamate–cysteine ligase, and heme oxygenase-1,[56] an early hypothesis emerged that indirect antioxidant activities of GTE upregulate Nrf2-dependent defence to protect against NAFLD-associated liver injury in an Nrf2-dependent manner. However, studies in Nrf2-knockout mice fed a high-fat diet demonstrated that Nrf2 deficiency exacerbates NAFLD but dietary GTE supplementation fully ameliorated hepatic inflammation and oxidative injury in association with improved NAFLD pathology independent of intact Nrf2 signalling.[36] Thus, GTE- or catechin-mediated decreases in NFκB-mediated NAFLD are independent of any antioxidant activity and likely occur by limiting receptor mediated-signalling that induces NFκB activation.[36]

Apart from their anti-inflammatory activity, green tea catechins may also protect against NAFLD by enhancing energy expenditure. Consistent with pharmacological efforts to target peroxisome proliferator-activated receptors (PPAR) due to their regulatory role in β-oxidation, dietary EGCG supplementation in high-fat-fed mice was found to decrease hepatic lipid accumulation while increasing PPAR-α expression and the expression of genes involved in cholesterol and fatty acid metabolism in the liver (*e.g.*, cholesterol 7α-hydroxylase, β-hydroxy β-methylglutaryl-CoA reductase, LDL receptor, and scavenger receptor B1).[57] Others have also reported that catechin supplementation in high-fat-fed rats decreased hepatic TAG accumulation and lipid peroxidation in association with normalizing the expression of adipose PPAR-γ and PPAR-δ, thereby inducing genes involved in fatty acid oxidation.[58] Increased physical activity is a key recommendation to lower NAFLD risk. At least in rodents fed a high-fat diet, GTE supplementation combined with exercise more effectively mitigates NAFLD than either therapy alone.[59] Not only were circulating alanine aminotransferase and hepatic lipid accumulation decreased, co-therapy most greatly increased the expression of genes related to mitochondrial biogenesis (*e.g.*, sirtuin-1, PPAR-γ coactivator 1α, nuclear respiratory factor 1, and transcription factor B1) and increasing faecal lipid content. Thus, GTE and its catechins appear to limit inflammation and normalize energy homeostasis to alleviate NAFLD, and exercise potentiates these benefits to limit NAFLD risk. However, studies in humans have yet to be reported examining whether such co-therapies are effective in individuals at risk of or with NAFLD.

4.3.3 Genetic Models of Obesity-associated NAFLD

Genetic models of obesity have provided important insights into the mechanism by which GTE and its catechins attenuate NAFLD risk. In Zucker (*fa/fa*) rats, which become obese due to hyperphagia attributed to the absence of intact leptin receptor signalling, dietary supplementation of green tea polyphenols was found to protect against hyperglycaemia, hyperinsulinemia, and hyperlipidaemia and to increase serum aminotransferases.[60] Supplementation also lowered hepatic TAG and the pro-inflammatory cytokines tumour necrosis factor alpha (TNFα) and interleukin (IL)-6 in the circulation while increasing the expression of hepatic AMPK, acetyl-CoA carboxylase, and SREBP-1c. This suggests that the anti-inflammatory activities of green tea polyphenols alleviate NAFLD by limiting hepatic lipogenesis,[60] but independent of leptin receptor signalling.

GTE supplementation has also been examined in leptin-deficient (*ob/ob*) mice that rapidly develop severe NAFLD. When these mice were provided powdered GTE, liver steatosis was nearly fully ameliorated without any benefit on circulating adiponectin or hepatic ascorbic acid concentrations.[61] Other studies in *ob/ob* mice have revealed that GTE protects against NAFLD consistent with a mechanism of decreasing the flux of adipose-derived free fatty acids to the liver where they would otherwise be esterified to TAG to promote liver steatosis.[62] GTE also upregulated cytoprotective antioxidant defences in association with reduced hepatic lipid peroxidation, suggesting that enhanced detoxification of ROS helps to alleviate NAFLD risk. GTE in *ob/ob* mice has also been shown to attenuate hepatic lipid peroxidation and protein nitration by limiting the expression of myeloperoxidase and inducible NO synthase and NADPH oxidase activity in association with reducing hepatic inflammatory infiltrates.[63] While these findings indicate an important role of green tea catechins in reducing the generation and/or enhancing the detoxification of reactive species to alleviate NAFLD risk, it appears that these benefits are entirely independent of leptin adequacy and signalling.

The adipokine adiponectin has an important function to regulate energy metabolism, including lipogenesis and β-oxidation.[64] In adiponectin-knockout mice, which readily develop obesity-associated NAFLD, dietary GTE supplementation lowered body weight, hepatic total cholesterol, and TAG, and increased energy expenditure and genes responsible for fatty acid oxidation and insulin signalling in the liver.[65] Similar studies in wild-type and SREBP-2-knockout rats fed a high-fat and high-cholesterol diet have been conducted to examine EGCG on the transcriptional regulation of SREBP-2 by sirtuin-1/forkhead box O1 (SIRT1/FOXO1).[66] Data show that EGCG in wild-type rats, but not SREBP-2-knockout rats, increases the expression of FOXO1 and decreases the expression of SREBP-2. In wild-type rats only, EGCG also decreased circulating cholesterol, histologic evidence of NAFLD, serum aminotransferases, and expression levels of hydroxymethylglutaryl-coenzyme A reductase, farnesyl diphosphate synthase, squalene synthase, and ATP binding cassette protein A1 while increasing acetyl-CoA acetyltransferase-2 expression. These findings suggest that EGCG protects against NAFLD by limiting cholesterol synthesis through a mechanism involving SIRT1/FOXO1-mediated suppression of SREBP-2 transcriptional activities. Taken together, studies in transgenic models have provided insights that green tea catechins likely protect against NAFLD, at least in part, by regulating lipid metabolism but independent of the known benefits of leptin or adiponectin signalling on energy metabolism.

4.4 Extrahepatic Benefits of Green Tea on NAFLD Risk

While the aforementioned sections of this chapter support that green tea or its catechins function directly in the liver to ameliorate NAFLD, accumulating evidence suggests that its hepatoprotective activities are mechanistically linked to health effects in the gut, adipose tissue, and skeletal muscle. For instance, adipose inflammation, which is known to promote lipolysis and subsequent hepatic lipid accumulation, is attenuated by GTE. Specifically, wild-type mice that were fed a high-fat diet supplemented with GTE were shown to have attenuated adiposity and serum alanine aminotransferase activity in association with lowered adipose TLR4/NFκB-mediated increases in pro-inflammatory genes (*e.g.*, TNF, monocyte chemoattractant protein-1, and inducible NO synthase), and expression levels of macrophage activation markers (*e.g.*, CD68 and F4/80).[67] Dysfunctional skeletal muscle metabolism, including increased lipid uptake and impaired glucose disposal, are implicated in NAFLD,[68] whereas EGCG supplementation in high-fat-fed mice increased the skeletal muscle expression of β-oxidation-related genes in association with decreasing liver steatosis, insulin resistance, and plasma alanine aminotransferase.[69]

Consistent with green tea catechins having poor bioavailability,[70] a significant proportion of the oral dose is not absorbed but instead directed to the lower gut. Here, GTE catechins attenuate "leaky gut" in association with improved NAFLD pathology by maintaining the expression levels of intestinal tight junction proteins to limit the translocation of inflammation-inducing gut-derived endotoxins to the portal circulation (Figure 4.3).[71,72] This is likely attributed to GTE catechins upregulating the transcriptional activities of hypoxia-inducible factor-1α in association with reducing mucosal inflammation that is otherwise induced by a high-fat diet in mice.[71] Others have also reported that EGCG limits epithelial barrier dysfunction *in vitro* that was induced by interferon-γ-mediated activation of signal transducer and activator of transcription-1.[73] These findings support evidence from high-fat-fed mice with NAFLD that GTE blocks TLR4/NFκB-mediated liver injury in association with lowered metabolic endotoxemia.[71] A separate line of evidence also suggests that GTE thwarts gut barrier dysfunction by alleviating mucosal inflammation downstream of TLR4/NFκB signalling.[67,71] Specifically, GTE supplementation prevented high-fat diet-induced increases in serum alanine aminotransferase while reducing the expression of TLR4 receptor complex genes in the ileum and colon along with lowering the expression of several pro-inflammatory mediators in the jejunum, ileum, and colon.[67] In a separate study, high-fat-fed mice supplemented with GTE, EGCG equivalent to its level in GTE, or (+)-catechin at a dose equivalent to EGCG were shown to have similar benefits on NAFLD pathology (Figure 4.4), insulin resistance, serum endotoxin, and expression levels of intestinal tight junction proteins; the effects of (+)-catechin were

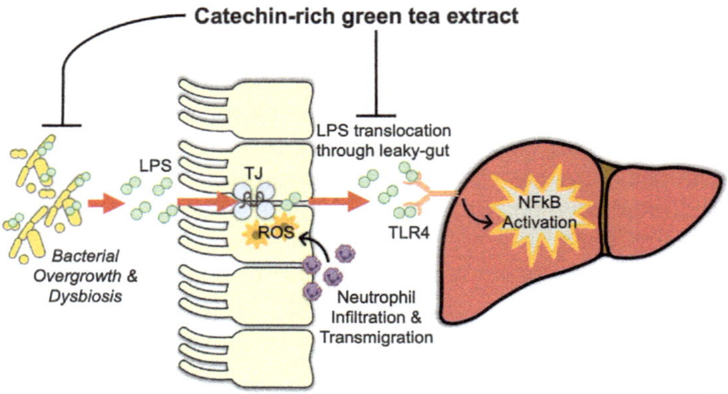

Figure 4.3 Gut-level activities of green tea catechins protect against endotoxemia-associated inflammation. NAFLD is associated with alterations in gut microbiota and intestinal barrier dysfunction that permits the translation of gut-derived endotoxins (*e.g.*, LPS) to the liver to activate pro-inflammatory TLR4/NFκB signalling. Dietary green tea catechins reduce endotoxin–TLR4–NFκB inflammation in the liver by maintaining intestinal TJ protein expression which is otherwise reduced in NAFLD. Prebiotic activities of green tea catechins likely contribute to these health benefits by increasing populations of commensal bacteria and health-promoting functions (*e.g.*, short-chain fatty acid biosynthesis) and decreasing pathogenic bacterial populations. Abbreviations: LPS, lipopolysaccharide; NAFLD, nonalcoholic fatty liver disease; NFκB, nuclear factor-kappa B; ROS, reactive oxygen species; TJ, tight junction; TLR4, toll-like receptor-4.

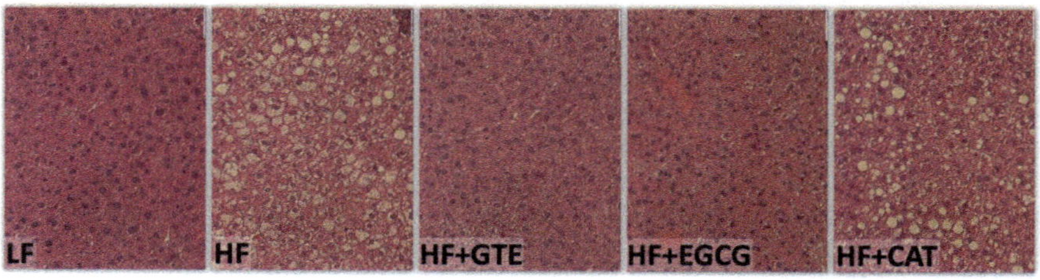

Figure 4.4 Differential effects of green tea catechins in mitigating NAFLD. Chronic feeding of high-fat diet results in hepatic steatosis, ballooning, and infiltration of inflammatory cells. Dietary supplementation of GTE and EGCG at the same concentration present within GTE ameliorates NAFLD to a similar extent. However, supplementation of CAT at a mass dose equivalent to EGCG mitigates NAFLD but not to the extent occurring from EGCG. Histopathological images are unpublished representative images of those reported in a study by the authors of this chapter (Dey *et al.*[71]). No permission required. Abbreviations: CAT, (+)-catechin; EGCG, epigallocatechin gallate; GTE, green tea extract; HF, high fat; LF, low fat; NAFLD, nonalcoholic fatty liver disease.

less profound for some endpoints.[71] However, faecal calprotectin concentrations were significantly lowered by all green tea-related treatments, which further supports a gut-level anti-inflammatory effect. Interestingly, when the gut microbiota composition of these mice was examined, it was revealed that only GTE and (+)-catechin prevented high-fat-diet induced loss of microbial diversity. Further, microbial metabolic functions were more similar between GTE and (+)-catechin. These findings suggest that the favourable effects of catechins on NAFLD likely occur through differential prebiotic and/or antimicrobial activities in the gut and that their protective effects may not be entirely mediated through shifts in microbiota composition.

Studies examining the effects of GTE and its catechins on gut microbiota composition and functions are especially important due to substantial evidence that NAFLD is driven, at least in part, by alterations in gut microbiota.[74] EGCG has been shown to exert antimicrobial activity in epithelial cells to limit bacterial translocation by improving barrier integrity.[75] Although a short-term study in adults indicated that green tea improved microbiota composition [*e.g.*, increased α-diversity, increased genera of short-chain fatty acid (SCFA) producers] and reduced microbial functions related to lipopolysaccharide biosynthesis,[76] long-term intervention trials in humans did not detect any measurable change in gut microbiota composition in adults supplemented with EGCG-enriched GTE.[77] The findings of this study are potentially attributed to the lack of dietary control of the human participants. Although instructed to follow their usual diets, no information was provided about intake of polyphenols or fibre nor was usual basal controlled of these individuals reported to understand whether dietary intake of placebo and GTE treatment groups differed. Nonetheless, well-controlled animal studies have provided strong insight that GTE or purified catechins favourably influence the composition and diversity of gut microbiota. For example, GTE and its catechins modulated gut microbial composition and increased bacterial diversity to limit weight gain and hepatic injury in rodents.[67,71,78] GTE in obese mice fed a high-fat diet also improved gut microbial diversity in association with decreasing microbial functions related to lipopolysaccharide biosynthesis and energy metabolism while increasing functions related to butyrate and propionate metabolism.[67] Further corroborating, green tea leaf powder supplementation increased intestinal SCFA, especially acetate and butyrate.[79] In mice, supplementation with GTE has been demonstrated to increase the abundance of several gut commensals and probiotic bacteria (*e.g.*, *Bifidobacterium*, *Akkermansia*, and *Lactobacillus*), and decrease the populations of pathobionts (*e.g.*, Proteobacteria).[67,78]

Taken together, the available evidence, especially that in rodent models, supports that the prebiotic and antimicrobial activities of green tea catechins in the gut likely contribute to reducing NAFLD risk. Their favourable effects that increase commensal bacterial populations likely drive the biosynthesis of SCFAs, especially butyrate, which are well-established to reduce intestinal permeability and prevent the excess absorption of inflammation-inducing endotoxins. Their antimicrobial activities are also expected to restrict the growth of pathogenic bacteria that are not only responsible for endotoxin biosynthesis but also provoking mucosal inflammation that contributes to the loss of gut barrier function. Critical to advancing green tea catechin-based recommendations for human health relative to NAFLD is the need for controlled clinical studies that can identify the optimal dose and duration of use of catechins to improve microbiota composition and functions, and the complex interactions with foodstuffs from the basal diet that has the potential to mask or potentiate the prebiotic and antimicrobial effects of the catechins. Another area that deserves attention is related to the interaction between green tea catechins and gut microbiota such that poorly absorbed catechins are metabolized in the lower gut to generate a number of microbial-derived metabolites that have an anti-inflammatory function. These metabolites, including valerolactones and valeric acids, are absorbed by intestinal epithelial cells. Studies of the hepatic metabolome in obese mice with NAFLD show that various microbial metabolites of catechins (*e.g.*, valerolactones) accumulate in the liver following GTE supplementation.[80] Experimental data show that EGCG-derived 5-(3′,4′,5′-trihydroxyphenyl)-γ-valerolactone display anti-cancer and anti-inflammatory activities.[81] Others have also reported that bacteria-derived catechin metabolites have antioxidant activities *in vitro*.[82,83] Phenyl-γ-valerolactones and their sulphated derivatives protect against oxidative stress in brown adipocytes without affecting their function or differentiation.[84] Importantly, valerolactones can be isomerized to form valeric acid and oxidized in the liver to generate phenylacetic, benzoic, hydroxypropionic, and hippuric acids to regulate host metabolic processes.[10,85] They may also favourably influence gut bacterial populations including *Bifidobacterium* while lowering the ratio of Firmicutes to Bacteroidetes,[86] an index that has been suggested to reflect gut dysbiosis.[87] Controlled experiments in obese mice support that catechin metabolites may protect against NAFLD.[88] Rather than supplementing catechin-rich GTE, these investigators prepared and supplemented bacteria-fermented GTE that is significantly depleted of catechins; catechin metabolites were not measured. Nonetheless, obese mice fed fermented GTE showed a reduction in obesity and NAFLD-associated metabolic complications.[88] This supports the possibility that catechins themselves along with their gut-derived metabolites may exert independent and/or additive effects to ameliorate NAFLD and stresses the importance of controlled human trials to unravel whether green tea catechins can improve microbiota composition and functions that promote the generation of anti-inflammatory metabolites.

4.5 Hepatoprotective Benefits of Green Tea: Clinical Evidence

NAFLD is most commonly associated with obesity,[89] which is why patients with NAFLD are strongly encouraged to engage in weight management strategies. Indeed, aggressive lifestyle intervention in obese individuals with NAFLD resulted in at least 10% weight loss which resolved nonalcoholic steatohepatitis in 90% of them, and fibrosis was regressed in 45% of cases.[90] Unfortunately, only a small proportion of patients from this study were able to achieve this weight loss goal, suggesting that alternative and/or complementary therapies are needed. Green tea has received considerable attention for its potential role in weight management. One meta-analysis of randomized placebo-controlled trials that examined green tea on weight loss indicated that supplementation was associated with a loss of body mass and reduced body mass index (BMI).[91] While a weight loss of 1.78 kg occurred for the entire population, subgroup analysis showed a greater benefit (2.53 kg weight loss) among those having a BMI of $\geq$30 kg m^{-2} compared with those having a BMI of 25–29.9 kg m^{-2}. Analysis of dose and duration of supplementation among the entire study population suggested that the greatest weight loss effects occurred when catechins were ingested up to 500 mg per day and for 12 weeks. A pooled analysis of six human trials with 921 persons randomized to placebo or green tea supplements found that long-term supplementation at a mean catechin dose of 564 mg per day was associated with a loss of body mass (1.69 kg), visceral adiposity (7.2%), and subcutaneous adiposity (10.5%).[92] GTE supplementation for 12 weeks (625 mg per day) has also been reported to enhance exercise-induced decreases in abdominal fat in overweight and obese persons with only a statistical tendency to reduce body mass.[93] However, there has not been any large-scale study examining whether green tea-mediated improvements in obesity, independent of or dependent on exercise, reverse NAFLD in humans and only a few trials have examined GTE to manage NAFLD-associated complications.

The few clinical trials that have examined GTE to treat NAFLD have been of a relatively small size and short duration. In a double-blind randomized study ($n = 45$), patients with confirmed NAFLD who received GTE (550 mg, 3 months) as part of a lifestyle intervention (diet and exercise education) had significantly reduced BMI, aspartate aminotransferase, and fasting glucose without any significant changes in body weight, plasma insulin, lipids, or alanine aminotransferase.[94] Unfortunately, no follow-up analysis was conducted to examine ultrasound or histologic evidence of NAFLD, thereby limiting the understanding of whether GTE improved liver pathology. Pezeshki *et al.*[95] conducted a double-blind, placebo-controlled, randomized clinical trial in 71 patients with NAFLD that was confirmed on the basis of elevated circulating liver enzymes and ultrasonography. The authors reported that 12 weeks of GTE supplementation (256 mg per day of total catechins) significantly reduced alanine and aspartate aminotransferases but those randomized to placebo also experienced a significant lowering of alanine aminotransferase and a tendency ($P = 0.07$) for reduced aspartate aminotransferase. The body weight in both groups was also decreased regardless of GTE supplementation. In a placebo-controlled trial, 80 patients with elevated aminotransferases and ultrasound evidence of NAFLD were randomized to GTE containing 512 mg total catechins daily for 12 weeks.[96] Among those receiving GTE, significant decreases in body mass, BMI, insulin resistance, circulating lipids (total cholesterol and TAG), and alanine and aspartate aminotransferases were observed, and these endpoints differed significantly from those receiving placebo where no within-group changes were detected. Ultrasonography-based measures at the end of the trial also suggested that NAFLD pathology regressed in 68% of GTE-treated patients compared with only 25% of those receiving placebo. In a comparative study, hypercholesterolemic adults were allocated to receive placebo, or catechin-rich beverages of oolong tea (640 mg catechins) or green tea (781 mg catechins) for 12 weeks. No changes in circulating liver enzymes were detected in response to either test beverage, although these liver function markers were within normal limits at study enrolment. However, the participants receiving either tea beverage but not placebo experienced a significant reduction in body weight and adiposity. Interestingly, there were 3 of 60 enrolled patients who had ultrasound-confirmed fatty infiltrates, and their liver pathology was normalized in response to daily ingestion of either of the catechin-rich teas.

Meta-analyses have also been conducted to assess whether green tea improves liver health in humans. In 440 903 regular green tea drinking cases and 385 246 irregular green tea drinking cases derived from 15 studies ($n = 9$ prospective cohort studies, 3 retrospective studies, and 4 cross-sectional studies), it was suggested that green tea consumption had an overall benefit to lower the incidence of liver disease.[33] This included a 35% lower relative risk for fatty liver disease among green tea drinkers in addition to protective effects on the risk for hepatitis, cirrhosis, and hepatocellular carcinoma. In a meta-analysis that focused on liver function biomarkers,[97] pooled analyses of 15 human trials suggested that GTE had no benefit on alanine aminotransferase, aspartate aminotransferase, or alkaline phosphatase. However, because not all trials had participants with liver enzymes that were elevated at study enrolment, subgroup analysis was considered to examine these endpoints among those with NAFLD. In NAFLD patients, GTE supplementation was associated with a significant reduction in alanine aminotransferase and aspartate aminotransferase, whereas GTE in healthy persons was associated with small increases in alanine aminotransferase ($P = 0.07$) and aspartate aminotransferase ($P = 0.001$). In a meta-analysis of six intervention studies that included only patients with NAFLD, GTE supplementation was associated with lowered levels of transaminases, BMI, total and LDL cholesterols, and TAG, but not insulin resistance.[98]

Undoubtedly, there is much heterogeneity in the clinical trials examining GTE on NAFLD risk, with apparent differences in GTE dose, catechin composition, duration of intervention, and lifestyle modification and reporting. While the few trials being generally promising for GTE to alleviate biomarkers of NAFLD and pathologic evidence to a more limited extent, additional intervention studies of higher quality are needed to examine dose-dependent effects in NAFLD patients. Ideally, these studies should enrol biopsy-confirmed patients due to known limitations of ultrasonography for the diagnosis and follow-up of NAFLD severity. Further, the basal diet requires reporting even in studies where no dietary modification has been employed to define how the overall dietary pattern impacts effective treatment. Nonetheless, the available evidence provides promise that GTE-based interventions can be effective for NAFLD management, although its benefits are likely connected at least in part to GTE-mediated decreases in body mass. Future studies, therefore, should assess whether GTE exerts its liver-centric benefits independent of or dependent on any changes in body mass, and aim to confirm its mechanism of action that has been detailed in preclinical studies.

4.6 Safety of Green Tea

Although consumption patterns vary widely among and between populations, the regular consumption of green tea is safe. In fact, it is not unusual for individuals in some Asian countries to consume >10 cups of green tea daily.[99] The intake of >5 cups per day has been associated with a 26% lower risk of mortality due to cardiovascular disease (CVD), whereas ≥10 cups daily was inversely related to liver injury biomarkers and circulating lipids.[99,100]

The popularity of dietary supplements containing high doses of GTE or purified catechins has heightened concerns for their potential adverse effects.[101] Case reports suggest that an inappropriate use of GTE supplements may cause acute hepatotoxicity, gastrointestinal distress, and neurological complications.[101–103] GTE-related hepatotoxicity potentially involves interactions between catechins and liver enzymes (*e.g.*, uridine 5′-diphospho-glucuronosyltransferase, cytochrome P450s, and mitochondrial enzymes), or immunological reactions.[104] The findings of a systemic review of 34 human trials suggest that adverse effects from GTE are rarely based on the documentation of 8 liver-related adverse effects (elevated liver enzymes) in 7 persons across 4 trials.[105] Most adverse effects were characterized as mild and no cases of severe adverse effects were reported. The European Food Safety Authority (EFSA) has also provided a scientific opinion concerning the safety of green tea catechins.[106] It recognized that EGCG consumption from tea infusions ranges from 90 to 300 mg per day, whereas supplements are estimated to provide upwards of 866 mg daily. This concludes that tea infusions or reconstituted beverages having a similar catechin level to conventionally prepared tea infusions are in general considered to be safe based on the assumption that intakes are similar to usual intakes in large populations. In addition, any adverse liver effects following tea infusion consumption are likely due to idiosyncratic reactions. Likewise, this authoritative panel concluded based on its review of clinical trials that intake of green tea supplements at doses ≥800 mg EGCG are linked to a heightened risk of liver enzymes. The United States Pharmacopeia (USP) has more recently conducted a similar comprehensive review concerning the hepatotoxicity of GTE.[107] Although it has previously been suggested that the hepatotoxic effects of GTE may be attributed to pesticide residues, pyrrolizidine alkaloids, and elemental impurities, this expert panel found no evidence to support this perspective. Rather, the USP concluded that severe hepatoxicity from GTE can occur in humans, but with a very low rate of occurrence, and these adverse events are likely attributed to EGCG or other catechins found in green tea. In addition, the concentration of the GTE or catechin dose, ingestion of large bolus doses and/or under fasting conditions, genetic susceptibility, and the presence of other underlying liver conditions may contribute to the observed hepatotoxic effects. Because of these factors, especially genetic or idiosyncratic causes, the panel refrained from setting a safe dose of consumption and indicated that a cautionary statement be included in the USP GTE monograph that states: "*Do not take on an empty stomach. Take with food. Do not use if you have a liver problem and discontinue use and consult a healthcare practitioner if you develop symptoms of liver trouble, such as abdominal pain, dark urine, or jaundice (yellowing of the skin or eyes)*".[107]

4.7 Conclusion

The hepatoprotective activities of green tea and its bioactive catechins to improve hepatocellular health related to NAFLD risk were discussed in this chapter. Studies *in vitro* and *in vivo* provide consistent evidence that GTE catechins exert anti-inflammatory and lipid-lowering effects in liver cell lines and animal models that could mitigate NAFLD risk. Many of these cellular studies attribute the effects of catechins to an antioxidant function that scavenges reactive species. However, the limited oral bioavailability of catechins in physiologically relevant models precludes their direct antioxidant function. This implies that green tea catechins, either directly or indirectly *via* their metabolites of microbial origin, likely regulate NAFLD-associated pathophysiological processes along the gut–liver–adipose axis. Indeed, GTE supplementation in rodent models of NAFLD demonstrated poor bioavailability of catechins, but significant anti-inflammatory activities in the intestine, liver, and adipose which likely contribute to ameliorating NAFLD. More studies are needed to clarify the independent and additive hepatoprotective effects of parental catechins and their microbial metabolites. In addition, evidence-based recommendations for green tea to improve liver health in humans are also precluded due to factors such as very few randomized controlled trials, heterogeneity of study outcomes, and high inter-individual effects of GTE and catechins. Potentially, these divergent effects are attributed to differences in gut microbiota composition between research participants that result in unique metabolic profiles of catechins. Nonetheless, such studies could be important for precision nutrition approaches that are increasingly being recognized to promote optimal human health. Additionally, although the chronic ingestion of green tea as a beverage is regarded as safe, controlled trials are needed that can better define the mechanisms by which high-dose boluses of green tea catechins (*e.g.*, supplements) adversely affect liver health in a small subset of persons. Such individuals would be contraindicated to utilize green tea supplements and/or should be monitored closely to ensure that co-ingestion of supplements with food does not induce acute hepatotoxicity. Nevertheless, green tea remains a popular beverage, and habitual intake at a controlled limit is likely to exert hepatoprotective benefits, especially in mitigating complications associated with NAFLD.

Acknowledgements

Support was provided by grants to RSB from USDA-NIFA (2019-67017-29259), USDA-HATCH (OHO01452-MRF), and the Ohio Agricultural Research and Development Center at The Ohio State University, Columbus, OH, USA.

References

1. L. Drew, The growth of tea, *Nature*, 2019, **566**, S2–S4.
2. P. Dey, G. Y. Sasaki and R. S. Bruno, Health benefits of green tea, in *Handbook of Nutraceuticals and Functional Foods*, ed. R. E. C. Wildman and R. S. Bruno, CRC Press, Boca Raton, FL, 2019, pp. 127–145.
3. S. Mitra, A. De and A. Chowdhury, Epidemiology of non-alcoholic and alcoholic fatty liver diseases, *Transl. Gastroenterol. Hepatol.*, 2020, **5**, 16.
4. J. Zhou, C.-T. Ho, P. Long, Q. Meng, L. Zhang and X. Wan, Preventive efficiency of green tea and its components on nonalcoholic fatty liver disease, *J. Agric. Food Chem.*, 2019, **67**, 5306–5317.
5. A. P. Neilson, R. J. Green, K. V. Wood and M. G. Ferruzzi, High-throughput analysis of catechins and theaflavins by high performance liquid chromatography with diode array detection, *J. Chromatogr. A*, 2006, **1132**, 132–140.
6. C. Musial, A. Kuban-Jankowska and M. Gorska-Ponikowska, Beneficial properties of green tea catechins, *Int. J. Mol. Sci.*, 2020, **21**, 1744.
7. S. M. Chacko, P. T. Thambi, R. Kuttan and I. Nishigaki, Beneficial effects of green tea: a literature review, *Chin. Med.*, 2010, **5**, 13.
8. M. E. Harbowy, D. A. Balentine, A. P. Davies and Y. Cai, Tea chemistry, *Crit. Rev. Plant Sci.*, 1997, **16**, 415–480.
9. H. N. Graham, Green tea composition, consumption, and polyphenol chemistry, *Prev. Med.*, 1992, **21**, 334–350.
10. S. Sang, J. D. Lambert, C.-T. Ho and C. S. Yang, The chemistry and biotransformation of tea constituents, *Pharmacol. Res.*, 2011, **64**, 87–99.
11. W. Koch, W. Kukula-Koch and K. Głowniak, Catechin composition and antioxidant activity of black teas in relation to brewing time, *J. AOAC Int.*, 2017, **100**, 1694–1699.
12. L. F. Wang, D. M. Kim and C. Y. Lee, Effects of heat processing and storage on flavanols and sensory qualities of green tea beverage, *J. Agric. Food Chem.*, 2000, **48**, 4227–4232.

13. Q. V. Vuong, J. B. Golding, C. E. Stathopoulos, M. H. Nguyen and P. D. Roach, Optimizing conditions for the extraction of catechins from green tea using hot water, *J. Sep. Sci.*, 2011, **34**, 3099–3106.
14. A. C. Nobre, A. Rao and G. N. Owen, L-theanine, a natural constituent in tea, and its effect on mental state, *Asia Pac. J. Clin. Nutr.*, 2008, **17**, 167–168.
15. N. P. Seeram, S. M. Henning, Y. Niu, R. Lee, H. S. Scheuller and D. Heber, Catechin and caffeine content of green tea dietary supplements and correlation with antioxidant capacity, *J. Agric. Food Chem.*, 2006, **54**, 1599–1603.
16. R. S. Bruno, J. A. Bomser and M. G. Ferruzzi, Antioxidant capacity of green tea (*Camellia sinensis*), in *Processing and Impact on Antioxidants in Beverages*, ed. V. Preedy, Academic Press, London, UK, 2014, pp. 33–39.
17. G.-D. Hung, P.-C. Li, H.-S. Lee, H.-M. Chang, C.-T. Chien and K.-L. Lee, Green tea extract supplementation ameliorates CCl4-induced hepatic oxidative stress, fibrosis, and acute-phase protein expression in rat, *J. Formos. Med. Assoc.*, 2012, **111**, 550–559.
18. X. Shi, J. Ye, S. Leonard, M. Ding, V. Vallyathan and V. Castranova, *et al.*, Antioxidant properties of (−)-epicatechin-3-gallate and its inhibition of Cr(VI)-induced DNA damage and Cr(IV)- or TPA-stimulated NF-κB activation, *Mol. Cell. Biochem.*, 2000, **206**, 125–132.
19. I. F. Benzie and Y. Szeto, Total antioxidant capacity of teas by the ferric reducing/antioxidant power assay, *J. Agric. Food Chem.*, 1999, **47**, 633–636.
20. S. M. Henning, C. Fajardo-Lira, H. W. Lee, A. A. Youssefian, V. L. Go and D. Heber, Catechin content of 18 teas and a green tea extract supplement correlates with the antioxidant capacity, *Nutr. Cancer*, 2003, **45**, 226–235.
21. S. Chandra and E. Gonzalez de Mejia, Polyphenolic compounds, antioxidant capacity, and quinone reductase activity of an aqueous extract of Ardisia compressa in comparison to mate (*Ilex paraguariensis*) and green (*Camellia sinensis*) teas, *J. Agric. Food Chem.*, 2004, **52**, 3583–3589.
22. D. Erba, P. Riso, A. Colombo and G. Testolin, Supplementation of Jurkat T cells with green tea extract decreases oxidative damage due to iron treatment, *J. Nutr.*, 1999, **129**, 2130–2134.
23. S. C. Forester and J. D. Lambert, Antioxidant effects of green tea, *Mol. Nutr. Food Res.*, 2011, **55**, 844–854.
24. N. Kager, F. Ferk, M. Kundi, K.-H. Wagner, M. Mišík and S. Knasmüller, Prevention of oxidative DNA damage in inner organs and lymphocytes of rats by green tea extract, *Eur. J. Nutr.*, 2010, **49**, 227–234.
25. Y.-W. Hsu, C.-F. Tsai, H.-C. Ting, W.-K. Chen and C.-C. Yen, Green tea supplementation in mice mitigates senescence-induced changes in brain antioxidant abilities, *J. Funct. Foods*, 2014, **7**, 471–478.
26. H. Sung, J. Nah, S. Chun, H. Park, S. Yang and W. Min, *In vivo* antioxidant effect of green tea, *Eur. J. Clin. Res.*, 2000, **54**, 527–529.
27. A. Basu, K. Sanchez, M. J. Leyva, M. Wu, N. M. Betts and C. E. Aston, *et al.*, Green tea supplementation affects body weight, lipids, and lipid

peroxidation in obese subjects with metabolic syndrome, *J. Am. College. Nutr.*, 2010, **29**, 31–40.

28. M. Benedict and X. Zhang, Non-alcoholic fatty liver disease: An expanded review, *World J. Hepatol.*, 2017, **9**, 715–732.
29. Y. L. Fang, H. Chen, C. L. Wang and L. Liang, Pathogenesis of non-alcoholic fatty liver disease in children and adolescence: From "two hit theory" to "multiple hit model", *World J. Gastroenterol.*, 2018, **24**, 2974–2983.
30. H. Tilg and A. R. Moschen, Evolution of inflammation in nonalcoholic fatty liver disease: the multiple parallel hits hypothesis, *Hepatology*, 2010, **52**, 1836–1846.
31. A. Ui, S. Kuriyama, M. Kakizaki, T. Sone, N. Nakaya and K. Ohmori-Matsuda, *et al.*, Green tea consumption and the risk of liver cancer in Japan: the Ohsaki Cohort study, *Cancer Causes Control*, 2009, **20**, 1939–1945.
32. C.-X. Ni, H. Gong, Y. Liu, Y. Qi, C.-L. Jiang and J.-P. Zhang, Green tea consumption and the risk of liver cancer: a meta-analysis, *Nutr. Cancer*, 2017, **69**, 211–220.
33. X. Yin, J. Yang, T. Li, L. Song, T. Han and M. Yang, *et al.*, The effect of green tea intake on risk of liver disease: a meta analysis, *Int. J. Clin. Exp. Med.*, 2015, **8**, 8339.
34. C. Masterjohn and R. S. Bruno, Therapeutic potential of green tea in nonalcoholic fatty liver disease, *Nutr. Rev.*, 2012, **70**, 41–56.
35. H. L. Jiao, P. Ye and B. L. Zhao, Protective effects of green tea polyphenols on human HepG2 cells against oxidative damage of fenofibrate, *Free Radic. Biol. Med.*, 2003, **35**, 1121–1128.
36. J. Li, T. N. Sapper, E. Mah, S. Rudraiah, K. E. Schill and C. Chitchumroonchokchai, *et al.*, Green tea extract provides extensive Nrf2-independent protection against lipid accumulation and NFκB proinflammatory responses during nonalcoholic steatohepatitis in mice fed a high-fat diet, *Mol. Nutr. Food Res.*, 2016, **60**, 858–870.
37. A. Shil, C. Davies, L. Gautam, J. Roberts and H. Chichger, Investigating the opposing effect of two different green tea supplements on oxidative stress, mitochondrial function and cell viability in HepG2 cells, *J. Diet. Suppl.*, 2022, **19**, 459–482.
38. T. Wu, Y. H. Liu, Y. C. Fu, X. M. Liu and X. H. Zhou, Direct evidence of sirtuin downregulation in the liver of non-alcoholic fatty liver disease patients, *Ann. Clin. Lab. Sci.*, 2014, **44**, 410–418.
39. X. Q. Deng, L. L. Chen and N. X. Li, The expression of SIRT1 in nonalcoholic fatty liver disease induced by high-fat diet in rats, *Liver Int.*, 2007, **27**, 708–715.
40. Y. Colak, O. Ozturk, E. Senates, I. Tuncer, E. Yorulmaz and G. Adali, *et al.*, SIRT1 as a potential therapeutic target for treatment of nonalcoholic fatty liver disease, *Med. Sci. Monit.*, 2011, **17**, HY5–HY9.
41. U.-J. Bae, J. Park, I. W. Park, B. M. Chae, M.-R. Oh and S.-J. Jung, *et al.*, Epigallocatechin-3-gallate-rich green tea extract ameliorates fatty liver and weight gain in mice fed a high fat diet by activating the sirtuin 1

and AMP activating protein kinase pathway, *Am. J. Chin. Med.*, 2018, **46**, 617–632.
42. C. A. Bursill and P. D. Roach, Modulation of cholesterol metabolism by the green tea polyphenol (−)-epigallocatechin gallate in cultured human liver (HepG2) cells, *J. Agric. Food Chem.*, 2006, **54**, 1621–1626.
43. J. J. Kim, Y. Tan, L. Xiao, Y.-L. Sun and X. Qu, Green tea polyphenol epigallocatechin-3-gallate enhance glycogen synthesis and inhibit lipogenesis in hepatocytes, *BioMed Res. Int.*, 2013, **2013**, 920128.
44. J. Zhou, B. L. Farah, R. A. Sinha, Y. Wu, B. K. Singh and B.-H. Bay, *et al.*, Epigallocatechin-3-gallate (EGCG), a green tea polyphenol, stimulates hepatic autophagy and lipid clearance, *PLoS One*, 2014, **9**, e87161.
45. C.-L. Lin, H.-C. Huang and J.-K. Lin, Theaflavins attenuate hepatic lipid accumulation through activating AMPK in human HepG2 cells, *J. Lipid Res.*, 2007, **48**, 2334–2343.
46. Z. Liu, Q. Li, J. Huang, Q. Liang, Y. Yan and H. Lin, *et al.*, Proteomic analysis of the inhibitory effect of epigallocatechin gallate on lipid accumulation in human HepG2 cells, *Proteome Sci.*, 2013, **11**, 1–11.
47. L. Baselga-Escudero, C. Blade, A. Ribas-Latre, E. Casanova, M. Suárez and J. L. Torres, *et al.*, Resveratrol and EGCG bind directly and distinctively to miR-33a and miR-122 and modulate divergently their levels in hepatic cells, *Nucleic. Acid. Res.*, 2014, **42**, 882–892.
48. H.-Y. Lin, Y.-L. Yang, P.-W. Wang, F.-S. Wang and Y.-H. Huang, The emerging role of microRNAs in NAFLD: Highlight of microRNA-29a in modulating oxidative stress, inflammation, and beyond, *Cells*, 2020, **9**, 1041.
49. P. Dongiovanni, M. Meroni, M. Longo, S. Fargion and A. L. Fracanzani, miRNA signature in NAFLD: A turning point for a non-invasive diagnosis, *Int. J. Mol. Sci.*, 2018, **19**, 3966.
50. H. J. Park, J.-Y. Lee, M.-Y. Chung, Y.-K. Park, A. M. Bower and S. I. Koo, *et al.*, Green tea extract suppresses NFκB activation and inflammatory responses in diet-induced obese rats with nonalcoholic steatohepatitis, *J. Nutr.*, 2012, **142**, 57–63.
51. M. Y. Chung, E. Mah, C. Masterjohn, S. K. Noh, H. J. Park and R. M. Clark, *et al.*, Green tea lowers hepatic COX-2 and prostaglandin E2 in rats with dietary fat-induced nonalcoholic steatohepatitis, *J. Med. Food*, 2015, **18**, 648–655.
52. J. Li, T. N. Sapper, E. Mah, M. V. Moller, J. B. Kim and C. Chitchumroonchokchai, *et al.*, Green tea extract treatment reduces NFκB activation in mice with diet-induced nonalcoholic steatohepatitis by lowering TNFR1 and TLR4 expression and ligand availability, *J. Nutr. Biochem.*, 2017, **41**, 34–41.
53. J. Xiao, C.-T. Ho, E. C. Liong, A. A. Nanji, T. M. Leung and T. Y. Lau, *et al.*, Epigallocatechin gallate attenuates fibrosis, oxidative stress, and inflammation in non-alcoholic fatty liver disease rat model through TGF/SMAD, PI3 K/Akt/FoxO1, and NF-kappa B pathways, *Eur. J. Nutr.*, 2014, **53**, 187–199.

54. H. Hou, W. Yang, S. Bao and Y. Cao, Epigallocatechin gallate suppresses inflammatory responses by inhibiting Toll-like receptor 4 signaling and alleviates insulin resistance in the livers of high-fat-diet rats, *J. Oleo. Sci.*, 2020, ess19303.
55. D. Xu, M. Xu, S. Jeong, Y. Qian, H. Wu and Q. Xia, *et al.*, The role of Nrf2 in liver disease: Novel molecular mechanisms and therapeutic approaches, *Front. Pharmacol.*, 2019, **9**, 1428.
56. H. K. Na and Y. J. Surh, Modulation of Nrf2-mediated antioxidant and detoxifying enzyme induction by the green tea polyphenol EGCG, *Food Chem. Toxicol.*, 2008, **46**, 1271–1278.
57. J. Huang, S. Feng, A. Liu, Z. Dai, H. Wang and K. Reuhl, *et al.*, Green tea polyphenol EGCG alleviates metabolic abnormality and fatty liver by decreasing bile acid and lipid absorption in mice, *Mol. Nutr. Food Res.*, 2018, **62**, 1700696.
58. J. Yan, Y. Zhao and B. Zhao, Green tea catechins prevent obesity through modulation of peroxisome proliferator-activated receptors, *Sci. China: Life Sci.*, 2013, **56**, 804–810.
59. W. Y. Khoo, B. J. Chrisfield, S. Sae-Tan and J. D. Lambert, Mitigation of nonalcoholic fatty liver disease in high-fat-fed mice by the combination of decaffeinated green tea extract and voluntary exercise, *J. Nutr. Biochem.*, 2020, **76**, 108262.
60. Y. Tan, J. Kim, J. Cheng, M. Ong, W.-G. Lao and X.-L. Jin, *et al.*, Green tea polyphenols ameliorate non-alcoholic fatty liver disease through upregulating AMPK activation in high fat fed Zucker fatty rats, *World J. Gastroenterol.*, 2017, **23**, 3805.
61. R. S. Bruno, C. E. Dugan, J. A. Smyth, D. A. DiNatale and S. I. Koo, Green tea extract protects leptin-deficient, spontaneously obese mice from hepatic steatosis and injury, *J. Nutr.*, 2008, **138**, 323–331.
62. H. J. Park, D. A. DiNatale, M.-Y. Chung, Y.-K. Park, J.-Y. Lee and S. I. Koo, *et al.*, Green tea extract attenuates hepatic steatosis by decreasing adipose lipogenesis and enhancing hepatic antioxidant defenses in ob/ob mice, *J. Nutr. Biochem.*, 2011, **22**, 393–400.
63. M. Y. Chung, H. J. Park, J. E. Manautou, S. I. Koo and R. S. Bruno, Green tea extract protects against nonalcoholic steatohepatitis in ob/ob mice by decreasing oxidative and nitrative stress responses induced by proinflammatory enzymes, *J. Nutr. Biochem.*, 2012, **23**, 361–367.
64. K. Khoramipour, K. Chamari, A. A. Hekmatikar, A. Ziyaiyan, S. Taherkhani and N. M. Elguindy, *et al.*, Adiponectin: Structure, physiological functions, role in diseases, and effects of nutrition, *Nutrients*, 2021, **13**, 1180.
65. L. T. Ferreira, C. P. B. de Sousa Filho, M. P. Marinovic, A. C. Rodrigues and R. Otton, Green tea polyphenols positively impact hepatic metabolism of adiponectin-knockout lean mice, *J. Funct. Foods*, 2020, **64**, 103679.
66. Y. Li and S. Wu, Epigallocatechin gallate suppresses hepatic cholesterol synthesis by targeting SREBP-2 through SIRT1/FOXO1 signaling pathway, *Mol. Cell. Biochem.*, 2018, **448**, 175–185.

67. P. Dey, G. Y. Sasaki, P. Wei, J. Li, L. Wang and J. Zhu, *et al.*, Green tea extract prevents obesity in male mice by alleviating gut dysbiosis in association with improved intestinal barrier function that limits endotoxin translocation and adipose inflammation, *J. Nutr. Biochem.*, 2019, **67**, 78–89.
68. S. Altajar and G. Baffy, Skeletal muscle dysfunction in the development and progression of nonalcoholic fatty liver disease, *J. Clin. Transl. Hepatol.*, 2020, **8**, 414–423.
69. S. Sae-Tan, K. A. Grove, M. J. Kennett and J. D. Lambert, (−)-Epigallocatechin-3-gallate increases the expression of genes related to fat oxidation in the skeletal muscle of high fat-fed mice, *Food Funct.*, 2011, **2**, 111–116.
70. K. Nakagawa, S. Okuda and T. Miyazawa, Dose-dependent incorporation of tea catechins, (−)-epigallocatechin-3-gallate and (−)-epigallocatechin, into human plasma, *Biosci., Biotechnol., Biochem.*, 1997, **61**, 1981–1985.
71. P. Dey, B. D. Olmstead, G. Y. Sasaki, Y. Vodovotz, Z. Yu and R. S. Bruno, Epigallocatechin gallate but not catechin prevents nonalcoholic steatohepatitis in mice similar to green tea extract while differentially affecting the gut microbiota, *J. Nutr. Biochem.*, 2020, **84**, 108455.
72. G. S. Jinhui, Li, P. Dey, C. Chitchumroonchokchai, A. N. Labyk, J. D. McDonald and J. B. Kim, *et al.*, Green tea extract protects against hepatic NFκB activation along the gut-liver axis in diet-induced obese mice with nonalcoholic steatohepatitis by reducing endotoxin and TLR4/MyD88 signaling, *J. Nutr. Biochem.*, 2017, **53**, 58–65.
73. J. L. Watson, S. Ansari, H. Cameron, A. Wang, M. Akhtar and D. M. McKay, Green tea polyphenol (−)-epigallocatechin gallate blocks epithelial barrier dysfunction provoked by IFN-γ but not by IL-4, *Am. J. Physiol.: Gastrointest. Liver Physiol.*, 2004, **287**, G954–G961.
74. J. K. Hodges, G. Y. Sasaki and R. S. Bruno, Anti-inflammatory activities of green tea catechins along the gut-liver axis in nonalcoholic fatty liver disease: lessons learned from preclinical and human studies, *J. Nutr. Biochem.*, 2020, **85**, 108478.
75. M. L. Wan, K. H. Ling, M. F. Wang and H. El-Nezami, Green tea polyphenol epigallocatechin-3-gallate improves epithelial barrier function by inducing the production of antimicrobial peptide pBD-1 and pBD-2 in monolayers of porcine intestinal epithelial IPEC-J2 cells, *Mol. Nutr. Food Res.*, 2016, **60**, 1048–1058.
76. X. Yuan, Y. Long, Z. Ji, J. Gao, T. Fu and M. Yan, *et al.*, Green tea liquid consumption alters the human intestinal and oral microbiome, *Mol. Nutr. Food Res.*, 2018, **62**, e1800178.
77. P. L. Janssens, J. Penders, R. Hursel, A. E. Budding, P. H. Savelkoul and M. S. Westerterp-Plantenga, Long-term green tea supplementation does not change the human gut microbiota, *PLoS One*, 2016, **11**, e0153134.
78. H. W. Jeong, J. K. Kim, A. Y. Kim, D. Cho, J. H. Lee and J. K. Choi, *et al.*, Green tea encourages growth of *Akkermansia muciniphila*, *J. Med. Food*, 2020, **23**, 841–851.

79. J. Wang, P. Li, S. Liu, B. Zhang, Y. Hu, H. Ma and S. Wang, Green tea leaf powder prevents dyslipidemia in high-fat diet-fed mice by modulating gut microbiota, *Food Nutr. Res.*, 2020, **64**, 3672.
80. G. Y. Sasaki, J. Li, M. J. Cichon, K. M. Riedl, R. E. Kopec and R. S. Bruno, Green tea extract treatment in obese mice with nonalcoholic steatohepatitis restores the hepatic metabolome in association with limiting endotoxemia-TLR4-NFκB-mediated inflammation, *Mol. Nutr. Food Res.*, 2019, **63**, e1900811.
81. J. D. Lambert, J. E. Rice, J. Hong, Z. Hou and C. S. Yang, Synthesis and biological activity of the tea catechin metabolites, M4 and M6 and their methoxy-derivatives, *Bioorg. Med. Chem. Lett.*, 2005, **15**, 873–876.
82. T. Unno, K. Tamemoto, F. Yayabe and T. Kakuda, Urinary excretion of 5-(3′,4′-dihydroxyphenyl)-γ-valerolactone, a ring-fission metabolite of (−)-epicatechin, in rats and its *in vitro* antioxidant activity, *J. Agric. Food Chem.*, 2003, **51**, 6893–6898.
83. A. Takagaki, S. Otani and F. Nanjo, Antioxidative activity of microbial metabolites of (−)-epigallocatechin gallate produced in rat intestines, *Biosci., Biotechnol., Biochem.*, 2011, **75**, 582–585.
84. P. Mena, L. Bresciani, N. Brindani, I. A. Ludwig, G. Pereira-Caro and D. Angelino, *et al.*, Phenyl-γ-valerolactones and phenylvaleric acids, the main colonic metabolites of flavan-3-ols: Synthesis, analysis, bioavailability, and bioactivity, *Nat. Prod. Rep.*, 2019, **36**, 714–752.
85. W. Yong Feng, Metabolism of green tea catechins: an overview, *Curr. Drug Metab.*, 2006, **7**, 755–809.
86. B. Liu, J. Zhang, P. Sun, R. Yi, X. Han and X. Zhao, Raw bowl tea (Tuocha) polyphenol prevention of nonalcoholic fatty liver disease by regulating intestinal function in mice, *Biomolecules*, 2019, **9**, 435.
87. R. E. Ley, P. J. Turnbaugh, S. Klein and J. I. Gordon, Microbial ecology: human gut microbes associated with obesity, *Nature*, 2006, **444**, 1022–1023.
88. D.-B. Seo, H. W. Jeong, D. Cho, B. J. Lee, J. H. Lee and J. Y. Choi, *et al.*, Fermented green tea extract alleviates obesity and related complications and alters gut microbiota composition in diet-induced obese mice, *J. Med. Food*, 2015, **18**, 549–556.
89. D. H. Ipsen, J. Lykkesfeldt and P. Tveden-Nyborg, Molecular mechanisms of hepatic lipid accumulation in non-alcoholic fatty liver disease, *Cell. Mol. Life Sci.*, 2018, **75**, 3313–3327.
90. E. Vilar-Gomez, Y. Martinez-Perez, L. Calzadilla-Bertot, A. Torres-Gonzalez, B. Gra-Oramas and L. Gonzalez-Fabian, *et al.*, Weight loss through lifestyle modification significantly reduces features of nonalcoholic steatohepatitis, *Gastroenterology*, 2015, **149**, 367–378.
91. Y. Lin, D. Shi, B. Su, J. Wei, M. A. Găman and M. Sedanur Macit, *et al.*, The effect of green tea supplementation on obesity: A systematic review and dose-response meta-analysis of randomized controlled trials, *Phytother. Res.*, 2020, **34**, 2459–2470.

92. M. Hibi, H. Takase, M. Iwasaki, N. Osaki and Y. Katsuragi, Efficacy of tea catechin-rich beverages to reduce abdominal adiposity and metabolic syndrome risks in obese and overweight subjects: a pooled analysis of 6 human trials, *Nutr. Res.*, 2018, **55**, 1–10.
93. K. C. Maki, M. S. Reeves, M. Farmer, K. Yasunaga, N. Matsuo and Y. Katsuragi, *et al.*, Green tea catechin consumption enhances exercise-induced abdominal fat loss in overweight and obese adults, *J. Nutr.*, 2009, **139**, 264–270.
94. S. M. Tabatabaee, S. M. Alavian, L. Ghalichi, S. M. Miryounesi, K. Mousavizadeh, S. Jazayeri and M. R. Vafa, Green tea in non-alcoholic fatty liver disease: a double blind randomized clinical trial, *Hepat. Mon.*, 2017, **17**, e14993.
95. A. Pezeshki, S. Safi, A. Feizi, G. Askari and F. Karami, The effect of green tea extract supplementation on liver enzymes in patients with non-alcoholic fatty liver disease, *Int. J. Prev. Med.*, 2016, **7**, 28.
96. M. Hussain, R. Habib Ur and L. Akhtar, Therapeutic benefits of green tea extract on various parameters in non-alcoholic fatty liver disease patients, *Pak. J. Med. Sci.*, 2017, **33**, 931–936.
97. M. Mahmoodi, R. Hosseini, A. Kazemi, R. Ofori-Asenso, M. Mazidi and S. M. Mazloomi, Effects of green tea or green tea catechin on liver enzymes in healthy individuals and people with nonalcoholic fatty liver disease: A systematic review and meta-analysis of randomized clinical trials, *Phytother. Res.*, 2020, **34**, 1587–1598.
98. F. Mansour-Ghanaei, A. Hadi, M. Pourmasoumi, F. Joukar, S. Golpour and A. Najafgholizadeh, Green tea as a safe alternative approach for nonalcoholic fatty liver treatment: A systematic review and meta-analysis of clinical trials, *Phytother. Res.*, 2018, **32**, 1876–1884.
99. K. Imai and K. Nakachi, Cross sectional study of effects of drinking green tea on cardiovascular and liver diseases, *Br. Med. J.*, 1995, **310**, 693–696.
100. S. Kuriyama, T. Shimazu, K. Ohmori, N. Kikuchi, N. Nakaya and Y. Nishino, *et al.*, Green tea consumption and mortality due to cardiovascular disease, cancer, and all causes in Japan: the Ohsaki study, *JAMA*, 2006, **296**, 1255–1265.
101. J. Hu, D. Webster, J. Cao and A. Shao, The safety of green tea and green tea extract consumption in adults–Results of a systematic review, *Regul. Toxicol. Pharmacol.*, 2018, **95**, 412–433.
102. A. M. Dostal, H. Samavat, S. Bedell, C. Torkelson, R. Wang and K. Swenson, *et al.*, The safety of green tea extract supplementation in postmenopausal women at risk for breast cancer: results of the Minnesota Green Tea Trial, *Food Chem. Toxicol.*, 2015, **83**, 26–35.
103. M. García-Cortés, M. Robles-Díaz, A. Ortega-Alonso, I. Medina-Caliz and R. J. Andrade, Hepatotoxicity by dietary supplements: a tabular listing and clinical characteristics, *Int. J. Mol. Sci.*, 2016, **17**, 537.
104. E. Palacio Sánchez, M. E. Ribero Vargas and J. C. Restrepo Gutiérrez, Hepatotoxicity due to green tea consumption (*Camellia sinensis*): A review, *Rev. Colomb. Gastroenterol.*, 2013, **28**, 46–52.

105. T. Isomura, S. Suzuki, H. Origasa, A. Hosono, M. Suzuki and T. Sawada, *et al.*, Liver-related safety assessment of green tea extracts in humans: a systematic review of randomized controlled trials, *Eur. J. Clin. Nutr.*, 2016, **70**, 1221–1229.
106. E. P. O. F. Additives, N. S. A. T. Food, M. Younes, P. Aggett, F. Aguilar and R. Crebelli, *et al.*, Scientific opinion on the safety of green tea catechins, *EFSA J.*, 2018, **16**, e05239.
107. H. A. Oketch-Rabah, A. L. Roe, C. V. Rider, H. L. Bonkovsky, G. I. Giancaspro and V. Navarro, *et al.*, United States Pharmacopeia (USP) comprehensive review of the hepatotoxicity of green tea extracts, *Toxicol. Rep.*, 2020, **7**, 386–402.

CHAPTER 5

The Effect of Consuming Spices on Vascular Function

ESTER S. OH,[a] CONNIE J. ROGERS,[b,c] KRISTINA S. PETERSEN[d] AND PENNY M. KRIS-ETHERTON*[b]

[a] Division of Renal Diseases and Hypertension, University of Colorado Anschutz Medical Campus, Aurora, CO, USA; [b] Department of Nutritional Sciences, The Pennsylvania State University, University Park, PA, USA; [c] Center for Molecular Immunology and Infectious Disease, The Pennsylvania State University, University Park, PA, USA; [d] Department of Nutritional Sciences, Texas Tech University, Lubbock, TX, USA
*Email: pmk3@psu.edu

5.1 Introduction

There were an estimated 422.7 million cases of cardiovascular disease (CVD) worldwide resulting in 17.9 million CVD deaths in 2015.[1] CVD continues to be the leading cause of mortality globally. Stiffening of the large elastic arteries and impaired vascular endothelial function are major contributors to CVD risk.[2,3] Pulse wave velocity (PWV) is used to assess arterial stiffness, which is a measure of pulse wave speed over a given distance (typically from the carotid to femoral artery).[4] Brachial artery flow-mediated dilation (FMD) is a noninvasive measure of vascular endothelial function that assesses endothelium-dependent dilation following a flow stimulus.[5,6] A high PWV and an impaired FMD are independently associated with an increased risk of CVD events.[7]

Numerous preclinical and clinical studies have shown that oxidative stress and inflammation have deleterious effects resulting in vascular dysfunction.[8]

Food Chemistry, Function and Analysis No. 39
Dietary Supplements with Antioxidant Activity: Understanding Mechanisms and Potential Health Benefits
Edited by Cesarettin Alasalvar, Fereidoon Shahidi and Chi-Tang Ho

Published by the Royal Society of Chemistry, www.rsc.org

Oxidative stress and inflammation induce structural (*e.g.*, collagen deposition) and functional (*e.g.*, smooth muscle tone-related) changes that contribute to arterial stiffness.[9] Moreover, oxidative stress and inflammation reduce nitric oxide (NO) bioavailability, a potent vasodilator,[10] in the vascular endothelium, contributing to vascular endothelial dysfunction.[10] Obesity, including central obesity and metabolic syndrome (MetS), is associated with elevated markers of oxidative stress and inflammation,[11] thereby increasing the risk of CVD. Thus, with the growing prevalence of obesity and related co-morbidities in the US and globally,[12] identifying novel interventions that reduce vascular dysfunction in individuals with obesity or other metabolic disorders is needed to reduce the initiation and/or progression of CVD.

Spices (and herbs) are recommended as a dietary strategy to decrease salt intake in many countries including the US, the UK, and Australia.[13–15] Spice consumption is associated with reduced incidence and/or progression of CVD and other chronic diseases including type 2 diabetes (T2D), rheumatoid arthritis (RA), and osteoarthritis.[16,17] Spices have a high oxygen radical absorbance capacity (ORAC) level on a per-gram basis[18] and are rich sources of phenolic compounds with high antioxidant capacity.[19] Moreover, spices and their bioactive metabolites inhibit inflammatory pathways that are related to vascular dysfunction and the progression of CVD.[20] Thus, spices may be a supplemental dietary component that can inhibit oxidative and inflammatory pathways and lower the risk of CVD.

The current review aims to comprehensively summarize the results from human intervention studies that have evaluated the effect of popular spices in the US on vascular function as well as biomarkers of oxidative stress and inflammation. The spices reviewed herein include cinnamon (*Cinnamomum* spp.), garlic (*Allium sativum*), ginger (*Zingiber officinale*), turmeric/curcumin (*Curcuma longa*), and spice blends. Moreover, the mechanism(s) of action that may underlie the effect of these spices on vascular function is(are) discussed.

5.2 Effect of Consuming Spices on Vascular Function

5.2.1 Cinnamon

Cinnamon (*Cinnamomum* spp.) is one of the most popular spices used not only for cooking but in modern as well as traditional medicines worldwide. It contains various compounds including cinnamaldehyde, cinnamyl acetate, cineole, eugenol, coumarin, linalool, humulene, ethyl cinnamate, β-caryophyllene, and τ-cadinol. Among them, cinnamaldehyde is known be the major bioactive compound of cinnamon.[20] In a randomized controlled trial (RCT) supplementation with 3 g of cinnamon powder delivered in a high-fat meal (634 kcal) had no effect on the postprandial reflective index (RI) and stiffness index (SI) in healthy subjects (Table 5.1).[21] These results are not unexpected considering that acute cinnamon consumption likely

Table 5.1 Evidence reporting the effect of spice consumption on vascular function, oxidative stress, and inflammation.

Study design	Characteristics	Intervention: Dose	Intervention: Duration	Results[a]	Ref.
Cinnamon					
RCT, SB, PC, and CO ($n = 19$)	Healthy (mean age = 26 years)	3 g per day of cinnamon powder added in a high-fat meal (634 kcal)	Single administration	↔ Postprandial SI ↔ Postprandial RI	21
In vitro cell culture	Rat aorta	10^{-7}, 10^{-6}, 10^{-5}, and 10^{-4} g mL^{-1} of cinnamaldehyde	—	↑ Vasodilation	22
In vivo	db/db mice	0.2% of cinnamaldehyde in the diet	12 weeks	↑ Endothelium-dependent vasodilation	23
Ginger					
In vitro cell culture	Diabetic rat aorta	0.3–10 μM of 6-gingerol	—	↓ Vasoconstriction	35
Garlic					
Systematic review (10 RCT; $n = 40$)	Healthy, obesity, hypertension, T2D, and CAD	0.4–2.4 g per day	Single administration, 2 weeks–3 months	Inconsistent results, but garlic has the potential to improve vascular function in subjects with cardiovascular risk	24
Turmeric/curcumin					
Meta-analysis (10 RCTs; $n = 765$)	Healthy, dyslipidaemia, T2D, allergy, and depression	0.15–2 g per day	8–24 weeks	↓ FMD ↔ PWV ↔ AI	37

RCT, DB, and PC ($n = 38$)	Healthy adults (mean age = 21 years)	0.2 g per day	8 weeks	↑ FMD	38
RCT, DB, PC, and CO ($n = 18$)	Healthy smokers (mean age = 65 years)	5 g per day	Single administration in a capsule	↔ Postprandial FMD	39
Spice blends					
RCT, DB, PC, and CO ($n = 18$)	Healthy men (mean age = 45 years)	23.5 g per day curry meal (clove, coriander, cumin, garlic, ginger, onion, red pepper, and turmeric)	Single administration	↑ Postprandial FMD	40
RCT and CO ($n = 18$)	Patients with T2D	11.25 g per day spice blend (clove, black pepper, paprika, and garlic powder) in a high-fat meal	Single administration	↑ Postprandial peripheral artery tonometry score	41
RCT and CO ($n = 12$)	Men with overweight or obesity	6 g per day spice blend (basil, bay leaf, black pepper, cinnamon, coriander, cumin, ginger, oregano, parsley, red pepper, rosemary, thyme, and turmeric) in high-saturated fat and high-carbohydrate meal	Single administration	↑ Postprandial FMD	42

[a] Results: ↑, ↓, and ↔ represent a significant increase, significant decrease, and non-significant effect, respectively.
Abbreviations: AI, augmentation index; CAD, coronary artery disease; CO, cross-over; DB, double-blind; FMD, flow-mediated dilation; PC, placebo-controlled; PWV, pulse wave velocity; RCT, randomized controlled trial; RI, reflective index; SB, single-blind; SI, stiffness index; T2D, type 2 diabetes.

does not result in a structural change in the blood vessels following a single meal challenge. However, additional studies are needed to evaluate the effect of longer-term cinnamon consumption on arterial stiffness. Several preclinical studies suggest a potential beneficial effect of cinnamaldehyde on vascular function. For example, an *in vitro* study demonstrated that treatment with 10^{-7} to 10^{-4} g mL^{-1} of cinnamaldehyde increased the vasodilation properties of isolated rat aorta.[22] Also, incorporation of 0.2% of cinnamaldehyde into the diet for 12 weeks in db/db mice increased acetylcholine-induced vasodilation compared to mice not given cinnamaldehyde in the diet.[23] The emerging evidence from these preclinical studies of cinnamon on vascular function justifies further studies to better understand the effectiveness and the underlying mechanism of action.

5.2.2 Garlic

Garlic (*Allium sativum*) has a reputation in numerous countries and cultures as a therapeutic medicinal plant and includes various compounds such as diallyl sulphide, diallyl disulphide, diallyl trisulphide, ajoene, allicin, alliin, methiin, *S*-allylcysteine, isoalliin, cycloalliin, and *S*-allylmercaptocysteine.[20] A systematic review of 10 RCTs examined the effect of garlic consumption on vascular function[24] (Table 5.1). In this systematic review, the effect of garlic consumption in the form of garlic powder and aged garlic extract (0.4–2.4 g per day for 2 weeks–3 months) was examined in five studies that assessed FMD,[25–29] three studies that assessed PWV,[30–32] and two studies that quantified other indices of vascular function, including SI[33] and RI.[34] The results compiled suggest that garlic has the potential to improve vascular function, particularly in subjects with cardiovascular risk factors. Thus, supplementation with 0.4–2.4 g of garlic daily for 2 weeks–3 months may improve vascular function, particularly in individuals at risk for CVD. However, additional studies are needed to confirm these results.

5.2.3 Ginger

Ginger (*Zingiber officinale*) has a long history of medicinal use and constitutes diverse compounds including 6-gingerol, 6-paradol, shogaol, 6-gingerdiol, gingerdione, zingiberene, citral, bisabolene, α-farnesene, β-phellandrene, cineole, and zingerone. The compound 6-gingerol is a major phenolic constituent of ginger.[20] To date, no clinical studies have examined the effect of consuming ginger on markers of vascular function (*e.g.*, FMD and PWV). However, several preclinical studies have demonstrated a beneficial effect of 6-gingerol on vascular function. For example, an *in vitro* study demonstrated that incubation of aorta isolated from diabetic rats with 6-gingerol (0.3–10 μM) for 30 minutes alleviated the exaggerated vasoconstriction in the aorta observed in this preclinical model of diabetes[35] (Table 5.1). Thus, ginger may have a potential beneficial effect on vascular function, but human intervention trials are needed to test this hypothesis.

5.2.4 Turmeric/Curcumin

Turmeric (*Curcuma longa*) has long been used both as a culinary spice and as a traditional medicinal plant in Asian countries. Curcumin is the main bioactive compound of turmeric.[20] A meta-analysis and two RCTs have examined the effect of curcumin consumption on vascular function (Table 5.1). A meta-analysis of 10 RCTs demonstrated that consuming curcumin (0.15–2 g per day for 8–24 weeks) increased FMD [weighted mean difference (WMD): 1.49%; 95% CI: 0.16, 2.82; $I^2 = 0\%$], but had no effect on PWV or the augmentation index, another indicator of arterial stiffness,[36] in healthy subjects and those with dyslipidaemia, T2D, perennial allergic rhinitis, or depression.[37] An RCT in healthy subjects demonstrated that consuming 0.2 g of curcumin daily for 8 weeks increased FMD.[38] However, a single administration of a 5 g curcumin capsule had no effect on postprandial FMD in healthy smokers.[39] Overall, these findings suggest that consuming 0.15–2 g of curcumin daily for 8–24 weeks may have a beneficial effect on FMD but may not affect PWV in humans. In addition, curcumin consumption for a duration of ≥8 weeks may be needed to improve FMD and assess its effects on the PWV.

5.2.5 Spice Blends

Three randomized crossover trials have evaluated the effect of consuming a blend of spices (6–23.5 g) delivered in a meal on postprandial FMD (Table 5.1). Nakayama *et al.*[40] reported that a single meal containing 23.5 g of a spice blend (clove, coriander, cumin, garlic, ginger, onion, red pepper, and turmeric) increased postprandial FMD in healthy men. Consumption of a high-fat meal containing 11.25 g of a spice blend (clove, black pepper, paprika, and garlic powder) also improved endothelial function, assessed by the peripheral artery tonometry score, in patients with T2D.[41] Moreover, consuming a much smaller amount (6 g) of a spice blend (basil, bay leaf, black pepper, cinnamon, coriander, cumin, ginger, oregano, parsley, red pepper, rosemary, thyme, and turmeric) delivered in a high-saturated-fat, high-carbohydrate meal increased postprandial FMD in men with overweight or obesity.[42] Therefore, incorporating 6–23.5 g of spices into a meal may improve postprandial FMD. However, additional studies are needed to test if longer-term spice consumption may also have a beneficial effect on FMD and PWV.

5.3 Possible Mechanisms Linking Spice Consumption and Vascular Function: Role of Oxidative Stress and Inflammatory Mediators

Vascular dysfunction, including stiffening of the large elastic arteries and impaired vascular endothelial function, is a critical contributor to elevated

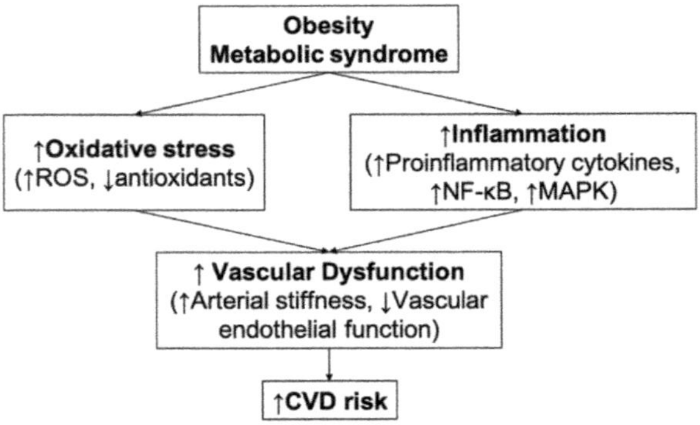

Figure 5.1 Mechanism underlying the association between obesity/metabolic syndrome and vascular dysfunction. Results: ↑ and ↓ represent a significant increase and significant decrease, respectively. Abbreviations: CVD, cardiovascular disease; MAPK, mitogen-activated protein kinase; NF-κB, nuclear factor-kappa B; ROS, reactive oxygen species.

CVD risk.[7] Since obesity and associated metabolic disorders are linked to both elevated markers of oxidative stress/inflammation and impaired vascular function,[8] oxidative stress and inflammation may, at least in part, mediate this association (Figure 5.1).

Several biomarkers are used to assess oxidative and inflammatory status in clinical studies. In subjects with obesity and MetS, an increase in the production of reactive oxygen species (ROS) generated by uncoupled endothelial nitric oxide synthase (eNOS), reduced antioxidant levels[43] [*e.g.*, superoxide dismutase (SOD), catalase (CAT), glutathione reductase (GR), and glutathione peroxidase (GPX)], and increased circulating malondialdehyde (MDA) levels[44] resulting from ROS-induced lipid peroxidation[45] has been reported. In addition, individuals with obesity and MetS have elevated biomarkers of inflammation, such as circulating levels of proinflammatory cytokines including interleukin (IL)-1β, IL-6, and tumour necrosis factor-alpha (TNF-α) and activation of signalling pathways that regulate a large array of genes encoding for proinflammatory cytokines [*e.g.*, nuclear factor-kappa B (NF-κB)[46] and mitogen-activated protein kinase (MAPK)[47]] (Figure 5.1).

This section summarizes the clinical evidence supporting the effect of spice consumption on biomarkers of oxidative stress and inflammation, which may underlie the relationship between spice consumption and improved vascular function.

5.3.1 Cinnamon

Numerous clinical trials have examined the effect of cinnamon on oxidative stress and inflammation (Table 5.2). A meta-analysis of 12 RCTs reported that cinnamon consumption (0.5–3 g per day for 10–112 days) reduced the

Table 5.2 Clinical evidence reporting the effect of spice consumption on oxidative stress and inflammation.

Study design	Subject characteristics	Intervention: Dose (g per day)	Intervention: Duration	Results[a]	Ref.
Cinnamon					
Meta-analysis (12 RCTs; $n = 696$)	Overweight or obesity, MetS, T2D, RA, PCOS, and NAFLD	1.5–3	10 days–12 weeks	↓ MDA ↑ TAC ↓ CRP ↓ IL-6	48
Meta-analysis (6 RCTs; $n = 285$)	Healthy, overweight or obesity, pre-diabetes, T2D, MetS, RA, and NAFLD	0.46–3	8–24 weeks	↓ CRP	49
Garlic					
Meta-analysis (7 RCTs; $n = 317$)	Healthy, obesity, T2D, postmenopausal osteoporotic women, and pregnant women	0.12–1.2	2–16 weeks	↓ MDA ↑ TAC	52
Meta-analysis (9 RCTs; $n = 363$)	Healthy, overweight or obesity, T2D, CAD, and pregnant women	0.25–3.6	2–48 weeks	↓ CRP	53
Meta-analysis (16 RCTs; $n = 743$)	Healthy, obesity, hypertension, hypercholesterolemia, MetS, T2D, CAD/CHD, postmenopausal osteoporotic women, and recalcitrant warts	0.8–3.6	2–52 weeks	↓ CRP ↓ IL-6 ↓ TNF-α	54
Meta-analysis (17 RCTs; $n = 912$)	Healthy, obesity, hypertension, hyperlipidaemia, MetS, T2D, CAD, postmenopausal osteoporotic women, osteoarthritis, and dialysis	0.4–4	1–48 weeks	↓ CRP ↔ IL-6 ↔ TNF-α	55
Meta-analysis (10 RCTs and 1 NRT; $n = 530$)	Healthy, obesity, hypertension, T2D, CAD, postmenopausal women, and osteoporosis ($n = 530$)	0.4–4.8	1–48 weeks	↔ CRP ↔ IL-6 ↔ TNF-α	56

Table 5.2 (*Continued*)

Study design	Subject characteristics	Intervention: Dose (g per day)	Intervention: Duration	Results[a]	Ref.
Ginger					
Meta-analysis (20 RCTs; $n=1077$)	Healthy, obesity, T2D, pulmonary tuberculosis, ARDS, increased colorectal cancer risk, breast cancer, osteoarthritis, NAFLD, UC, and PD	0.12–3	10 days–12 weeks	↓ MDA ↑ TAC ↓ CRP ↓ TNF-α ↓ IL-6	57
Meta-analysis (16 RCTs; $n=1010$)	Overweight, T2D, low back pain, osteoarthritis, NAFLD, and PD	1–3	4–12 weeks	↓ CRP ↓ hs-CRP ↓ TNF-α ↔ IL-6	58
Meta-analysis (9 RCTs and 1 NRT; $n=449$)	Healthy, obesity, hyperlipidaemia, T2D, and PD	1–3	45 days–12 weeks	↓ CRP	59
Turmeric/curcumin					
Meta-analysis (17 RCTs; $n=1279$)	MetS, T2D, osteoarthritis, PUD, infertile oligoasthenospermia, CKD, haemodialysis, CG, β-thalassemia, CP, chronic pulmonary complications, cancer, anxiety, and fatigue	0.08–3	1 week–4 months	↓ MDA ↑ SOD ↑ CAT ↑ GPX ↔ TAC ↔ GR	61
Meta-analysis (8 RCTs; $n=626$)	T2D, MetS, osteoarthritis, ESRD, and chronic pruritic skin lesions	0.6–1.5	4 weeks–4 months	↓ MDA ↑ SOD ↔ GPX (erythrocyte)	62
Meta-analysis (15 RCTs; $n=1223$)	Obesity, MetS, T2D, diabetic nephropathy, and CAD	0.294–4	3 days–36 weeks	↓ MDA ↓ IL-6 ↓ hs-CRP ↔ SOD ↔ TNF-α	63

Meta-analysis (6 RCTs; $n = 342$)	Healthy, obesity, osteoarthritis, SM intoxication, OLP, and coronary artery bypass grafting	0.08–6	4 weeks–3 months	↓ CRP	64
Meta-analysis (8 RCTs; $n = 562$)	Healthy, obesity, dyslipidaemia, osteoarthritis, MI, cancer, SM intoxication, and OLP	0.08–6	2–12 weeks	↓ CRP	65
Meta-analysis (8 RCTs and 1 NRT; $n = 609$)	Obesity, MetS, T2D, osteoarthritis, SM intoxication, chronic pruritic skin lesions, and OLP	0.2–6	4 weeks–8 months	↓ IL-6	66
Meta-analysis (8 RCTs; $n = 549$)	Overweight or obesity, T2D, MetS, osteoarthritis, depression, and SM intoxication	0.03–1.5	4 weeks–2 months	↓ TNF-α	74
Meta-analysis (18 RCTs; $n = 356$)	Overweight or obesity, hyperlipidaemia, MetS, T2D, RA, diabetic nephropathy, CAD, osteoarthritis, haemodialysis, and systemic lupus erythematosus	0.06–2.4	4 weeks–4 months	↔ CRP ↔ IL-1 ↔ IL-6 ↔ TNF-α	67
Spice blends					
RCT and SB ($n = 26$)	Women with polycystic ovary syndrome (mean age = 46 years)	0.7 g spice blend (cinnamon, ginger, spearmint, and sweet orange) per day	3 months	↓ MDA ↑ CAT ↑ GPX ↑ SOD	68
RCT, DB, and PC ($n = 64$)	Subjects with obesity (mean age = ∼39 years)	3 g spice blend capsule (ajwain, caraway, common rue, and oregano) and 500 kcal removal from diet per day	8 weeks	↓ TNF-α	69

Table 5.2 *(Continued)*

Study design	Subject characteristics	Intervention: Dose (g per day)	Intervention: Duration	Results[a]	Ref.
RCT, DB, and PC ($n = 60$)	Patients with RA (≥18 years)	Spice blend capsule (black pepper, curcumin, and ginger)	8 weeks	↔ CRP	70
RCT and CO ($n = 18$)	Patients with T2D	11.25 g spice blend (clove, black pepper, paprika, and garlic powder) in a high-fat meal	Single administration	↓ MDA (plasma and urine)	71
RCT and CO ($n = 6$)	Men with overweight (mean age = 44 years)	14 g spice blend (black pepper, cinnamon, clove, garlic, ginger, oregano, paprika, rosemary, and turmeric) in a high-fat meal	Single administration	↑ FRAP ↑ hydrophilic ORAC ↔ CRP	73

RCT and CO ($n = 12$)	Men with overweight or obesity	6 g spice blend (basil, bay leaf, black pepper, cinnamon, coriander, cumin, ginger, oregano, parsley, red pepper, rosemary, thyme, and turmeric) in a high-saturated fat and high-carbohydrate meal	Single administration	↓IL-1β, IL-8 and TNF-α secretion from LPS stimulated PBMC	72
RCT, DB, PC, and CO ($n = 18$)	Healthy men (mean age = 45 years)	23.5 g curry meal (clove, coriander, cumin, garlic, ginger, onion, red pepper, and turmeric)	Single administration	↔ MDA modified LDL ↔ 8-Isoprostane (urine) ↔ CRP	40

[a] Results: ↑, ↓, and ↔ represent a significant increase, significant decrease, and non-significant effect, respectively.

Abbreviations: ARDS, acute respiratory distress syndrome; CAD, coronary artery disease; CAT, catalase; CG, chronic gastritis; CHD, coronary heart disease; CKD, chronic kidney disease; CO, cross-over; CP, chronic pruritus; CRP, c-reactive protein; DB, double-blind; ESRD, end-stage renal disease; FRAP, ferric reducing antioxidant pathway; GPX, glutathione peroxidase; GR, glutathione reductase; hs-CRP, high sensitivity c-reactive protein; IL, interleukin; LDL, low-density lipoprotein; LPS, lipopolysaccharide; MDA, malondialdehyde; MetS, metabolic syndrome; MI, myocardial infarction; NAFLD, non-alcoholic fatty liver disease; NRT, non-randomized trial; OLP, oral lichen planus; ORAC, oxygen radical absorbance capacity; PBMC, peripheral blood mononuclear cell; PC, placebo-controlled; PCOS, polycystic ovary syndrome; PD, peritoneal dialysis; PUD, peptic ulcer disease; RA, rheumatoid arthritis; RCT, randomized controlled trial; SB, single-blind; SM, sulphur mustard; SOD, superoxide dismutase; TAC, total antioxidant capacity; TNF-α, tumour necrosis factor-alpha; UC, ulcerative colitis.

circulating levels of MDA [WMD: −0.97 mmol L^{-1}; 95% confidence interval (CI): −1.66, −0.27; $I^2 = 89\%$], C-reactive protein (CRP) (WMD: −2.22 mg L^{-1}, 95% CI: −3.74, −0.69; $I^2 = 89\%$), and IL-6 (WMD: −1.48 pg mL^{-1}; 95% CI: −2.96, −0.01; $I^2 = 79\%$) and increased total antioxidant capacity (TAC) (WMD: 0.34 mmol L^{-1}; 95% CI: 0.04, 0.64; $I^2 = 97\%$) in a heterogeneous population [*e.g.*, subjects with MetS, non-alcoholic fatty liver disease (NAFLD), RA, T2D, or healthy subjects].[48] Results from a meta-analysis including six RCTs demonstrated that cinnamon consumption (0.46–3 g per day for 8–24 weeks) reduced circulating CRP levels (WMD: −0.81 mg dL^{-1}; 95% CI: −1.36, −0.26; $I^2 = 79\%$) in a population with diverse health conditions (*e.g.*, healthy subjects and those with overweight/obesity, prediabetes, T2D, MetS, RA, and NAFLD).[49] A subgroup analysis showed that cinnamon supplementation with both doses of <1.5 g per day and ≥1.5 g per day significantly reduced CRP levels.[49] However, CRP levels were significantly reduced only when the duration of the intervention was >12 weeks (WMD: −0.42 mg dL^{-1}; 95% CI: −0.65, −0.20; $I^2 = 0\%$),[49] suggesting that longer-term consumption of cinnamon may be necessary to observe a change in CRP levels. Furthermore, db/db mice supplemented with 0.02% cinnamaldehyde in the diet daily for 12 weeks had greater acetylcholine-induced relaxation of the aorta *via* inhibition of ROS production and increased NO production, and elevated nuclear factor erythroid 2-related factor 2 (Nrf2, a regulator of antioxidant response[50]) compared to controls.[51] Collectively, these findings suggest that cinnamon may alleviate vascular dysfunction *via* changes in oxidative and inflammatory pathways. However, well-designed human intervention trials are needed to explore this relationship (Figure 5.2).

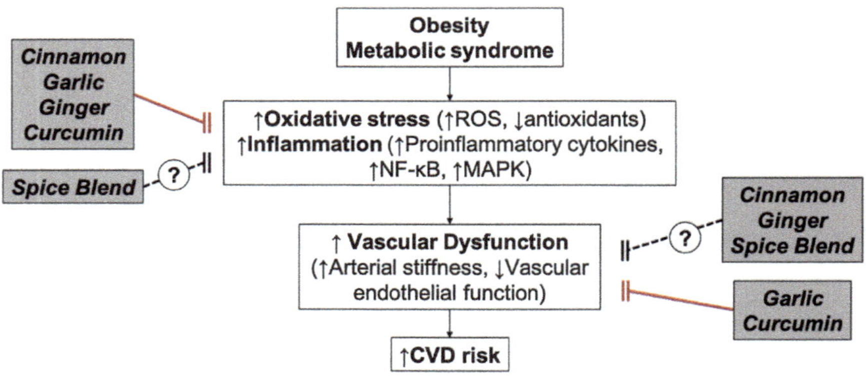

Figure 5.2 The effect of spices on vascular function, oxidative stress, and inflammation. Results: ↑ and ↓ represent a significant increase and significant decrease, respectively. Abbreviations: CVD, cardiovascular disease; MAPK, mitogen-activated protein kinase; NF-κB, nuclear factor-kappa B; ROS, reactive oxygen species.

5.3.2 Garlic

Five meta-analyses have investigated the effect of garlic on biomarkers of oxidative stress and inflammation (Table 5.2). A meta-analysis of seven RCTs reported that garlic consumption (0.12–1.2 g per day for 2–16 weeks) reduced circulating MDA levels (WMD: -1.88 mmol L^{-1}; 95% CI: -0.33, -0.46; $I^2 = 98\%$) and increased TAC levels (WMD: 11.03 mmol L^{-1}; 95% CI: 4.78, 17.28; $I^2 = 100\%$) in healthy subjects, postmenopausal osteoporotic women, those with obesity and T2D, and pregnant women.[52] A subgroup analysis in this study revealed that there was a greater increase in TAC levels when the duration of the trial was ≥8 weeks.[52] A meta-analysis of nine RCTs reported a reduction in CRP levels (WMD: -0.8 mg L^{-1}; 95% CI: -1.5, -0.1; $I^2 = 98\%$) with garlic consumption (0.25–3.6 g per day for 2–48 weeks) in a heterogeneous population [healthy subjects, pregnant women, and those with overweight or obesity, T2D, coronary artery disease (CAD), heart scan score >30 (intermediate to high risk for CAD)].[53] However, a reduction in CRP levels was greater when garlic supplementation was ≥1.2 g per day as compared to <1.2 g per day.[53] A meta-analysis of 16 RCTs reported that garlic consumption (0.8–3.6 g per day for 2–52 weeks) reduced circulating CRP (WMD: -0.61 mg L^{-1}; 95% CI: -1.12, -0.11; $I^2 = 77\%$), IL-6 (WMD: -0.73 ng L^{-1}; 95% CI: -1.06, -0.40; $I^2 = 0\%$), and TNF-α (WMD: -0.26 ng L^{-1}; 95% CI: -0.41, -0.12; $I^2 = 0\%$) levels in healthy subjects and those with obesity, MetS, hypertension, hypercholesterolemia, T2D, CAD/coronary heart disease (CHD), and postmenopausal osteoporotic women.[54] Furthermore, the IL-6 level was reduced only when the garlic dose given was >1.2 g per day.[54] Similarly, a meta-analysis of 17 RCTs demonstrated that garlic consumption (0.4–4 g per day for 1–48 weeks) reduced CRP levels (WMD: -0.53 mg L^{-1}; 95% CI: -0.93, -0.13; $I^2 = 48\%$), but not IL-6 and TNF-α levels, in a population with diverse health conditions (*e.g.*, healthy subjects and those with obesity, hypertension, hyperlipidemia, MetS, T2D, CAD, and osteoarthritis, postmenopausal osteoporotic women, and patients on dialysis).[55] However, a reduction in IL-6 and TNF-α was observed only when garlic was consumed as an aged garlic extract, whereas garlic powder, tablets, and capsules had no significant effect. A meta-analysis of 10 RCTs and one non-randomized trial (NRT) reported that consuming garlic (0.4–4.8 g per day for 1–48 weeks) had no effect on CRP, IL-6 and TNF-α levels;[56] however, TNF-α, CRP, and IL-6 levels were reduced when the duration of garlic supplementation was >8 weeks, >6 weeks, and ≥4 weeks, respectively.[56]

Numerous clinical studies have demonstrated an antioxidative and anti-inflammatory effect of garlic consumption, which may in part underlie the beneficial effect on vascular function. However, given the results of numerous subgroup analyses in the aforementioned studies, future human intervention trials should consider the duration of intervention, dose of garlic, and type of garlic preparation (*e.g.*, raw, powder, oil, and aged extract) when evaluating the effect of garlic consumption on vascular function (Figure 5.2).

5.3.3 Ginger

Three meta-analyses have examined the anti-oxidative and anti-inflammatory effects of ginger consumption (Table 5.1). A meta-analysis of 20 RCTs reported that consumption of ginger (0.12–3 g per day for 10 days–11 weeks) increased TAC levels [mean difference (MD): 0.99; 95% CI: 0.65, 1.33; $I^2 = 95\%$] and reduced concentrations of MDA (MD: −0.65; 95% CI: −1.28, −0.03; $I^2 = 83\%$), CRP (MD: −1.03; 95% CI: −1.53, −0.53; $I^2 = 86\%$), IL-6 (MD: −1.02, 95% CI: −1.71, −0.34; $I^2 = 73\%$), and TNF-α (MD: −0.95; 95% CI: −1.59, −0.31; $I^2 = 88\%$) in healthy subjects and those with obesity, T2D, pulmonary tuberculosis, acute respiratory distress syndrome (ARDS), increased colorectal cancer risk, breast cancer, osteoarthritis, non-alcoholic fatty liver disease (NAFLD), and ulcerative colitis (UC), and patients on peritoneal dialysis (PD).[57] Furthermore, a significant reduction in CRP and TNF-α levels was observed when ginger was consumed at doses of 2–3 g per day and 1–2 g per day, respectively. Similarly, a meta-analysis of 16 RCTs reported that consuming ginger (1–3 g per day for 4–12 weeks) reduced the concentrations of CRP [standardized MD (SMD): −5.11; 95% CI: −7.91, −2.30; $I^2 = 98\%$], hs-CRP (SMD: −0.88; 95% CI: −1.63, −0.12; $I^2 = 91\%$), and TNF-α (SMD: −0.85; 95% CI: −1.48, −0.21; $I^2 = 89\%$), but not IL-6, in subjects with overweight, T2D, low back pain, osteoarthritis, and NAFLD and patients on PD.[58] A meta-analysis of nine RCTs and one NRT demonstrated that ginger consumption (1–3 g per day for 45 days–12 weeks) reduced CRP levels (WMD: −0.84 $mg\,L^{-1}$; 95% CI: −1.38, −0.31; $I^2 = 56\%$) in healthy subjects, those with obesity, hyperlipidaemia, and T2D, and patients on PD.[59]

Although no clinical studies to date have investigated the effect of ginger consumption on markers of vascular function, numerous clinical studies have demonstrated an anti-oxidative and anti-inflammatory effect of ginger, which may confer benefits on vascular function. Moreover, *in vitro* treatment of isolated porcine coronary arteries with ginger extract (1, 3, 10, 30, or 100 $\mu g\,mL^{-1}$) for 30 min increased potassium chloride-induced vasodilation by increasing nitric oxide (NO) production, as well as inhibited eNOS activity and the cyclooxygenase-2 (COX-2) inflammatory pathway.[60] These findings demonstrate a beneficial effect of ginger on vascular function that is mediated *via* a reduction in oxidative stress and inflammation. However, additional well-designed human intervention trials are needed to explore this hypothesis (Figure 5.2).

5.3.4 Turmeric/Curcumin

Nine meta-analyses have examined the effect of curcumin/turmeric consumption on biomarkers of oxidative stress and inflammation (Table 5.2). In a meta-analysis of 17 RCTs, supplementation with turmeric/curcumin (0.08–3 g per day for 1 week–4 months) reduced MDA levels (SMD: −0.46; 95% CI: −0.68, −0.25; $I^2 = 97\%$) and increased the concentrations of SOD (SMD: 0.82; 95% CI: 0.27, 1.38; $I^2 = 98\%$), CAT (SMD: 10.26; 95% CI: 0.92,

19.61; $I^2 = 99\%$), and GPX (SMD: 8.90; 95% CI: 6.62, 11.19, $I^2 = 100\%$)[61] in a heterogeneous population [*e.g.*, subjects with MetS, T2D, osteoarthritis, infertile oligoasthenospermia, peptic ulcer disease (PUD), solid cancer, chronic kidney disease (CKD), chronic gastritis (CG), β-thalassemia, chronic pruritus (CP), chronic pulmonary complications, anxiety, and fatigue, and patients on haemodialysis]. Similarly, a meta-analysis of eight RCTs reported that curcumin consumption (0.6–1.5 g per day for 4 weeks–4 months) decreased MDA concentrations (SMD: −0.769; 95% CI: −1.06, −0.48; $I^2 = 60\%$) and increased SOD levels (SMD: 1.08; 95% CI: 0.49, 1.68; $I^2 = 84\%$) in subjects with T2D, MetS, osteoarthritis, end-stage kidney disease (ESRD), and chronic pruritic skin lesions.[62] In a meta-analysis of 15 RCTs, supplementation with curcumin (0.294–4 g for 3 days–36 weeks) reduced the levels of MDA (SMD: −3.14; 95% CI: −4.76, −1.53; $I^2 = 97\%$), IL-6 (SMD: −2.08; 95% CI: −3.90, −0.25; $I^2 = 97\%$), and high sensitivity-CRP (hs-CRP) (SMD: −0.65; 95% CI: −1.20, −1.0; $I^2 = 93\%$) in subjects with obesity, MetS, T2D, diabetic nephropathy, and CAD.[63] A meta-analysis of six RCTs demonstrated a reduction in CRP levels (WMD: −6.44 $mg\,L^{-1}$; 95% CI: −10.77, −2.11; $I^2 = 97\%$) with curcumin supplementation (0.08–6 g per day for 4 weeks–3 months) in a diverse population.[64] A meta-analysis of eight RCTs also reported that curcumin consumption (0.08–6 g per day for 2–12 weeks) reduced CRP levels (WMD: −2.20 $mg\,L^{-1}$; 95% CI: −3.96, −0.44; $I^2 = 95\%$) in a heterogeneous population [*e.g.*, healthy subjects and those with obesity, dyslipidaemia, osteoarthritis, myocardial infarction (MI), solid tumors receiving chemotherapy, sulphur mustard (SM) intoxication, and oral lichen planus (OLP)].[65] A meta-analysis of eight RCTs and one NRT reported that curcumin consumption (0.2–6 g per day for 4 weeks–8 months) decreased IL-6 levels (WMD: −0.60 $pg\,mL^{-1}$; 95% CI: −1.06, −0.14) in subjects with obesity, MetS, T2D, osteoarthritis, SM intoxication, chronic pruritic skin lesions, and OLP.[66] A meta-analysis of eight RCTs demonstrated that curcumin consumption (0.03–1.5 g per day for 4 weeks–2 months) reduced TNF-α concentrations (WMD: −4.69 $pg\,mL^{-1}$; 95% CI: −7.10, −2.28) in a heterogeneous population (*e.g.*, overweight or obesity, T2D, MetS, osteoarthritis, depression, and SM intoxication). In contrast, a recent meta-analysis of 18 RCTs demonstrated no effects of curcumin consumption on CRP, IL-1β, IL-6, and TNF-α levels in patients with chronic inflammatory diseases.[67] However, there was a greater reduction in CRP in subjects on the trial for >2 months compared to those on the trial for ≤2 months.[67]

The anti-oxidative and anti-inflammatory effects of turmeric/curcumin consumption are well-established in clinical trials including subjects with a wide array of health conditions. Given that turmeric/curcumin consumption also improves vascular function, a reduction in oxidative stress and inflammation after turmeric/curcumin supplementation may in part mediate this association (Figure 5.2).

5.3.5 Spice Blends

Seven RCTs have investigated the effect of consuming spice blends (either acutely or chronically) on markers of oxidative stress and inflammation (Table 5.2). Three studies examined the effect of longer-term consumption of a spice blend (0.7–3 g per day for 8 weeks–3 months), and four studies examined the postprandial effect of a single administration of a spice blend (6–23.5 g).

5.3.5.1 Longer-term Consumption

Ainehchi *et al.*[68] reported that supplementation with 0.7 g of a spice blend (cinnamon, ginger, spearmint, and sweet orange) daily for 3 months reduced circulating MDA levels and increased circulating CAT, GPX, and SOD levels in women with polycystic ovary syndrome. Supplementation with 3 g of a spice blend capsule (ajwain, caraway, common rue, and oregano) daily for 8 weeks decreased circulating TNF-α levels in subjects with obesity.[69] However, Hemmati *et al.*[70] reported no change in circulating CRP levels following supplementation with a spice blend capsule (black pepper, curcumin, and ginger) daily for 8 weeks in subjects with rheumatoid arthritis.

5.3.5.2 Single Administration

Li *et al.*[71] demonstrated that consuming a high fat meal containing 11.25 g of a spice blend (clove, black pepper, paprika, and garlic powder) reduced MDA levels in plasma and urine. Skulas-Ray *et al.*[73] examined the postprandial effect of consuming 14 g of a spice blend (black pepper, cinnamon, clove, garlic, ginger, oregano, paprika, rosemary, and turmeric) delivered in a high-fat meal on lipid peroxidation in men with overweight in a randomized crossover trial. After consuming the meal containing the spice blend, postprandial levels of ferric reducing ability of plasma (FRAP) and hydrophilic ORAC were decreased.[42] We examined the effect of consuming a high-saturated-fat, high-carbohydrate meal alone (0 g of a spice blend), or with 2 g and 6 g of a spice blend (basil, bay leaf, black pepper, cinnamon, coriander, cumin, ginger, oregano, red pepper, rosemary, thyme, and turmeric) on postprandial inflammatory responses in men with overweight or obesity.[72] After consuming the high-saturated fat, high-carbohydrate meal containing 6 g of a spice blend, a reduction in postprandial secretion of IL-1β, IL-8, and TNF-α from lipopolysaccharide (LPS)-stimulated peripheral blood mononuclear cells was observed compared to when the meal was consumed without the spice blend.[72] However, Nakayama *et al.*[40] reported no change in circulating MDA, modified low-density lipoprotein (LDL), urinary 8-isoprostane, and circulating CRP levels following consumption of a curry meal containing 23.5 g of a spice blend (clove, coriander, cumin, garlic, ginger, onion, red pepper, and turmeric) in healthy men. The change in the markers of oxidative stress after the curry meal might have been too small to be detected. Other studies utilized a high-fat meal,[71–73] which may exaggerate the postprandial oxidative stress/inflammation and enable the detection of a significant effect on the outcomes of interest.

Clinical intervention studies that have examined the effect of a spice blend on vascular function have focused mainly on acute effects (*e.g.*, increased postprandial FMD).[40–42] Thus, there is a need for additional longer-term human intervention studies to examine the possible vascular effects of consuming a blend of spices. Furthermore, both short (single administration) and longer (8 weeks–3 months) duration studies of spice consumption have beneficial effects on oxidative stress and vascular function, which supports the hypothesis that spice consumption is cardioprotective (Figure 5.2).

5.4 Conclusion

In this review, we presented evidence linking spice consumption (cinnamon, ginger, garlic, turmeric/curcumin, and a spice blend) to changes in vascular function. A possible mechanism of action that may underlie the beneficial effect of spice consumption on vascular function has been discussed. The clinical evidence available to date suggests a beneficial effect of several individual spices (cinnamon, garlic, ginger, and curcumin), as well as a spice blend (as generally occurs in a meal), on markers of vascular function, oxidative stress, and inflammation. However, additional studies are needed to (1) examine the potential benefits of consuming a spice blend *vs.* individual spices; (2) determine the dose and duration of spice consumption needed to achieve improvements in vascular function; and (3) evaluate in which populations spice consumption might be most beneficial. Nonetheless, the available evidence clearly indicates that including spices in the diet confers health benefits on key mechanisms that improve vascular function.

References

1. G. A. Roth, C. Johnson, A. Abajobir, F. Abd-Allah, S. F. Abera and G. Abyu, *et al.*, Global, regional, and national burden of cardiovascular diseases for 10 causes, 1990 to 2015, *J. Am. Coll. Cardiol.*, 2017, **70**, 1–25.
2. J. Yeboah, Brachial flow-mediated dilation predicts incident cardiovascular events in older adults. The Cardiovascular Health Study, *Circulation*, 2004, **109**, 613–619.
3. M. Shechter, A. Issachar, I. Marai, N. Koren-Morag, D. Freinark and Y. Shahar, *et al.*, Long-term association of brachial artery flow-mediated vasodilation and cardiovascular events in middle-aged subjects with no apparent heart disease, *Int. J. Cardiol.*, 2009, **134**, 52–58.
4. H.-L. Kim and S.-H. Kim, Pulse wave velocity in atherosclerosis, *Front. Cardiovasc. Med.*, 2019, **6**, 41.
5. D. R. Seals, K. L. Jablonski and A. J. Donato, Aging and vascular endothelial function in humans, *Clin. Sci.*, 2011, **120**, 357–375.

6. S. N. Doshi, K. K. Naka, N. Payne, C. J. Jones, M. Ashton and M. J. Lewis, *et al.*, Flow-mediated dilatation following wrist and upper arm occlusion in humans: the contribution of nitric oxide, *Clin. Sci.*, 2001, **101**, 629–635.
7. R. M. Bruno, E. Bianchini, F. Faita, S. Taddei and L. Ghiadoni, Intima media thickness, pulse wave velocity, and flow mediated dilatio, *Cardiovasc. Ultrasound*, 2014, **12**, 1–9.
8. D. P. Hajjar and A. M. Gotto Jr, Biological relevance of inflammation and oxidative stress in the pathogenesis of arterial diseases, *Am. J. Pathol.*, 2013, **182**, 1474–1481.
9. V. Cachofeiro, M. Goicochea, S. G. De Vinuesa, P. Oubiña, V. Lahera and J. Luño, Oxidative stress and inflammation, a link between chronic kidney disease and cardiovascular disease: new strategies to prevent cardiovascular risk in chronic kidney disease, *Kidney Int.*, 2008, **74**, S4–S9.
10. J. Wu, S. Xia, B. Kalionis, W. Wan and T. Sun, The role of oxidative stress and inflammation in cardiovascular aging, *BioMed Res. Int.*, 2014, 615312.
11. A. Fernández-Sánchez, E. Madrigal-Santillán, M. Bautista, J. Esquivel-Soto, Á. Morales-González and C. Esquivel-Chirino, *et al.*, Inflammation, oxidative stress, and obesity, *Int. J. Mol. Sci.*, 2011, **12**, 3117–3132.
12. World Health Organization (WHO), Obesity and overweight. Published online at: https://www.who.int/news-room/fact-sheets/detail/obesity-and-overweight (accessed June 12, 2021).
13. National Health and Medical Research Council (NHMRC), *Australian Dietary Guidelines*, National Health and Medical Research Council, Canberra, Australia, 2013.
14. U. S. Department of Agriculture (USDA), *Dietary Guidelines for Americans, 2020–2025*, U. S. Department of Agriculture, Washington, DC, 9th edn, 2020.
15. Public Health England, *The Eatwell Guide*, England, United Kingdom, 2016.
16. D. Bukvicki, D. Gottardi, S. Prasad, M. Novakovic, P. D. Marin and A. K. Tyagi, The healing effects of spices in chronic diseases, *Curr. Med. Chem.*, 2020, **27**, 4401–4420.
17. L. C. Tapsell, I. Hemphill, L. Cobiac, C. S. Patch, D. R. Sullivan and M. Fenech, *et al.*, Health benefits of herbs and spices: the past, the present, the future, *Med. J. Aust.*, 2006, **185**, S1–S24.
18. D. B. Haytowitz and S. Bhagwat, *USDA Database for the Oxygen Radical Absorbance Capacity (orac) of Selected Foods, Release 2*, U. S. Department of Agriculture, Washinton, DC, 2010.
19. B. Shan, Y. Z. Cai, M. Sun and H. Corke, Antioxidant capacity of 26 spice extracts and characterization of their phenolic constituents, *J. Agric. Food Chem.*, 2005, **53**, 7749–7759.
20. A. B. Kunnumakkara, B. L. Sailo, K. Banik, C. Harsha, S. Prasad and S. C. Gupta, *et al.*, Chronic diseases, inflammation, and spices: how are they linked?, *J. Transl. Med.*, 2018, **16**, 1–25.

21. O. Markey, C. M. McClean, P. Medlow, G. W. Davison, T. R. Trinick and E. Duly, *et al.*, Effect of cinnamon on gastric emptying, arterial stiffness, postprandial lipemia, glycemia, and appetite responses to high-fat breakfast, *Cardiovasc. Diabetol.*, 2011, **10**, 78.
22. Y.-L. Xue, H.-X. Shi, F. Murad and K. Bian, Vasodilatory effects of cinnamaldehyde and its mechanism of action in the rat aorta, *Vasc. Health. Risk. Manag.*, 2011, 7, 273–280.
23. X. Guo, W. Sun, L. Huang, L. Wu, Y. Hou and L. Qin, *et al.*, Effect of cinnamaldehyde on glucose metabolism and vessel function, *Med. Sci. Mon. Int. Med. J. Exp. Clin. Res.*, 2017, **23**, 3844–3853.
24. H. Emamat, H. Tangestani, A. S. Totmaj, H. Ghalandari and J. Nasrollahzadeh, The effect of garlic on vascular function: A systematic review of randomized clinical trials, *Clin. Nutr.*, 2020, **39**, 3563–3570.
25. M. Mahdavi-Roshan, P. Mirmiran, M. Arjmand and J. Nasrollahzadeh, Effects of garlic on brachial endothelial function and capacity of plasma to mediate cholesterol efflux in patients with coronary artery disease, *Anatolian J. Cardiol.*, 2017, **18**, 116–121.
26. M. J. Williams, W. H. Sutherland, M. P. McCormick, D. J. Yeoman and S. A. de Jong, Aged garlic extract improves endothelial function in men with coronary artery disease, *Phytother. Res.*, 2005, **19**, 314–319.
27. C. Womack, D. Lawton, L. Redmond, M. Todd and T. Hargens, The effects of acute garlic supplementation on the fibrinolytic and vasoreactive response to exercise, *J. Int. Soc. Sports Nutr.*, 2015, **12**, 1–8.
28. N. Weiss, N. Ide, T. Abahji, L. Nill, C. Keller and U. Hoffmann, Aged garlic extract improves homocysteine-induced endothelial dysfunction in macro-and microcirculation, *J. Nutr.*, 2006, **136**, 750S–754S.
29. D. Gómez-Arbeláez, V. Lahera, P. Oubiña, M. Valero-Muñoz, N. De las Heras and Y. Rodríguez, *et al.*, Aged garlic extract improves adiponectin levels in subjects with metabolic syndrome: a double-blind, placebo-controlled, randomized, crossover study, *Mediators Inflammation*, 2013, **2013**, e285795.
30. B. Turner, C. Mølgaard and P. Marckmann, Effect of garlic (*Allium sativum*) powder tablets on serum lipids, blood pressure and arterial stiffness in normo-lipidaemic volunteers: a randomised, double-blind, placebo-controlled trial, *Br. J. Nutr.*, 2004, **92**, 701–706.
31. K. Ried, N. Travica and A. Sali, A, The effect of Kyolic aged garlic extract on gut microbiota, inflammation, and cardiovascular markers in hypertensives: the GarGIC trial, *Front. Nutr.*, 2018, **5**, 122.
32. K. Ried, N. Travica and A. Sali, The effect of aged garlic extract on blood pressure and other cardiovascular risk factors in uncontrolled hypertensives: The AGE at Heart trial, *Integr. Blood Pressure Control*, 2016, **9**, 9–21.
33. M. Szulińska, M. Kręgielska-Narożna, J. Świątek, P. Styś, B. Kuźnar-Kamińska and H. Jakubowski, *et al.*, Garlic extract favorably modifies markers of endothelial function in obese patients–randomized double blind placebo-controlled nutritional intervention, *Biomed. Pharmacother.*, 2018, **102**, 792–797.

34. M. Atkin, D. Laight and M. H. Cummings, The effects of garlic extract upon endothelial function, vascular inflammation, oxidative stress and insulin resistance in adults with T2Dat high cardiovascular risk. A pilot double blind randomized placebo controlled trial, *J. Diabetes Its Complications*, 2016, **30**, 723–727.
35. S. A. Ghareib, H. M. El-Bassossy, A. A. Elberry, A. Azhar, M. L. Watson and Z. M. Banjar, 6-Gingerol alleviates exaggerated vasoconstriction in diabetic rat aorta through direct vasodilation and nitric oxide generation, *Drug Des., Dev. Ther.*, 2015, **9**, 6019.
36. I. B. Wilkinson, H. MacCallum, L. Flint, J. R. Cockcroft, D. E Newby and D. J. Webb, The influence of heart rate on augmentation index and central arterial pressure in humans, *J. Physiol.*, 2000, **525**, 263–270.
37. J. Hallajzadeh, A. Milajerdi, F. Kolahdooz, E. Amirani, H. Mirzaei and Z. Asemi, The effects of curcumin supplementation on endothelial function: A systematic review and meta-analysis of randomized controlled trials, *Phytother. Res.*, 2019, **33**, 2989–2995.
38. J. M. Oliver, L. Stoner, D. S. Rowlands, A. R. Caldwell, E. Sanders and A. Kreutzer, *et al.*, Novel form of curcumin improves endothelial function in young, healthy individuals: a double-blind placebo controlled study, *J. Nutr. Metab.*, 2016, **2016**, e1089653.
39. N. Barber-Chamoux, D. Milenkovic, M. A. Verny, V. Habauzit, B. Pereira and C. Lambert, *et al.*, Substantial variability across individuals in the vascular and nutrigenomic response to an acute intake of curcumin: a randomized controlled trial, *Mol. Nutr. Food Res.*, 2018, **62**, e1700418.
40. H. Nakayama, N. Tsuge, H. Sawada, N. Masamura, S. Yamada and S. Satomi, *et al.*, A single consumption of curry improved postprandial endothelial function in healthy male subjects: a randomized, controlled crossover trial, *Nutrition*, 2014, **13**, 1–7.
41. Z. Li, S. Henning, Y. Zhang, N. Rahnama, A. Zerlin and G. Thames, *et al.*, Decrease of postprandial endothelial dysfunction by spice mix added to high-fat hamburger meat in men with Type 2 diabetes mellitus, *Diabet. Med.*, 2013, **30**, 590–595.
42. K. S. Petersen, C. J. Rogers, S. G. West, D. N. Proctor and P. M. Kris-Etherton, The effect of culinary doses of spices in a high-saturated fat, high-carbohydrate meal on postprandial lipemia and endothelial function: a randomized, controlled, crossover pilot trial, *Food Funct.*, 2020, **11**, 3191–3200.
43. E. Birben, U. M. Sahiner and C. Sackesen, S. Erzurum and O. Kalayci, Oxidative stress and antioxidant defense, *World Allergy Organ. J.*, 2012, **5**, 9–19.
44. F. Moreto, E. P. de Oliveira, R. M. Manda and R. C. Burini, The higher plasma malondialdehyde concentrations are determined by metabolic syndrome-related glucolipotoxicity, *Oxid. Med. Cell. Longevity*, 2014, e505368.
45. E. Gianazza, M. Brioschi, A. M. Fernandez and C. Banfi, Lipoxidation in cardiovascular diseases, *Redox Biol.*, 2019, **23**, e101119.

46. L. Catrysse and G. van Loo, Inflammation and the metabolic syndrome: the tissue-specific functions of NF-κB, *Trends Cell Biol.*, 2017, **27**, 417–429.
47. K. C. Nandipati, S. Subramanian and D. K. Agrawal, Protein kinases: mechanisms and downstream targets in inflammation-mediated obesity and insulin resistance, *Mol. Cell. Biochem.*, 2017, **426**, 27–45.
48. C. Zhu, H. Yan, Y. Zheng, H. O. Santos, M. S. Macit and K. Zhao, Impact of cinnamon supplementation on biomarkers of inflammation and oxidative stress: A systematic review and meta-analysis of randomized controlled trials, *Complement. Ther. Med.*, 2020, **53**, e102517.
49. N. Vallianou, C. Tsang, M. Taghizadeh, A. Davoodvandi and S. Jafarnejad, Effect of cinnamon (*Cinnamomum zeylanicum*) supplementation on serum C-reactive protein concentrations: A meta-analysis and systematic review, *Complementary Ther. Med.*, 2019, **42**, 271–278.
50. S. Vomund, A. Schäfer, M. J. Parnham, B. Brüne and A. Von, Knethen, Nrf2, the master regulator of anti-oxidative responses, *Int. J. Mol. Sci.*, 2017, **18**, e2772.
51. P. Wang, Y. Yang, D. Wang, Q. Yang, J. Wan and S. Liu, *et al.*, Cinnamaldehyde ameliorates vascular dysfunction in diabetic mice by activating Nrf2, *Am. J. Hypertens.*, 2020, **33**, 610–619.
52. S. P. Moosavian, A. Arab, Z. Paknahad and S. Moradi, The effects of garlic supplementation on oxidative stress markers: A systematic review and meta-analysis of randomized controlled trials, *Complementary Ther. Med.*, 2020, **50**, e102385.
53. M. Taghizadeh, Z. Hamedifard and S. Jafarnejad, Effect of garlic supplementation on serum C-reactive protein level: A systematic review and meta-analysis of randomized controlled trials, *Phytother. Res.*, 2019, **33**, 243–252.
54. M. Darooghegi Mofrad, A. Milajerdi, F. Koohdani, P. J. Surkan and L. Azadbakht, Garlic supplementation reduces circulating C-reactive protein, tumor necrosis factor, and Interleukin-6 in adults: a systematic review and meta-analysis of randomized controlled trials, *J. Nutr.*, 2019, **149**, 605–618.
55. F. Mirzavandi, M. Mollahosseini, A. Salehi-Abargouei and H. Mozaffari-Khosravi, Effects of garlic supplementation on serum inflammatory markers: A systematic review and meta-analysis of randomized controlled trials, *Diabetes Metab. Syndr.*, 2020, **14**, 1153–1161.
56. M. Koushki, N. Amiri-Dashatan, Y. Pourfarjam and A. H. Doustimotlagh, Effect of garlic intake on inflammatory mediators: a systematic review and meta-analysis of randomised controlled trials, *Postgrad. Med. J.*, 2021, **97**, 156–163.
57. M. Jalali, M. Mahmoodi, S. P. Moosavian, R. Jalali, G. Ferns and A. Mosallanezhad, *et al.*, The effects of ginger supplementation on markers of inflammatory and oxidative stress: A systematic review and meta-analysis of clinical trials, *Phytother. Res.*, 2020, **34**, 1723–1733.
58. M. Morvaridzadeh, S. Fazelian, S. Agah, M. Khazdouz, M. Rahimlou and F. Agh, *et al.*, Effect of ginger (*Zingiber officinale*) on inflammatory

markers: A systematic review and meta-analysis of randomized controlled trials, *Cytokine*, 2020, **135**, e155224.
59. M. Mazidi, H.-K. Gao, P. Rezaie and G. A. Ferns, The effect of ginger supplementation on serum C-reactive protein, lipid profile and glycaemia: a systematic review and meta-analysis, *Food. Nutr. Res.*, 2016, **60**, e32613.
60. H. C. Wu, C. T. Horng, S. C. Tsai, Y. L. Lee, S. C. Hsu and Y. J. Tsai, *et al.*, Relaxant and vasoprotective effects of ginger extracts on porcine coronary arteries, *Int. J. Mol. Med.*, 2018, **41**, 2420–2428.
61. M. Alizadeh and S. Kheirouri, Curcumin reduces malondialdehyde and improves antioxidants in humans with diseased conditions: A comprehensive meta-analysis of randomized controlled trials, *Biomedicines*, 2019, **9**, e2019090423.
62. S. Qin, L. Huang, J. Gong, S. Shen, J. Huang and Y. Tang, *et al.*, Meta-analysis of randomized controlled trials of 4 weeks or longer suggest that curcumin may afford some protection against oxidative stress, *Nutr. Res.*, 2018, **60**, 1–12.
63. R. Tabrizi, S. Vakili, M. Akbari, N. Mirhosseini, K. B. Lankarani and M. Rahimi, *et al.*, The effects of curcumin-containing supplements on biomarkers of inflammation and oxidative stress: A systematic review and meta-analysis of randomized controlled trials, *Phytother. Res.*, 2019, **33**, 253–262.
64. A. Sahebkar, Are curcuminoids effective C-reactive protein-lowering agents in clinical practice? Evidence from a meta-analysis, *Phytother. Res.*, 2014, **28**, 633–642.
65. Y. Panahi, M. S. Hosseini, N. Khalili, E. Naimi, M. Majeed and A. Sahebkar, Antioxidant and anti-inflammatory effects of curcuminoid-piperine combination in subjects with metabolic syndrome: a randomized controlled trial and an updated meta-analysis, *Clin. Nutr.*, 2015, **34**, 1101–1108.
66. G. Derosa, P. Maffioli, L. E. Simental-Mendia, S. Bo and A. Sahebkar, Effect of curcumin on circulating interleukin-6 concentrations: a systematic review and meta-analysis of randomized controlled trials, *Pharmacol. Res.*, 2016, **111**, 394–404.
67. C. M. White, V. Pasupuleti, Y. M. Roman, Y. Li and A. V. Hernandez, Oral turmeric/curcumin effects on inflammatory markers in chronic inflammatory diseases: A systematic review and meta-analysis of randomized controlled trials, *Pharmacol. Res.*, 2019, **146**, e104280.
68. N. Ainehchi, A. Khaki, A. Farshbaf-Khalili, M. Hammadeh and E. Ouladsahebmadarek, The effectiveness of herbal mixture supplements with and without clomiphene citrate in comparison to clomiphene citrate on serum antioxidants and glycemic biomarkers in women with polycystic ovary syndrome willing to be pregnant: a randomized clinical trial, *Biomolecules*, 2019, **9**, e215.
69. E. Valizadeh, H. Akbari, F. Ghalichi and A. Ostadrahimi, The "Mohazell" herbal formula in combination with a calorie-restricted diet can improve

systemic inflammation in obesity: A randomized double-blind, clinical trial, *Prog. Nutr.*, 2019, **21**, 341–347.
70. A. A. Hemmati, E. Rajaee, G. Houshmand, M. Fakhroddin, M. Dargahi-MalAmir and S. Hesam, *et al.*, Study the effects of anti-inflammatory curcumex capsules containing three plants (ginger, curcumin and black pepper) in patients with active rheumatoid arthritis, *Int. J. Pharm. Clin. Res.*, 2016, 7, 389–392.
71. Z. Li, S. M. Henning, Y. Zhang, A. Zerlin, L. Li and K. Gao, *et al.*, Antioxidant-rich spice added to hamburger meat during cooking results in reduced meat, plasma, and urine malondialdehyde concentrations, *Am. J. Clin. Nutr.*, 2010, **91**, 1180–1184.
72. E. S. Oh, K. S. Petersen, P. M. Kris-Etherton and C. J. Rogers, Spices in a high-saturated-fat, high-carbohydrate meal reduce postprandial proinflammatory cytokine secretion in men with overweight or obesity: A 3-period, crossover, randomized controlled trial, *J. Nutr.*, 2020, **150**, 1600–1609.
73. A. C. Skulas-Ray, P. M. Kris-Etherton, D. L. Teeter, C.-Y. Oliver Chen, J. P. Vanden Heuvel and S. G. West, A high antioxidant spice blend attenuates postprandial insulin and triglyceride responses and increases some plasma measures of antioxidant activity in healthy, overweight men, *J. Nutr.*, 2011, **141**, 1451–1457.
74. A. Sahebkar, A. F. G. Cicero, L. E. Simental-Mendía, B. B. Aggarwal and S. C. Gupta, Curcumin downregulates human tumor necrosis factor-α levels: A systematic review and meta-analysis of randomized controlled trials, *Pharmacol. Res.*, 2016, **107**, 234–242.

CHAPTER 6

Phytoceutical-based Traditional Weight Loss Strategies for Management of Body Recomposition: Common Misconceptions and Novel Technological Breakthroughs

DEBASIS BAGCHI,*[a,b] BERNARD W. DOWNS,[c] MANASHI BAGCHI,[d] SAMUDRA P. BANIK,[e] STEVE KUSHNER,[f] BRUCE MORRISON,[g] SANJOY CHAKRABORTY[h] AND STEPHEN HESSON[i]

[a] College of Pharmacy and Health Sciences, Texas Southern University, Houston, TX, USA; [b] Department of Biology, College of Arts and Sciences, Adelphi University, Garden City, NY, USA; [c] Victory Nutrition International, Inc., Department of R&D, Bonita Springs, FL, USA; [d] Dr. Herbs LLC, Concord, CA, USA; [e] Department of Microbiology, Maulana Azad College, Kolkata, India; [f] ALM R&D, Oldsmar, FL, USA; [g] Morrison Family and Sports Medicine, Huntingdon Valley, PA, USA; [h] Department of Biological Sciences, New York City College of Technology/CUNY, Brooklyn, NY, USA; [i] PFHIX Inc., Neenah, WI, USA
*Email: debasisbagchi@gmail.com

Food Chemistry, Function and Analysis No. 39
Dietary Supplements with Antioxidant Activity: Understanding Mechanisms and Potential Health Benefits
Edited by Cesarettin Alasalvar, Fereidoon Shahidi and Chi-Tang Ho

Published by the Royal Society of Chemistry, www.rsc.org

6.1 Introduction

According to the Center for Disease Control and Prevention, the prevalence of obesity in the years 2017–2018 was 42.4%.[1,2] Obesity-related conditions include diverse cardiovascular disease (CVD) and dysfunctions, stroke, type 2 diabetes (T2D), and certain types of cancer that are some of the leading causes of preventable, premature death.[3] The estimated annual medical cost of obesity in the US was $147 billion in 2008. The medical cost for individuals with obesity was $1429 higher than that for individuals of normal weight.[4] As a result of the alarming rise in the global obesity epidemic, the terminologies "weight loss" and "weight management" have become common descriptors, even in the medical and scientific communities. However, the correct scientific terminology should more accurately refer to express changes in body composition and size, especially in regard to a reduction in fat mass, rather than weight reduction alone.[5] In fact, focusing on "weight" as an appropriate measurement criterion constitutes a contradiction and a misconception to the natural sequence of metabolic events in the body recomposition process. Overall, the term "weight loss" is an unreliable metric as it automatically places the primary focus on the heaviness of the body and does not provide a correct and accurate perspective for evaluating healthy changes in metabolism, body composition, or size.[6] For example, losing a small amount of fat while increasing muscle density, strength, and performance could result in a small increase in weight, as muscle is heavier than fat. It is especially important to highlight that fat is the lightest of pertinent macromolecules in the human body, lighter than water, muscle mass, and bone. Moreover, fat is usually "the last to go" in the body recomposition process. Therefore, creating short-term expectations in weight loss is erroneous and maybe even fraudulent.[6] This chapter provides a comprehensive account of the general misconceptions and ideas that exist among people to get rid of excess body fat and simultaneously discusses a remedial phytoceutical-based approach to gradually eradicate obesity and establish a healthy body recomposition.

6.2 Fat Metabolism and Shortcomings of Available Weight Management Techniques

Obesity can be classified as a metabolic disorder arising out of loss in body homeostasis and enhanced dependence on anaerobic pathways, which culminate in a decline of the basal/resting metabolic rate and the rate of lipid oxidation. Fat metabolism is regulated by diverse factors ranging from genetics, hormonal influences, microbiome composition, neurotransmitters, and lifestyle, as well as a disruption in energy metabolism and homeostasis.[7–12] Available "weight management" techniques for the most part have greatly failed to provide successful means to achieve sustainable healthy body composition and improve healthy fat loss. Commercialized "weight loss" programs, even medically supervised versions, do not consider the "bi-phasic" nature of genetically regulated set point "defence response" mechanisms that mandate preservation/conservation of body fat stores against famine and survival threats simulated, and even induced by aggressive weight loss tactics during phase 1. A logical expectation would be that, with all the commercial and medical expenditures on weight management, the incidence and severity of obesity would be diminishing.[8,11,12] However, the evidence is overwhelming – the natural products industry, medical institutions, and the totality of the weight loss industry with their products, programs, expertise, and financial resources have neither been able to even put a dent in the incidence of obesity nor have consistently reversed its pathological progression. It continues to rise. The irrefutable evidence, therefore, is that there is something significantly wrong with current perspectives, paradigms, approaches, and treatments of obesity.

6.3 Misconception

The first significant misconception overlooks a crucial biological factor. Biological systems maintain compensatory mechanisms of homeostatic counterbalancing as close to ideal as possible *via* a highly accurate and complex system of crosstalk, interconnectivity, and feedback signalling.[8–15] The medical/scientific and commercial marketers have not adhered to Einstein's axiom, "If you want to find a solution, you must first be able to (accurately) state the problem". The medical/scientific and commercial marketers' use of deprivation and physical exertion tactics to induce weight loss asserts that obesity is due to a person overeating and not exercising enough. While these are contributory and exacerbating influences, they are not the primary causes. Admittedly, nutritional deficits are an important contributing factor.[8–10]

Etiological patterns of obesity indicate that its escalating incidence is occurring because of the rebound retaliation to the early phases of weight loss using variations of deprivation, elimination, stimulation, and physical exertion-based tactics.[8–10] Rapid weight loss tactics are probably the greatest contributor to the obesity epidemic, greater even than conventional low quality food choices that could have started the obesogenic cascade. Moreover, food technologies have made many foodstuffs better tasting and more appealing, which often results in weight loss food deprivation-induced excessive self-medicating intake of not-so-healthy comfort foods.[11–16] This all-too-common behaviour contributes to rebound weight gain, dysbiosis, and related health consequences.

6.4 Human Physiology and Energy Homeostasis: Salient Features

The body is genetically pre-programmed to survive and has three forms of survival insurance (fat, sugar as glycogen, and water).[7,17–22] When the body is put into a survival mode, *e.g.*, "survival panic", it upregulates survival insurance storage of these substances and retards energy expenditure in favour of energy conservation.

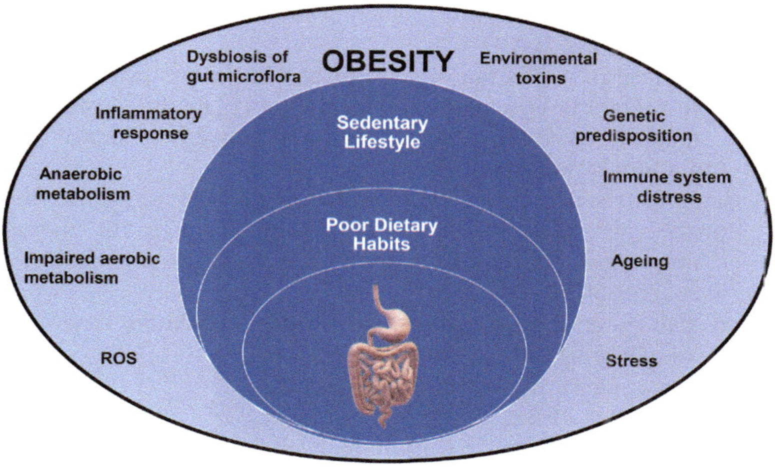

Figure 6.1 Obesity and its underlying factors. Abbreviation: ROS, reactive oxygen species.

Chronic disorders are anaerobic pathologies that induce survival panic.[8–10,17,18] There are numerous factors that contribute to obesity. Figure 6.1 shows a pictorial representation of the intricate factors involved in the pathogenesis of obesity. In addition to numbers 1 and 2, other important contributory factors include, but are not limited to:

1. Poor dietary habits – overfed and undernourished (*e.g.*, nutrient-deficient high calorie intake).[8,9]
2. Insufficient or inadequate physical activity.[8,9]
3. Excess exposure to environmental toxins (*e.g.*, polychlorinated biphenyls, gasoline exhaust and other aerial pollutants, chemical outgassing of synthetic materials, and many other factors),[9,10] agribusiness treatments (*e.g.*, chemical fertilizers, growth enhancers, pesticides, herbicides, fungicides, genetically modified organisms, irradiation, gassing, and colouring, among others),[9–12] livestock treatments (*e.g.*, hormonal growth enhancers, antibiotics, and genetic engineering, to name a few),[10–12] and food overprocessing/diverse use of food additives, flavour enhancers, preservatives, and colouring agents.[12]
4. Digestive malfunction and "dysbiosis": a manifestation of this is impaired disintegration and dissolution of food stuffs, especially proteins, contributing to leaky gut syndrome and inflammatory bowel disorders.[11,12] This can be a major cause of obesity and metabolic disorders. A breakdown of gut wall/barrier integrity, common in society, can allow toxins to permeate into the bloodstream causing a wide range of problems (*e.g.*, inducing survival panic) with a major problem being an inability to "let go" of fat reserves. Gut dysbiosis also has an adverse effect on neurotransmission integrity (*via* gut–brain interconnectivity).

5. Overburdening imposition on anaerobic metabolism (cannot effectively use O_2 and H_2O).
6. An increase in inflammatory conditions: inflammation plays a major role in metabolic syndrome (MetS) including insulin resistance.[12–14]
   - ✓ Unhealthy levels of inflammation can cause "stubborn" fat accumulation around the belly.[8,9,15]
   - ✓ This fat tissue acts as an organ unto itself sending out chemical messengers that inhibit healthy fat metabolism and increase fat storage (and can retard protein conversion into muscle) for energy conservation.[8,9,22–24]
7. Increase in reactive oxygen species (ROS), oxidative stress, and excessive oxidative damage to tissues.[25–27]
8. Immune system distress and exhaustion.[19]
9. Stress (further inducing "survival panic").[17,20]
10. Survival-induced genetic responsivity leading to energy conservation, increased stress intolerance, impaired cognition, and increased sadness.[11,12]
11. Aging (*e.g.*, accelerated premature aging consequences caused by diverse factors discussed earlier).[2,3]
12. Impaired aerobic metabolism, inducing anaerobic glycolysis and metabolism, can lead to a deficiency of adenosine-5-monophosphate-activated protein kinase (AMPK) that can cause abnormal fat deposition, especially in the gut. AMPK is an enzyme found in every cell in the body and declines with advancing age. One of the key actions of AMPK is to trigger burning of stored fat for energy. Anaerobic glycolysis and metabolism, inducing energy conservation, can reduce circulating AMPK, which is a major contributor to abdominal fat formation and storage.[28,29]

Moreover, out of previous 12 extensive factors, the "weight loss" industry primarily addresses only overeating/unhealthy food intake and inadequate exercise. Selected nutraceutical researchers are starting to state that they address genetic factors but have truly little supporting evidence of any success. While there is significant scientific evidence regarding the role of epigenetics and the use of nutrigenomics (*e.g.*, KB220Z, leptin/ghrelin optimization, *etc.*) to influence fat metabolomics, the scientific support for allegations by weight loss researchers is primarily rhetorical hyperbole.[6]

There are two therapeutic perspectives (involving numerous strategies) for improving fat metabolism:

1. Avoiding the bad factors, such as sugar intake, gluten, and fast food consumption.
2. Indulging in the good factors, such as healthy diet, lifestyle (including modest exercise), and evidence-based supplementation, to improve cellular metabolism (*e.g.*, optimal energy production, management, and cellular waste removal).

Incorporating these two perspectives enables a much more successful strategy to reduce excess fat stores, improve musculature, and achieve a desirable body composition and optimal health than either one alone.

6.5 Systematic Approach

A more comprehensive and accurate approach is, therefore, to fundamentally restore healthy and efficient aerobic metabolism, improving oxygen utilization, and management that enables competent cellular waste removal. Adversarial factors to this objective involve nutrient deprivation (the most common factor employed by weight loss products and programs), the imposition of various types of stimulants inducing, *via* feedback, regulatory energy-conserving retaliation in the form of even greater survival insurance storage, and triggering the "yo–yo" rebound weight regain. Moreover, conventional weight loss tactics seem to ignore the biological, genetic, and metabolic consequences that occur with aging. Diminishing hormonal potencies are crucially important considerations that need to be addressed in the obesity epidemic. Cellular oxygen deprivation-induced anaerobic pathologies exacerbate and amplify the effects of diminished hormone functionality and contribute to an increasingly sluggish metabolism with increased fat, glycogen, and water storage. This also increases the detrimental strain on the cardiovascular system, kidneys, gastrointestinal (GI) tract, and microbiome, to name a few. The subsequent sections of this chapter elaborate the specific causes of obesity as well as novel therapeutic interventions to minimize uncontrolled weight gain.

6.6 Factors Adversely Affecting Fat Metabolism

Intrinsic misconceptions, malpractices, and wrong philosophies of leading a healthy lifestyle, all have together contributed and aggravated the scenario of rising incidences of obesity and failure to keep up to a healthy aerobic lifestyle. The following section delves into the specific causes arising out of both environmental influence as well as dysfunction of the body physiology which are accountable for inflicting significant damage towards management of body recomposition.

6.6.1 Adverse Influence of Environmental Toxins on Fat Metabolism

Diverse environmental risk factors including occupational exposures to polluted air, soil, water, gasoline exhaust, hazardous heavy metal wastes/pesticides/chemical wastes, and even mental stress significantly contribute to survival panic and the development of numerous chronic non-communicable diseases, as well as immensely affecting fat metabolism. These environmental stressors are greatly involved in increased risks for numerous metabolic, neuronal dysfunctions and neurotoxicity, attention deficit disorders, autism spectrum disorders, dementia, CVD, and dysfunctions including mitochondrial dysfunction, disruption in redox balance and inflammatory response, and depression, digestive disorders, anxiety disorders, diabetes, stroke, arrythmia, atrial fibrillation, myocardial infarction, and hypertension.[9] In fact, diverse environmental toxins and pollutants are potent obesogens, carcinogens, and autogens, which contribute to and even induce diverse autoimmune, inflammatory responses, and cardiovascular events and mortality. Several of these environmental toxins and pollutants are well documented to induce weight loss at high exposure levels; however, interestingly chronic exposure to lower concentrations of these toxins and pollutants exert a significant increase in fat storage-promoting actions. Interestingly, the scientific community commercially exploited these properties to synthesize growth hormones that fatten livestock.[9,10]

6.6.2 Adverse Influence of Digestive Malfunction and Gut Dysbiosis on Fat Metabolism

The human GI tract, the largest source of non-self-antigens, contains numerous microbes of more than 1000 different species, which immensely contribute to 3.3 million unique microbial genes.[11] In fact, the GI tract functions as the major immunological organ to strategize and preserve tolerance to commensal and dietary antigens, and, in addition, protect the body when encountering pathogenic stimuli and insults. Once the balance is disrupted, an array of inflammatory sequela are initiated, leading to host cell damage and/or autoimmunity.[11,12] The literature demonstrates that the compositional aspects of the intestinal microbiota can modulate the susceptibility to chronic diseases/dysfunctions of the intestinal tract including ulcerative colitis, Crohn's disease, celiac disease, and irritable bowel syndrome. Moreover, several systemic diseases including obesity, type 1 diabetes (T1D) and T2D, and various cancers, among others, can result from a disruption in intestinal microbiota. Unhealthy diet is causally linked with the increased incidence of several inflammatory diseases and distress.[12]

6.6.3 Adverse Influence of Anaerobic Pathologies on Fat Metabolism

An array of chronic diseases take place during anaerobic pathologies, which are characterized by an increased inability to effectively utilize oxygen, resulting in anaerobic metabolism and lactate accumulation, among other sequelae.[13] An impairment of oxidative pathways during lactate production contributes to a net gain of H^+ with increasing cellular acid burden, thereby decreasing the blood pH. Therefore, this progressive inability of the human body to effectively use cellular oxygen, inducing progressive acidemia, a metabolic shift toward cellular anaerobic glycolysis, and a compensatory expenditure of alkalinizing histidine molecules from the heme protein of deconjugated hemoglobin, which releases iron. These events cause an increase in the anaerobic/acidic cellular environment, especially in the blood, and can manifest in several symptoms and pathological manifestations including altered fat metabolism, vaso-occlusive incidences, cardiovascular dysfunctions, and an array of inflammatory diseases.[14] However, use of properly designed nutraceuticals and functional foods may have the ability to reverse these adverse conditions.

6.6.4 Adverse Effects of Inflammation on Fat Metabolism

It is well demonstrated that chronic inflammation is associated with anaerobic sequela and consequential insulin resistance, which in turn is linked with nutrient metabolism in the peripheral tissues. Several inflammation and macrophage-associated genes are remarkably upregulated in white adipose tissue (WAT) in mouse models of genetic and high-fat diet-induced obesity. Rapid upregulation occurred in the WAT of mice following high-fat diet-induced obesity. Treatment with an insulin-sensitizing drug, *via* histological analysis, was shown to down-regulate macrophage derived genes.[15] Xu *et al.*[15] concluded that macrophages in WAT play a vital mechanistic role in morbid obesity, while macrophage-associated inflammatory responses contribute to the pathogenesis of obesity-induced insulin resistance. In other words, obesity-associated insulin resistance plays a major role in chronic inflammatory diseases and dysfunctions initiated in the adipose tissue.

6.6.5 Adverse Effects of ROS-induced Oxidative Damage on Fat Metabolism

A potential link exists between adipose tissue hypoxia, ROS, oxidative stress, mitochondrial dysfunction, and low-grade inflammation induced by adipocyte hypertrophy, macrophage infiltration, and production of inflammatory mediators.[15,16] Particularly, metabolic incompetence in WAT in overweight or obese people immensely contributes to the pathogenesis of insulin resistance and the development of T2D. It is important to interject that insulin resistant impairment of the oxidative respiratory metabolism of glucose into energy causes a shift to anaerobic glycolysis and an increase in pathogenic anaerobic sequela. Therefore, compromised insulin homeostasis in adipocytes is intricately associated with elevated lipolysis and elevated free fatty acids, which ultimately leads to ectopic fat deposition in hepatic tissues, skeletal muscles, as well as other chronic pathologies. The literature demonstrates that continued adipose tissue hypoxia (*e.g.*, anaerobic state) induces dysfunctions of adipocytes and elevated oxidative stress, and inhibits the production of adiponectin, which is involved in regulating glucose levels and the breakdown of fatty acids. However, the time-dependent effects of hypoxia on WAT and the generation of oxidative stress in adipocytes are very devastating.[17]

6.6.6 Adverse Effects of Immune Distress on Fat Metabolism

It has been well demonstrated that inflammatory response is activated early in adipose expansion as well as during chronic obesity, permanently compromising the immune system to a proinflammatory phenotype.[18] This also activates an impact on innate and adaptive immunity. Metabolic incompetence is escorted by chronic low-grade inflammation and has fundamentally modified the viewpoints of researchers on the underlying causes and progression of obesity and MetS.[19] Furthermore, inflammatory responses play a major role in fibrosis and angiogenesis. Inflammation also has an impact on altering lipid metabolism, immune cell activation, and immune distress. The obesity induced disbalance in immune cell plasticity, therefore, leads to gross alterations in the metabolism of fatty acids and cholesterol.[18,19]

6.6.7 Adverse Effects of Stress on Fat Metabolism

A universal observation is that when confronted with an incredibly stressful situation, an individual tends to sigh heavily. This is a natural response to improve oxygen intake and cellular oxygenation, as stress exerts significant demands on cells. Cortisol, the stress hormone, stimulates fat and carbohydrate metabolism and insulin sensitization leading to an increase in and maintenance of blood sugar levels. This causes an increase in appetite and cravings for sweet, high-fat, and salty foods. However, stress alleviation using various and appropriate nutraceutical supplements can rapidly improve metabolic competence and, thereby, reduce sugar and food cravings.[20]

6.6.8 Reversing Adversarial Gene Expression of Neurotransmitters and Hormones Involved in Fat Metabolism

The underlying cause of dysfunction in body metabolism and homeostasis has been linked primarily to the altered secretion and/or expression patterns of neurotransmitters and hormones. Increased appetite of obese individuals has been almost always attributable to abnormal hormone levels of the hypothalamic–pituitary axes. Alongside, neurotransmitters also play a critical role in regulating energy balance and controlling appetite. The mode of action of many anti-obesity drugs involves altered metabolism of neurotransmitters. The following are key molecules implicated in regulation of appetite and ensuring effective maintenance of body homeostasis.

6.6.8.1 Leptin

Leptin deficiency or resistance can result in and/or exacerbate profound obesity, T1D, T2D, infertility, and Rabson–Mendenhall syndrome in humans, among others.[21–23] Adipose tissues, cells, and enterocytes in the small intestine are greatly involved in synthesizing the hormone "leptin". Leptin, a 167 amino acid adipokine (or adipocytokine), plays a critical role in energy homeostasis, metabolism, endocrine function, and body composition. In fact, leptin modulates hunger and appetite, and, if appropriately regulated, it would downregulate fat storage in adipocytes.[21,22] It plays important physiological roles in the development of leptin-based therapeutics for other prevalent disorders including obesity-associated non-alcoholic fatty liver disease, depression, and dementia. Leptin orchestrates complex biological effects through its receptors, expressed both centrally and peripherally. An understanding of diverse biological functions of leptin has expanded from anti-obesity to broad effects on reproduction, hematopoiesis, angiogenesis, blood pressure, bone mass, lymphoid organ homeostasis, and T lymphocyte systems. Leptin therapy is also adopted in conditions where leptin levels are high.[22,23]

6.6.8.2 Ghrelin

Ghrelin has been described as the "hunger hormone" as ghrelin release increases appetite. Until recently, this known function remained the exclusive focus of therapeutic perspectives with the objective of reducing the appetite-increasing effects of ghrelin. However, additional mechanistic actions of ghrelin include as a growth hormone releasing catalyst, a potent antioxidant, protecting the central nervous system (CNS), the functional integrity of the GI tract, hepatic and pancreatic tissues, and the brain, inhibition of apoptosis, and protecting biological tissues.[24] The question that must be asked is how does increasing appetite to pathological obesogenic levels reconcile with what seem to be a highly beneficial antioxidant and protective function? A plausible answer, requiring further research, is that ghrelin is released in response to a state of genetically-mediated "survival panic". When a risk to survival is genetically determined, for example, to reverse nutritional deficiencies, a state of "survival panic" triggers an increase in appetite and the various tissue protective measures depicted above, to better ensure survival. Anaerobic/hypoxic metabolic events can induce an obesogenic momentum that can abolish satiety signalling. Conventional corrective strategies are targeted at reducing/blocking ghrelin release and upregulating leptin release.[25] A more foundational and first step strategy would be to restore aerobic energy metabolism to terminate "survival panic" signalling.

6.6.8.3 Influence of Reproductive Hormones on Obesity

In the neuronal system, energy homeostasis is regulated by the hypothalamus. To be more specific, two of the components, namely, (a) the ventrolateral region of the ventral medial nucleus (VMN) and (b) the arcuate, respond to hormones as well as other signals to regulate energy homeostasis, particularly energy intake and energy expenditure.[26] A disruption or injury in the hypothalamus, which includes damage to both VMN and the arcuate and also induces survival panic, causes animals (and humans) to eat more, which subsequently reduces energy expenditure, increases energy conservation (genetically perceived "survival panic"), and increases fat storage leading to obesity. The hypothalamus also contains two oestrogen receptors, *viz.* oestrogen receptor (ER)α and ERβ. The oestrogens in the VMN and arcuate are involved in regulating food intake and body weight. Activation of ERα reduces food intake and, in turn, increases fat storage. It is important to emphasize that oestrogen receptors are also found in several peripheral tissues including adipose tissue.[26,27] Oestrogens in the adipose tissue exert anti-inflammatory activities in the periphery and in the CNS, which protects women from diverse inflammatory diseases. A thorough understanding of the regulatory mechanisms on how oestrogens modulate body weight and anti-inflammatory efficacy will help us to unveil more effective therapeutic modalities.[26–28]

6.6.8.4 Thyroid Hormones and Neurotransmitters

Thyroid glands produce four hormones as given below:

(i) triiodothyronine (T3), which plays vital roles in metabolic homeostasis, cardiovascular function, digestive function, muscle control, neuronal development and function, and the maintenance of bones;
(ii) tetraiodothyronine (thyroxine or T4), the prime hormone secreted by the thyroid gland into the bloodstream, which has diverse functions in digestion, cardiac and muscle functions, neuronal development, and maintenance of bones;
(iii) 3,5-diiodothyronine (3,5-T2), an active thyroid hormone that stimulates the thyroid hormone receptor (TR)-β, which upregulates energy expenditure;[29,30] 3,5-T2 promotes thyromimetic effects in myocardial tissue and pituitary which results in the suppression of thyroid stimulating hormone (TSH) release;[31,32] it is an allosteric regulator/modulator of the cytochrome *c* oxidase, complex IV of the electron transport chain; it increases its activity by preventing the interaction of adenosine triphosphate (ATP) as an allosteric inhibitor;[33] and
(iv) calcitonin, a hormone secreted by the thyroid, which is involved in lowering blood calcium.[28,35]

T3 and T4 are the best-known hormones pharmaceutically used to influence thyroid function. Using nutraceuticals to promote T2 production and function may offer new, novel, and safe strategies to modulate cellular energy production more effectively. T4 is a 32-amino acid peptide hormone known to oppose the action of parathyroid hormone, which stimulates the release of calcium from bones for transfer into the bloodstream and reduces the loss of calcium in the urine to maintain an appropriate balance of calcium in the bloodstream and in tissues. Immunohistochemical detection of brain T3 showed that the hormone was concentrated in both noradrenergic and adrenergic centers.[34,35] During neuronal development, neurotransmitters function as growth regulators. In the adult brain (and elsewhere, *e.g.*, the GI tract), neurotransmitters play their roles *via* synthesis, transport, reception, and degradation for disposal or reuptake, as well as maintain neuronal plasticity and other growth-related processes.

6.6.8.5 Neurotransmitters: Serotonin and Dopamine

Dysfunction in feeding signals by neuronal cells plays a vital role in obesity. Neurotransmitters, also known as chemical messengers, transmit a signal/message from a neuron across the synapse to diverse target cells including different neurons, muscle cells, or gland cells.[35]

Serotonin is involved in the transmission of nerve impulses. It binds to the receptor proteins within the postsynaptic cell, inducing a change in the electrical state of the cells including modulation of the electrical state, which can either excite the neuronal cells transmitting the chemical message or inhibit them. Serotonergic neurons have axons that are abundant in different neuronal tissues and affect diverse behavioural aspects. Excess serotonin molecules are taken back by the presynaptic cell and reprocessed.[35]

A strong link exists between the serotonin and dopamine systems. Serotonin is the first step in the "Brain Reward Cascade" [*e.g.*, serotonin, enkephalin, endorphins, gamma aminobutyric acid (GABA), cannabinoids, and dopamine] that culminates in modulation of dopamine synthesis.[36] Dopamine is released by neurons to transmit an electrical signal from one neuron to the next to facilitate interactive communications to and from the CNS. It is intricately involved in regulating feelings of reward satisfaction that includes modulating food intake. A dopamine deficit was found in obese individuals characterized by increased food consumption behaviours.[37] Dopamine modulates food cravings, motivation, and reward circuits, and thus dopamine deficiency in obese individuals may perpetuate pathological eating to compensate for decreased activation of these circuits.[37]

6.6.9 Reversing the Adverse Effects of Premature Aging on Fat Metabolism

Aside from the natural decline of genetic competence associated with reproduction and normal aging, an increase in anaerobic glycolysis, cellular lactic acid production, and ROS are hallmarks of impaired cellular metabolism and characteristic of chronic degenerative diseases that also accelerate premature aging. These anaerobic/hypoxic pathogenic events raise the demand for and expenditure of alkalinizing buffers, *e.g.*, bicarbonate, various ions, and histidine from the heme protein in hemoglobin, among others.

Restoring aerobic glycolysis is an important anti-aging objective to reduce the anaerobic cellular requirement for significantly greater amounts of sugar (as glucose) that result in the concomitant increase in anaerobic-induced lactic acid production. Additionally, improving insulin sensitivity and glucose utilization contributes to a more efficient energy dynamic and reduces insulin resistance and the demand for significantly greater amounts of sugar. A healthy diet, moderate exercise, and sufficient restful sleep are foundational strategies to retard premature aging.

6.6.10 Reversing the Adverse Effects of AMPK Deficiency on Fat Metabolism

AMPK is an enzyme found in every cell in the body that declines with age. It regulates whole-body and cellular energy balance in response to energy supply from food and beverage intake as well as cellular energy demands. AMPK enhances fat oxidation by stimulating fat burning, especially of stored fat for energy production, and enhances glucose uptake while inhibiting fat and cholesterol synthesis.[38–40] It is well established that ATP serves as cellular energy, while AMPK senses and responds to low ATP levels. When AMPK is activated, it turns up processes in cellular mitochondria that enhance conversion of ingested food molecules into ATP, which helps to restore and boost energy levels. A deficit in AMPK increases the conversion of sugar into fat with a concomitant increase in fat storage.[41,42] Therefore, rejuvenating AMPK synthesis and metabolism is an important objective.

6.7 Phytoceutical Intervention for Effective Management of Body Weight and Body Recomposition

Phytochemicals in combination with small molecule therapeutics and/or exercise have been effective in efficient weight management.[43] It is believed from the evidence regarding the mode of action of phytoceuticals obtained thus far that most of them act by repressing the action of catechol-*O*-methyltransferase, the enzyme that mediates degradation of nor-epinephrine. Nor-epinephrine is responsible for boosting lipolysis and, therefore, inhibition of catechol-*O*-methyltransferase would promote enhanced fatty acid oxidation. However, an optimal formulation of the right constituents together with enhanced bioavailability to bolster body recomposition had by and large eluded scientists. Although there have been several reports of plant derived individual components as bioactives or dietary supplements in effective management of obesity (Table 6.1),[44–55] almost all of these constituents have failed to make a significant impact as authentic and reliable therapies for ensuring holistic body recomposition and transformation from an obese individual to a healthy person.

Table 6.1 Popular bioactive components and dietary supplements for effective body weight management obtained from plant sources.

Bioactive components/dietary supplements	Source	Function	Ref.
Chitosan	Mushrooms	Chitosan by virtue of its cationic nature can sequester negatively charged lipids, thus preventing their gastrointestinal uptake.	44
Raspberry (*Rubus idaeus*) ketone	Fruits of red raspberry	RK mimics the structures of capsaicin and synephrine, compounds with anti-obese properties and alters lipid metabolism.	45
Green coffee (*Coffea canephora robusta*)	Green coffee beans	Inhibition of fat absorption and activation of fat metabolism in the liver.	46
Spirulina (*Arthrospira platensis*)	A type of cyanobacteria found in both marine and fresh water environments	Spirulina turns on the AMP-activated protein kinase signalling pathway which results in repression of lipid synthesizing genes, also reduces triacylglycerol levels, and blocks the synthesis of fatty acids. Additionally, it also alters gut microbiota for decreasing body lipids.	47
Korean pine nut (*Pinus koraiensis*) oil	Korean pine nuts	The free fatty acids from these nuts bolster the release of satiety hormones such as cholecystokinin, which delays gastric emptying and subsequently decreases appetite.	48
Palm (*Arecaceae* sp.) oil and oat (*Avena sativa*) oil	Palm and oat seeds	Similar to Korean pine nuts.	49
Malabar tamarind (*Garcinia cambogia*) extract	Hydroxycitric acid extract of pericarp of dried fruit	It induces weight loss by suppressing *de novo* fatty acid synthesis and promotes lipid oxidation.	50
Tea (*Camellia sinensis*) catechins	Leaves of tea plants	They block formation of small-intestine micelle α-glucosidase activity, which would lead to a decrease in carbohydrate absorption.	51
Bitter orange (*Citrus aurantium*) extract	Bitter orange plant	The alkaloids present in this plant, such as *p*-octopamine and synephrine, promote β-adrenergic response.	52
Khat (*Catha edulis*)	Leaves of khat plant	Cathinone and cathine (active ingredients) resemble amphetamine-like compounds that exert their effects on the hypothalamus, thus reducing appetite.	53

Table 6.1 (*Continued*)

Bioactive components/dietary supplements	Source	Function	Ref.
		Cathinone also increases sympathomimetic activity leading to a delay in gastric emptying.	
Hoodia plant (*Hoodia parviflora*)	Leaves of hoodia plant (originally found in Kalahari Desert, Africa)	It causes activation of β adrenergic receptors to enhance lipid oxidation.	54
Guar gum (*Cyamopsis tetragonoloba*)	Seeds of the drought tolerant plant (guar)	It reduces blood cholesterol by promoting excretion of bile acids in faeces. This leads to a fall in enterohepatic bile acid so that the levels are replenished by production of bile acids from cholesterol and thus hepatic free cholesterol concentration is reduced.	55

Abbreviations: AMP, adenosine monophosphate; RK, raspberry ketone.

Prodosome is a novel encapsulation technology designed to promote efficient absorption and maximize bioavailability. The prodosomed antioxidant is a proprietary formulation composed of alpha-glyceryl phosphoryl choline (GPC) and five botanical extracts rich in a diverse range of phytoceutical components, including phytosaccharides, theobromine (other dimethylxanthines, *viz.* paraxanthine and theophylline), amino acids, gallic acid, guanidine, isobutyric acid, B-vitamins, vitamin C, nitric oxide, chlorogenic acid, ellagic acid, triterpenes and pentacyclic triterpenoid acids, and saponins (such as gypenosides), as well as ursolic acid (0.7–1%, a pentacyclic triterpene), and amyrin esters (up to 0.5%), betulinic acid, bioactive phenolic compounds including anthocyanins (cyanidin-3-*O*-glucoside and delphinidin-3-*O*-glucoside), flavonols (myricetin and quercetin), ellagic acid, ellagitannins, and proanthocyanidins, as well as carotenoids (lutein, carotene, violaxanthin, and luteoxanthin). All of these components can substantially rejuvenate metabolic turnover in the body and alleviate obesity and related ailments. The following are the main constituents in TRCAP21 formulation, which play their roles in boosting positive body metabolism. A schematic representation of their mode of action is presented in Figure 6.2.

6.7.1 D-Ribose Nicotinamide (DRN) Complex

This is a proprietary combination of the DRN complex and has been demonstrated to be superior to the nicotinamide riboside (NR) complex that is known to clinically improve the salvage and availability of nicotinamide

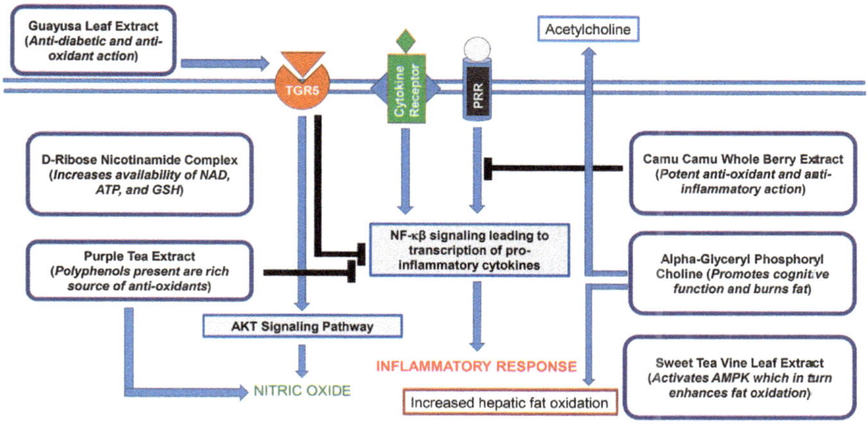

Figure 6.2 Schematic representation of the gross mechanism of action of different phytoceuticals on reducing obesity and establishing healthy body homeostasis by curbing release of pro-inflammatory cytokines and inducing synthesis of nitric oxide. Abbreviations: AKT, AK strain transforming; AMPK, adenosine monophosphate activated protein kinase; ATP, adenosine triphosphate; GSH, glutathione; NAD, nicotinamide adenine dinucleotide; NF-κB, nuclear factor-kappa B; PRR, pattern recognition receptor; TGR5, Takeda G protein coupled receptor 5.

adenine dinucleotide (NAD).[56] In a pilot study, it was reported that human blood NAD+ can rise as much as 2.7-fold with a single oral dose of NR, and that oral NR elevates mouse hepatic NAD+ with distinct and superior beneficial bioactivity to nicotinic acid and nicotinamide alone. It was further shown that single doses of 100, 300, and 1000 mg of NR produce dose-dependent increases in the blood NAD+ metabolome in the first clinical trial of NR pharmacokinetics in humans.[57] Nicotinic acid adenine dinucleotide (NAAD) was not known to be involved in the bioconversion of NR to NAD+. However, it is formed from NR and it was discovered that the rise in NAAD is a highly sensitive biomarker of effective NAD+ repletion. The highest concentrations of NAD, a coenzyme in redox reactions and a donor of adenosine diphosphate (ADP)-ribose moieties in ADP-ribosylation reactions, are in the mitochondria and are important to optimize cellular energy production.

The DRN complex provides effective benefits in all tissues. Pharmacodynamics of the DRN complex reveal broad advantages over nicotinamide riboside alone because it increases the availability of NAD, ATP, and glutathione (GSH) simultaneously, including improvements in muscle and brain performance. Following DRN complex administration (0, 100, 300, 900, and 2700 mg) in animals, NAD levels increased steadily at all dosages, in a dose dependent manner.[58] It was observed that NAD levels reach their plateau after 4 days of supplementation. A positive dose–response relationship exists between the contents of the DRN complex and NAD metabolites in the liver, muscle, and brain tissues.

A single-center, randomized, double-blind, comparator controlled, crossover study was conducted in 18 healthy males and females (age = 35–65 years) to investigate the safety and efficacy of the DRN complex.[57] Significant increases were observed in both NAD+ and nicotinamide adenine dinucleotide phosphate (NADP+) by 15% and ATP and ADP levels by 7.3%. The GSH level increased by 11% and redox balance was maintained. The DRN complex increased insulin response and glucose tolerance significantly with a steady reduction observed in salivary cortisol. Furthermore, it improved the quality of life significantly, especially reducing fatigue, while both concentration and motivation levels increased significantly.

6.7.2 Sweet Tea Vine (*Gynostemma pentaphyllum*) Leaf Extract

Sweet tea vine is enriched in potent antioxidants and has been demonstrated to activate AMPK which regulates whole-body and cellular energy balance in response to energy supply from food and beverage intake and cellular energy demands. It enhances fat oxidation by stimulating fat burning for energy production and glucose uptake while inhibiting fat and cholesterol synthesis.[38,40] It is well established that ATP serves as cellular energy, while AMPK senses and responds to low ATP levels. When AMPK is activated, it turns up processes in cellular mitochondria that enhance conversion of ingested food molecules into ATP, which helps to restore and boost energy levels.

Sweet tea vine has been reported to stimulate catabolic pathways, *e.g.*, the breakdown of large complex molecules into smaller simpler ones, such as through glycolysis [*e.g.*, the breakdown (lysis) and uptake of sugars as glucose (glycol) and fatty acid oxidation (*e.g.*, the burning of fat)] that produces ATP and releases energy.[39–42]

Sweet tea vine also reduces cytokine production and activates nuclear factor-kappa B (NF-κB) and signal transducer and activator of transcription 3 (STAT3) signalling in lipopolysaccharide-induced macrophages, without affecting their viability.[59] Cytokines are important in cell signalling which induce the sequence of inflammatory events in the immune process. The ability of sweet tea vine to reduce cytokine production is a unique property in reducing the need for inflammation in the tissue repair and restoration process. Total extracts or saponins from sweet tea vine have also been shown to exert a wide range of beneficial effects such as maintaining healthy cholesterol and blood glucose levels, strengthening immunity, boosting the strength of DNA, and inhibiting cell mutations.[60–63]

Of great interest is the finding that sweet tea vine was shown to increase hemoglobin, liver glycogen, and muscle glycogen concentrations, and significantly extend the amount of time rats could swim until exhaustion, as well as decrease blood lactic acid levels and blood urea nitrogen concentrations. This indicates that sweet tea vine increases oxygen utilization and aerobic metabolism providing significant anti-fatigue effects demonstrated by extended muscle tissue output.

A randomized, double-blind, placebo-controlled, clinical study was conducted to investigate the effect of sweet tea vine (450 mg extract per day) on body weight, fat loss, and metabolic markers in 80 subjects [body mass index (BMI) ≥ 25 kg m^{-2} and waist–hip ratio (WHR) ≥ 0.90 for males or WHR ≥ 0.85 for females] over a period of 12 weeks.[64] Anthropometric parameters, blood lipid profiles, abdominal fat distribution, and adverse events were monitored. Total abdominal fat area, body weight, body fat mass, percent body fat, and BMI were significantly reduced in the treatment group as compared to the placebo group. No adverse events were reported.

6.7.3 Camu Camu (*Myrciaria dubia*) Whole-berry Extract

Camu camu contains a significant amount of vitamin C (3 g of vitamin C per 100 g), which exhibits potent antioxidant efficacy. In fact, this cherry-sized fruit contains more vitamin C than 70 oranges. Research studies have demonstrated that it is a potent antioxidant, protects DNA integrity, and boosts immune competence by reducing interleukin (IL)-6 and IL-8, exerts anti-inflammatory effects by reducing cytokine production and C-reactive proteins (CRP), accelerates tissue reconstruction and repair, and protects against allergens and mutagens.[65,66]

In mice, the orally administered extract significantly suppressed water retention (as the formation of edema) and the release of nitric oxide from mouse macrophage (immune) cells *in vitro*.[67] Camu camu is a rich source of betulinic acid, which is well known as a potent antioxidant and anti-inflammatory agent and plays an important role in promoting immune competence. Another randomized controlled trial demonstrated that compared to the same dose of vitamin C as that contained in camu camu, in 20 male smokers, it had a significantly greater ability to decrease free radical damage and inflammatory markers, and improve immune responsivity than straight vitamin C.[68] It is reported that some people have used camu camu to "increase energy" and maintain healthy gums, eyes, and skin. Importantly, these mechanistic benefits translate to improving metabolic competence, a crucially important factor in restoring and maintaining healthy body composition.

In addition to betulinic acid and vitamin C, the fruit is rich in cyanidin 3-glucoside, delphinidin 3-glucoside, ellagic acid, kaempferol, myricetin, quercetin, quercitrin, and rutin, among others.[65] It is rich in nutrients such as vitamins (in addition to vitamin C, beta-carotene, riboflavin, thiamine, and niacin), minerals (sodium, potassium, calcium, iron, phosphorus, zinc, magnesium, manganese, and copper), fatty acids, and protein. Moreover, camu camu exhibited potent antioxidant efficacy and broad-spectrum safety,[69,70] all of which are important to maintain metabolic homeostasis.

6.7.4 Guayusa (*Ilex guayusa* Loes.) Leaf Extract

Guayusa is an evergreen dioecious medicinal plant enriched in a significant amount of phenolic antioxidants. This plant is native to the Amazon

Rainforest between 200 and 2000 meters of elevation.[71–73] It contains a small amount of caffeine (1.73–3.48%), theobromine, L-theanine, gallic acid, guanidine, isobutyric acid (contributing to microbiome homeostasis), nicotinic acid, ascorbic acid, riboflavin, choline, pyridoxine, triterpenes as well as ursolic acid (0.7–1%; a pentacyclic triterpene acid), chlorogenic acid, and amyrin esters (up to 0.5%), among others. Guayusa-derived ursolic acid plays a vital role in the activation of Takeda G protein-coupled receptor 5 (TGR5) ($EC_{50} = 4.7$ μM), a nuclear receptor involved in the prevention and management of diabetes and MetS *via* modulation of energy expenditure and insulin sensitization.[74] García-Ruiz *et al.*[72] identified, characterized, and quantified a total of 14 phenolic constituents, chlorogenic acid, quercetin-3-*O*-hexose, flavonols, and carotenoids, including considerable amounts of lutein in guayusa. Both oxygen radical absorbance capacity (ORAC) and 2,2-diphenyl-1-picrylhydrazyl (DPPH) assays revealed high antioxidant capacity. However, fermentation lowered these bioactive constituents and exhibited the lowest antioxidant capacity. Guayusa exhibited a broad safety spectrum with no genotoxicity, making it an important contributory factor to reducing physiological distress and achieving healthy body composition.

A 90 day sub-chronic toxicity study at 1200, 2500, and 5000 mg per kg body weight per day of guayusa concentrate, and a caffeine-positive control (150 mg per kg body weight per day) induced a lowering of body weights, food efficiency, triacylglycerols (TAG), fat pad weights, and an elevation in blood chemistry, *viz.*, serum aspartate aminotransferase, serum alanine aminotransferase, and cholesterol, as well as adaptive salivary gland hypertrophy.[75] Guayusa also obtained Food and Drug Administration (FDA)-notified generally recognized as safe (GRAS) status in 2019.[76] Dried leaves of guayusa also received a novel food notification for authorization of a traditional food from a third country under regulation of European Union (2015/2283).[77]

Schustera and Mitchell[78] critically summarized the psychopharmacology of the guayusa constituents including methylxanthines, polyphenols, and L-theanine, which exhibited mood elevation and enhanced cognition. It has been reported by Adhikary and Mandal[79] that L-theanine maintains normal sleep, potentiates memory function, protects against neurotoxins, and exhibits potential beneficial effects in diabetes, CVD, hypertension, stress alleviation, tumour suppression, menstruation, and hepatic injury. The authors highlighted the pharmacokinetics data and the biosynthetic pathology. Trujillo *et al.*[80] demonstrated the antiviral potential of guayusa against acyclovir-resistant herpes simplex virus. Guayusa exhibited outstanding antimicrobial activity on microorganisms including *Porphyromonas gingivalis* ATCC 33277 and *Prevotella intermedia* associated with periodontal disease.[81] Once again, achieving these mechanistic benefits contributes to the multifaceted systems biology approach in alleviating metabolic duress to reduce or eliminate survival panic-induced reduction in energy expenditure and energy storage as fat, and promoting fat oxidation.

Krieger *et al.*[82] conducted a double-blind, randomized, crossover clinical investigation in 12 adult males to receive 200 mg of caffeine from a green coffee extract, a guayusa, and a synthetic control to evaluate their comparative safety, absorption, and effect on neurotransmitters. No statistically significant changes in blood pressure or heart rate were observed. Guayusa exerted a significantly lower increase in epinephrine as compared to the control. Broad spectrum safety of guayusa (dose: 118 g containing 200 mg caffeine or 2.5 mg per kg body weight on average) was reported in this investigation. These researchers demonstrated that the guayusa caffeine source potentially differs from the synthetic and green coffee-based caffeine sources with respect to exerting an insignificant impact on epinephrine, which may signify a decreased incidence of adrenaline-related side effects experienced by other caffeine sources. The neurotransmitter data showed that guayusa did not induce statistically significant changes in dopamine, norepinephrine, or GABA levels. Essentially, guayusa did not exert any caffeine stimulatory effects.

Overall, guayusa exhibits broad-spectrum safety (GRAS affirmed, antioxidant efficacy, and antimicrobial activity), elevates mood, focus, and memory functions, enhances cognition and alleviates stress, protects against neurotoxins, lowers the serum TAG level, and reduces the incidence of MetS. Ursolic acid, an integral constituent in guayusa, plays a vital role in the activation of TGR5 ($EC_{50} = 4.7\ \mu M$), a nuclear receptor involved in the prevention and management of diabetes and MetS *via* modulation of energy expenditure and insulin sensitization. Human studies demonstrated that guayusa, the only naturally occurring caffeine source that reduces the adrenaline-induced side effects, exhibits broad spectrum safety and boosts energy, stamina, and endurance without caffeine stimulatory effects.

6.7.5 Purple Tea (*Camellia sinensis*) Extract

The purple tea, a unique "purple" variation of *Camellia sinensis*, was developed by the Tea Research Foundation of Kenya. The purple tea leaves and shoots used in the purple tea are grown on a pesticide-free plantation in Kenya and have been tested for and are free of residual agricultural chemicals, acute toxicity, and mutagenicity. The tea is cultivated at mountain areas with 1500 and 2500 meters above sea level and exposed to strong UV rays from the sun. This more intense sun exposure increases the protective properties of the plant. The young leaves and shoots get their purple colour from their high anthocyanin (especially delphinidin) content and have potent antioxidant activity significantly higher than green, oolong, and black teas.

Additionally, the purple tea contains a unique polyphenol, 1,2-di-*O*-galloyl-4,6-*O*-(*S*)-hexahydroxydiphenoyl-β-D-glucose (GHG), which is the main polyphenolic ellagitannin in this novel purple tea extract, and exhibits potent antioxidant efficacy. Incidentally, GHG is not found in other tea varieties. GHG has been demonstrated to contribute significantly to body weight management, dermal health, and blood flow.[83] In addition, purple tea is a rich source of theobromine, an alkaloid that is close in action to caffeine, but does not overstimulate the central nervous system which results in jitters and also provides other valuable health benefits. Purple tea contains 16.5% of polyphenols, which is significantly high as compared to green tea (9.1%), oolong tea (7.4%), and black tea (10.1%), while it contains lower amounts of caffeine. It exhibits several benefits including recovery from muscle damage, improvements in blood flow, perfusion, and recovery.

In vitro studies have demonstrated its ability to improve antioxidant efficacy and nitric oxide production, promote exercise recovery, and mitigate muscle damage through antioxidant and anti-inflammatory mechanisms. Harty *et al.*[84] further demonstrated its impact on exercise recovery by reducing creatine kinase, oxidative stress, and exercise-induced soreness. These improvements provide another facet in the systems biology approach to improve metabolic competence, increase oxygen utilization, and enhance fat and glucose oxidation contributing to healthy body composition.

Shimoda *et al.*[85] reported that purple tea administration (200 mg per kg) significantly suppressed body weight gain, liver weight, abdominal fat, and TAG in serum and hepatic tissues. Expression of carnitine palmitoyltransferase 1A was also enhanced. The purple tea administration (100 mg per kg) also suppressed fat absorption. These investigators demonstrated that 4 week daily consumption of purple tea improved obesity parameters including body weight and BMI by suppressing fat absorption and increasing hepatic fat metabolism.[85]

Cesareo *et al.*[83] examined the effects of purple tea (100 mg per day) in exercising humans after stressful and damaging exercise. A randomized, placebo-controlled, double-blind, crossover study was conducted in 30 healthy men over a period of 8 consecutive days. Following 5 and 8 days of supplementation, changes in muscle oxygenation, body composition, reactive hyperemia, visual analog responses, exercise performance, and muscle damage markers were assessed. Interestingly, lactate dehydrogenase, a marker of muscle injury, was significantly reduced after eight days of supplementation and exercise performance challenge. Moreover, lower body muscle endurance performance was significantly improved.

6.7.6 Alpha-glycerylphosphorylcholine (α-GPC)

Alpha-GPC, a novel safe non-stimulating endurance-enhancing ergogenic antioxidant supplement, is enzymatically converted to phosphorylcholine and increases the synthesis of and enhances the release of the neurotransmitter acetylcholine. This promotes improvements in learning, focus and cognition, and brain function, among others.[86] Moreover, α-GPC prevents exercise-induced reductions in choline levels and boosts endurance, performance, and growth hormone secretion. Plasma free choline levels significantly increased at 60 and 120 min after α-GPC administration. Plasma growth hormone secretion was increased significantly 60 min after taking α-GPC, whereas no significant change was observed with the placebo.[87,88] The research by Kawamura *et al.*[87] also demonstrated that serum free fatty acid was increased 120 min after α-GPC ingestion, but no changes were seen with the placebo. In addition, serum acetoacetate and 3-hydroxybutyrate levels, which are indices of hepatic fat oxidation, were increased at 120 min after taking α-GPC, whereas the placebo had no effect. These findings suggest that a single dose of α-GPC increases growth hormone secretion and hepatic fat oxidation, with concomitant increases in choline levels, in young adults. Parnetti *et al.*[86] demonstrated that, in addition to learning, α-GPC was shown to enhance mood and physiological performance in athletes. Importantly, α-GPC was shown to exert its beneficial effects regardless of body mass or age.[89]

6.8 Molecular Basis of Action of Phytoceuticals as Effective Anti-obesity Formulations

It is important to address the two main factors to which obesity has been historically attributed (*e.g.*, overeating and lack of exercise). It is important to have a healthy lifestyle that includes a wholesome and nutritious plant-based dietary plan with modest amounts of fish, poultry, or red meats (optional), sufficient restful sleep, and at least a modest amount of exercise. Implementing this sort of plan enables even greater beneficial results in conjunction with the novel prodosomed phytoceutical antioxidant formulation technology.

The severity of obesity is proportional to the number, intensity, and impact of concurrent factors (*e.g.*, genetics, poor lifestyle choices, stress, and toxic environmental exposures, among others) and a foundational understanding of the factors involved in the genetically-mediated "drive to survive". This is of the utmost importance in formulating strategies to rally numerous biological systems and events to harmoniously and homeostatically reduce survival-induced energy conservation and upregulate cellular energy expenditure, especially from oxidized fat. Inflammation is induced because of hypoxic/anaerobic events and pathologies such as trauma, allergen/antigen insult, infection, stress, or any factors irritating/damaging connective tissues and/or causing or leading to tissue hypoxia and genetically induced survival panic.[90]

With the preceding information as a background, the prodosomed nutraceutical technology was formulated to effectively address the other 10 factors previously identified. The most important priority is to reverse anaerobic metabolic events and pathologies to optimize oxygen utilization and hydration, thereby reducing survival panic signalling and the generation of ROS.[14] The prodosomed phytoceutical antioxidant formulation strategically designed with six research-affirmed ingredients at research validated doses was shown to reduce inflammation,[59] known to increase with excess adiposity.

The DRN complex exhibited increases in the availability of NAD, ATP, and GSH simultaneously, including improvements in muscle and brain performance.[57] In a randomized, double-blind, placebo-controlled, cross-over clinical study, significant increases were observed in both NAD+ and NADP+ by 15%, and ATP and ADP levels by 7.3%. The glutathione level increased by 11% and redox balance was maintained. Insulin response and glucose tolerance increased significantly. A steady reduction was observed in salivary cortisol.[57] This study demonstrates important benefits for the DRN complex as a potent antioxidant that also enhances cellular energy production.

Sweet tea vine was shown to improve AMPK synthesis, muscle, and liver glycogen as well as insulin sensitivity and glucose utilization[39,40] and increase antioxidant potential, cellular energy production, and metabolic activity.[38–40,91,92] Collectively, these events demonstrate an improvement in cellular aerobic metabolism (*e.g.*, aerobic glycolysis, lipolysis, and oxidative respiration), crucial to reverse anaerobic pathologies and genetically induced survival panic.[14] Moreover, among its other components, sweet tea vine is high in potent antioxidants quercetin and rutin, both of which are potent antioxidants that are highly beneficial, including exerting neuroprotective effects and modulation of gut microbiota.[93]

Camu camu has been shown to boost immunity as evidenced by decreases in IL-6 and IL-8, among other markers.[66,67] Moreover, it has been demonstrated that camu camu protects against oxidative stress, cell mutations, and damage as well as protecting genomic DNA integrity as evidenced by reducing levels of urinary 8-hydroxy-deoxyguanosine.[67–69]

Guayusa-derived ursolic acid plays a vital role in the activation of a nuclear receptor, TGR5 ($EC_{50} = 4.7\ \mu M$), a member of G-protein-coupled bile acid receptor superfamily, involved in the prevention and management of diabetes and MetS *via* modulation of energy expenditure and insulin sensitization.[73] Moreover, guayusa contains a small amount of caffeine that, unlike other caffeine sources, did not induce statistically significant changes in dopamine, norepinephrine (pre-adrenaline), or GABA levels.[79] Guayusa has also been shown to elevate mood, reduce stress, and enhance memory and cognition.[81]

Purple tea, in an evidence-based dosage amount in the prodosomed phytoceutical antioxidant formulation, was shown to significantly reduce lactate dehydrogenase (an anaerobic marker of muscle damage) and increase muscle oxygen saturation, thereby improving aerobic cellular energy output and endurance.[85] The rich botanical source of antioxidants in the prodosomed phytoceutical formulation helps to neutralize ROS, mitigate the consequences of exposure to environmental toxins, reduce oxidative stress and cellular/tissue damage while strengthening connective tissues, and improve cellular oxygen utilization. Reducing ROS also reduces immune burden and GI tract dysbiosis.[93]

Finally, in addition to increasing fat oxidation, α-GPC has been shown to promote an increase in growth hormone, among others, and improvements in strength and athletic performance, including endurance, regardless of body mass or age.[85,89]

6.9 Conclusion

Studies reporting the additive effect of phytochemicals in conjunction with exercise on fatty acid oxidation are limited. For most of the plant-based chemical compounds and extracts, additional experiments are needed to determine optimal doses as well as efficacy involved in long-term use along with exercise to enhance fatty acid oxidation. From a practical standpoint, the additive effects of combining phytochemicals with exercise on fatty acid oxidation can provide new insights for developing strategies that facilitate fat loss and improve weight management in obese patients. The prodosomed formulation is a proprietary formulation that collectively harnesses the goodness of all constituent phytochemicals to restore aerobic metabolism and normal physiological homeostasis, thus mitigating obesogenic factors. Most of the traditionally exploited herbal supplements and bioactive compounds had failed to cater for a long-standing need of the health industry to develop a formulation of body weight management, chiefly arising out of their limited bioavailability coupled with poor shelf life. Therefore, the need of the hour for the scientific fraternity is to culminate traditional knowledge with state-of-the-art technological breakthroughs for development of formulations for holistic management of obesity, the most significant reason behind human ailments.

References

1. C. M. Hales, M. D. Carroll, C. D. Fryar and C. L. Ogden, NCHS Data Brief No. 360 February 2020, U.S. Department of Health and Human Services, Centers for Disease Control and Prevention, National Center for Health Statistics; Prevalence of Obesity and Severe Obesity Among Adults: United States, 2017–2018, Published online at: https://www.cdc.gov/nchs/data/databriefs/db360-h.pdf (accessed January 4, 2022).
2. Adult Obesity Facts, Obesity is a Common, Serous and Costly Disease, Feb 11, 2021, Published online at: https://www.cdc.gov/obesity/data/adult.html (accessed January 4, 2022).
3. Managing Overweight and Obesity in Adults, Systematic Evidence Review from the Obesity Expert Panel, 2013, U.S. Department of Health and Human Services, Published online at: https://www.nhlbi.nih.gov/health-topics/managing-overweight-obesity-in-adults (accessed December 28, 2021).
4. E. A. Finkelstein, J. G. Trogdon, J. W. Cohen and W. Dietz, Annual medical spending attributable to obesity: Payer-and service-specific estimates, *Health Aff.*, 2009, **28**, w822–w831.
5. M. C. Bastawrous, C. Piernas, A. Bastawrous, J. Oke, D. Lasserson and W. Mathenge, *et al.*, Reference values for body composition and associations with blood pressure in Kenyan adults aged ≥50 years old, *Eur. J. Clin. Nutr.*, 2019, **73**, 558–565.

6. B. W. Downs, A. L. C. Chen, T. J. H. Chen, R. L. Waite, E. R. Braverman and M. Kerner, *et al.*, Nutrigenomic targeting of carbohydrate craving behavior: can we manage obesity and aberrant craving behaviors with neurochemical pathway manipulation by immunological compatible substances (nutrients) using a Genetic Positioning System (GPS) map?, *Med. Hypotheses*, 2009, **73**, 427–434.
7. G. D. Miller, Appetite regulation: hormones, peptides, and neurotransmitters and their role in obesity, *Am. J. Lifestyle Med.*, 2019, **13**, 586–601.
8. S. F. L. Kirk, T. L. Penney, T. L. F. McHugh and A. M. Sharma, Effective weight management practice: a review of the lifestyle intervention evidence, *Int. J. Obes.*, 2012, **36**, 178–185.
9. P. F. Baillie-Hamilton, Chemical toxins: a hypothesis to explain the global obesity epidemic, *J. Altern. Complement. Med.*, 2002, **8**, 185–192.
10. M. A. Hyman, Environmental toxins, obesity, and diabetes: an emerging risk factor, *Altern. Ther. Health Med.*, 2010, **16**, 56–58.
11. K. Brown, D. DeCoffe, E. Molcan and D. L. Gibson, Diet-induced dysbiosis of the intestinal microbiota and the effects on immunity and disease, *Nutrients*, 2012, **4**, 1095–1119.
12. C. L. Boulangé, A. L. Neves, J. Chilloux, J. K. Nicholson and M. E. Dumas, Impact of the gut microbiota on inflammation, obesity, and metabolic disease, *Genome Med.*, 2016, **8**, 42.
13. B. Phypers and T. Pierce, Lactate physiology in health and disease, *Contin. Educ. Anaesth. Crit. Care Pain*, 2006, **6**, 128–132.
14. J. R. Corbier, B. W. Downs, S. Kushner, T. Aloisio, D. Bagchi and M. Bagchi, VMP35 MNC, a novel iron-free supplement, enhances cytoprotection against anemia in human subjects: a novel hypothesis, *Food Nutr. Res.*, 2019, **63**, 3410.
15. F. Zatterale, M. Longo, J. Naderi, G. A. Raciti, A. Desiderio and C. Miele, *et al.*, Chronic adipose tissue inflammation linking obesity to insulin resistance and type 2 diabetes, *Front. Physiol.*, 2020, **10**, 1607.
16. N. Netzer, H. Gatterer, M. Faulhaber, M. Burtscher, S. Pramsohler and D. Pesta, Hypoxia, oxidative stress and fat, *Biomolecules*, 2015, **5**, 1143–1150.
17. S. Furukawa, T. Fujita, M. Shimabukuro, M. Iwaki, Y. Yamada and Y. Nakajima, *et al.*, Increased oxidative stress in obesity and its impact on metabolic syndrome, *J. Clin. Invest.*, 2004, **114**, 1752–1761.
18. A. R. Saltiel and J. M. Olefsky, Inflammatory mechanisms linking obesity and metabolic disease, *J. Clin. Invest.*, 2017, **127**, 1–4.
19. M. J. Hubler and A. J. Kennedy, Role of lipids in the metabolism and activation of immune cells, *J. Nutr. Biochem.*, 2016, **34**, 1–7.
20. C. Rabasa and S. L. Dickson, Impact of stress on metabolism and energy balance, *Curr. Opin. Behav. Sci.*, 2016, **9**, 71–77.
21. Y. Zhang and S. Chua Jr., Leptin function and regulation, *Compr. Physiol.*, 2017, **8**, 351–369.
22. F. Zhang, Y. Chen, M. Heiman and R. Dimarchi, Leptin: structure, function and biology, *Vitam. Horm.*, 2005, **71**, 345–372.

23. G. A. Triantafyllou, S. A. Paschou and C. S. Mantzoros, Leptin and hormones: Energy homeostasis, *Endocrinol. Metab. Clin. North Am.*, 2016, **45**, 633–645.
24. R. Akki, K. Raghay and M. Errami, Potentiality of ghrelin as antioxidant and protective agent, *Redox Rep.*, 2021, **26**, 71–79.
25. S. Yarnell, M. Oscar-Berman, N. M. Avena, K. Blum and M. S. Gold, Pharmacotherapies for overeating and obesity, *J. Genet. Syndr. Gene Ther.*, 2013, **4**, 131.
26. B. Leeners, N. Geary, P. N. Tobler and L. Asarian, Ovarian hormones and obesity, *Hum. Reprod. Update*, 2017, **23**, 300–321.
27. N. M. Lainez and D. Coss, Obesity, neuroinflammation, and reproductive function, *Endocrinology*, 2019, **160**, 2719–2736.
28. L. M. Brown, L. Gent, K. Davis and D. J. Clegg, Metabolic impact of sex hormones on obesity, *Brain Res.*, 2010, **1350**, 77–85.
29. F. Goglia, The effects of 3,5-diiodothyronine on energy balance, *Front. Physiol.*, 2014, **5**, 528.
30. A. Lombardi, R. Senese, R. De Matteis, R. A. Busiello, F. Cioffi and F. Goglia, *et al.*, 3,5-Diiodo-L-thyronine activates brown adipose tissue thermogenesis in hypothyroid rats, *PLoS One*, 2015, **10**, e0116498.
31. A. S. Padron, R. A. Neto, T. U. Pantaleão, M. C. de Souza dos Santos and R. L. Araujo, *et al.*, Administration of 3,5-diiodothyronine (3,5-T2) causes central hypothyroidism and stimulates thyroid-sensitive tissues, *J. Endocrinol.*, 2014, **221**, 415–427.
32. W. Jonas, J. Lietzow, F. Wohlgemuth, C. S. Hoefig, P. Wiedmer and U. Schweizer, *et al.*, 3,5-Diiodo-L-thyronine (3,5-T2) exerts thyromimetic effects on hypothalamus-pituitary-thyroid axis, body composition, and energy metabolism in male diet-induced obese mice, *J. Endocrinol.*, 2015, **156**, 389–399.
33. S. Arnold, F. Goglia and B. Kadenbach, 3,5-Diiodothyronine binds to subunit Va of cytochrome-c oxidase and abolishes the allosteric inhibition of respiration by ATP, *Eur. J. Biochem.*, 1998, **252**, 325–330.
34. M. B. Dratman and J. T. Gordon, Thyroid hormones as neurotransmitters, *Thyroid*, 1996, **6**, 639–647.
35. G. D. Miller, Appetite regulation: Hormones, peptides, and neurotransmitters and their role in obesity, *Am. J. Lifestyle Med.*, 2019, **13**, 586–601.
36. K. Blum, M. Febo and R. D. Badgaiyan, Fifty years in the development of a glutaminergic-dopaminergic optimization complex (KB220) to balance brain reward circuitry in reward deficiency syndrome: A pictorial, *Austin Addict. Sci.*, 2016, **1**, 1006.
37. G. J. Wang, N. D. Volkow, J. Logan, N. R. Pappas, C. T. Wong and W. Zhu, *et al.*, Brain dopamine and obesity, *Lancet*, 2001, **357**, 354–357.
38. D. Carling, The AMP-activated protein kinase cascade; a unifying system for energy control, *Trends Biochem. Sci.*, 2004, **29**, 18–24.
39. S. Fogarty and D. G. Hardie, Development of protein kinase activators: AMPK as a target in metabolic disorders and cancer, *Biochim. Biophys. Acta*, 2010, **1804**, 581–591.

40. D. G. Hardie, AMPK: a key regulator of energy balance in the single cell and the whole organism, *Int. J. Obes.*, 2008, **32**(Suppl. 4), 7–12.
41. A. Woods, K. Dickerson, R. Heath, S. P. Hong, M. Momcilovic and S. R. Johnstone, *et al.*, Ca21/calmodulin-dependent protein kinase kinase-beta acts upstream of AMP-activated protein kinase in mammalian cells, *Cell Metab.*, 2005, **2**, 21–33.
42. B. B. Zhang, G. Zhou and C. Li, AMPK: an emerging drug target for diabetes and the metabolic syndrome, *Cell Metab.*, 2009, **9**, 407–416.
43. S. A. J. Trammell, M. S. Schmidt, B. J. Weidemann, P. Redpath, F. Jaksch and R. W. Dellinger, *et al.*, Nicotinamide riboside is uniquely and orally bioavailable in mice and humans, *Nat. Comm.*, 2016, 7, 12948.
44. V. R. Trivedi, M. C. Satia, A. Deschamps, V. Maquet, R. B. Shah and P. H. Zinzuwadia, *et al.*, Single-blind, placebo controlled randomised clinical study of chitosan for body weight reduction, *Nutr. J.*, 2016, **15**, 3.
45. C. Morimoto, Y. Satoh, M. Hara, S. Inoue, T. Tsujita and H. Okuda, Anti-obese action of raspberry ketone, *Life Sci.*, 2005, 77, 194–204.
46. H. Shimoda, E. Seki and M. Aitani, Inhibitory effect of green coffee bean extract on fat accumulation and body weight gain in mice, *BMC Complement. Altern. Med.*, 2006, **6**, 9.
47. J. J. DiNicolantonio, A. G. Bhat and J. OKeefe, Effects of spirulina on weight loss and blood lipids: a review, *Open Heart*, 2020, 7, e001003.
48. W. J. Pasman, J. Heimerikx, C. M. Rubingh, R. Van den Berg, M. O'Shea and L. Gambelli, *et al.*, The effect of Korean pine nut oil on *in vitro* CCK release, on appetite sensations and on gut hormones in post-menopausal overweight women, *Lipids Health Dis.*, 2008, 7, 10.
49. A. A. Burns, M. B. Livingstone, R. W. Welch, A. Dunne and I. R. Rowland, Dose–response effects of a novel fat emulsion (Olibra) on energy and macronutrient intakes up to 36 h post-consumption, *Eur. J. Clin. Nutr.*, 2002, **56**, 368–377.
50. M. McCarty and M. Majeed, The pharmacology of citrin, in *Citrin®. A Revolutionary, Herbal Approach to Weight Management*, ed. M. Majeed, R. Rosen, M. McCarty, A. Conte, D. Patil and E. Butrym, New Editions Publishing, Burlingame, CA, 1994, pp. 34–52.
51. K. Muramatsu, M. Fukuyo and Y. Hara, Effect of green tea catechins on plasma cholesterol level in cholesterol-fed rats, *J. Nutr. Sci. Vitaminol.*, 1986, **32**, 613–622.
52. J. Park, H. L. Kim, Y. Jung, K. S. Ahn, H. J. Kwak and J. Y. Um, Bitter orange (*Citrus aurantium* Linné) improves obesity by regulating adipogenesis and thermogenesis through AMPK activation, *Nutrients*, 2019, **11**, 1988.
53. A. M. Lemieux, B. Li and M. Al'Absi, Khat use and appetite: an overview and comparison of amphetamine, khat and cathinone, *J. Ethnopharmacol.*, 2015, **160**, 78–85.
54. S. Perna, V. Infantino, G. Peroni, C. Gasparri, M. A Faliva and M. Naso, *et al.*, Effects of *Hoodia parviflora* on satiety, abdominal obesity and

weight in a group of overweight subjects: a randomized, blinded, placebo-controlled trial, *Minerva Gastroenterol. Dietol.*, 2020, **66**, 211–218.
55. M. H. Pittler and E. Ernst, Guar gum for body weight reduction: meta-analysis of randomized trials, *Am. J. Med.*, 2001, **110**, 724–730.
56. A. K. Hopp, P. Grüter, O. Michael and M. O. Hottiger, Regulation of glucose metabolism by NAD+ and ADP-ribosylation, *Cells*, 2019, **8**, 890.
57. A. Xue, *Pharmacodynamics Study Summary of RiaGev, American Pre-clinical Services*, Bioenergy Life Science, Inc., Minneapolis, MN, 2022, unpublished.
58. J. H. Kim and Y. Park, Combined effects of phytochemicals and exercise on fatty acid oxidation, *J. Exerc. Nutr. Biochem.*, 2016, **20**, 20–26.
59. W. Y. Wong, M. L. L. Lee, B. D. Chan, V. W. S. Ma, W. Zhang and T. T. C. Yip, *et al.*, *Gynostemma pentaphyllum* saponins attenuate inflammation *in vitro* and *in vivo* by inhibition of NF-κB and STAT3 signaling, *Oncotarget*, 2017, **8**, 87401–87414.
60. S. Megalli, N. M. Davies and B. D. Roufogalis, Anti-hyperlipidemic and hypoglycemic effects of *Gynostemma pentaphyllum* in the Zucker fatty rat, *J. Pharm. Pharm. Sci.*, 2006, **9**, 281–291.
61. J. Yeo, Y. J. Kang, S. M. Jeon, U. J. Jung, M. K. Lee and H. Song, *et al.*, Potential hypoglycemic effect of an ethanol extract of *Gynostemma pentaphyllum* in C57BL/KsJ-db/db mic, *J. Med. Food*, 2008, **11**, 709–716.
62. H. F. Lu, Y. S. Chen, J. S. Yang, J. C. Chen, K. W. Lu and T. H. Chiu, *et al.*, Gypenosides induced G0/G1 arrest *via* inhibition of cyclin E and induction of apoptosis *via* activation of caspases-3 and 29 in human lung cancer A-549 cells, *In Vivo*, 2008, **22**, 215–221.
63. V. Razmovski-Naumovski, T. H. W. Huang, V. H. Tran, G. Q. Li, C. C. Duke and B. D. Roufogalis, Chemistry and pharmacology of *Gynostemma pentaphyllum*, *Phytochem. Rev.*, 2005, **4**, 197–219.
64. S. H. Park, T. L. Huh, S. Y. Kim, M. R. Oh, T. Pichiah and S. W. Chae, *et al.*, Antiobesity effect of *Gynostemma pentaphyllum* extract (Actiponin): A randomized, double-blind, placebo-controlled trial, *Obesity*, 2014, **22**, 63–71.
65. P. C. Langley, J. V. Pergolizzi Jr., R. Taylor Jr. and C. Ridgway, Antioxidant and associated capacities of Camu camu (*Myrciaria dubia*): a systematic review, *J. Altern. Complement. Med.*, 2015, **21**, 8–14.
66. D. Fracassetti, C. Costa, L. Moulay and F. A. Tomás-Barberán, Ellagic acid derivatives, ellagitannins, proanthocyanidins and other phenolics, vitamin C and antioxidant capacity of two powder products from camu-camu fruit (*Myrciaria dubia*), *Food Chem.*, 2013, **139**, 578–588.
67. K. Yazawa, K. Suga, A. Honma, M. Shirosaki and T. Koyama, Anti-inflammatory effects of seeds of the tropical fruit camu-camu (*Myciaria dubia*), *J. Nutr. Sci. Vitaminol.*, 2011, **57**, 104–107.
68. T. Inoue, H. Komoda, T. Uchida and T. Node, Tropical fruit camu-camu (*Myrciaria dubia*) has anti-oxidative and anti-inflammatory properties: A randomized controlled trial, *J. Cardiol.*, 2008, **52**, 127–132.

69. F. C. da Silva, A. Arruda, A. Ledel, C. Dauth, N. F. Romao and R. N. Viana, *et al.*, Antigenotoxic effect of acute, subacute and chronic treatments with Amazonian camu-camu, *Food Chem. Toxicol.*, 2012, **50**, 2275–2281.
70. F. C. da Silva, J. N. Picada, N. F. Romão, F. D. O. S. Sorbal, D. Lemos and S. de Vargas Schons, *et al.*, Antigenotoxic and antimutagenic effects of *Myrciaria dubia* juice in mice submitted to ethanol 28-day treatment, *J. Toxicol. Environ. Health, Part A*, 2019, **82**, 956–968.
71. J. F. Dueñas, C. Jarrett, I. Cummins and E. Logan–Hines, Amazonian guayusa (*Ilex guayusa* Loes.): A historical and ethnobotanical overview, *Economic Botany*, 2016, **70**, 85–91.
72. A. García-Ruiz, N. Baenas, A. M. Benítez-González, C. M. Stinco, A. J. Meléndez-Martínez and D. A. Moreno, *et al.*, Guayusa (*Ilex guayusa* L.) new tea: Phenolic and carotenoid composition and antioxidant capacity, *J. Sci. Food Agric.*, 2017, **97**, 3929–3936.
73. J. Brinckmann and T. Brendler, Herb profile: Guayusa (*Ilex guayusa* family: Aquifoliaceae), *Herbalgram*, 2019, **124**, 6–13.
74. G. Chianesea, S. D. Golin-Pachecob, O. Taglialatela-Scafatia, J. A. Colladoc, E. Munoz and G. Appendino, *et al.*, Bioactive triterpenoids from the caffeine-rich plants guayusa and mate, *Food Res. Int.*, 2019, **115**, 504–510.
75. R. W. Kapp Jr., O. Mendes, S. Roy, R. S. McQuate and R. Kraska, General and genetic toxicology of Guayusa concentrate (*Ilex guayusa*), *Int. J. Toxicol.*, 2016, **35**, 222–242.
76. GRAS Notice (GRN), Gras Notice No. 835 from the Food Additive Safety, Published online at: https://www.fda.gov/food/generally-recognized-safe-gras/gras-notice-inventory (accessed January 28, 2022).
77. G. Wise and D. Santander, 2018 Technical Report, Published online at: https://www.researchgate.net/publication/329687006 (accessed December 14, 2021).
78. J. Schustera and E. S. Mitchell, More than just caffeine: psychopharmacology of methylxanthine interactions with plant-derived phytochemicals, *Prog. Neuropsychopharmacol. Biol. Psychiatry*, 2019, **89**, 263–274.
79. R. Adhikary and V. Mandal, L-Theanine: A potential multifaceted natural bioactive amide as health supplement, *Asian Pac. J. Trop. Biomed.*, 2017, 7, 842–848.
80. D. C. Trujillo, I. Dai Prá, I. V. Farias, F. H. Reginatto, I. T. da Silva and G. M. Costa, Antiviral activity of *Ilex guayusa* Loes against acyclovir-resistant Herpes Simplex Virus, in *7th Brazilian Conference on Natural Products/XXXIII RESEM Proceedings*, Rio de Janeiro, RJ, Brazil, 2019, pp. 10–13.
81. F. Gamboa, C. C. Muñoz, G. Numpaque, L. G. Sequeda-Castañeda, S. J. Gutierrez and N. Ellez, Antimicrobial activity of *Piper marginatum* Jacq and *Ilex guayusa* Loes on microorganisms associated with periodontal disease, *Int. J. Microbiol.*, 2018, **2018**, 4147383.
82. D. R. Krieger, D. S. Kalman, S. Feldman, L. Arnillas, D. Goldberg and O. Gisbert, *et al.*, The safety, pharmacokinetics, and nervous system

effects of two natural sources of caffeine in healthy adult males, *Clin. Transl. Sci.*, 2016, **9**, 246–251.

83. K. Cesareo, T. Ziegenfuss, B. Raub, J. Sandrock and H. Lopez, Effects of purple tea on muscle hyperemia and oxygenation, serum markers of nitric oxide production and muscle damage, and exercise performance, *J. Exer. Nutr.*, 2020, **3**, 2640–2572.
84. P. S. Harty, H. A. Zabriskie, J. L. Erickson, P. E. Molling, C. M. Kerksick and A. R. Jagim, Multi-ingredient pre-workout supplements, safety implications, and performance outcomes: a brief review, *J. Int. Soc. Sports Nutr.*, 2018, **15**, 41.
85. H. Shimoda, S. Hitoe, S. Nakamura and H. Matsuda, Purple tea and its extract suppress diet-induced fat accumulation in mice and human subjects by inhibiting fat absorption and enhancing hepatic carnitine palmitoyltransferase expression, *Int. J. Biomed. Sci.*, 2015, **11**, 67–75.
86. L. Parnetti, G. Abate, L. Bartorelli, D. Cucinotta, M. Cuzzupoli and M. Maggioni, *et al.*, Multicentre study of l-alpha-glyceryl-phosphorylcholine *vs.* ST200 among patients with probable senile dementia of Alzheimer's type, *Drugs Aging*, 1993, **3**, 159–164.
87. T. Kawamura, T. Okubo, K. Sato, S. Fujita, K. Goto and T. Hamaoka, *et al.*, Glycerophosphocholine enhances growth hormone secretion and fat oxidation in young adults: a clinical trial, *Nutrition*, 2012, **28**, 1122–1126.
88. T. Ziegenfuss, J. Landis and J. Hofheins, Acute supplementation with alpha-glycerylphosphorylcholine augments growth hormone response to, and peak force production during resistance exercise, *J. Int. Soc. Sports Nutr.*, 2008, **5**(Suppl. 1), 1–15.
89. L. Pfleger, M. Gajdošík, P. Wolf, S. Smajis, P. Fellinger and A. Kuehne, *et al.*, Absolute quantification of phosphor-containing metabolites in the liver using 31P MRSI and hepatic lipid volume correction at 7T suggests no dependence on body mass index or age, *J. Magn. Reason. Imag.*, 2019, **49**, 597–607.
90. P. Bortolotti, E. Faure and E. Kipnis, Inflammasomes in tissue damages and immune disorders after trauma, *Front. Immunol.*, 2018, **9**, 1900.
91. B. W. Downs, J. R. Corbier, N. Speight, S. Kushner, T. Aloisio and M. Bagchi, *et al.*, Anemia: Influence of dietary fat, sugar, and salt on hemoglobin and blood health, in *Dietary Sugar, Salt, and Fat in Human Health*, ed. H. G. Preuss and D. Bagchi, Elsevier/Academic Press, Amsterdam, The Netherlands, 2020, pp. 103–128.
92. S. Imai and J. Yoshino, The importance of NAMPT/NAD+/SIRT1 in the systemic regulation of metabolism and ageing, *Diabetes, Obes. Metab.*, 2013, **15**(Suppl. 3), 26–33.
93. J. Xia, W. Song, X. Liang, Q. Zhang, Y. Shi and W. Liu, *et al.*, Protective effect of quercetin on streptozotocin-induced diabetic peripheral neuropathy rats through modulating gut microbiota and reactive oxygen species level, *Biomed. Pharmacother.*, 2020, **127**, 110147.

CHAPTER 7

Dietary Lipid Supplements

FEREIDOON SHAHIDI*[a] AND JIANKANG WANG[b]

[a] Department of Biochemistry, Memorial University of Newfoundland, St. John's, NL, Canada; [b] School of Food and Biological Engineering, Shaanxi University of Science and Technology, Xi'an, P. R. China
*Email: fshahidi@mun.ca

7.1 Introduction

Lipids are essential macronutrients in food and serve as a major source of energy. Their diverse biological functions are believed to be exerted by their unique chemical compositions, including fatty acid profiles with respect to the essential polyunsaturates, fat-soluble vitamins, and other minor components.[1] Food lipids in the unprocessed form are principally triacylglycerols (TAG), phospholipids, and sterols which are found naturally in most biological materials. They serve as important ingredients and processing media for the food industry. Lipids, especially TAG, are a key caloric source, an important source of essential fatty acids, and a carrier for fat-soluble vitamins.[2] Moreover, they occur as emulsions or as free oil/fat in a food matrix and contribute desirable properties, including attributes of flavour and texture in foods.[3,4] It has been proven that a balanced diet with various nutrients is essential to maintain biological function in the body, and hence the intake of essential lipids is imperative in a balanced diet.

Dietary supplementation of lipids, especially those supplements containing functional fatty acids, such as omega-3 and omega-6 fatty acids, as well as other bioactive compounds have all strengthened the development of the nutrient supplement industry. For instance, linoleic acid and α-linolenic

Food Chemistry, Function and Analysis No. 39
Dietary Supplements with Antioxidant Activity: Understanding Mechanisms and Potential Health Benefits
Edited by Cesarettin Alasalvar, Fereidoon Shahidi and Chi-Tang Ho

Published by the Royal Society of Chemistry, www.rsc.org

acid are essential fatty acids that need to be obtained from dietary sources, and thus their adequate intake is imperative for human health (Figure 7.1). Omega-3 fatty acids, especially long chain ones, as well as tocopherols and tocotrienols, sterols, phenolics, squalene, and carotenoids, are of interest to the lipid supplement industry.[5,6] These lipids are commonly used as food supplements since their constituents are expected to offer biological function in the body, in addition to their basic nutritional value. For instance, consumption of fatty fish is relatively low in Western countries, and thus sufficient intake of polyunsaturated fatty acids (PUFA) rich in those fish can be achieved by taking dietary lipid supplements, such as fish oil, algal oil, krill oil as well as oils from certain botanical sources (Figure 7.2).[7,8] Edible lipids have been used for a variety of purposes throughout history, with their integration into food and pharmaceutical products.[9–11] There has been growing interest in both the promotion and research of dietary lipid supplements for their diverse health benefits, especially the management and treatment of different health conditions in the last several decades. This chapter aims to provide an account of different lipid supplement classes and their chemical compositions and health benefits.

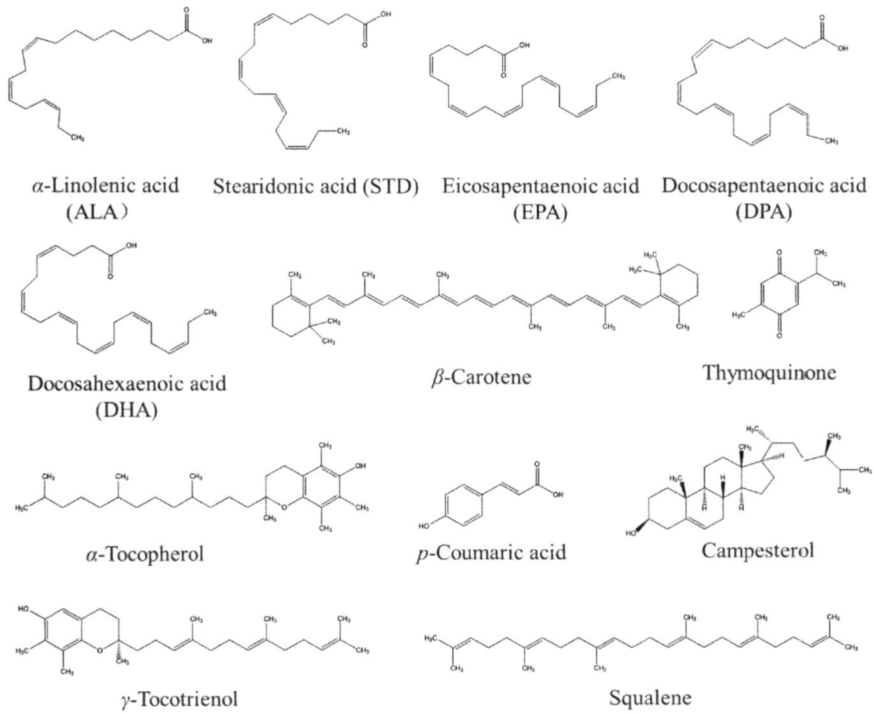

Figure 7.1 Chemical structures of the key bioactive compounds present in dietary lipid supplements.

Figure 7.2 Marine and terrestrial sources of dietary lipid supplements.

7.2 Health Benefits of Dietary Lipid Supplements

The use of dietary supplements to maintain normal development and biological function of human beings has been a common practice in the modern society. For dietary lipid supplements, they can be obtained from marine origin (including fish oil, seal blubber oil, krill oil, and algal oil) and terrestrial plant origin (including evening primrose oil, borage oil, black cumin seed oil, pumpkin seed oil, and berry seed oils). Dietary supplementation of lipids is particularly attractive to people who cannot obtain those lipids from their diet or require enhanced intake to maintain or improve health (Table 7.1). Most of the specialized plant oils are rich in antioxidant phenolics, especially tocopherols, which act as powerful antioxidants and render them a protection against oxidation.[1,28] Moreover, these antioxidants render a wide range of biological functions in the human body.

Table 7.1 Health benefits of dietary lipid supplements.

Dietary lipids	Health outcomes	Ref.
Fish oil	Improves insulin sensitivity and reduces plasma and liver lipid levels Decreases mean cognitive subscale scores Increases Mini-Mental State Examination scores	8, 12
Seal blubber oil	Decreases the rate of CHD Increases the contents of EPA, DHA, DPA, and non-esterified fatty acids in serum phospholipids Prevents CVD and thrombotic diseases	13–15
Krill oil	Prevents cognitive impairment	16
Algal oil	Alleviates human metabolic disorders and anti-inflammation	17, 18
Evening primrose oil	Reduces serum TAG Increases HDL-C	19
Borage oil	Prevents CVD	20
Black cumin seed oil	Reduces fasting blood glucose Increases the insulin level	21
Pumpkin seed oil	Inhibits cellular damage and apoptosis Improves semen parameters	22, 23
Berry seed oils	Prevents CVD	24, 25
Sea buckthorn seed oil	Decreases blood pressure Reduces blood cholesterol	26, 27

Abbreviations: CHD, coronary heart disease; CVD, cardiovascular disease; DHA, docosahexaenoic acid; DPA, docosapentaenoic acid; EPA, eicosapentaenoic acid; HDL-C, high-density lipoprotein cholesterol; TAG, triacylglycerols.

7.2.1 Dietary Lipid Supplements from Marine Sources

7.2.1.1 Fish Oils

Fish oils are among the most widely-consumed dietary lipid supplements globally, which are commonly available in pharmacies and grocery stores. More than 30% of regular dietary supplement consumers take fish oil supplements in the United States.[29] Fish oils can be obtained from the body of fatty fish such as herring, mackerel, menhaden, and salmon, as well as the liver of lean fish such as cod and halibut.[30] Fish oils consist of saturated fatty acids (SFA), monounsaturated fatty acids (MUFA), and PUFA, such as omega-3 PUFA, mainly eicosapentaenoic acid (EPA), docosapentaenoic acid (DPA), and docosahexaenoic acid (DHA). The high content of omega-3 PUFA in fish oils is suggested to be a consequence of cold temperature adaptation in which omega-3 PUFA remains liquid at room temperature. The fatty acid profile of fish oils generally depends on fish species, diet, and environmental factors.[31–33]

There is compelling evidence indicating that the intake of sufficient fish oil leads to a wide range of health benefits, including preventing cardiovascular disease (CVD) and cognitive deficits, alleviating diabetes, maintaining normal growth of the brain and eyes for children, and enhancing cognitive function, among others.[34] Haimeur *et al.*[34] investigated the effect of fish oil on risk factors for CVD, including platelet aggregation, dyslipidemia, and oxidative stress in a rat model study. They found that dietary supplementation of fish oil inhibited the development of adiposity and the prothrombotic effect, improved insulin sensitivity, and reduced plasma and liver lipid levels. Therefore, fish oil showed a promising function in the prevention of CVD. The effect of fish oil on cognitive impairment was assessed in a clinical trial conducted by Daiello *et al.*[12] The results of the study indicated that dietary supplementation of fish oil leads to significantly lower mean cognitive subscale and higher Mini-Mental State Examination scores. In addition, less atrophy in one or more brain regions of interest was observed in the group administered with fish oil.

7.2.1.2 Seal Blubber Oil

Marine oils have long been perceived as nutraceuticals since their function of preventing and treating various health conditions have been well elucidated. Traditionally used for industrial purposes, seal blubber oil is an important marine source of omega-3 fatty acids, including EPA, DPA, and DHA.[13,35] The total content of omega-3 fatty acids in seal blubber oil is about 15.6–20%. Seal blubber oil is chemically different from fish oil with its high level of DPA (4.7%) and the dominant distribution of PUFA in the terminal positions compared to the dominant location of these fatty acids in the middle position in fish oil. The positional distribution of omega-3 fatty acids in the TAG backbone of seal blubber oil is viewed as favourable as fatty acids are released from the terminal positions of the TAG backbone upon hydrolysis in the digestive system, while fatty acids in the middle position remain esterified, which is carried to the circulation system in the form of chylomicrons.

Approximately 29% of the total carcass weight of seals is made of blubber from where lipids are mainly extracted. Seal blubber oil can be commonly seen in some pharmacy stores in the form of capsules. Seal bubbler oil has been traditionally consumed as an essential dietary supplement in different regions, such as Canada, Norway, and Denmark, for its health benefits. Back to the early 1970s, it had been reported that there was a low rate of coronary heart disease (CHD) in Eskimos located in Greenland, which was associated with the high consumption of omega-3 fatty acid rich food from seafoods, seal meat, and blubber.[13,36] The effect of consuming seal blubber oil on cardiovascular and thrombotic risk factors as well as fatty acid profiles of serum phospholipid and nonesterified fatty acid was evaluated in a clinical study.[14] It was revealed that dietary supplementation of seal oil leads to increased contents of EPA, DHA, DPA, and nonesterified fatty acids in serum phospholipids in addition to a modest beneficial function on fibrinogen and protein C levels. The inhibitory effect of seal blubber oil on cardiovascular and thrombotic diseases was later confirmed by Bonefeld-Jrgensen *et al.*[15]

Moreover, it was reported that seal blubber oil may be more effective in improving plasma lipid profiles and lowering thrombotic risk than fish oils due to its unique fatty acid composition, structure, and minor components. The effect of dietary supplementation of seal blubber oil on the diabetic sensorimotor polyneuropathy in type 1 diabetes (T1D) was evaluated in a clinical trial, and the results indicated that long-term dietary supplementation of seal blubber oil increased corneal nerve fibre length in T1D individuals, which is viewed as Class IV evidence.

7.2.1.3 Krill Oil

Krill oil is extracted from Antarctic krill (*Euphausia superba*) which are small crustacean species native to the Antarctic Ocean.[37] With an estimate of 300 000 million metric tons from the Antarctic Ocean, krill is considered as the biggest biomass on earth and indeed provides enough material for the production of krill oil. It has been proven to be effective in preventing or treating a number of common health conditions, including hyperlipidemia, inflammation, and arthritis.[37] With a substantial level of total EPA and DHA, krill oil is viewed as an important alternative to fish oils. EPA accounts for 19% (w/w), while DHA accounts for about 10.9% (w/w) in krill oil. The total omega-3 fatty acids account for 34.1% (w/w) in its fatty acid profiles.[38] The signature chemical composition of krill oil is that its omega-3 fatty acids including EPA and DHA are bound in phospholipids such as phosphatidylcholine.

The difference in the chemical binding of EPA and DHA in krill and fish oils has been suggested to affect their bioavailability. The product labelling and commercials commonly pose a claim that the bioavailability of krill oil is much higher than that of fish oils, and thus the daily supplement of krill oil is smaller than that of fish oils to achieve the same level of health benefits. The capsules of commercial krill oil products are small in size compared to those of fish oils. A large number of studies have been conducted to evaluate the bioavailability of krill oil and different fish oils, but the results have not been conclusive.[37,38] Köhler *et al.*[39] assessed the long-term bioavailability of krill oil and fish oil on fatty acid compositions of plasma TAG and phospholipids in 15 healthy subjects, in which a randomized, single-dose, single-blind, cross-over, active-reference trial was conducted in individuals who ingested krill oil and fish oil that contained about 1700 mg EPA and DHA. The results of the study showed that EPA and DHA in krill oil had a higher 72 hour bioavailability than fish oil, which proved that the bioavailability of EPA and DHA bound to phospholipids was better taken in the human body than that in their TAG forms. Yurko-Mauro *et al.*[40] reported that there was no significant difference in the plasma and red blood cell levels of EPA + DHA in 66 administered healthy adults after receiving the same doses of fish oil and krill oil supplements for 4 weeks in a double blind, randomized, three-treatment, multi-dose and parallel study, which suggested similar oral bioavailability of krill oil and fish oil. Therefore, it is necessary to conduct further studies on the longer-term effect of supplementing these oils on fatty acid composition in tissues to better understand the bioavailability difference between krill oil and fish oils. In addition to its fatty acid profiles, the high phospholipid content of krill oil may be responsible for its better oxidative stability compared to fish oils as phospholipids may act as potential antioxidants or synergists. Furthermore, krill oil contains a high amount of carotenoids, mainly xanthophylls, which are effective antioxidants and these not only protect the oil from oxidation, but may also render their own beneficial health effects.[37,38]

7.2.1.4 Algal Oil

Microalgae (including cyanobacteria) have been confirmed to be exceptional producers of supplementary fatty acids such as EPA and DHA which render a number of biological functions, and algal oils are suitable for vegetarians (Table 7.1).[41] Lipids are key structural components in the cells of microalgae and account for 1–40% of the dry weight of the cells. Fatty acids from algal sources normally have a chain length between 12 and 22 carbon atoms. Myristic and palmitic acids are two major SFA in algal oils, while oleic acid is the key MUFA in these oils.[42] Algal oils are rich sources of omega-3 and omega-6 fatty acids, including α-linolenic acid, EPA, DHA, linoleic acid, γ-linolenic acid, and arachidonic acid, and most of these fatty acids have been commercialized as nutrition enhancers.

Algal oils have been shown to alleviate human metabolic disorders.[17] The effect of DHA rich algal oil on colonic inflammation and intestinal microbiota was evaluated in a dextran sulphate sodium (DSS) induced colitis mouse model study, and the results showed that dietary supplementation of algal oil exerted inhibitory effects on the progress of DSS induced colitis in tested animals *via* modulation of the intestinal microbiota and metabolites as well as the repair of the intestinal barrier, which indicated a potential use of DHA rich algal oil in people with intestinal inflammation. Growing evidence also indicates that omega-3 PUFA have anti-melanoma activity, but their action mechanism remains unclear. Tan *et al.*[18] proved that the intake of algal oil significantly increased in omega-3 PUFA in lung metastatic tissues in tested C57BL/6 mice that received injection of B16F10 melanoma cells. The results from their study showed that supplementation of algal oil suppressed pulmonary metastases and outgrowth of melanoma cells, and led to a decrease in p62 accumulation and secretion of proinflammatory tricytokine interleukin-1β, all of which suggested the anti-tumorigenic activity of algal oil through autophagy-mediated p62 elimination and anti-inflammatory properties.

Algal sources contain many carotenoids, including fucoxanthin, among others, which have many considerable antioxidant effects and a variety of health benefits and are known to have anti-diabetic, anti-cancer, anti-bacterial, and anti-obesity effects. For instance, Karpiński and Adamczak[43] investigated the anti-bacterial activity of fucoxanthin using various bacteria, and the results showed that the growth of more than half of them was inhibited. Moreover, fucoxanthin has been confirmed to play a role in anti-obesity by inhibiting fat accumulation in abdominal white adipose tissue through the up-regulation of the thermogenesis and down-regulation of the expression of pro-inflammatory adipokines involved in insulin resistance.[44]

7.3 Dietary Lipid Supplements from Terrestrial Plants

7.3.1 Evening Primrose Oil

Evening primrose (*Oenothera biennis* L.) is a plant belonging to the *Onagraceae* family.[45] Evening primrose oil is commonly obtained from the seeds of the plant through cold-pressing. Some 98% of the cold-pressed evening primrose oil is in the form of TAG. Thirteen TAG molecules have been identified in the oil, of which the one with the highest content is linoleic–linoleic–linoleic (40%), while the linoleic–linoleic–γ-linolenic content is approximately 15% of all species. The dominant fatty acid species in evening primrose oil is linoleic acid (70–74%), and the fatty acid with the second highest content is γ-linolenic acid (8–10%).

Regarding the minor components in evening primrose oil, phospholipids comprise a small portion of the oil, about 0.05%, and the key phospholipids identified include phosphatidic acid, phosphatidylcholine, phosphatidylinositol, phosphatidylglycerol, and phosphatidylethanolamine.[46] In addition, aliphatic alcohols (798 $mg\,kg^{-1}$) have been found to be another group of minor components in the oil, namely 1-tetracosanol and 1-hexacosanol. Other minor components identified in evening primrose oil include triterpenes (β-amyrin and squalene), tocopherols (α-tocopherol, γ-tocopherol, and δ-tocopherol), polyphenols (hydroxytyrosol, vanillic acid, *p*-coumaric acid, and ferulic acid), and sterols.[28]

The fatty acid composition and minor components present in evening primrose oil offer various biological functions, and oral intake of the oil has been proven to render a wide range of health benefits. For instance, Khorshidi *et al.*[19] confirmed the effect of evening primrose oil supplementation on lipid profiles in a systematic review and meta-analysis of randomized clinical trials. The findings of related studies were searched from various databases and search engines, and a meta-analysis was conducted using a model that involves random effects. After analysing a total of 926 publications, it was found that evening primrose oil supplementation at a dose of up to 4 g per day showed a significant effect on the reduction of serum TAG levels and the increase of high-density lipoprotein levels in hyperlipidemic individuals. The meal from evening primrose contains very important phenolic antioxidant components and this by-product of the oil may provide important health effects which could be related to the antioxidant potential of the compounds present.[47]

7.3.2 Borage Oil

Borage (*Borago officinalis* L.) is a plant native to the Mediterranean region, known since ancient times as a culinary and medicinal material.[9] There has been growing interest in borage seeds since it was confirmed to be the richest known source of γ-linolenic acid. The oil from borage seeds contains a remarkable amount of γ-linolenic acid, which has important nutritional value as a dietary supplementation of omega-6 essential fatty acids, and has attracted potential interest in the treatment of various health conditions. The fatty acid composition of borage oil is dominated by linoleic acid (40.80%) and γ-linolenic acid (22.90%), the latter being two to three times more than that in evening primrose oil.[48] Regarding key minor components, the tocopherol composition of borage oil is dominated by γ-tocopherol,[49] and the major phytosterols include Δ5-avenasterol, campesterol, and β-sitosterol.[50]

Borage oil has been promoted to treat different pathologies.[20,51] Al-Okbi *et al.*[20] studied the preventive effect of borage oil on dyslipidemia, oxidative stress, inflammation, and fatty liver which are common risk factors for CVD in a rat model. It was revealed that borage oil protected the tested animals from CVD and fatty liver disease to different degrees. γ-Linolenic acid has been reported to have the potential to treat cancers and chronic skin conditions, and such activities can be partially explained by the incorporation of these fatty acids into phospholipids and ceramides.[51] The inhibitory effect of borage oil on epidermal hyperproliferation was confirmed in guinea pigs.[51] Borage oil showed higher anti-proliferative efficacy compared to evening primrose oil, which may be due to its higher distribution of γ-linolenic acid in the middle sn-2 position. Therefore, the bioactivity of γ-linolenic acid rich oils is not solely determined by their γ-linolenic acid content and the stereospecificity of fatty acids also plays a significant role in their biological function.

The deoiled meal from borage also serves as a viable source of phenolic compounds with strong antioxidant potential and this might be important for further exploitation by the supplement industry.[52]

7.3.3 Black Cumin Seed Oil

Black cumin seed oil is extracted from the seeds of *Nigella sativa*, an annual flowering plant that is native to southwest Asia. As one of the popular herbal supplements, black cumin seed oil has been marketed as a "miracle supplement".[10] With health benefits in treating asthma, diabetes, hypertension, weight loss, and other conditions, it is used for both topical applications and oral consumption. One of its key components is thymoquinone, a natural antioxidant.[53–56] The major components of fatty acids are linoleic and oleic acids.

Mahdavi *et al.*[55] investigated the effects of black cumin seed oil with a low-calorie diet on cardiometabolic risk factors in a double-blind randomized controlled clinical trial of 90 obese women, 25–50 years old. The results of the study showed that the weight and waist circumference decreased significantly in the group with intervention compared with the placebo group at the end of the trial. Moreover, the group administered with black cumin seed oil compared to the placebo group had a lower amount of very low-density lipoproteins. Therefore, black cumin seed oil coupled with a low-calorie diet may reduce cardiometabolic risk factors in obese women, although additional clinical trials may be required for further confirmation. Black cumin seed oil has also been reported to have beneficial effects in the treatment of diabetic complications. In an attempt to better understand such an effect, Heba *et al.*[21] conducted an animal model study, from which fasting blood glucose, insulin, and lipid profile levels were determined in mature male Wistar rats. It was revealed that administration of 2 mL kg^{-1} of black cumin seed oil led to a reduction of fasting blood glucose and an increase of the insulin level in tested animals.

7.3.4 Pumpkin Seed Oil

Pumpkin (*Cucurbita pepo* L.) is cultivated all over the world.[57,58] Its seed oil is non-refined edible oil from the seeds of *Cucurbita pepo* L. The total content of unsaturated fatty acids in pumpkin seed oil is about 80.1%, with 62.4% as linoleic acid. The oil renders desirable biological functions of maintaining healthy blood vessels, nerves, and tissues.[59] The total content of SFA accounts for 19.9% and the major one is palmitic acid (17.5%). Pumpkin seed oil contains high levels of α- and γ-tocopherols with up to 282 and 800 mg kg^{-1} of oil, respectively, which contribute to its strong free radical quenching capacity.[57,60] A number of other phenolic compounds have also been identified in pumpkin seed oil, including vanillic, caffeic, *trans*-cinnamic, and *p*-coumaric acids.

Pumpkin seed oil can improve membrane fluidity and osmosis as well as enhance intracellular and extracellular gaseous exchange. Hamdi[22] investigated the protective effect of pumpkin seed oil on testicular dysfunction induced by alumina nanoparticles (Al_2O_3 NPs) in a rat model. The results of the study revealed the protective function of pumpkin seed oil by inhibiting cellular damage and apoptosis induced by Al_2O_3 nanoparticles in the tested animals. In another study, Akang *et al.*[23] investigated the effect of pumpkin seed oil on the testis and semen parameters in an adult male Sprague-Dawley rat model, where a slight improvement in semen parameters was found in the group administered with pumpkin seed oil.

7.3.5 Berry Seed Oils

Berry seed oils are by-products from berry juice production. They have gained great commercial interest and include blackberry, black raspberry, blueberry, sea buckthorn, marionberry, boysenberry, and caneberry seed oils that can be sold as supplements in pharmacy stores.[1,61–64]

These oils contain oleic, linoleic, and α-linolenic acids and thus are considered as rich sources of essential fatty acids.[1] Le *et al.*[1] analysed the fatty acid profiles of the cold-pressed seed oils of several commonly known berries including blackberries, black raspberries, and blueberries, and the results showed that palmitic, stearic, oleic, linoleic, and α-linolenic acids were major fatty acids in oils. In addition, high amounts of total PUFA ranging from 65.8% to 87.2% were detected in these oils, of which black raspberry seed oil showed the highest content of PUFA (87.2%), while blueberry seed oil had the highest contents of MUFA (22.2%) and SFA (6.5%). The most abundant fatty acid in blackberry, black raspberry, and blueberry seed oils was linoleic acid, ranging from 41.3% to 63.4%. A substantial content of α-linolenic acid was also detected in these oils and this contributes to a favourable omega-6/omega-3 ratio, ranging from 4.7 to 1.5,[1] and the highest ratio was observed for blackberry seed oil, while the lowest ratio was for blueberry seed oil. Such n-6/n-3 ratios make these berry seed oils more favourable than the majority of vegetable oils as the lower ratio of omega-6/omega-3 fatty acids leads to a number of health benefits, such as prevention of cancer and CVD as well as bone health improvement.[24,25,65] The supplementation of these oils in the diet can balance the ratio of omega-6/omega-3 fatty acids. These berry seed oils serve as rich sources of natural antioxidants, tocopherols, and tocotrienols.[1] Le *et al.*[1] found that there were considerable amounts of total tocopherols and tocotrienols in non-stripped blackberry, black raspberry, and blueberry seed oils. For instance, blueberry seed oil contained 1245 $mg\,kg^{-1}$ of γ-tocotrienol, but a low level of α-tocopherol was observed in the same oil. The non-stripped black raspberry seed oil also contained a high concentration of γ-tocopherol (449 $mg\,kg^{-1}$ of oil). Non-stripped blueberry seed oil showed the highest concentration of total tocols among blackberry seed, black raspberry seed, and blueberry seed oils and non-stripped blackberry seed oil contained high contents of both γ- and δ-tocopherols compared to non-stripped blueberry seed oil. The high contents of essential fatty acids and tocols as well as lower ratios of omega-6/omega-3 fatty acids make these berry seed oils excellent dietary lipid supplements.

7.3.6 Sea Buckthorn Seed Oil

Sea buckthorn (*Hippophae rhamnoides* L.) is a shrub that is native to Eurasia. The leaves, berries, and seeds of this plant have been used for their pharmaceutical properties.[66] The seeds of sea buckthorn contain 8–10% of oil that is rich in oleic, linoleic, α-linolenic, and palmitic acids, of which oleic acid has the highest content among detected fatty acids.[67] The minor components include carotenoid and tocopherols, which contribute to the antioxidant capacity of the oil.[68] The medicinal properties of sea buckthorn seed oil have been reported in the literature.[26,27] Vashishtha *et al.*[26] investigated the effect of sea buckthorn seed oil supplementation on CVD, hypertension, and systolic blood pressure. It was confirmed that supplementation of sea buckthorn seed oil leads to decreased blood pressure in hypertensive individuals as well as reduction of cholesterol, oxidized-low-density lipoprotein, and TAG in hypercholesterolemic individuals in a randomized, controlled, and double-blind longitudinal study. In addition, supplementation of carotenoid- and tocopherol-rich sea buckthorn seed oil improved the circulatory antioxidant status in both normal and hypertensive individuals. Hao *et al.*[27] studied the effect of sea buckthorn seed oil on blood cholesterol and the gut microbiota in hypercholesterolemic hamsters and revealed that sea buckthorn seed oil supplementation effectively reduced blood cholesterol in hypercholesterolemic hamsters due to increased intestinal cholesterol excretion and improved growth of short chain fatty acid-producing bacteria.

7.4 Mechanisms of the Health Benefits of Dietary Lipid Supplements

Lipids play a variety of important biological roles, including being structural components of membranes, storage and transport forms of metabolic fuels, cell signalling, protective coatings, and carriers of fat-soluble vitamins.[69] Moreover, the diverse health benefits of dietary lipid supplements have been reported in a large body of literature and their mechanisms have been investigated in the last several decades. For instance, lipids are involved in cell signalling, which is thought to be qualitatively different from other classical signalling paradigms as they can freely diffuse through cell membranes.[70] A number of lipid categories have been identified as signalling molecules and cellular messengers.[71] For instance, sphingosine-1-phosphate, a sphingolipid derived from ceramide, is a potent messenger molecule involved in regulating calcium mobilization, cell growth and apoptosis.[72] Diacylglycerols and the phosphatidylinositol phosphates are involved in calcium-mediated activation of protein kinase C.[73] Prostaglandins, which are one type of fatty-acid derived eicosanoid, are involved in inflammation and immunity.[74] Omega-3 PUFA, especially EPA and DHA, offer a wide range of health benefits and play a critical role in many bodily functions. Their roles in alleviating cancer, CVD, psychiatric disorders, Parkinson's disease, and inflammatory ailments have been well demonstrated.[75–78] For instance, the role of DHA in supporting the normal development of the brain, the eyes and the nerves, especially in infants and fetuses, has been well elucidated. Therefore, DHA has been an imperative ingredient in the diet of pregnant and lactating women as well as infant formula.[79] Omega-3 PUFA function mainly by changes in membrane composition and eicosanoid production and in regulating gene expression.[80] Omega-3 PUFA can profoundly alter the biochemical makeup of lipid rafts and caveolae membrane microdomains, thereby influencing cellular signalling.[81,82] EPA and arachidonic acid serve as precursors of the secondary signalling intermediate – eicosanoid. SFA play several beneficial roles in the human body although they are considered as bad fat to the public. For instance, SFA constitute at least 50% of the cell membranes and help maintain the necessary stiffness and integrity of the cells. Medium chain fatty acids are also quick energy suppliers.[83] They can be readily hydrolysed by pancreatic lipases in the small intestine and their metabolites are able to enter the circulatory system with sufficient water solubility. They travel rapidly to the liver where they can be quickly oxidized for energy. Furthermore, short chain fatty acids have antibacterial effects and therefore their presence in the colon prevents the establishment of pathogenic bacteria.[84]

It has been well-elucidated that PUFA are prone to oxidation, which leads to the deterioration of their nutritional values.[7] However, it was reported that lower levels of omega-3 fatty acids may possess antioxidant potential in the human body.[85,86] The intake of DHA and EPA can significantly inhibit the content of plasma lipid hydroperoxide levels in individuals with mild cognitive impairment, which suggested the capacity to reduce the oxidative stress of these omega-3 fatty acids.[85] In addition, it was revealed that DHA may exert an inhibitory effect toward Nox-4 in the tested individuals after the designed intake of the fatty acids.[86]

Fat-soluble minor components from dietary lipids exert their biological function through various mechanisms. For instance, astaxanthin, an important minor component in krill oil, which is a fat-soluble carotenoid, occurs at approximately 100 ppm (w/w) in commercially produced krill oil. It is the source of the dark red appearance of krill oil, which is from the algal diet of krill. Astaxanthin is a natural antioxidant that cannot only protect krill oil from oxidation, but also renders various health benefits through its free radical-quenching ability.[37,38] Kim *et al.*[16] investigated the protective effects of krill oil against high fat diet-induced cognitive impairment in mice. It was revealed that administration of krill oil improved spatial memory ability, learning, and memory abilities as well as reduced the contents of reactive oxygen species and nitric oxide compared with the control group, which indicated a reduced oxidative stress and potential function of preventing cognitive impairment. Pumpkin seed oil may inhibit testicular toxicity induced by Al_2O_3 NPs due to the potent antioxidant activity of its minor components, including vitamin A, vitamin B, carotenes, and tannins.

Other minor components that occur in edible oils and have a significant effect on their stability and possibly antioxidant potential have been mentioned briefly in each section, but it is important to emphasize the role of phospholipids as synergists with antioxidant potential.[87] In addition, certain non-tocopherol/tocotrienol constituents may exist in some oils such as tyrosol and hydroxytyrosol in olive oil which have many beneficial health effects.[88,89]

Some derivatives of fatty acids with phenolic compounds (lipophenols) have also been produced which serve as antioxidants for use in fats and oils as stabilizing supplements/additives and have received approval.[90] As an example, the palmitate ester of epigallocatechin gallate may be mentioned which has been approved in China and more recently in the United States.[7]

7.5 Safety Concerns of Dietary Lipid Supplements

There has always been public concern that dietary supplements should be safe and of good quality. Dietary lipid supplements, especially those from marine sources, can be contaminated by toxic materials from their sources or during manufacturing, transportation, and storage. Toxic chemicals such as methylmercury, dibenzodioxins, polychlorinated biphenyls, and chlordane that are concentrated in fish and marine mammals high in omega-3 PUFA may negate the beneficial effects of omega-3 PUFA, if present.[91] The aforementioned toxic substances are present at low levels in fresh waters and oceans, but they bioconcentrate in the aquatic food chain such that their levels are the highest in older, larger, predatory fish such as tuna and marine mammals such as seals that may reside in certain locations.[92] Oils obtained from these marine animals caught from such waters may contain high levels of toxic contaminants that are associated with several diseases. For instance, methylmercury can be transported freely across the blood–brain barrier and the placenta throughout the body. Studies have indicated that methylmercury is linked to subtle developmental deficits in children exposed *in utero* such as decreased memory function and attention deficits.[93] Methylmercury exposure in adults has been associated with an increased risk of CVD.[94] Therefore, contaminant levels should be evaluated in oils originally from suspicious areas of water.

The side effects of dietary lipid supplements should also be thoroughly evaluated to ensure their safety for human consumption. It was reported that fatty acid concentrates from marine sources may cause blood thinning and decreased platelet aggregation.[95] Therefore, individuals taking anticoagulant/antiplatelet drugs (blood thinners) may take extra caution and receive advice from practitioners before taking fish oil supplements due to the increased risk of bleeding. Low toxicity of black cumin seed oil has also been reported by different research groups.[96–98] Zaoui *et al.*[98] investigated the toxicity of black cumin seed fixed oil in mice and rats through determination of lethal dose 50 (LD_{50}) values and examination of possible biochemical, haematological, and histopathological changes. The results indicated that oral administration of fixed black cumin seed oil for 12 weeks decreased platelet counts by approximately 15–35% compared to the control. Individuals with known allergies to crustaceans or shellfish should also pay attention to krill oil as the presence of any residual protein material in the oil may cause allergy.[37] Moreover, dietary lipid supplements are rich in PUFA, which makes them extremely susceptible to oxidation.[99,100] This is why most fish oil supplements contain antioxidants such as mixed tocopherols. However, the content of oxidation products in the supplement should be monitored to ensure their quality and safety for human consumption. Furthermore, the purity and authenticity of dietary lipid supplements should also be strictly regulated by governmental agencies.[29]

7.6 Conclusion

Lipids are widely present in food and biological systems and their irreplaceable role in nutritional and health benefits has been well understood. Lipids containing PUFA and other bioactives are required for normal functioning of physiological systems in humans. With the growing public awareness of the nutritional, diet, and health benefits of specialty lipids, dietary lipid supplements are on demand. Both the benefits and risks of these supplements should be well understood and the safety of consuming oils from marine origin as well as those from terrestrial sources should be further evaluated in areas of environmental pollution or those with potential health risks.

References

1. Q. Q. Li, J. K. Wang and F. Shahidi, Chemical characteristics of cold-pressed blackberry, black raspberry and blueberry seed oils and the role of the minor components in their oxidative stability, *J. Agric. Food Chem.*, 2016, **64**, 5410–5416.
2. J. E. Kinsella, Food lipids and fatty acids: importance in food quality, nutrition, and health, *Food Technol.*, 1988, **42**, 129.
3. M. K. Youssef and S. Barbut, Effects of protein level and fat/oil on emulsion stability, texture, microstructure and color of meat batters, *Meat Sci.*, 2009, **82**, 228–233.
4. D. A. Forss, Role of lipids in flavors, *J. Agric. Food Chem.*, 1969, **17**, 681–685.
5. D. Boskou, Edible old pressed oils and their biologically active components, *J. Exp. Food Chem.*, 2017, **3**, 1–2.
6. W. Y. Oh, Y. Chioub, P. Leeb, M. Panb and F. Shahidi, Lipophilized resveratrol affects the generation of reactive nitrogen species in murine macrophages and cell viability of human cancer cell lines, *J. Food Bioact.*, 2019, **7**, 73–77.
7. J. Wang, H. Han, D. Wang, Y. Sun, J. Huang and F. Shahidi, Stability and stabilization of omega-3 oils: A review, *Trends Food Sci. Technol.*, 2021, **118**, 17–35.
8. C. Alasalvar, S. K. Chang, B. Bolling, W. Y. Oh and F. Shahidi, Specialty seeds: Nutrients, bioactives, bioavailability, and health benefits: A comprehensive review, *Compr. Rev. Food Sci. Food Saf.*, 2021, **20**, 2382–2427.
9. L. Casas-Cardoso, C. Mantell, S. Obregón, C. Cejudo-Bastante, A. Alonso-Moraga and E. J. M. de la Ossa, *et al.*, Health-promoting properties of borage seed oil fractionated by supercritical carbon dioxide extraction, *Foods*, 2021, **10**, 2471.
10. J. Wang, H. Han, D. Wang, P. Li and F. Shahidi, Conjugated fatty acids in muscle food products and their potential health benefits: a review, *J. Agric. Food Chem.*, 2020, **68**, 13530–13540.

11. Z. Yu, J. Wang and F. Shahidi, Two step-production of aclyglycerols containing a high proportion of docosapentaenoic acid from marine omega-3 oils and their oxidative stability, *J. Food Bioact.*, 2018, **4**, 123–129.
12. L. A. Daiello, A. Gongvatana, S. Dunsiger, R. A. Cohen and B. R. Ott, Association of fish oil supplement use with preservation of brain volume and cognitive function, *Alzheimer's Dementia*, 2015, **11**, 226–235.
13. P. K. J. P. D. Wanasundara, F. Shahidi and U. N. Wanasundara, Seal blubber oil: a novel source of ω3 fatty acids, *J. Food Lipids*, 1996, **3**, 293–306.
14. J. A. Conquer, L. A. Cheryk, E. Chan, P. A. Gentry and B. J. Holub, Effect of supplementation with dietary seal oil on selected cardiovascular risk factors and hemostatic variables in healthy male subjects, *Thromb. Res.*, 1999, **96**, 239–250.
15. E. C. Bonefeld-Jrgensen, S. M. Mller and S. M. Hansen, Modulation of atherosclerotic risk factors by seal oil: a preliminary assessment, *Int. J. Circumpolar Health*, 2001, **60**, 25–33.
16. J. H. Kim, H. J. Seo, Q. Q. Pang, Y. R. Kwon, J. Kim and E. J. Cho, Protective effects of krill oil on high fat diet-induced cognitive impairment by regulation of oxidative stress, *Free Radical Res.*, 2021, **55**, 1–11.
17. Z. Xu, H. Tang, F. Huang, Z. Qiao, X. Wang and C. Yang, *et al.*, Algal oil rich in n-3 PUFA alleviates DSS-induced colitis via regulation of gut microbiota and restoration of intestinal barrier, *Front. Microbiol.*, 2020, **11**, 615404.
18. R. Tan, F. Wang, C. Fan, X. Zhang, J. Zhao and J. Zhang, *et al.*, Algal oil rich in n-3 polyunsaturated fatty acids suppresses B16F10 melanoma lung metastasis by autophagy induction, *Food Funct.*, 2018, **9**, 6179–6186.
19. M. Khorshidi, M. Zarezadeh, O. M. Moghaddam, M. R. Emami, H. Kord-Varkaneh and S. M. Mousavi, *et al.*, Effect of evening primrose oil supplementation on lipid profile: A systematic review and meta-analysis of randomized clinical trials, *Phytother. Res.*, 2020, **34**, 2628–2638.
20. S. Y. Al-Okbi, S. M. El-Qousy, S. El-Ghlban and H. F. Moawa, Role of borage seed oil and fish oil with or without turmeric and alpha-tocopherol in prevention of cardiovascular disease and fatty liver in rats, *J. Oleo Sci.*, 2018, **67**, 1551–1562.
21. H. M. A. Abdelrazek, O. E. Kilany, M. A. A. Muhammad, H. M. Tag and A. M. Abdelazim, Black seed thymoquinone improved insulin secretion, hepatic glycogen storage, and oxidative stress in streptozotocin-induced diabetic male wistar rats, *Oxid. Med. Cell. Longevity*, 2018, **2018**, 8104165.
22. H. Hamdi, Testicular dysfunction induced by aluminum oxide nanoparticle administration in albino rats and the possible protective role of the pumpkin seed oil, *J. Basic Appl. Zool.*, 2020, **81**, 1–12.

23. E. N. Akang, A. A. Oremosu, O. O. Dosumu, C. C. Noronha and A. O. Okanlawon, The effect of fluted pumpkin (*Telferia occidentalis*) seed oil (FPSO) on testis and semen parameters, *Agric. Biol. J. North Am.*, 2011, **1**, 697–703.
24. B. A. Watkins, Y. Li, K. G. Allen, W. E. Hoffmann and M. F. Seifert, Dietary ratio of (n-6)/(n-3) polyunsaturated fatty acids alters the fatty acid composition of bone compartments and biomarkers of bone formation in rats, *J. Nutr.*, 2000, **130**, 2274–2284.
25. F. B. Hu, L. Bronner, W. C. Willett, M. J. Stampfer, K. M. Rexrode and C. M. Albert, *et al.*, Fish and omega-3 fatty acid intake and risk of coronary heart disease in women, *J. Am. Med. Assoc.*, 2002, **287**, 1815–1821.
26. V. Vashishtha, B. Barhwal, A. Kumar, S. K. Hota, O. P. Chaurasia and B. Kumar, Effect of seabuckthorn seed oil in reducing cardiovascular risk factors: a longitudinal controlled trial on hypertensive subjects, *Clin. Nutr.*, 2017, **36**, 1231–1238.
27. W. J. Hao, Z. Y. He, H. Y. Zhu, J. H. Liu, K. Kwek and Y. M. Zhao, *et al.*, Sea buckthorn seed oil reduces blood cholesterol and modulates gut microbiota, *Food Funct.*, 2019, **10**, 5669–5681.
28. S. Montserrat-de la Paz, M. A. Fernandez-Arche, M. Angel-Martin and M. D. Garcia-Gimenez, Phytochemical characterization of potential nutraceutical ingredients from evening primrose oil (Oenothera biennis L.), *Phytochem. Lett.*, 2014, **8**, 158–162.
29. A. A. Jairoun, M. Shahwan and H. Z. Sa'ed, Fish oil supplements, oxidative status, and compliance behaviour: Regulatory challenges and opportunities, *PLoS One*, 2020, **15**, e0244688.
30. C. Alasalvar, F. Shahidi and P. Quantick, Food and health applications of marine nutraceuticals: A review, in *Seafoods Quality, Technology and Nutraceutical Applications*, ed. C. Alasalvar and T. Taylor, Springer-Verlag, Berlin, Germany, 2002, pp. 175–204.
31. F. Shahidi and P. Ambigaipalan, Omega-3 polyunsaturated fatty acids and their health benefits, *Annu. Rev. Food Sci. Technol.*, 2018, **9**, 345–381.
32. J. Wang and F. Shahidi, Effect of enzymatic randomization on positional distribution and stability of seal blubber and menhaden oils, *J. Agric. Food Chem.*, 2011, **59**, 4232–4237.
33. J. Wang, E. Reyes Suárez, J. Kralovec and F. Shahidi, Effect of chemical randomization on positional distribution and stability of omega-3 oil triacylglycerols, *J. Agric. Food Chem.*, 2010, **58**, 8842–8847.
34. A. Haimeur, N. Meskini, V. Mimouni, L. Ulmann, H. Messaouri and F. Pineau-Vincent, *et al.*, A comparative study on the effect of argan oil *versus* fish oil on risk factors for cardio-vascular disease in high-fat-fed rats, *Nutrition*, 2019, **57**, 32–39.
35. J. Wang and F. Shahidi, Oxidative stability of marine oils as affected by added wheat germ oil, *Int. J. Food Prop.*, 2017, **20**(Suppl. 3), S3334–S3344.
36. H. O. Bang, J. Dyerberg and N. Hjorne, The composition of food consumed by Greenland Eskimos, *Acta Med. Scand.*, 1976, **200**, 59–73.

37. J. M. Kwantes and O. Grundmann, A brief review of krill oil history, research, and the commercial market, *J. Diet. Suppl.*, 2015, **12**, 23–35.
38. S. M. Ulven, B. Kirkhus, A. Lamglait, S. Basu, E. Elind and T. Haider, Metabolic effects of krill oil are essentially similar to those of fish oil but at lower dose of EPA and DHA, in healthy volunteers, *Lipids*, 2011, **46**, 37–46.
39. A. Köhler, E. Sarkkinen, N. Tapola, T. Niskanen and I. Bruheim, Bioavailability of fatty acids from krill oil, krill meal and fish oil in healthy subjects – a randomized, single-dose, cross-over trial, *Lipids Health Dis.*, 2015, **14**, 19.
40. K. Yurko-Mauro, J. Kralovec, E. Bailey-Hall, V. Smeberg, J. G. Stark and N. Salem Jr, Similar eicosapentaenoic acid and docosahexaenoic acid plasma levels achieved with fish oil or krill oil in a randomized double-blind four-week bioavailability study, *Lipids Health Dis.*, 2015, **15**, 1–9.
41. K. Lane, E. Derbyshire, W. L. Li and C. Brennan, Bioavailability and potential uses of vegetarian sources of omega-3 fatty acids: a review of the literature, *Crit. Rev. Food Sci. Nutr.*, 2014, **54**, 572–579.
42. Â. P. Matos, The impact of microalgae in food science and technology, *J. Am. Oil Chem. Soc.*, 2017, **94**, 1333–1350.
43. T. M. Karpiński and A. Adamczak, Fucoxanthin – an antibacterial carotenoid, *Antioxidants*, 2019, **8**, 239.
44. K. Miyashita and M. Hosokawa, Fucoxanthin in the management of obesity and its related disorders, *J. Funct. Foods*, 2017, **36**, 195–202.
45. M. Timoszuk, K. Bielawska and E. Skrzydlewska, Evening primrose (*Oenothera biennis*) biological activity dependent on chemical composition, *Antioxidants*, 2018, 7, 108.
46. W. W. Christie, The analysis of evening primrose oil, *Ind. Crop. Prod.*, 1999, **10**, 73–83.
47. F. Shahidi, Antioxidant factors in plant foods and selected oilseeds, *BioFactors*, 2008, **13**, 179–185.
48. R. H. Foster, G. Hardy and R. G. Alany, Borage oil in the treatment of atopic dermatitis, *Nutrition*, 2010, **26**, 708–718.
49. N. A. M. Eskin, Borage and evening primrose oil, *Eur. J. Lipid Sci. Technol.*, 2008, **110**, 651–654.
50. D. Fabrikova, J. L. Guil-Guerrerob, M. J. González-Fernándezb, I. Rodríguez-Garcíac, F. Gómez-Mercadod and M. Urrestarazu, *et al.*, Borage oil: Tocopherols, sterols and squalene in farmed and endemic-wild Borago species, *J. Food Compos. Anal.*, 2019, **83**, 103299.
51. S. Chung, K. Kong, K. Seong and Y. Cho, γ-Linolenic acid in borage oil reverses epidermal hyperproliferation in guinea pigs, *J. Nutr.*, 2002, **132**, 3090–3097.
52. C. Soto, R. Chamy and M. E. Zúñiga, Enzymatic hydrolysis and pressing conditions effect on borage oil extraction by cold pressing, *Food Chem.*, 2007, **102**, 834–840.

53. J. Heshmati and N. Namazi, Effects of black seed (*Nigella sativa*) on metabolic parameters in diabetes mellitus: A systematic review, *Complementary Ther. Med.*, 2015, **23**, 275–282.
54. S. Kheirouri, V. Hadi and M. Alizadeh, Immunomodulatory effect of *Nigella sativa* oil on T lymphocytes in patients with rheumatoid arthritis, *Immunol. Invest.*, 2016, **45**, 271–283.
55. R. Mahdavi, N. Namazi, M. Alizadeh and S. Farajnia, Effects of *Nigella sativa* oil with a low-calorie diet on cardiometabolic risk factors in obese women: a randomized controlled clinical trial, *Food Funct.*, 2015, **6**, 2041–2048.
56. A. Sahebkar, G. Beccuti, L. E. Simental-Mendia, V. Nobili and S. Bo, *Nigella sativa* (black seed) effects on plasma lipid concentrations in humans: a systematic review and meta-analysis of randomized placebo-controlled trials, *Pharmacol. Res.*, 2016, **106**, 37–50.
57. T. Potočnik, M. R. Cizej and I. J. Košir, Influence of seed roasting on pumpkin seed oil tocopherols, phenolics and antiradical activity, *J. Food Compos. Anal.*, 2018, **69**, 7–12.
58. S. N. Nakić, D. Rade, D. Škevin, D. Štrucelj, Ž. Mokrovčak and M. Bartolić, Chemical characteristics of oil from naked and husk seeds of *Cucurbita pepo* L., *Eur. J. Lipid Sci. Technol.*, 2006, **108**, 936–943.
59. C. B. Samue, K. K. D. Barine and E. E. Joy, Comparative assessment of the physicochemical properties and fatty acid profile of fluted pumpkin seed oil with some commercial vegetable oils in rivers state, Nigeria, *Res. J. Food Nutr.*, 2018, **2**, 32–40.
60. O. G. Fruhwirth and A. Hermetter, Production technology and characteristics of Styrian pumpkin seed oil, *Eur. J. Lipid Sci. Technol.*, 2008, **110**, 637–644.
61. J. Parry, L. Su, K. Zhou, M. P. Yurawecz, P. Whittaker and L. Yu, Fatty acid composition and antioxidant properties of cold-pressed marionberry, boysenberry, red raspberry, and blueberry seed oils, *J. Agric. Food Chem.*, 2005, **53**, 566–573.
62. B. S. Bushman, B. Phillips, T. Isbell, B. Ou, J. M. Crane and S. J. Knapp, Chemical composition of caneberry (Rubus spp.) seeds and oils and their antioxidant potential, *J. Agric. Food Chem.*, 2004, **52**, 7982–7987.
63. V. Van Hoed, N. De Clercq, C. Echim, M. Andjelkovic, E. Leber and K. Dewettinck, Berry seeds: A source of specialty oils with high content of bioactives and nutritional value, *J. Food Lipids*, 2009, **16**, 33–49.
64. M. E. Maitin, E. Grao-Cruces, M. C. Millan-Linares and S. Montserrat-de la Paz, Grape (*Vitis vinifera* L.) seed oil: a functional food from the winemaking industry, *Foods*, 2020, **9**, 1360.
65. W. J. Aronson, J. A. Glaspy, S. T. Reddy, D. Reese, D. Heber and D. Bagga, Modulation of omega-3/omega-6 polyunsaturated ratios with dietary fish oils in men with prostate cancer, *Urology*, 2001, **58**, 283–288.
66. Z. P. Li, J. Wang, Y. T. Xiong, Z. W. Li and S. W. Feng, The determination of the fatty acid content of sea buckthorn seed oil using near

infrared spectroscopy and variable selection methods for multivariate calibration, *Vib. Spectrosc.*, 2016, **84**, 24–29.
67. M. Kaushal and P. C. Sharma, Nutritional and antimicrobial property of seabuckthorn (Hippophae sp.) seed oil, *J. Sci. Ind. Res.*, 2011, **70**, 1033–1036.
68. A. Zielińska and I. Nowak, Abundance of active ingredients in sea-buckthorn oil, *Lipids Health Dis.*, 2017, **16**, 1–11.
69. F. Shahidi and S. P. J. N. Senanyake, Nutraceutical and specialty lipids, in *Nutralceutical and Specialty Lipids and their Co-products*, ed. F. Shahidi, CRC Press, Boca Raton, FL, 2006, pp. 1–26.
70. S. Ghosh, J. C. Strum and R. M. Bell, Lipid biochemistry: functions of glycerolipids and sphingolipids in cellular signaling, *FASEB J.*, 1997, **11**, 45–50.
71. K. M. Eyster, The membrane and lipids as integral participants in signal transduction, *Adv. Physiol. Educ.*, 2007, **31**, 5–16.
72. V. Hinkovska-Galcheva, S. M. Van Way, T. P. Shanley and R. G Kunkel, The role of sphingosine-1-phosphate and ceramide-1-phosphate in calcium homeostasis, *Curr. Opin. Invest. Drugs*, 2008, **9**, 1192–1205.
73. C. Klein and A. N. Malviya, Mechanism of nuclear calcium signaling by inositol 1,4,5-trisphosphate produced in the nucleus, nuclear located protein kinase C and cyclic AMP-dependent protein kinase, *Front. Biosci.*, 2008, **13**, 1206–1226.
74. J. A. Boyce, Eicosanoids in asthma, allergic inflammation, and host defense, *Curr. Mol. Med.*, 2008, **8**, 335–349.
75. H. Nemets, B. Nemets, A. Apter, Z. Bracha and R. H. Belmaker, Omega-3 treatment of childhood depression: A controlled, double-blind pilot study, *Am. J. Psychiatry*, 2006, **163**, 1098–1100.
76. F. Shahidi, Functional foods: their role in health promotion and disease prevention, *J. Food Sci.*, 2006, **64**, 146–149.
77. C. C. Wang, W. S. Harris, M. Chung, A. H. Lichtenstein, E. M. Balk and B. Kupelnick, *et al.*, n-3 Fatty acids from fish or fish-oil supplements, but not alpha-linolenic acid, benefit cardiovascular disease outcomes in primary- and secondary-prevention studies: a systematic review, *Am. J. Clin. Nutr.*, 2006, **84**, 5–17.
78. J. Wang and F. Shahidi, Acidolysis of *p*-coumaric acid with omega-3 oils and antioxidant activity of phenolipid products in *in vitro* and biological model systems, *J. Agric. Food Chem.*, 2013, **62**, 454–461.
79. J. Wang and F. Shahidi, Oxidative stability of marine oils as affected by added wheat germ, *Int. J. Food Prop.*, 2018, **20**, 3334–3344.
80. H. Sampath, Polyunsaturated fatty acid regulation of genes of lipid metabolism, *Annu. Rev. Nutr.*, 2005, **25**, 317–340.
81. D. W. L. Ma, S. Jeongmin, K. C. Switzer, Y. Fan, D. N. McMurray and J. R. Lupton, *et al.*, Chapkin n-3 PUFA and membrane microdomains: a new frontier in bioactive lipid research, *J. Nutr. Biochem.*, 2004, **15**, 700–706.

82. J. Wang and F. Shahidi, Antioxidant activity of monooleyl and dioleyl *p*-coumarates in *in vitro* and biological model systems, *Eur. J. Lipid Sci. Technol.*, 2014, **116**, 370–379.
83. B. Brinkmann, Quality criteria of industrial frying oils and fats, *Eur. J. Lipid Sci. Technol.*, 2000, **102**, 539–541.
84. M. Bonhoff, C. P. Miller and W. R. Martin, Resistance of the mouse's intestinal tract to experimental salmonella infection: 1. factors which interfere with the initiation of infection by oral inoculation, *J. Exp. Med.*, 1964, **120**, 805–816.
85. L. K. Lee, S. Shahar, N. F. Rajab, N. A. M. Yusoff, R. A. Jamal and S. M. Then, The role of long chain omega-3 polyunsaturated fatty acids in reducing lipid peroxidation among elderly patients with mild cognitive impairment: a case-control study, *J. Nutr. Biochem.*, 2013, **24**, 803–808.
86. E. Giordano and F. Visioli, Long-chain omega 3 fatty acids: Molecular bases of potential antioxidant actions, *Prostaglandins, Leukotrienes Essent. Fatty Acids*, 2013, **90**, 1–4.
87. E. B. Burlakova, N. M. Storozhok and N. G. Khrapova, Synergistic effect of antioxidants and phospholipids during oxidation of natural lipids, *Vopr. Pitan.*, 1990, **4**, 53–58.
88. A. K. Etherton and S. T. Omaye, Oxidation of olive oil fortified with quercetin, caffeic acid, tyrosol and hydroxytyrosol, *Nutr. Food Sci.*, 2015, **45**, 493–508.
89. A. Cañuelo, B. Gilbert-López, P. Pacheco-Liñán, E. Martínez-Lara, E. Siles and A. Miranda-Vizuete, Tyrosol, a main phenol present in extra virgin olive oil, increases lifespan and stress resistance in *Caenorhabditis elegans*, *Mech. Ageing Dev.*, 2012, **133**, 563–574.
90. Y. Zhong and F. Shahidi, Lipophilised epigallocatechin gallate (EGCG) derivatives and their antioxidant potential in food and biological systems, *Food Chem.*, 2012, **131**, 22–30.
91. L. Hooper, R. L. Thompson and R. A. Harrison, Risks and benefits of omega 3 fats for mortality, cardiovascular disease, and cancer: systematic review, *Br. Med. J.*, 2006, **332**, 752–760.
92. P. M. Kris-Etherton, W. S. Harris and L. J. Appel, Fish consumption, fish oil, omega-3 fatty acids, and cardiovascular disease, *Circulation*, 2002, **106**, 2747–2757.
93. G. Carrier, M. Bouchard and M. Caza, A toxicokinetic model for predicting the tissue distribution and elimination of organic and inorganic mercury following exposure to methyl mercury in animals and humans. II. Application and validation of the model in humans, *Toxicol. Appl. Pharm.*, 2001, **171**, 50–60.
94. J. T. Salonen, K. Seppänen, K. Nyyssönen, H. Korpela, J. Kauhanen and M. Kantola, *et al.*, Intake of mercury from fish, lipid peroxidation, and the risk of myocardial infarction and coronary, cardiovascular, and any death in eastern Finnish men, *Circulation*, 1995, **91**, 645–655.

95. P. Saravanan, N. C. Davidson, E. B. Schmidt and P. C. Calder, Cardiovascular effects of marine omega-3 fatty acids, *Lancet*, 2010, **376**, 540–550.
96. B. H. Ali and G. Blunden, Pharmacological and toxicological properties of *Nigella sativa*, *Phytother. Res.*, 2003, **17**, 299–305.
97. N. W. Liles, E. E. Page, A. L. Liles, S. K. Vesely, G. E. Raskob and J. N. George, Diversity and severity of adverse reactions to quinine: a systematic review, *Am. J. Hematol.*, 2016, **91**, 461–466.
98. A. Zaoui, Y. Cherrah, N. Mahassini, K. Alaoui, H. Amarouch and M. Hassar, Acute and chronic toxicity of *Nigella sativa* fixed oil, *Phytomedicine*, 2002, **9**, 69–74.
99. F. Shahidi, J. Wang and U. N. Wanasundara, Methods for measuring oxidative rancidity in fats and oils, in *Food Lipids: Chemistry, Nutrition, and Biotechnology*, ed. C. C. Akoh, CRC Press, Boca Raton, FL, 2017, pp. 519–538.
100. Z. X. Wu, Y. X. Liu, M. M. Yu, F. W. Yin, Z. Q. Liu and A. Li, *et al.*, Effect of chitosan coatings incorporated with antioxidants of bamboo leaves and potassium sorbate on lipid oxidation and hydrolysis of scallop (*Chlamys farreri*) during refrigerated storage, *J. Food Bioact.*, 2019, **8**, 66–73.

CHAPTER 8

Mechanism of the Anti-cancer Action of Natural Dietary Compounds

CHING-SHU LAI* AND ANWAR CHOIRUL

Department of Seafood Science, National Kaohsiung University of Science and Technology, Kaohsiung, Taiwan
*Email: chlai@nkust.edu.tw

8.1 Introduction

Despite considerable decline in mortality rates over the past few years, cancer (also known as malignant tumour or neoplasm) remains the second leading cause of death globally. The epidemiological data estimated 19.3 million new cancer cases in 2020, which is expected to rise to 28.4 million cases by 2040.[1] The global cancer incidence is still growing and continued efforts to strengthen prevention, diagnosis, therapeutics, and management are critical for the reduction in cancer incidence. Cancer is a complicated and multifactorial disease resulting from a strong interplay between genetic and environmental factors, which induce genetic mutations and genome instability in cells. The gene alterations drive cellular transformation and further evolve into abnormally proliferating cells with malignant features such as growth and metastasis. This process is extremely complex and involves genetic and epigenetic interactions for the dysregulation of gene expression to disrupt various cellular functions. The complexity of cancer is presented with its six hallmarks proposed by Hanahan and Weinberg[2]

Food Chemistry, Function and Analysis No. 39
Dietary Supplements with Antioxidant Activity: Understanding Mechanisms and Potential Health Benefits
Edited by Cesarettin Alasalvar, Fereidoon Shahidi and Chi-Tang Ho

Published by the Royal Society of Chemistry, www.rsc.org

in 2000: self-sufficiency in growth signals, insensitivity to anti-growth signals, limitless replicative potential, evasion of apoptosis, sustained angiogenesis, tissue invasion, and metastasis. These six hallmarks conceptualize the pathogenesis of cancer and its biology and serve as potential targets for disease therapeutics. Two new emerging hallmarks, avoiding immune destruction and deregulating cellular energetics, and two enabling characteristics, tumour-promoting inflammation and genome instability and mutation, have also now been included.[3] Advances in knowledge regarding the hallmarks of cancer and its enabling characteristics offer new approaches in cancer treatment. Recent research has revealed that tumours are heterogeneous and composed of a wide variety of cell types with distinct phenotypic and functional signatures in response to diverse environmental factors. This heterogeneity boosts the clonal evolution of tumour cells to sustain their growth and resist clinical therapy, leading to poor prognosis. Cancer is viewed as an ecosystem that involves multiple cellular communication processes not only between cancer cells and surrounding tissue, but also among organs in the body.[4] Although our understanding of tumorigenesis has progressed, the therapeutic challenge of cancer continues to exist because of its diverse nature. Preventive and therapeutic strategies should be designed and improved based on the recognition of cancer as a dynamic and systemic disease.

Some dietary natural compounds, namely resveratrol, curcumin, tetrahydrocurcumin, and (−)-epigallocatechin-3-gallate (EGCG), which exhibit anticancer potential through their antioxidative effects, and the underlying molecular mechanisms of their actions are summarized and discussed.

8.2 Reactive Oxygen Species in Tumorigenesis

Accumulating evidence has revealed that reactive oxygen species (ROS) are implicated in many hallmarks of cancer. ROS play a vital role in each stage of tumorigenesis by acting as signal transduction messengers or through interaction with macromolecules, thus leading to cellular transformation and tumorigenesis.[5] Excessive ROS cause gene mutations due to oxidative DNA damage, result in genome instability and mutation of oncogenes and tumour suppressor genes, which ultimately lead to oncogenic transformation for cancer initiation. Mutation of tumour suppressor genes, such as p53, phosphatase and tensin homolog (PTEN), and retinoblastoma protein, results in loss of anti-growth signals and apoptosis, which allows the transformation of normal cells to a malignant phenotype.[6] In addition to inducing gene mutations, increased ROS activate proliferative signals by modifying numerous proteins and enzymes, such as mitogen-activated protein kinase (MAPK), phosphoinositide 3-kinase (PI3K)/Akt, and IκB kinase/nuclear-factor kappa B (IKK/NF-κB), which lead to the survival and uncontrolled proliferation of cancer cells. High levels of ROS also cause terminal oxidation of proteins (sulfonic acid and carbonyl), which usually leads to complete protein dysfunction and degradation.[7] ROS also induce lipid peroxidation by damaging polyunsaturated fatty acids (PUFA) in the cellular membrane or organelle membrane, initiating a self-perpetuating chain reaction that generates a variety of lipid peroxides and breakdown products (*e.g.*, aldehydes, malondialdehyde, and 4-hydroxynonenal) to exacerbate the oxidative damage of macromolecules. In addition, oxidative degradation of lipids alters the structural and functional properties of cellular membranes and increases the mutational load of transformed cells to facilitate their malignant phenotype and proliferation.

To facilitate tumour growth, cancer cells elevate ROS production to sustain the signalling pathways necessary for transformation, survival, proliferation, and hypoxia adaption.[6] Elevation of ROS in cancer cells might be caused by chronic inflammation, aberrant metabolism, mitochondrial dysfunction, and hypoxia. Among others, chronic inflammation is a hallmark of various human diseases and cancers. Cancer cells and tumour-infiltrating inflammatory/immune cells create a sustained inflammatory/oxidative microenvironment by producing numerous cytokines, growth factors, ROS, and reactive nitrogen species (RNS) that favours further oxidative stress and deleterious inflammation to promote tumour growth.[8] In addition, mutations in several mitochondrial DNA (mtDNA)-encoded components involved in the electron transport chain impair electron transfer, further leading to electron leakage, generation of superoxide, and subsequent oxidative stress.[9] High levels of mitochondrial ROS further induce widespread oxidative damage to macromolecules and aberrant signalling pathways, such as hypoxia-inducible factor (HIF), that facilitate angiogenesis, invasion, and metastasis.[9] To adapt to oxidative stress, cancer cells evolve their antioxidant mechanisms or induce autophagy to reduce the oxidative damage of macromolecules and organelles. Upregulation of antioxidant defence in cancer cells favours not only tumorigenesis but also resistance to chemotherapy by enhancing survival and apoptosis.[10] Mitochondrial and nicotinamide adenine dinucleotide phosphate (NADPH) oxidase 4 (NOX4)-derived ROS are linked to cell migration and invasion by the upregulation of matrix metalloproteinases (MMPs).[11]

Metabolic dysregulation is another important mechanism involved in the effects of ROS on tumour growth. Several important metabolic pathways are elevated in cancer cells, contributing to macromolecular synthesis, including glycolysis, the pentose phosphate pathway, glycerol synthesis, one-carbon metabolism, and the hexosamine pathway.[12–14] ROS-mediated inactivation of the tricarboxylic acid cycle (TCA) and mitochondrial components directly and indirectly impairs mitochondrial oxidative phosphorylation and ROS detoxification, inducing HIF activation and metabolic switch.[9] The inactivation of the TCA enzymes results in the accumulation of metabolic intermediates, which further disrupts the TCA cycle, thus flipping the metabolic switch in the cell to glycolysis. Moreover, citrates are used in the biosynthesis of lipids, which play important roles in cell proliferation.[15] The metabolic switch in cancer cells not only satisfies the anabolic demands of macromolecular biosynthesis but also maintains redox homeostasis.

Studies have demonstrated that the pentose phosphate pathway generates NADPH, and glutaminolysis-derived glutathione (GSH) is a compensatory mechanism for ROS detoxification that benefits cancer cell survival and tumour progression.[16,17]

Collectively, the abovementioned studies reveal that increased ROS production is necessary for boosting malignant transformation, proliferation, and tumorigenesis by modulating protein function, enzyme activity, signal transduction, gene transcription, and membrane integrity. Elevated ROS also play a central role in the metabolic reprogramming of cancer cells to adapt to nutritional and hypoxic stresses. These observations also provide a preventive and therapeutic strategy against cancer with the use of nutraceuticals and pharmaceuticals by targeting oxidative stress.

8.3 Anticancer Potential of Dietary Natural Compounds: Targeting Oxidative Stress

Although the role of ROS in cancer cells remains controversial, disruption of redox homeostasis and signalling is a common characteristic in many types of cancers. High levels of ROS influence a wide array of cellular responses and participate in almost all hallmarks of cancer; thus, the regulation of oxidative stress has been a promising strategy for cancer prevention and treatment.[10,18] Natural products have been attractive for drug discovery in the past several decades because of their structural diversity and a broad range of bioactivities, particularly their antioxidative properties.[19] Many natural dietary antioxidants exert beneficial effects against various diseases by influencing redox homeostasis.[20–22] They diminish oxidative stress through multiple mechanisms, including direct scavenging of ROS and RNS, interruption of oxidative reactions, inactivation of ROS-generating enzymes, upregulation of redox enzymes and proteins, and interaction with various redox signalling pathways. This chapter provides a summary of selected natural dietary compounds identified in recent years, which exert anticancer effects targeting the modulation of redox homeostasis (Figure 8.1).

8.3.1 Resveratrol

The antioxidative activity of resveratrol (3,4′,5-trihydroxystilbene) usually contributes to its chemopreventive effect for the suppression of tumorigenesis. In addition to directly scavenging ROS and RNS, resveratrol restrains the generation of free radicals and their damaging reactions induced by exogenous toxicants. A report on the Epstein–Barr virus (EBV)-transformed B-lymphoblastoid cell line showed that resveratrol was effective in reducing phorbol 12-myristate 13-acetate (PMA), a tumour promoter, stimulated H_2O_2 production. Because NADPH oxidase 2 (NOX2) is considered a major source of ROS production in B cells, the effect of resveratrol on PMA-stimulated H_2O_2 generation was attributed to the suppression of

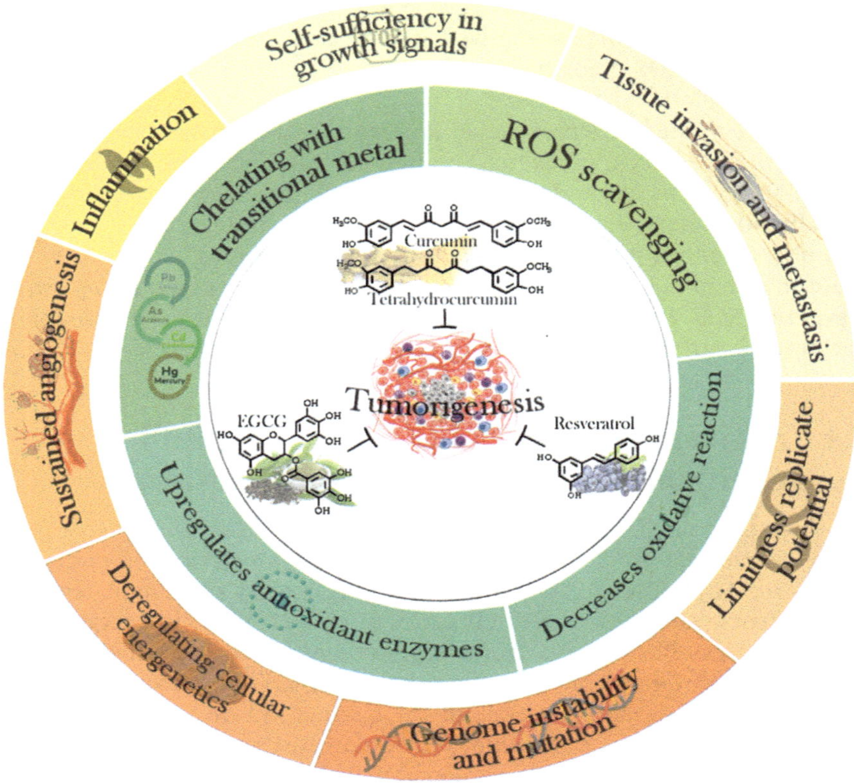

Figure 8.1 Proposed antioxidative mechanism of natural dietary compounds for anti-tumorigenesis. Abbreviations: EGCG, (−)-epigallocatechin-3-gallate; ROS, reactive oxygen species.

NOX2[23] (Table 8.1). Moreover, xanthine/xanthine oxidase-dependent superoxide generation was decreased by resveratrol, but the mechanism was not investigated.[23] However, a previous study showed that resveratrol directly scavenged superoxide anions and also inhibited xanthine oxidase activity in the xanthine/xanthine oxidase system.[24] These results revealed the ability of resveratrol to block endogenous ROS production by enzymes such as NOXs and xanthine oxidase. Reduction in mitochondrial ROS by resveratrol may inhibit malignant transformation. The Bhas 42 cells (v-Ha-ras-transfected mouse embryo fibroblast cells) were incubated with benzo[*a*]pyrene (B[*a*]P), a well-known carcinogen, which induced H_2O_2 and mitochondrial superoxide generation, accompanied by increased cell proliferation and colony formation, suggesting a neoplastic transformation. Resveratrol significantly reduced B[*a*]P-induced cell transformation by decreasing ROS production, and gene expression responded to mitochondrial oxidative stress and inflammation such as tumour necrosis factor-alpha (TNF-α). B[*a*]P was found to cause mitochondrial dysfunction by downregulating the mitochondrial

Table 8.1 Anticancer mechanism of resveratrol targeting oxidative stress.

Targeting hallmarks in cancer	Experimental model	Treatment	Molecular mechanisms for antioxidative activity	Biological outcomes	Ref.
↓ Genome instability and mutation	• PMA exposed B-LCL • Xanthine/xanthine oxidase	0.5–100 μM	↓ NOX2-dependent H_2O_2 production ↓ O_2 production	—	23
	H_2O_2 exposed porcine IPEC-J2	20 and 50 μM	↑ Nrf2 ↑ SOD, CAT, GPx, and HO-1	↑ Cell viability ↑ Tight junction protein	26
	Nicotine exposed male Wistar rats	20 mg kg^{-1}, 4 weeks, and oral	↓ Lipid peroxidation ↓ H_2O_2 and NO production ↓ XO and myeloperoxidase ↑ SOD, CAT, and GPx	↓ Lung pathology	27
↓ Genome instability and mutation ↓ Inflammation	Carbon black nanoparticle exposed human lung epithelial cells (A549)	1–10 μM	↓ H_2O_2 production ↓ NOX2 protein ↓ PKCα protein ↓ iNOS and COX-2	↓ Nuclear condensation ↑ Cell viability	28
	DEN/Fe-NTA exposed male Wistar rats	30 mg kg^{-1}, 24 weeks, and oral	↓ Lipid peroxidation ↓ XO ↑ SOD, CAT, and GPx ↑ Nrf2 and HO-1 ↓ NF-κB and STAT3 ↓ TGF-β1, IL-6, and TNF-α	↑ Renal function ↓ Kidney pathology (clear cells and lymphoid aggregates)	29
↓ Limitless replicative potential	B[*a*]P-exposed mouse embryo-fibroblasts (Bhas 42)	5 μM	↓ H_2O_2 production ↓ Mitochondrial O_2 –	↑ Mitochondrial content and function ↓ Cell proliferation ↓ Colonies	25

Table 8.1 (*Continued*)

Targeting hallmarks in cancer	Experimental model	Treatment	Molecular mechanisms for antioxidative activity	Biological outcomes	Ref.
↓ Tissue invasion and metastasis	Human pancreatic cancer cells (BxPC-3, Panc-1) under hypoxia	12.5–50 μM	↓ H_2O_2 production ↓ HIF-1α	↓ Migration and invasion ↓ uPA and MMP-2	31
	Human pancreatic stellate cells	50 μM	↓ H_2O_2 production	↓ Stellate cell activation ↓ Migration and invasion ↓ Glycolysis	32
	Human pancreatic stellate cells under hypoxia	50 μM	↓ HIF-1α	↓ Stellate cell activation ↓ Invasion and EMT	33
	High glucose exposed human pancreatic cancer cells (Panc-1)	50 μM	↓ H_2O_2 production	↓ Proliferation ↓ Migration and invasion ↓ p-ERK, p38, and p-NF-κB	34

Abbreviations: B[*a*]P, benzo[*a*]pyrene; B-LCL, B-lymphoblastoid cell line; CAT, catalase; COX-2, cyclooxygenase-2; DEN, diethylnitrosamine; EMT, epithelial–mesenchymal transition; ERK, extracellular signal-regulated kinase; Fe-NTA, ferric nitrilotriacetate; GPx, glutathione peroxidase; HIF-1α, hypoxia-inducible factor-1alpha; HO-1, heme oxygenase-1; IL-6, interleukin-6; iNOS, inducible nitric oxide synthase; IPEC-J2, intestinal porcine enterocyte cell line; MMP-2, matrix metalloproteinase-2; NF-κB, nuclear factor-kappa B; NO, nitric oxide; NOX2, NADPH oxidase 2; Nrf2, nuclear factor erythroid 2-related factor 2; PKCα, protein kinase C alpha; PMA, phorbol 12-myristate 13-acetate; p38, phospho-38; SOD, superoxide dismutase; STAT3, signal transducer and activator of transcription 3; TGF-β, transforming growth factor-beta; TNF-α, tumour necrosis factor-alpha; uPA, urokinase-type plasminogen activator.

content and Sirtuin 1 activity, followed by increased mitochondrial superoxide generation, leading to cell transformation. However, the impaired mitochondrial function and increased mitochondrial superoxide were ameliorated by resveratrol, thus contributing to the reduction of B[*a*]P-induced malignant transformation.[25] Upregulation of detoxifying enzymes is another mechanism by which resveratrol reduces oxidative stress and cellular damage. Resveratrol protected against H_2O_2-induced disruption of barrier intestinal function and apoptosis in an intestinal porcine enterocyte cell line (IPEC-J2). The reduced markers of tight junctions, including claudin-1, occludin, and zonula occludens-1 (ZO-1), caused by H_2O_2 in IPEC-J2 cells, were upregulated by resveratrol. This cytoprotective effect of resveratrol against ROS-induced intestinal barrier dysfunction was mediated by the induction of nuclear factor erythroid 2-related factor 2 (Nrf2) and its downstream antioxidant enzymes, superoxide dismutase (SOD), catalase (CAT), and glutathione peroxidase (GPx).[26] A similar result was observed in Wistar rats with nicotine-induced lung damage. Oral administration of resveratrol (30 $mg\,kg^{-1}$) for 4 weeks effectively improved nicotine-induced histopathological changes in lung tissue, such as scattered congestion of peribronchiolar, severe congestion of the alveolar lung, and inflammatory infiltrates. In addition to suppression of xanthine oxidase and myeloperoxidase, and thus further ROS production, as well as lipid peroxidation, resveratrol exerts its protective effect against lung injury caused by nicotine, as suggested by the elevation of SOD, CAT, and GPx levels.[27]

Resveratrol also attenuated carbon black nanoparticle-induced H_2O_2 generation and cytotoxicity in human lung epithelial A549 cells. The inhibitory effect of resveratrol is exerted *via* downregulation of protein kinase C (PKC) and NOX2, as evidenced by the use of specific inhibitors for both.[28] NOX-dependent ROS production has been implicated in lung inflammation. In A549 cells, carbon black nanoparticles upregulated inflammatory inducible nitric oxide synthase (iNOS) and cyclooxygenase-2 (COX-2) was reduced by resveratrol, *N*-acetylcysteine (NAC), and diphenyleneiodonium (NOX inhibitor), indicating that the antiinflammatory activity of resveratrol may be mediated by eliminating ROS.[28] In a diethylnitrosamine (DEN)/ferric nitrilotriacetate (Fe-NTA)-induced renal carcinogenesis model, Wistar rats orally treated with resveratrol at 30 $mg\,kg^{-1}$ for 24 weeks displayed improved renal function, as evidenced by decreased blood urea, serum creatinine, an increase in creatinine clearance, and a significant decrease in clear cells and lymphoid aggregates in the renal cortex, suggesting an ameliorative effect on clear cell renal cell carcinoma. The authors reported that this ameliorative effect may be because of the antioxidative properties of resveratrol, not only by inhibiting lipid peroxidation and xanthine oxidase, but also by upregulating Nrf2 and downstream heme oxygenase-1 (HO-1), SOD, CAT, and GPx, which exerts a renal protective effect.[29] In the same study, resveratrol also downregulated NF-κB and signal transducer and activator of transcription 3 (STAT3), as well as the downstream inflammatory cytokines transforming growth factor-β (TGF-β), interleukin (IL)-6, and TNF-α in renal carcinoma tissue from rats treated with DEN/Fe-NTA, followed by reduced lipid peroxidation and increased antioxidant enzymes. Fe-NTA is known to induce lipid peroxidation and promote renal carcinogenesis.[30] Therefore, downregulation of inflammatory mediators in tumour tissue by resveratrol is likely to attenuate lipid peroxidation caused by Fe-NTA.[29]

Tumour hypoxia-derived ROS are associated with cancer malignancy by enhancing invasion, angiogenesis, and metastasis. In human pancreatic cancer cell lines (BxPC-3 and Panc-1), hypoxia-induced H_2O_2 production, invasion, and migration were reduced by resveratrol, together with downregulation of hypoxia-inducible factor-1alpha (HIF-1α), urokinase-type plasminogen activator (uPA), and matrix metalloproteinase-2 (MMP-2). NAC also suppressed ROS production, invasion, and migration in both BxPC-3 and Panc-1 cells, suggesting that the inhibitory effect of resveratrol in pancreatic cancer cells was mediated by the downregulation of ROS under hypoxia.[31] Activation of pancreatic stellate cells produces ROS, and cytokines play an important role in pancreatic fibrosis and contribute to the malignant behaviour of cancer cells. Resveratrol alleviated H_2O_2-induced ROS production in human pancreatic stellate cells along with their activation, which manifested as elevated lipid content and a decrease in alpha-smooth muscle actin (α-SMA) protein. H_2O_2-induced migration and invasion of pancreatic stellate cells was suppressed by resveratrol. A similar result was found in cells treated with NAC, suggesting the improvement effect of resveratrol on malignant progression of pancreatic cancer by reducing ROS-induced pancreatic stellate cell activation, migration, and invasion.[32] Hypoxia caused the activation of pancreatic stellate cells but was inhibited by resveratrol, followed by the downregulation of HIF-1α, vascular endothelial growth factor (VEGF), and IL-6. Resveratrol also attenuated the invasive ability of pancreatic stellate cells under hypoxia, accompanied by a decrease in the epithelial–mesenchymal transition (EMT) markers, N-cadherin and vimentin. In addition, the culture medium of pancreatic stellate cells under hypoxic conditions facilitated pancreatic cancer Panc-1 and Mia Paca-2 cell invasion and enhanced the EMT markers. Hypoxia-induced pancreatic stellate cell activation and invasion, as well as increased invasiveness of pancreatic cancer cells, were abolished by HIF-1α and small interfering RNA (siRNA), indicating that the inhibitory effect of resveratrol on pancreatic stellate cells is most likely due to the targeting of HIF-1α. This study also demonstrated the anticancer potential of resveratrol in the disruption of the crosstalk between pancreatic cancer cells and pancreatic stellate cells to reduce malignancy by regulating HIF-1α.[33] Another study showed that resveratrol inhibited high glucose (25 mM)-induced proliferation, invasion, and migration of pancreatic cancer Panc-1 cells, followed by decreased ROS production and uPA, and increased E-cadherin. High glucose levels activated extracellular signal-regulated kinase (ERK), p38 mitogen-activated protein kinase (p38 MAPK), and NF-κB signalling, which further promoted invasion and migration in Panc-1 cells but was attenuated by resveratrol. Specific kinase inhibitors and NAC confirmed the role of ERK, p38 MAPK, and ROS in the proliferative, invasive, and migratory ability of pancreatic cancer cells. This study suggested that resveratrol diminished ROS-mediated proliferation and malignant progression of pancreatic cancer cells caused by high glucose levels, which implies that it may serve as a potential candidate for suppressing the malignant progression of pancreatic cancer by targeting hyperglycemia-driven ROS.[34]

8.3.2 Curcumin

The antioxidative effect is one of the potential mechanisms by which curcumin works against cancer. Curcumin is a typical antioxidant that exerts scavenging activity against various radicals and reactive oxidants. The phenolic hydroxyl groups and unique β-diketone moiety contribute to the direct radical scavenging ability of curcumin.[35,36] In addition, the β-diketone moiety is responsible for the metal-chelating activity of curcumin.[35] In rat splenocytes, curcumin (5 and 10 μM) treatment increased cell viability and reduced apoptosis caused by copper exposure, followed by amelioration of oxidative DNA damage with decreased micronuclei and comet formation (Table 8.2). The mechanism by which curcumin prevents copper-induced spleen dysfunction has not been fully discussed, but may be attributed to its ability to chelate copper, which has been widely investigated.[37] An animal study revealed that male Wistar rats administered with a single dose (5 mg kg^{-1}) of mercuric chloride ($HgCl_2$) exhibited marked H_2O_2 production in the erythrocytes. $HgCl_2$ treatment increased the level of protein carbonyl and reduced total thiol in the erythrocyte membrane and blood plasma of the Wistar rats. However, pretreatment of Wistar rats with curcumin for 21 days significantly protected them against $HgCl_2$-induced oxidative stress by ameliorating the generation of H_2O_2 and protein carbonyl, and increasing total thiol.[38]

Many studies have revealed that curcumin attenuates lipid peroxidation and upregulates antioxidant enzymes to protect cells under oxidative stress. Aflatoxin B1 is a mycotoxin and hepatocarcinogen produced by the *Aspergillus* fungi. It is usually converted to a highly reactive electrophilic intermediate (aflatoxin B1 exo-8,9-epoxide) by cytochrome P450 (CYP450) and further forms a DNA adduct in the liver cells, which contributes to hepatic carcinogenesis. In fatal hepatocytes, curcumin pre-treatment significantly reduced aflatoxin B1-induced cytotoxicity. The decreased level of aflatoxin B1 metabolites in the culture medium was caused by curcumin treatment with reduced cytochrome P450 3A (CYP3A) enzyme activity, indicating that curcumin mitigated its metabolism. In addition, lipid peroxidation was elevated in fatal hepatocytes by aflatoxin B1 but was markedly diminished by curcumin, followed by the induced NAD(P)H quinone dehydrogenase-1 (NQO-1) antioxidant enzyme activity. This suggests that curcumin exhibits hepatoprotective effects by not only mitigating the bioactivation of the toxicant but also elevating antioxidant defence.[39] Ultraviolet B (UVB) is known to trigger the overproduction of ROS that is involved in many skin disorders and is

Table 8.2 Anticancer mechanism of curcumin targeting oxidative stress.

Targeted hallmarks in tumorigenesis	Experimental model	Treatment	Molecular mechanisms for antioxidative activity	Biological outcomes	Ref.
↓ Genome instability and mutation	Copper exposed splenocytes	5 and 10 µM	↓ Oxidative DNA damage	↑ Viability ↓ Apoptosis	37
—	$HgCl_2$ exposed male Wistar rats	340 mg kg^{-1}, 21 days, and oral	↓ H_2O_2 production ↓ Protein oxidation ↑ Total thiol (–SH)	↓ Oxidative stress	38
—	Aflatoxin B1 exposed bovine fetal hepatocyte-derived cells (BFH12)	2.5–40 µM	↑ Antioxidant enzyme (NQO-1) ↓ Phase I enzyme (CYP3A) ↓ Lipid peroxidation	↑ Cell viability ↓ Aflatoxin B1 metabolites	39
—	UVB exposed human skin organ	6.8–27.1 mM (micro-emulsion)	↑ Radical scavenging ↑ Antioxidant enzyme	↓ Apoptosis	40
↓ Self-sufficiency in growth signals ↓ Limitless replicative potential	Human non-small cell lung cancer cells (A549)	2.5–80 µM	↓ H_2O_2 production ↑ Antioxidant enzyme (SOD and γ-GCS)	↓ Proliferation ↓ Wnt/β-catenin signalling	42
↓ Limitless replicative potential	Mice with Ehrlich solid tumours	25 and 50 mg kg^{-1}	↓ Lipid peroxidation ↑ Antioxidant enzyme (CAT and SOD)	↓ Tumour growth	43
—	• Cisplatin-resistant epithelial cancer cells (HONE1) • Cisplatin-resistant HONE1 xenograft mouse model	• 2 and 3.5 µM • 5 mg kg^{-1}, twice weekly for 32 days	↓ NOX5	↓ Cell viability ↓ Tumour growth	44
↓ Deregulating cellular energetics	Female Wistar-Furth rats fed ethanol and ω-3 PUFA	150 mg kg^{-1}, 8 weeks, and oral	↓ Lipid peroxidation ↑ GSH	↓ Body weight ↓ Dyslipidaemia ↓ Hepatic steatosis and injury	45

—	Streptozotocin treated male SD rats	1.5 g kg^{-1}, 21 days, and oral	↓ Lipid peroxidation ↑ Antioxidant enzyme ↑ GSH	↓ Blood glucose ↑ Hepatic glycogen	46
—	Human lung (H1299), breast (MCF-7), cervical (HeLa), and prostate (PC3) cancer cells	2.5–20 μM	↓ HIF-1α	↓ Glycolysis ↓ Pyruvate kinase M2 ↓ Cell viability	47
↓ Inflammation	IL-1β exposed primary hepatocytes	5–25 μM	↓ NO production ↓ iNOS	↓ Proinflammatory cytokines ↓ NF-κB activation ↓ Akt signalling	48
—	CCl_4 exposed male C57BL/6 mice	200 mg kg^{-1}, 4 weeks, and oral	↓ Inflammatory infiltration ↓ Kupffer cell activation ↓ iNOS	↓ Liver inflammation and fibrosis ↓ Proinflammatory cytokines and chemokines	49
—	L-Arginine exposed male ICR mice	25 and 50 mg kg^{-1}, 3 days	↓ Inflammatory infiltration ↓ Lipid peroxidation ↓ Myeloperoxidase	↓ Pancreatitis ↓ Pancreatic inflammation, edema, and necrosis ↓ Apoptotic acinar cells	50
↓ Tissue invasion and metastasis	H_2O_2 exposed pancreatic cancer cells (BxPC-3 and Panc-1)	5–40 μM	↓ H_2O_2 production	↓ ERK and NF-κB signalling ↓ MMP-2 and MMP-9 ↓ Migration and invasion	51
—	SOD-stimulated pancreatic cancer cells (BxPC-3 and Panc-1)	20 μM	↓ H_2O_2 production	↓ Mesenchymal phenotype and markers	52

Table 8.2 (*Continued*)

Targeted hallmarks in tumorigenesis	Experimental model	Treatment	Molecular mechanisms for antioxidative activity	Biological outcomes	Ref.
				↓ Akt and NF-κB signalling ↓ Migration and invasion	
↓ Sustained angiogenesis ↓ Tissue invasion and metastasis	HCC cells (HepG2), human hepatic stellate cells (HSCs), and human umbilical vein endothelial cells (HUVECs)	50 μM	↓ H_2O_2 production ↓ HIF-1α ↑ Nrf2 and GSH	↓ Angiogenesis ↓ VEGF and CTGF ↓ Mesenchymal marker ↓ Invasion and MMP-9	53
	• TGF-β1 exposed mouse embryonic hepatocytes (BNL CL.2 cells) • CCl_4 exposed SD rats	• 10–30 μM • 100, 200, and 400 mg kg^{-1}, 8 weeks (every other day), and oral	↓ H_2O_2 production ↑ GSH ↑ PPARα	↓ Hepatic fibrosis ↓ Mesenchymal and fibrogenesis marker ↓ TGF-β/Smad signalling ↑ Autophagy	54

Abbreviations: CAT, catalase; CCl_4, carbon tetrachloride; CTGF, connective tissue growth factor; CYP3A, cytochrome P450 3A; ERK, extracellular signal-regulated kinase; γ-GCS, gamma-glutamylcysteine synthetase; GSH, glutathione; HIF-1α, hypoxia-inducible factor-1alpha; ICR, Institute of Cancer Research; IL-1β, interleukin-1beta; iNOS, inducible nitric oxide synthase; MMP-2, matrix metalloproteinase-2; NF-κB, nuclear factor-kappa B; NO, nitric oxide; NOX5, NADPH oxidase 5; NQO-1, NAD(P)H quinone dehydrogenase-1; Nrf2, nuclear factor erythroid 2-related factor 2; PPARα, peroxisome proliferator-activated receptor-alpha; ω-3 PUFA, omega-3 polyunsaturated fatty acid; UVB, ultraviolet B; SD, Sprague-Dawley; Smad, mothers against decapentaplegic homolog; SOD, superoxide dismutase; TGF-β, transforming growth factor-beta; VEGF, vascular endothelial growth factor; Wnt, wingless-type MMTV integration site family.

responsible for the development of skin cancer. In an *ex vivo* study, the human skin, pretreated with a curcumin-loaded microemulsion, but not free curcumin, showed significantly reduced epidermal apoptosis by UVB-irradiation. This may result from the elevation of HO-1, an antioxidant enzyme, and thus exhibited the protective effect of the curcumin microemulsion against UVB-induced skin injuries and damages.[40] The mechanism by which curcumin upregulates antioxidant enzymes is linked to the activation of Nrf2, an important transcription factor for maintaining redox balance by regulating the gene expression of phase II detoxifying/antioxidant enzymes including SOD, glutathione reductase (GR), HO-1, and NQO-1.[41] Alleviation of tumour cell proliferation potentiated by oxidative stress is involved in the anticancer mechanism of curcumin. Curcumin decreased the viability and proliferation of human non-small cell lung cancer cells (A549), elevated SOD and gamma-glutamylcysteine synthetase (γ-GCS), and reduced H_2O_2 levels. NAC pre-treatment mitigated a curcumin-mediated decrease in viability following an increase in ROS production and a decrease in antioxidant enzymes, confirming the role of curcumin in regulating oxidative stress in A549 cell death. Curcumin induced reduction of A549 cell proliferation by suppressing wingless-type MMTV integration site family (Wnt)/β-catenin signalling and downregulating cyclin D1. The inhibitory effect of curcumin on Wnt/β-catenin signalling was blocked by NAC, clearly demonstrating that curcumin inhibited A549 cell proliferation by targeting oxidative stress-mediated Wnt/β-catenin signalling.[42] Curcumin has been shown to mitigate tumour growth by reducing lipid peroxidation. In mice with Ehrlich solid tumours, tumour growth was accompanied by increased lipid peroxidation in the liver, kidneys, and testis. However, mice with tumours were subcutaneously treated with 25 and 50 mg kg^{-1} curcumin, which suppressed tumour progression by alleviating lipid peroxidation.[43] Cisplatin resistance is a major challenge in the clinical treatment of various cancers. Cancer cells upregulate NADPH oxidase to generate ROS, which potentiates cisplatin resistance. Chronic cisplatin exposure significantly upregulated NOX5 and increased Akt signalling in the nasopharyngeal carcinoma cell line (HONE1). However, liposomal curcumin sensitized cisplatin to induce a decrease in cell viability *via* downregulation of NOX5 in cisplatin-resistant HONE1 cells. The chemosensitizing effect also occurred in the cisplatin-resistant HONE1 xenograft model, as evidenced by the reduced tumour volume and tumour weight.[44]

Some studies have shown that curcumin improves the dysregulated cellular metabolism to counteract tumorigenesis. In female Wistar-Furth rats with chronic omega-3 (ω-3) PUFA and ethanol exposure, oral treatment with 150 mg kg^{-1} curcumin for 8 weeks significantly reduced body weight gain and improved dyslipidemia. Curcumin administration also reduced liver weight, liver injury, and hepatosteatosis caused in chronic ω-3 PUFA and ethanol-fed rats. Curcumin abrogated the deleterious effects of chronic ethanol with ω-3PUFA on liver injury, which is linked to the decreased lipid peroxidation and elevated GSH.[45] The elevating effect of the antioxidant enzyme, caused by curcumin, is also found in diabetes, which is characterized by hyperglycemia and is linked to the risk of cancer due to oxidative stress. The type 1 diabetic male Sprague-Dawley (SD) rats, which were orally administered with 1.5 mg kg^{-1} curcumin for 21 days, showed reduced blood glucose. The enzymes involved in glucose metabolism, such as glucose-polymerizing enzyme and glucose synthetase, were markedly upregulated by curcumin, thus resulting in increased hepatic glycogen and leading to blood glucose homeostasis. In addition to the modulating effect on glucose metabolizing enzymes, oxidative stress was alleviated by curcumin treatment in diabetic rats, including elevated hepatic GPx and GSH. The plasma lipid peroxidation was also reduced by curcumin, followed by the increase in SOD. The elevation of antioxidants by curcumin was attributed to improved redox imbalance in type 1 diabetic rats.[46] Curcumin decreased glucose uptake and lactate production in human lung, breast, cervical, and prostate cancer cell lines, followed by reduced cell viability. The inhibition was dependent on suppression of pyruvate kinase M2 because its overexpression reversed the decreasing effect of curcumin on glycolysis. Curcumin downregulated pyruvate kinase M2 through the reduction of HIF-1α, known as a critical transcriptional activator for cancer cells under hypoxic conditions. This study implied that the ability of curcumin alters glucose metabolism in cancer cells to attenuate tumour progression.[47]

iNOS is a pro-inflammatory enzyme that produces excessive nitric oxide (NO) and is involved in tissue damage and tumorigenesis. Primary rat hepatocytes stimulated with IL-1β induced NO production and upregulated iNOS, but were attenuated by curcumin (5–25 μM) without exerting any cytotoxic effects. Other pro-inflammatory cytokines, such as TNF-α and IL-6, were also downregulated by curcumin. The inhibitory effect of curcumin on IL-1β-induced iNOS expression was mediated by the inactivation of NF-κB by blocking its nuclear translocation. Reduction of the PI3K/Akt signalling has been proposed as another mechanism for curcumin in the suppression of NF-κB activation.[48] In carbon tetrachloride (CCl_4)-injected rats, curcumin (200 mg kg^{-1}) treatment significantly reduced liver injury, inflammation, and fibrosis, followed by decreased leukocyte, monocyte, and macrophage infiltration. In addition, curcumin markedly lowered the infiltration of $Ly6C^{hi}$ monocytes, the pro-fibrotic classical monocyte subset, into the liver during fibrosis. Curcumin reduced $Ly6C^{hi}$ monocyte infiltration, accompanied by decreased levels of monocyte chemokines, such as monocyte chemoattractant protein-1 and chemokine C–C motif ligand 7. More importantly, curcumin suppressed the activation of Kupffer cells, the specific liver macrophages, which were shown to reduce the cell markers iNOS and cluster of differentiation 86 (CD86). This result indicates that curcumin attenuated $Ly6C^{hi}$ monocyte infiltration by blocking the activation of Kupffer cells.[49] Curcumin was also effective in reducing pancreatic inflammation in L-arginine-induced acute pancreatitis in mice. Male Institute of Cancer Research (ICR) mice pre-treated with 50 and 200 mg kg^{-1} curcumin for 3 days significantly alleviated pancreatitis with reduced neutrophilic infiltration, interstitial edema, mesenteric fat necrosis, and apoptotic acinar cells in the pancreatic tissue. Curcumin not only attenuated neutrophilic infiltration, but also enhanced their activities, as shown by the reduced level of myeloperoxidase, a peroxidase enzyme most abundantly expressed in neutrophils. In addition, the level of a marker of lipid peroxidation, 4-hydroxynonenal, was reduced by curcumin in the pancreatic tissue. This study suggested that curcumin reduced inflammation and oxidative stress to alleviate L-arginine-induced acute pancreatitis in mice.[50]

Several studies have demonstrated that curcumin inhibits cell migration and invasion through the modulation of ROS production and antioxidants. Curcumin dose-dependently inhibited H_2O_2-induced proliferation of human pancreatic cancer (BxPC-3) and Panc-1 cells, followed by a decrease in intracellular H_2O_2 levels. The H_2O_2-mediated cell proliferation was blocked by NAC. In addition, H_2O_2-induced cell migration and invasion were attenuated by curcumin and NAC, with decreased levels of MMP-2 and MMP-9. The activation of ERK and NF-κB signalling in pancreatic cancer cells by H_2O_2 was also alleviated by curcumin, proving that curcumin inhibited H_2O_2-induced cancer cell invasion and migration *via* suppression of the ROS/ERK/NF-κB pathway.[51] A similar effect was observed in pancreatic cancer cells with SOD-dependent H_2O_2 production. SOD-driven H_2O_2 production triggered ROS production in BxPC-3 and Panc-1 cells, but was reduced by curcumin with downregulation of the Akt and NF-κB signalling. The migration and invasion of both cell lines were also attenuated by curcumin treatment. SOD-driven cell migration and invasion were blocked by CAT and Akt inhibitors, suggesting the role of ROS and PI3K/Akt signalling in H_2O_2-mediated cell migration and invasion. SOD-mediated H_2O_2 production caused a phenotype transition from epithelial to mesenchymal in pancreatic cancer cells, suggesting SOD-driven H_2O_2 induced epithelial–mesenchymal transition (EMT) progression, which was inhibited by curcumin. Curcumin diminished the EMT progression caused by SOD, followed by downregulation of N-cadherin and vimentin, but upregulation of E-cadherin.[52] In human umbilical vein endothelial cell (HUVEC) tubule formation assays, HUVECs cultured with the medium from the hepatocellular carcinoma (HCC) cell line (HepG2) and human hepatic stellate cell (HSC) line significantly increased tubule formation compared to only the medium from HepG2 cells, indicating that HSCs enhanced HepG2-induced angiogenesis. However, NAC abrogated tubule formation, indicating the role of oxidative stress in HSC-mediated angiogenesis in HCC cells. This also reflected the mechanism by which curcumin suppresses angiogenesis, which may be similar to NAC, and was further supported by the upregulation of Nrf2 and GSH in HepG2 cells. The study further clarified that knocking down HIF-1α in HepG2 cells significantly blocked ROS production and failed to enhance HUVEC tubule formation when cultured with the medium from HepG2 cells and HSCs. In contrast, curcumin and NAC had no effect on ROS production in HepG2 cells with HIF-1α knockdown. Moreover, curcumin and NAC markedly attenuated HSC-induced invasion and migration ability in HepG2 cells as well as related markers such as VEGF, IL-6, MMP-9 (invasion), and vimentin (EMT), while the inhibition was eliminated by HIF-1α knockdown. This indicates that the inhibitory effect of curcumin appears to be dependent on the expression of HIF-1α. Upregulation of HIF-1α is an adaptive response to hypoxic conditions in cancer cells and is regulated by ROS, suggesting that curcumin may eliminate oxidative stress by upregulating Nrf2 and GSH, which reduces the stabilization of HIF-1α, leading to the suppression of HSC-mediated angiogenesis, invasion, and migration of HepG2 cells.[53] Another study demonstrated that different mechanisms of curcumin suppress EMT. In CCl_4-treated rats, curcumin administration significantly attenuated liver fibrosis, reduced EMT markers, and upregulated autophagy. Further *in vitro* studies using mouse embryonic hepatocytes (BNL CL.2 cells) showed that curcumin abrogated transforming growth factor-betta (TGF-β)-induced EMT by downregulating vimentin. The fibrogenic marker (α-SMA) was also decreased by curcumin, indicating that curcumin suppressed the differentiation of hepatocytes into myofibroblasts. In addition to suppressing TGF-β/Smad signalling, upregulation of peroxisome proliferator-activated receptor-alpha (PPAR-α) by curcumin is required to eliminate ROS, leading to autophagy activation and inhibition of hepatocyte EMT.[54]

8.3.3 Tetrahydrocurcumin

Tetrahydrocurcumin (THC) is a superior antioxidant and exerts cancer chemopreventive and therapeutic effects.[55] More importantly, THC is not pro-oxidative and is less cytotoxic than curcumin because of the absence of conjugated bonds in the central seven-carbon chain.[56] THC has been reported to exhibit better 2,2′-azinobis-(3-ethylbenzothiazoline-6-sulfonic acid) (ABTS) radical scavenging activity than curcumin. Furthermore, THC was more potent than curcumin in protecting against H_2O_2-induced cytotoxicity in HepG2 cells through the reduction of lipid peroxidation and elevation of antioxidant enzymes such as SOD and CAT[57] (Table 8.3). The hepatoprotective efficacy of THC has been widely demonstrated in various animal models. In male ICR mice fed with a high-fat and high-fructose diet, oral administration of 100 and 200 $mg\,kg^{-1}$ THC for 8 weeks markedly reduced the serum levels of alanine transaminase (ALT) and aspartate transaminase (AST), as well as improved glucose tolerance. Hepatic lipid peroxidation was also attenuated by THC treatment. The hepatoprotective effect of THC is presumably mediated by the downregulation of NADPH–CYP450 reductase, a source of ROS generation *via* oxidation of NADPH.[58] THC displayed antiinflammatory activity *via* downregulation of proinflammatory cytokines such as TNF-α, IL-1β, and IL-6 in lipopolysaccharide (LPS)-stimulated mouse splenocytes[57] and carrageenan-treated mice.[59] Although the mechanism has not been fully investigated, it was shown that THC attenuated carrageenan-induced paw edema by interrupting the activation of NF-κB.[59] In osteosarcoma cells, THC suppressed migration and invasion through the

Table 8.3 Anticancer mechanism of tetrahydrocurcumin targeting oxidative stress.

Targeted hallmarks in tumorigenesis	Experimental model	Treatment	Molecular mechanisms for antioxidative activity	Biological outcomes	Ref.
↓ Genome instability and mutation	• ABTS radical • H_2O_2 exposed HepG2 cells	0.1, 1, and 10 $\mu g\,mL^{-1}$	↑ Radical scavenging ↓ Lipid peroxidation ↑ Antioxidant enzymes	↑ Cell viability	57
	High-fat/high-fructose diet fed mice	100 and 200 $mg\,kg^{-1}$, 8 weeks, and oral	↓ Lipid peroxidation ↓ NADPH–CYP450 reductase activity	↓ Glucose tolerance ↓ Hepatic injury	58
↓ Inflammation	Lipopolysaccharide-stimulated mouse splenocytes	5, 7.4, and 10 $\mu g\,mL^{-1}$	—	↓ Inflammatory cytokines	57
	Carrageenan-exposed mouse paw	40 $mg\,kg^{-1}$ and oral	↓ NF-κB signalling	↓ Edema ↓ Inflammatory cytokines	59
↓ Limitless replicative potential ↓ Sustained angiogenesis ↓ Tissue invasion and metastasis	• Osteosarcoma cells (U-2OS and MG-63) • Xenograft model	• 20 and 50 μM • 100 $mg\,kg^{-1}$, 8 weeks	↓ HIF-1α	↓ Proliferation ↓ Invasion and migration ↓ Angiogenesis ↓ Lung metastasis ↑ MET process ↑ Autophagy	60

Abbreviations: ABTS, 2,2′-azinobis-3-ethylbenzothiazoline-6-sulfonic acid; HIF-1α, hypoxia-inducible factor-1 alpha; MET, mesenchymal–epithelial transition; NADPH–CYP450, nicotinamide adenine dinucleotide phosphate–cytochrome P450; NF-κB, nuclear factor-kappa B.

induction of the mesenchymal–epithelial transition (MET) process, followed by attenuation of mesenchymal molecules, including N-cadherin, vimentin, Snail, Twist, and β-catenin. The THC-mediated MET process was dependent on the downregulation of HIF-1α. THC suppressed HIF-1α to induce autophagy and MET, leading to reduced cell proliferation, migration, invasion, and angiogenesis under hypoxia. The *in vivo* efficacy of THC in suppressing lung metastasis was also observed. Male Balb/c nude mice were intravenously administered with sarcoma osteogenic-2 (SaOS-2) cells to induce lung metastasis for one week. After that, mice intraperitoneally injected with 100 $mg\,kg^{-1}$ THC for 8 weeks showed fewer and smaller metastatic nodules in the lungs. The mechanism was not further investigated in the *in vivo* study.[60]

8.3.4 Epigallocatechin Gallate (EGCG)

EGCG is recognized as a potent antioxidant with mechanisms against carcinogenesis; however, its anticancer effect is also attributed to its pro-oxidant activity due to catechol oxidation.[61,62] This pro-oxidative effect of EGCG derived from ROS production plays an important role in triggering apoptosis in cancer cells[61] while still exhibiting harmful effects on normal cells.[62] Although the adverse effect of EGCG has been reported due to its pro-oxidative properties, the concentration effect might be a major factor as it occurs at higher doses, whereas it is an effective antioxidant at low doses. However, the antioxidative and pro-oxidative effects are also influenced by various experimental conditions, such as exposure duration, cellular stress, and cell type.[62]

The excellent antioxidant properties of EGCG are attributed to its ability to directly scavenge free radicals because of the hydroxyl groups in the gallate group. In addition to 2,2-diphenyl-1-picrylhydrazyl (DPPH) radicals, EGCG scavenges methylglyoxal, a highly reactive dicarbonyl intermediate for the formation of advanced glycation end products (AGEs). In an *in vitro* glycosylation assay, EGCG significantly reduced $FeCl_3$/glucose-mediated oxidation of tryptophan and protein carbonylation of human serum albumin and hemoglobin. The formation of pentosidine, a well-known AGE, was also decreased by EGCG, suggesting its ability to protect against glucose-mediated oxidative damage[63] (Table 8.4). A similar effect was also reported in another *in vitro* glycosylation assay conducted using methylglyoxal. The methylglyoxal-induced AGE formation in bovine serum albumin was attenuated by EGCG and may result from the direct methylglyoxal scavenging effect of EGCG.[64] These studies show that EGCG protects against hyperglycemia-induced cellular oxidative damage. *N*-Nitrosodiethylamine (NDEA), also known as dimethylnitrosamine, is a well-known hepatic carcinogen that is metabolized by CYP450 to form an active ethyl radical metabolite. The radical not only acts as an alkylating agent that reacts with DNA and proteins, but also triggers ROS production, leading to oxidative damage to macromolecules.[65] In NDEA-treated male Wistar rats, the administration

Table 8.4 Anticancer mechanism of EGCG targeting oxidative stress.

Targeted hallmarks in tumorigenesis	Experimental model	Treatment	Molecular mechanisms for antioxidative activity	Biological outcomes	Ref.
—	• DPPH radical • Iron/glucose-induced protein glycation	0.7 μM	↑ Radical scavenging ↓ Protein glycation	↓ AGE (carbonylation and pentosidine)	63
—	• Methylglyoxal • Methylglyoxal-induced protein glycation	0.25–2.5 mM	↑ Methylglyoxal scavenging ↓ Protein glycation	↓ AGE (carbonylation)	64
↓ Genome instability and mutation	*N*-Nitrosodiethylamine exposed male Wistar rats	5 and 15 mg mL^{-1}, 21 days, in drinking water	↓ Lipid peroxidation ↓ Protein glycation	↓ Liver injury ↓ Degeneration of hepatocytes ↓ Oxidative DNA damage	66
	• ABTS radical • Heavy metal exposed rat adrenal pheochromocytoma cells (PC12)	5–320 μg mL^{-1}	↑ Radical scavenging ↓ H_2O_2 production	↓ Apoptosis ↓ Necrosis	67
↓ Self-sufficiency in growth signals	Nickel nanoparticle exposed mouse epidermal cells (JB6)	10 μM	↓ H_2O_2 and O_2 production	↓ Apoptosis ↓ AP-1 and NF-κB transcription activity ↓ MAPK signalling	69
—	$NaAsO_2$ exposed male SD rats	50 mg kg^{-1}, 30 days, and oral	↓ H_2O_2 production ↓ Lipid peroxidation ↑ Antioxidant	↓ Liver injury ↓ Histopathological changes of hepatotoxicity	70
↓ Deregulating cellular energetics	High glucose exposed HCC cells (HepG2) and primary hepatocytes	0.25–2.5 mM	↓ H_2O_2 and O_2 production	↑ Glycogen content ↓ JNK signalling ↑ Insulin signalling	71

Table 8.4 (*Continued*)

Targeted hallmarks in tumorigenesis	Experimental model	Treatment	Molecular mechanisms for antioxidative activity	Biological outcomes	Ref.
↓ Deregulating cellular energetics	• HCC cells (HCC-LM3 and HepG2) • HCC xenograft model	• 25–100 μM • 10 mg kg^{-1}, 30 days	↓ Phosphofructokinase level and activity	↓ Glycolysis ↑ Cell cycle arrest and apoptosis ↓ Tumour growth	72
↓ Deregulating cellular energetics	• Breast cancer cells (4T1) • Xenograft model	• 10–320 μM • 5, 10, and 20 mg kg^{-1}, 14 days	↓ HIF-1α ↓ Phosphofructokinase, lactic dehydrogenase, and hexokinase activity	↓ Glycolysis ↓ ATP ↑ Apoptosis and autophagy ↓ Tumour growth	73
↓ Tissue invasion and metastasis	Oxalate exposed renal tubular cells (MDCK)	25 μM	↓ H_2O_2 production ↑ Nrf2 and CAT	↓ EMT process ↑ Epithelial marker ↓ Mesenchymal marker	74

Abbreviations: ABTS, 2,2′-azinobis-(3-ethylbenzothiazoline-6-sulfonic acid); AGE, advanced glycation end product; AP-1, activator protein-1; ATP, adenosine triphosphate; CAT, catalase; DPPH, 2,2-diphenyl-1-picrylhydrazyl; EGCG, (−)-epigallocatechin-3-gallate; EMT, epithelial–mesenchymal transition; HCC, hepatocellular carcinoma; HIF-1α, hypoxia-inducible factor-1alpha; JNK, c-Jun N-terminal kinase; MAPK, mitogen-activated protein kinase; $NaAsO_2$, sodium arsenite; NF-κB, nuclear factor-kappa B; Nrf2, nuclear factor erythroid 2-related factor 2.

of EGCG significantly inhibited the degeneration of hepatocytes and reduced liver injury, as evidenced by the decreased serum levels of ALT, AST, alkaline phosphatase (ALP), and lactate dehydrogenase (LDH). The markers of oxidative stress, such as lipid peroxidation and protein glycation, were also attenuated by EGCG, followed by decreased oxidative DNA damage, which was shown as micronucleus formation and comet tail in hepatocytes. This study revealed that the action of EGCG against NDEA-induced hepatotoxicity is by alleviating oxidative stress.[66]

ECGC efficiently chelates metals and thus protects cells from oxidative damage. Heavy metals, including cadmium chloride ($CdCl_2$), $HgCl_2$, cobalt chloride ($CoCl_2$), and lead chloride ($PbCl_2$) decreased the viability of rat adrenal pheochromocytoma PC12 cells following increased apoptosis and necrosis. ECGC significantly attenuated the cellular toxicity, which was attributed to the inhibition of ROS production by heavy metals. In addition to scavenging ABTS radicals, this study implied that ECGC protected against heavy metal-induced cellular damage through its metal-chelating ability.[67] The application of nickel nanoparticles (Ni NPs) in various fields, such as biomedicine, catalysis, and adsorption of industrial waste has increased;[68] however, its toxicity is still a concern for human health. An *in vitro* study showed that exposure to Ni NPs (2.5–10 $\mu g\, cm^{-2}$) caused a cytotoxic effect in mouse epidermal JB6 cells, but was significantly abrogated by EGCG (10 μM). Ni NP treatment induced apoptosis in JB6 cells, while EGCG reduced apoptotic cells only in the lower dose of Ni NPs. ROS production increased in a dose-dependent manner in Ni NP-treated JB6 cells, followed by upregulation of MAPK and the transcriptional activity of activator protein-1 (AP-1) and NF-κB, which were markedly decreased after EGCG treatment. This study suggested that EGCG attenuated Ni NP-induced cytotoxicity by reducing intracellular ROS production and thus modulating MAPK signalling to reduce apoptosis.[69] In sodium arsenite ($NaAsO_2$) (5 mg per kg per day)-exposed male SD rats, oral administration of 50 $mg\, kg^{-1}$ EGCG markedly attenuated liver injury, as evidenced by the reduced serum levels of ALT, AST, ALP, and LDH. Hepatic histopathological changes, such as steatosis, inflammatory infiltration, and hepatocyte balloon degeneration, were also inhibited by EGCG. EGCG alleviated $NaAsO_2$ caused hepatotoxicity, presumably by reducing ROS production and elevating antioxidant enzymes, SOD and CAT, as well as GSH levels in the liver.[70]

The effect of EGCG on reprogramming glucose metabolism in cancer cells is documented. In HepG2 cells treated with high glucose, ROS production increased, followed by a reduced glycogen content, indicating impaired glucose metabolism caused by insulin resistance, which was proved by dysregulation of the insulin receptor substrate 1/Akt signalling. EGCG effectively reduced ROS production and improved glycogen content, which mitigated ROS-mediated c-Jun N-terminal kinase (JNK) signalling and thus improved insulin signalling in HepG2 cells.[71] Most cancer cells show elevated ROS levels partially due to a higher metabolic rate, which is recognized as one of the hallmarks of cancer. Upregulation of aerobic glycolysis promotes tumour growth and the malignant phenotype. Compared to normal liver cells (QSG-7701 and LO2), hepatocellular carcinoma (HCC) cells, such as HCC-LM3 and HepG2, exhibited higher mRNA and protein levels as well as the activity of phosphofructokinase, a key enzyme involved in glycolysis. EGCG inhibited glucose uptake and lactate production in HCC cells, leading to the suppression of proliferation, induction of the cell cycle, and apoptosis. In addition, the inhibitory effect of EGCG on glycolysis was primarily attributed to the downregulation of phosphofructokinase activity by stabilizing the inactive dimers of the enzyme, thus leading to the reduction of fully active tetramers. EGCG also enhanced apoptosis in HCC-LM3 and HepG2 cells treated with sorafenib, a standard therapeutic agent for advanced HCC, by suppressing phosphofructokinase levels. This was further confirmed using a mouse xenograft model with HCC-LM3 cells. Male Balb/c nude mice with HCC tumours treated with sorafenib (10 $mg\,kg^{-1}$) and intraperitoneal injection of 10 $mg\,kg^{-1}$ EGCG for 30 days showed decreased tumour volume and higher apoptotic cells than those with sorafenib treatment alone. This suggests the potential of EGCG, which targets phosphofructokinase to enhance the therapeutic efficacy of patients with HCC.[72] It also suppresses glycolysis and inhibits breast tumour growth. EGCG induced apoptosis and autophagy in breast cancer 4T1 cells with reduced glucose consumption, lactate production, and ATP levels. HIF-1α was also downregulated by EGCG, which interferes with glycolysis mainly by reducing the transcription and activity of phosphofructokinase, lactic dehydrogenase, and hexokinase. In the Balb/c xenograft model, mice bearing tumours from 4T1 cells and intraperitoneally injected with 5, 10, and 20 $mg\,kg^{-1}$ EGCG for 14 days showed significantly decreased tumour weight. Reduced glucose and lactate levels in tumour tissues were observed after 20 $mg\,kg^{-1}$ EGCG administration. This study demonstrated that EGCG exerted its anticancer effect by suppressing glycolytic enzymes to suppress glucose metabolism.[73]

EGCG targets oxidative stress to attenuate the malignant progression of cancer cells. In the Madin–Darby Canine Kidney (MDCK) cell line, renal tubular cells, EGCG pre-treatment attenuated oxalate-induced EMT, which was confirmed by revealing epithelial-like morphology and upregulated related markers, including E-cadherin, occludin, cytokeratin, and zona occludens-1. The mesenchymal markers, vimentin and fibronectin, induced by oxalate were also decreased by EGCG. In addition, oxalate significantly induced ROS production in renal tubular cells, which was diminished by EGCG treatment. This inhibitory effect might be partially mediated by the activation of Nrf2 and downstream levels of CAT because Nrf2 siRNA abrogated the downregulating effect of EGCG against oxalate-induced EMT.[74] Although EGCG is water-soluble compared to lipophilic resveratrol and curcumin, it shows low oral bioavailability due to poor intestinal stability, absorption, and membrane permeability. Therefore, improving the oral bioavailability of EGCG is necessary for its application in nutraceuticals and pharmaceuticals. Structural modifications, such as esterification, nano-based delivery, and co-administration with other bioactive ingredients, have been reported to improve the bioavailability of EGCG together with its biological activity, but require further validation.[75]

8.4 Conclusion

Cancer is one of the major challenges for human health, and therefore elucidation of the causes, pathology, and molecular mechanisms involved in tumorigenesis is critical to identify an effective approach for its prevention and intervention. The concept of ROS in cancer development has been known for a long time. ROS play physiological roles in organisms, but they are often recognized as oncogenic drivers. Oxidative stress derived from excessive ROS production or defective antioxidants enables initiated cells to survive by attacking macromolecules, interfering with signalling networks, deregulating gene transcription and protein function, and exerting further effects on cellular function and metabolic homeostasis leading to malignancy. Oxidative stress also implies stress-related adaptation of cancer cells to favour their malignant behaviour. Natural dietary compounds are long-term candidates for anticancer agents because of their structural diversity and a variety of biological effects, including antioxidative properties. Natural dietary compounds exert their antioxidative effects through multiple mechanisms, such as radical scavenging, metal chelation, inhibition of ROS production, attenuation of oxidative reactions and damage, and upregulation of antioxidant defence, thus preventing and suppressing oxidative stress-driven malignant transformation and progression of cancer cells. Although evidence from current research suggests the cancer chemopreventive and therapeutic potential of dietary natural compounds *via* targeting the modulation of oxidative stress, most of them are cell-based studies. Many issues, such as *in vivo* and clinical evidence, effective dosage, and bioavailability, require further investigation. In addition, dietary natural compounds are generally considered as nontoxic or having low toxicity, but the pro-oxidant effect of some natural compounds is still a concern for their application. Furthermore, the combination of different dietary natural compounds may result in synergistic effects that may increase bioavailability and anticancer efficacy *via* multiple molecular targets. Additional extensive research on dietary natural compounds is required for their potential application as anticancer agents.

References

1. H. Sung, J. Ferlay, R. L. Siegel, M. Laversanne, I. Soerjomataram and A. Jemal, *et al.*, Global cancer statistics 2020: GLOBOCAN estimates of incidence and mortality worldwide for 36 cancers in 185 countries, *Ca-Cancer J. Clin.*, 2021, **71**, 209–249.
2. D. Hanahan and R. A. Weinberg, The hallmarks of cancer, *Cell*, 2000, **100**, 57–70.

3. D. Hanahan and R. A. Weinberg, The hallmarks of cancer, *Cell*, 2011, **144**, 646–674.
4. N. McGranahan and C. Swanton, Clonal heterogeneity and tumor evolution: past, present, and the future, *Cell*, 2017, **168**, 613–628.
5. T. Fiaschi and P. Chiarugi, Oxidative stress, tumor microenvironment, and metabolic reprogramming: a diabolic liaison, *Int. J. Cell Biol.*, 2012, 762825.
6. C. R. Reczek and N. S. Chandel, The two faces of reactive oxygen species in cancer, *Annu. Rev. Cancer Biol.*, 2017, **1**, 79–98.
7. A. Corcoran and T. G. Cotter, The two faces of reactive oxygen species in cancer, *FEBS J.*, 2013, **280**, 1944–1965.
8. S. Reuter, S. C. Gupta, M. M. Chaturvedi and B. B. Aggarwal, Oxidative stress, inflammation, and cancer: how are they linked?, *Free Radic. Biol. Med.*, 2010, **49**, 1603–1616.
9. S. S. Sabharwal and P. T. Schumacker, Mitochondrial ROS in cancer: initiators, amplifiers or an Achilles' heel?, *Nat. Rev. Cancer*, 2014, **14**, 709–721.
10. B. Perillo, D. M. Di, A. Pezone, Z. E. Di, P. Giovannelli and G. Galasso, *et al.*, ROS in cancer therapy: the bright side of the moon, *Exp. Mol. Med.*, 2020, **52**, 192–203.
11. K. Mori, T. Uchida, T. Yoshie, Y. Mizote, F. Ishikawa and M. Katsuyama, *et al.*, A mitochondrial ROS pathway controls matrix metalloproteinase 9 levels and invasive properties in RAS-activated cancer cells, *FEBS J.*, 2019, **286**, 459–478.
12. J. Kim, J. Kim and J. S. Bae, ROS homeostasis and metabolism: a critical liaison for cancer therapy, *Exp. Mol. Med.*, 2016, **48**, e269.
13. A. C. Newman and O. D. K. Maddocks, One-carbon metabolism in cancer, *Br. J. Cancer*, 2017, **116**, 1499–1504.
14. N. M. Akella, L. Ciraku and M. J. Reginato, Fueling the fire: emerging role of the hexosamine biosynthetic pathway in cancer, *BMC Biol.*, 2019, **17**, 52.
15. P. Icard, L. Poulain and H. Lincet, Understanding the central role of citrate in the metabolism of cancer cells, *Biochim. Biophys. Acta*, 2012, **1825**, 111–116.
16. L. Jin and Y. Zhou, Crucial role of the pentose phosphate pathway in malignant tumors, *Oncol. Lett.*, 2019, **17**, 4213–4221.
17. L. Yang, S. Venneti and D. Nagrath, A hallmark of cancer metabolism, *Annu. Rev. Biomed. Eng.*, 2017, **19**, 163–194.
18. A. S. John, M. K. Ankem and C. Damodaran, Oxidative stress: a promising target for chemoprevention, *Curr. Pharmacol. Rep.*, 2016, **2**, 73–81.
19. A. G. Atanasov, S. B. Zotchev, V. M. Dirsch and C. T. Supuran, Natural products in drug discovery: advances and opportunities, *Nat. Rev. Drug Discovery*, 2021, **20**, 200–216.
20. S. B. Nimse and D. Pal, Free radicals, natural antioxidants, and their reaction mechanisms, *RSC Adv.*, 2015, **5**, 27986–28006.

21. R. M. Seifried, E. Harrison and H. E. Seifried, Antioxidants in health and disease, in *Nutrition in the Prevention and Treatment of Disease*, ed. A. M. Coulston, C. J. Boushey, M. G. Ferruzzi and L. M. Delahanty, Academic Press Inc., London, UK, 4th edn, 2017, pp. 321–346.
22. S. Li, G. Chen, C. Zhang, M. Wu, S. Wu and Q. Liu, Research progress of natural antioxidants in foods for the treatment of diseases, *Food Sci. Hum. Wellness*, 2014, **3**, 110–116.
23. S. Whitehouse, P. L. Chen, A. L. Greenshields, M. Nightingale, D. W. Hoskin and K. Bedard, Resveratrol, piperine and apigenin differ in their NADPH-oxidase inhibitory and reactive oxygen species-scavenging properties, *Phytomedicine*, 2016, **23**, 1494–1503.
24. Z. Jia, H. Zhu, B. R. Misra, J. E. Mahaney, Y. Li and H. P. Misra, EPR studies on the superoxide-scavenging capacity of the nutraceutical resveratrol, *Mol. Cell. Biochem.*, 2008, **313**, 187–194.
25. K. Omidian, H. Rafiei and B. Bandy, Increased mitochondrial content and function by resveratrol and select flavonoids protects against benzo[*a*]pyrene-induced bioenergetic dysfunction and ROS generation in a cell model of neoplastic transformation, *Free Radic. Biol. Med.*, 2020, **152**, 767–775.
26. Y. Zhuang, H. Wu, X. Wang, J. He, S. He and Y. Yin, Resveratrol attenuates oxidative stress-induced intestinal barrier injury through PI3K/Akt-mediated Nrf2 signaling pathway, *Oxid. Med. Cell. Longevity*, 2019, 7591840.
27. R. Z. Hamza and N. S. El-Shenawy, Anti-inflammatory and antioxidant role of resveratrol on nicotine-induced lung changes in male rats, *Toxicol. Rep.*, 2017, **4**, 399–407.
28. H. T. Hsu, Y. T. Tseng, W. J. Wong, C. M. Liu and Y. C. Lo, Resveratrol prevents nanoparticles-induced inflammation and oxidative stress *via* downregulation of PKC-alpha and NADPH oxidase in lung epithelial A549 cells, *BMC Complementary Altern. Med.*, 2018, **18**, 211.
29. A. M. Kabel, A. Atef and R. S. Estfanous, Ameliorative potential of sitagliptin and/or resveratrol on experimentally-induced clear cell renal cell carcinoma, *Biomed. Pharmacother.*, 2018, **97**, 667–674.
30. M. Athar and M. Iqbal, Ferric nitrilotriacetate promotes N-diethylnitrosamine-induced renal tumorigenesis in the rat: implications for the involvement of oxidative stress, *Carcinogenesis*, 1998, **19**, 1133–1139.
31. W. Li, L. Cao, X. Chen, J. Lei and Q. Ma, Resveratrol inhibits hypoxia-driven ROS-induced invasive and migratory ability of pancreatic cancer cells *via* suppression of the Hedgehog signaling pathway, *Oncol. Rep.*, 2016, **35**, 1718–1726.
32. B. Yan, L. Cheng, Z. Jiang, K. Chen, C. Zhou and L. Sun, *et al.*, Resveratrol inhibits ROS-promoted activation and glycolysis of pancreatic stellate cells *via* suppression of miR-21, *Oxid. Med. Cell. Longevity*, 2018, 1346958.
33. Y. Xiao, T. Qin, L. Sun, W. Qian, J. Li and W. Duan, *et al.*, Resveratrol ameliorates the malignant progression of pancreatic cancer by

inhibiting hypoxia-induced pancreatic stellate cell activation, *Cell Transplant.*, 2020, **29**, 963689720929987.
34. L. Cao, X. Chen, X. Xiao, Q. Ma and W. Li, Resveratrol inhibits hyperglycemia-driven ROS-induced invasion and migration of pancreatic cancer cells *via* suppression of the ERK and p38 MAPK signaling pathways, *Int. J. Oncol.*, 2016, **49**, 735–743.
35. K. I. Priyadarsini, The chemistry of curcumin: from extraction to therapeutic agent, *Molecules*, 2014, **19**, 20091–20112.
36. A. Barzegar and A. A. Moosavi-Movahedi, Intracellular ROS protection efficiency and free radical-scavenging activity of curcumin, *PLoS One*, 2011, **6**, e26012.
37. R. Mandil, A. Prakash, A. Rahal, S. P. Singh, D. Sharma and R. Kumar, *et al.*, *In vitro* and *in vivo* effects of flubendiamide and copper on cyto-genotoxicity, oxidative stress and spleen histology of rats and its modulation by resveratrol, catechin, curcumin and alpha-tocopherol, *BMC Pharmacol. Toxicol.*, 2020, **21**, 29.
38. P. Singh, S. Singh, G. Garg, A. K. Singh and S. I. Rizvi, Curcumin has protective effects on ROS production and redox imbalance in an experimental oxidative-stressed model of rat, *J. Biol. Act. Prod. Nat.*, 2020, **10**, 484–494.
39. M. Pauletto, M. Giantin, R. Tolosi, I. Bassan, A. Barbarossa and A. Zaghini, *et al.*, Curcumin mitigates AFB1-induced hepatic toxicity by triggering cattle antioxidant and anti-inflammatory pathways: a whole transcriptomic *in vitro* study, *Antioxidants*, 2020, **9**, 1509.
40. G. M. Ben Yehuda, M. Frusic-Zlotkin, Y. Soroka, S. S. Ben, R. Bitton and H. Bianco-Peled, *et al.*, Curcumin protects skin against UVB-induced cytotoxicity *via* the Keap1-Nrf2 pathway: The use of a microemulsion delivery system, *Oxid. Med. Cell. Longevity*, 2017, 5205471.
41. M. Ashrafizadeh, Z. Ahmadi, R. Mohammadinejad, T. Farkhondeh and S. Samarghandian, Curcumin activates the Nrf2 pathway and induces cellular protection against oxidative injury, *Curr. Mol. Med.*, 2020, **20**, 116–133.
42. J. Y. Wang, X. Wang, X. J. Wang, B. Z. Zheng, Y. Wang and X. Wang, *et al.*, Curcumin inhibits the growth *via* Wnt/beta-catenin pathway in non-small-cell lung cancer cells, *Eur. Rev. Med. Pharmacol. Sci.*, 2018, **22**, 7492–7499.
43. M. Nisari, S. Yilmaz, T. Ertekin, D. Ceylan, N. Inanc and O. Al, *et al.*, Effects of curcumin on lipid peroxidation and antioxidant enzymes in kidney, liver, brain and testis of mice bearing ehrlich solid tumor, *Proceedings*, 2017, **1**, 994.
44. S. Chen, W. Gao, M. J. Zhang, J. Y. Chan and T. S. Wong, Curcumin enhances cisplatin sensitivity by suppressing NADPH oxidase 5 expression in human epithelial cancer, *Oncol. Lett.*, 2019, **18**, 2132–2139.
45. R. Varatharajalu, M. Garige, L. C. Leckey, K. Reyes-Gordillo, R. Shah and M. R. Lakshman, Protective role of dietary curcumin in the prevention of the oxidative stress induced by chronic alcohol with respect to hepatic

injury and antiatherogenic markers, *Oxid. Med. Cell. Longevity*, 2016, 5017460.

46. Z. Xie, X. Zeng, X. Li, B. Wu, G. Shen and Q. Wu, *et al.*, Curcumin attenuates oxidative stress in liver in Type 1 diabetic rats, *Open Life Sci.*, 2017, **12**, 452–459.
47. F. A. Siddiqui, G. Prakasam, S. Chattopadhyay, A. U. Rehman, R. A. Padder and M. A. Ansari, *et al.*, Curcumin decreases Warburg effect in cancer cells by down-regulating pyruvate kinase M2 *via* mTOR-HIF1alpha inhibition, *Sci. Rep.*, 2018, **8**, 8323.
48. R. Nakatake, H. Hishikawa, H. Matushima, Y. Nakamura, M. Ishizaki and K. Matsui, *et al.*, Curcumin protects liver inflammation by suppressing expression of inducible nitric oxide synthase in primary cultured rat hepatocytes, *Funct. Foods Health Dis.*, 2017, 7, 716–734.
49. X. A. Zhao, G. Chen, Y. Liu, Y. Chen, H. Wu and Y. Xiong, *et al.*, Curcumin reduces Ly6C(hi) monocyte infiltration to protect against liver fibrosis by inhibiting Kupffer cells activation to reduce chemokines secretion, *Biomed. Pharmacother.*, 2018, **106**, 868–878.
50. P. Siriviriyakul, T. Chingchit, N. Klaikeaw, M. Chayanupatkul and D. Werawatganon, Effects of curcumin on oxidative stress, inflammation and apoptosis in L-arginine induced acute pancreatitis in mice, *Heliyon*, 2019, **5**, e02222.
51. L. Cao, J. Liu, L. Zhang, X. Xiao and W. Li, Curcumin inhibits H_2O_2-induced invasion and migration of human pancreatic cancer *via* suppression of the ERK/NF-kappaB pathway, *Oncol. Rep.*, 2016, **36**, 2245–2251.
52. W. Li, Z. Jiang, X. Xiao, Z. Wang, Z. Wu and Q. Ma, *et al.*, Curcumin inhibits superoxide dismutase-induced epithelial-to-mesenchymal transition *via* the PI3K/Akt/NF-kappaB pathway in pancreatic cancer cells, *Int. J. Oncol.*, 2018, **52**, 1593–1602.
53. S. Shao, W. Duan, Q. Xu, X. Li, L. Han and W. Li, *et al.*, Curcumin suppresses hepatic stellate cell-induced hepatocarcinoma angiogenesis and invasion through downregulating CTGF, *Oxid. Med. Cell. Longevity*, 2019, 8148510.
54. D. Kong, Z. Zhang, L. Chen, W. Huang, F. Zhang and L. Wang, *et al.*, Curcumin blunts epithelial–mesenchymal transition of hepatocytes to alleviate hepatic fibrosis through regulating oxidative stress and autophagy, *Redox Biol.*, 2020, **36**, 101600.
55. C.-S. Lai, C.-T. Ho and M.-H. Pan, The cancer chemopreventive and therapeutic potential of tetrahydrocurcumin, *Biomolecules*, 2020, **10**, 831.
56. B. B. Aggarwal, L. Deb and S. Prasad, Curcumin differs from tetrahydrocurcumin for molecular targets, signaling pathways and cellular responses, *Molecules*, 2014, **20**, 185–205.
57. M. K. Trivedi, P. Panda, K. K. Sethi, M. Gangwar and S. C. Mondal, *et al.*, Solid and liquid state characterization of tetrahydrocurcumin using XRPD, FT-IR, DSC, TGA, LC-MS, GC–MS, and NMR and its biological activities, *J. Pharm. Anal.*, 2020, **10**, 334–345.

58. K. Jarukamjorn, W. Chatuphonprasert, N. Jearapong, C. Punvittayagul and R. Wongpoomchai, Tetrahydrocurcumin attenuates phase I metabolizing enzyme-triggered oxidative stress in mice fed a high-fat and high-fructose diet, *J. Funct. Foods*, 2019, **55**, 117–125.
59. Z. B. Zhang, D. D. Luo, J. H. Xie, Y. F. Xian, Z. Q. Lai and Y. H. Liu, *et al.*, Curcumin's metabolites, tetrahydrocurcumin and octahydrocurcumin, possess superior anti-inflammatory effects *in vivo* through suppression of TAK1-NF-kappaB pathway, *Front. Pharmacol.*, 2018, **9**, 1181.
60. Y. Zhang, Y. Liu, J. Zou, L. Yan, W. Du and Y. Zhang, *et al.*, Tetrahydrocurcumin induces mesenchymal–epithelial transition and suppresses angiogenesis by targeting HIF-1alpha and autophagy in human osteosarcoma, *Oncotarget*, 2017, **8**, 91134.
61. J. D. Lambert and R. J. Elias, The antioxidant and pro-oxidant activities of green tea polyphenols: a role in cancer prevention, *Arch. Biochem. Biophys.*, 2010, **501**, 65–72.
62. J. Ouyang, K. Zhu, Z. Liu and J. Huang, Prooxidant effects of epigallocatechin-3-gallate in health benefits and potential adverse effect, *Oxid. Med. Cell. Longevity*, 2020, 9723686.
63. H. Ozyurt, C. Luna and M. Estevez, Redox chemistry of the molecular interactions between tea catechins and human serum proteins under simulated hyperglycemic conditions, *Food Funct.*, 2016, **7**, 1390–1400.
64. M. Nikoo, J. M. Regenstein and G. H. Ahmadi, Antioxidant and antimicrobial activities of (-)-epigallocatechin-3-gallate (EGCG) and its potential to preserve the quality and safety of foods, *Compr. Rev. Food Sci. Food Saf.*, 2018, **17**, 732–753.
65. J. George, M. Tsuchishima and M. Tsutsumi, Molecular mechanisms in the pathogenesis of *N*-nitrosodimethylamine induced hepatic fibrosis, *Cell Death Dis.*, 2019, **10**, 18.
66. Y. H. Siddique and F. Ali, Protective effect of epigallocatechin gallate against *N*-nitrosodiethylamine (NDEA) induced toxicity in rats, *Cogent Biol.*, 2016, **2**, 1141451.
67. L. Yang, K. Shen and D. Ji, Natural compounds attenuate heavy metal-induced PC12 cell damage, *J. Int. Med. Res.*, 2020, **48**, 300060520930847.
68. N. D. Jaji, H. L. Lee, M. H. Hussin, H. M. Akil, M. R. Zakaria and M. B. H. Othman, Advanced nickel nanoparticles technology: From synthesis to applications, *Nanotechnol. Rev.*, 2020, **9**, 1456–1480.
69. Y. Gu, Y. Wang, Q. Zhou, L. Bowman, G. Mao and B. Zou, *et al.*, Inhibition of nickel nanoparticles-induced toxicity by epigallocatechin-3-gallate in JB6 cells may be through down-regulation of the MAPK signaling pathways, *PLoS One*, 2016, **11**, e0150954.
70. X. D. Han, Y. Y. Zhang, K. L. Wang, Y. P. Huang, Z. B. Yang and Z. Liu, The involvement of Nrf2 in the protective effects of (-)-epigallocatechin-3-gallate (EGCG) on NaAsO2-induced hepatotoxicity, *Oncotarget*, 2017, **8**, 65302–65312.

71. S. B. Ma, R. Zhang, S. Miao, B. Gao, Y. Lu and S. Hui, *et al.*, Epigallocatechin-3-gallate ameliorates insulin resistance in hepatocytes, *Mol. Med. Rep.*, 2017, **15**, 3803–3809.
72. S. Li, L. Wu, J. Feng, J. Li, T. Liu and R. Zhang, *et al.*, *In vitro* and *in vivo* study of epigallocatechin-3-gallate-induced apoptosis in aerobic glycolytic hepatocellular carcinoma cells involving inhibition of phosphofructokinase activity, *Sci. Rep.*, 2016, **6**, 28479.
73. R. Wei, L. Mao, P. Xu, X. Zheng, R. M. Hackman and G. G. Mackenzie, *et al.*, Suppressing glucose metabolism with epigallocatechin-3-gallate (EGCG) reduces breast cancer cell growth in preclinical models, *Food Funct.*, 2018, **9**, 5682–5696.
74. R. Kanlaya, S. Khamchun, C. Kapincharanon and V. Thongboonkerd, Protective effect of epigallocatechin-3-gallate (EGCG) *via* Nrf2 pathway against oxalate-induced epithelial mesenchymal transition (EMT) of renal tubular cells, *Sci. Rep.*, 2016, **6**, 30233.
75. W. Dai, C. Ruan, Y. Zhang, J. Wang, J. Han and Z. Shao, *et al.*, Bioavailability enhancement of EGCG by structural modification and nano-delivery: A review, *J. Funct. Foods*, 2020, **65**, 103732.

CHAPTER 9

Anti-diabetic Activity of Antioxidative Teas and Tea Polyphenol Supplements

ZISHENG HAN,[a] LIANG ZHANG,[b,c] XIAOCHUN WAN[b,c] AND CHI-TANG HO*[a,c]

[a] Department of Food Science, Rutgers University, New Brunswick, NJ, USA; [b] State Key Laboratory of Tea Plant Biology and Utilization, Anhui Agricultural University, Hefei, China; [c] International Joint Laboratory on Tea Chemistry and Health Effects of Ministry of Education, Anhui Agricultural University, Hefei, China

*Email: ctho@sebs.rutgers.edu

9.1 Introduction

Tea is a beverage consumed by many people in the world. With the improvement of people's living standard, consumers pay more and more attention to healthy diets. As a health promoting product, tea contains a variety of bioactive components, such as tea polyphenols, L-theanine, and caffeine. These compounds possess many health functions such as anti-neurodegenerative diseases, prevention of metabolic syndrome (MetS), and prevention of cancer, among others. Tea polyphenols account for the largest proportion of tea dry weight, while possessing the most abundant structures and most active properties. Green tea contains appreciable amounts of phytochemicals, especially catechins. According to the literature, dry tea leaves

Food Chemistry, Function and Analysis No. 39
Dietary Supplements with Antioxidant Activity: Understanding Mechanisms and Potential Health Benefits
Edited by Cesarettin Alasalvar, Fereidoon Shahidi and Chi-Tang Ho

Published by the Royal Society of Chemistry, www.rsc.org

usually contain 10–25% polyphenols which belong to various classes, *e.g.*, flavanols and flavonols. Green tea catechins are further composed of different chemical entities that include epigallocatechin-3-gallate (EGCG), epicatechin (EC), epicatechin-3-gallate (ECG), and epigallocatechin (EGC). Among these catechins, EGCG is present in the highest amount and considered to be an effective antioxidant with appreciable radical scavenging ability.

Diabetes is a typical chronic disease, which could be classified into two types: one is the insulin secretion deficiency caused by destruction of pancreatic β-cells. The other is type 2 diabetes (T2D), which accounts for about 90% of all diabetic patients and is caused by insulin resistance. Among the contributory factors of obesity, dietary habits are considered one of the major reasons for its increase. During the last few years, excessive consumption of carbohydrates and animal fat has contributed to increased incidence of atherosclerosis induced cardiovascular complications and diabetes mellitus.[1] T2D is the main risk for many people globally, including in some Asian countries, where the number of T2D patients has rapidly grown in the past decades.[2]

According to the processing technology and fermentation degree, fresh tea leaves can be processed into at least six types of tea. For example, when fresh tea leaves were not fermented, but actually oxidized by polyphenol oxidase, they retain most of the polyphenol compounds from the tea plant and their compositions of flavan-3-ols and other flavonoids or flavonoid glycosides are similar. If tea is fermented (enzymatic oxidation), most of flavan-3-ol monomers are oxidized into *o*-quinone and condensed into theaflavins or other oxidation products, most of which are tea pigments.[3] Therefore, although there have been many reports about the hypoglycemic effects of various types of tea, the bioactive compounds are different in non-fermented and fermented teas.

The main tea polyphenols in green tea include EGCG, EGC, EC, ECG, and some (2*S*,3*R*)-catechins. In black tea, there are mainly four theaflavins and large molecular weight thearubigins. The main bioactive compounds of tea are given in Figure 9.1. To control or prevent diabetes, there are some known mechanisms of tea, such as inhibiting carbohydrate hydrolysis and the absorption and transportation of glucose, inhibiting the biosynthesis of glucose, promoting the biosynthesis of glycogen, enhancing the energy expenditure of glucose, improving insulin sensitivity, and promoting the uptake of glucose in muscle or target tissue. This chapter discusses the health benefits of tea and its bioactive components, among which anti-diabetic properties and mechanisms are the major topics.

9.2 Health Benefits of Tea and Tea Polyphenols

With the change of lifestyle and dietary habits, such as increased consumption of carbohydrates and animal fat, as well as the lack of physical exercise, more and more cases of MetS have become a chronic epidemic in both developed and developing countries. MetS refers to a group of risk

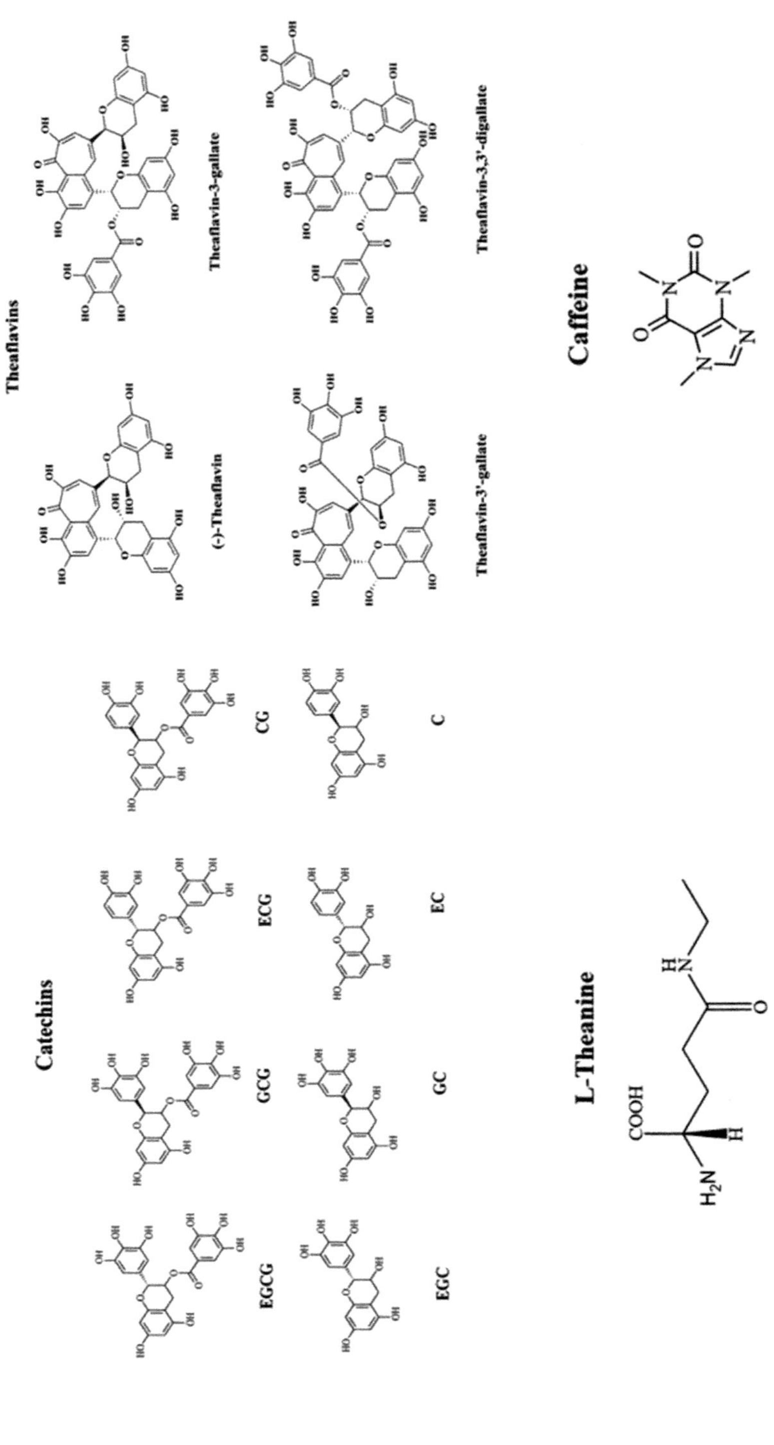

Figure 9.1 The main bioactive compounds of tea.

factors related to coronary heart disease (CHD) and T2D. It has been defined according to the presence of any three of the following five risk factors: central obesity (excess weight around the stomach), high triacylglycerols (TAG) in the blood, low levels of high-density lipoprotein (HDL) cholesterol in the blood, high blood pressure (HBP), and insulin resistance.[4] One of the representative outcomes of MetS is T2D and is often associated with other diseases, such as obesity. While the social economy and life quality of people have improved, concerns and demands for health and healthy foods have intensified. As a popular daily drink, especially in Asian countries, tea and its associated health promoting properties have attracted much public attention. Over decades, there have been thousands of papers on the biological activities of tea and tea polyphenols. It was found that tea and its bioactive components possess potential health promoting properties, such as anti-carcinogenic, anti-inflammatory, and antioxidative capacities.[5] In addition, tea and tea polyphenols, for example, EGCG, contribute to the prevention and improvement of several MetS ailments, including T2D.[6] Numerous studies have validated the health benefits of tea and its bioactive components towards T2D.[7] Among these studies, epidemiological studies on the distribution and determinants of T2D and health conditions in type 2 diabetic population have been carried out in order to study strategies for preventing and treating T2D and promoting health. A large number of *in vitro* and *in vivo* animal studies have also been conducted to assess the overall effect of tea and its bioactive components on T2D and relative indices. In addition, human intervention studies are another characteristic type of population-based investigation, which contribute to a more comprehensive analysis of the health benefits of tea and its bioactive components. In this chapter, we intend to summarize the development of these research activities within the last decade. Table 9.1 lists the anti-diabetic effects of tea and tea compounds reported in population-based investigations.

9.2.1 Epidemiological Studies

There have been hundreds of publications about population-based investigations of tea, including intervention experiments, double-blind, randomized trials, pre-clinical and clinical investigations, and some epidemiological studies. Based on these reports, there are also many cohort or evidence-based statistical analyses. However, the health benefits of tea in patients with various diseases remain equivocal because of the lifestyle, background dietary factor, disease types, and dose applied. Furthermore, because tea, cocoa, coffee, and polyphenol-rich red wine are the most consumed beverages or light-alcohol drinks, comparison among these has also emerged in recent years.[8] Therefore, before epidemiological analysis of tea, it is worth comparing the different results of epidemiological studies and interventional trials, the latter of which mainly focuses on the diabetic patients, while epidemiological studies may involve lots of normal people with regular physiological parameters. In most epidemiological studies, tea has been

Table 9.1 Anti-diabetic effect of tea or tea compounds reported in population-based investigations.

Samples	Dosage	Subjects	Duration	Results	Ref.
Total polyphenols	560 mg per day	19 F and 8 M patients	20 weeks	No changes of glycaemic parameters, renal function markers, or main antioxidant capacity markers	14
White tea	Tea infusion 150 mL per day	13 F patients	6 months	Decrease of weight, BMI, fat%, glucose, insulin, LDL cholesterol, and TAG	16
Green tea	1500 mg per day	96 patients with DPN	16 weeks	Symptoms of DPN, sensorineural pain, and neural dysfunction were improved	17
Green tea	1000 mg per day	120 overweight F	12 weeks	Decrease of fasting glucose, cholesterol, and LDL cholesterol	18
Green tea	≥1 cup per week	100 diabetic retinopathy patients	1 year	Protective effect on retinopathy in women, not in men	19
Green tea/green tea extract	4 cups or 2 capsules per day	13 and 10 adults with metabolic syndrome	8 weeks	Improvement of antioxidant capacity and whole blood glutathione	9

Abbreviations: BMI, body mass index; DPN, diabetic peripheral neuropathy; F, female; LDL, low-density lipoprotein; M, male; TAG, triacylglycerols.

shown to have the ability to reduce the risk of T2D, while a few studies have indicated that tea drinking is not associated with T2D risk, and fewer have suggested it may increase the risk of T2D. In intervention trials, the positive effects of tea and tea polyphenols on blood glucose and lipids are limited, but those on some antioxidant index and immunological index could be observed. Therefore, various teas and their ingredients are more recognized as health beverages or functional foods, rather than herbal medicines.

In a population-based investigation, green tea infusion and tea polyphenols can significantly increase plasma antioxidant capacity by improving the level of glutathione, but significantly reduce the plasma iron level.[9] The iron chelating ability of polyphenols may be one of the important mechanisms of regulating antioxidant activity. In a cohort study of 38 176 people, tea consumption was associated with lowered risk of T2D. Moreover, when tea consumption was at least 3 cups per day, the lowest hazard ratio was observed, but if the consumption of tea was >7 cups per day it did not show any further benefit on the hazard ratio.[10] In a prospective cohort study conducted in Singapore, black and oolong tea consumption was not significantly associated with glycaemic markers, while green tea was significantly associated with lower C-reactive protein (CRP) but there was no significant association between tea consumption (cups) and metabolic markers.[11] However, the association between green tea consumption and diabetes risk was not consistent among many studies.[12]

A large population-based cohort study was used to assess the association between sleep duration and the risk of T2D incidence on male subjects. The results showed that longer sleep duration, particularly among individuals already exhibiting factors linked to poor quality of sleep, was associated with diabetes. The association between sleep duration and diabetes may be modified by tea drinking, especially in older men or men with more sleep-related factors.[13]

Furthermore, there are some population investigations about the effects of tea on the diabetic index and complications. In a randomized, double-blind, controlled trial, 27 T2D patients consuming 560 mg green tea extract for 20 weeks did not show any changes in terms of glycaemic parameters, renal function markers, and other antioxidant capacity markers.[14] It was suggested that the diabetes duration may be the reason for inconsistent results in population investigations. If diabetes was diagnosed for more than 5 years, or the diabetes presented more severe pathological conditions, the benefits of tea extract may be out of the scope of medical treatment.[15] Recently, white tea was also chosen to conduct a randomized trial in patients with T2D. In this study, it was obvious that white tea can decrease the glucose level, insulin level, low-density lipoprotein (LDL) cholesterol, and TAG for six months, but the exact dose was not calculated. If white tea consumption (tea infusion) was together with aerobic training, the effects may be enhanced compared to either tea consumption or aerobic training alone. Interestingly, tea consumption had no effect on energy metabolism, but decreased fat percentage and body mass index (BMI).[16]

When diabetic patients had peripheral neuropathy, green tea extract showed an improvement in both clinical and neurophysiological parameters, including the Toronto clinical scoring system, visual analogue scale, and vibration perception thresholds. It also significantly decreased fasting blood glucose.[17] Recently, a randomized trial involving 120 overweight women indicated that green tea can reduce the levels of fasting glucose and total and LDL cholesterol.[18] It is important to note that the overweight people that participated in the study were not diabetic.

9.2.2 Human Intervention Studies Including Meta-analysis

A recent randomized controlled trial showed no significant effect on obese male subjects, but their diastolic blood pressure (DBP) was improved.[20] As mentioned in the review on tea chemistry and biological activities, it was found that when tea or tea polyphenols were consumed by obese or diabetic patients or population, the observed protective effects were limited.[3] Some meta-analyses also indicated that green tea supplementation significantly decreased fasting blood glucose by 1.44 mg dL^{-1}, in which a high content of catechins ($\geq$500 mg dL^{-1}) led to a significant decrease in fasting blood glucose.[21] Decaffeinated green tea exerted effects on decreasing insulin and homeostatic model assessment for insulin resistance (HOMA-IR), and was suggested to be associated with the decrease of T2D risk, but the blood glucose level was not affected.[22] The administration of 3 cups per day (600 mL, high intake) or 1 cup per day (200 mL, low intake) of black tea in 36 patients with T2D (18 men and 18 women) for 12 weeks showed certain improvements related to the disease in the high-intake patients. Reduced haemoglobin A1c and total serum cholesterol levels, along with increased regulatory T cells cluster of differentiation (CD)3 + CD4 + CD25 + forkhead box protein 3, CD3 + CD4 + interleukin (IL)-10 + cells (an immunosuppressive phenotype), reduced (pro-inflammatory) CD3 + CD4 + IL-17 + cells, and reduced T helper 1 cell-associated CD3 + CD4 + interferon gamma + cells. The preparation of tea was as follows: one tea bag (2.5 g) of dry black tea as sold was infused in 200 mL of water brought to boil, then steeped for 3 min.[23] In addition, tea products, such as instant tea and matcha supplementation, had beneficial effects on the regulation of blood glucose and gut microbiota through reversing the changes in microbiota caused by alloxan injection.[24] A systematic review and meta-analysis indicated that green tea significantly reduced the serum levels of CRP in patients with T2D. However, no significant effect was observed for green tea on the total antioxidant capacity and serum levels of malondialdehyde in patients with T2D.[25]

Dyslipidemia is common in T2D patients and contributes to an increased risk of cardiovascular disease (CVD) and mortality. A systematic review and meta-analysis revealed that the supplementary intake of green tea extracts may improve the lipid profile by reducing the serum concentrations of TAG in patients with T2D. In addition, with long term intervention of green tea extracts, the concentrations of total serum cholesterol were reduced.[26] Two randomized clinical trials also revealed the association between the blood pressure of T2D patients and green tea consumption. Consumption of four cups of green tea (each 2.5 g/200 mL) per day decreased the systolic blood pressure (SBP), body weight, BMI, and waist circumference.[27] Similarly, decreases in SBP and DBP were detected in 100 mildly hypertensive patients (age ranging between 30 and 60 years) that received a green tea infusion three times per day 2 hours after each meal for 4 weeks.[28] In another case study, green tea infusion was prepared by boiling 3 gram green tea bags in 150 mL of water for 5 min. The administration of a green tea extract (500 mg in one capsule 30 min after meals three times daily for 16 weeks) to 92 patients with T2D (20–65 years) decreased TAG levels, HOMA-IR, and increased the levels of high HDL cholesterol, adiponectin, apolipoprotein (Apo) A1, Apo B100, and glucagon-like peptide 1.[15]

Diabetes mellitus is often accompanied by various metabolic disturbances that may cause a wide range of long-term complications including ophthalmic, nervous, and renal disturbances. In a randomized, double-blind, placebo-controlled clinical trial, 42 diabetic patients with a urinary albumin–creatinine ratio (UACR) >30 mg g^{-1} received either four capsules of green tea polyphenols (corresponding to 800 mg of EGCG, $n=21$) or placebo ($n=21$) for 12 weeks; a 41% decrease in UACR in the green tea polyphenol group was observed.[29] It was not surprising that, in another study, green tea did not alter blood glucose levels, but it decreased the risk of diabetic retinopathy through declining the level of occlusion, the early marker of glial fibrillary acidic protein. The mechanism may be associated with counteracting glutamine/glutamate impairment and improvement of oxidative stress.[30] Green tea also ameliorated the nephropathy of diabetes by decreasing the renal accumulation of collagen IV and renal oxidative stress through down-regulation of nicotinamide adenine dinucleotide phosphate oxidase 4 (Nox4). A human intervention study demonstrated that patients with diabetic retinopathy who had regularly drunk Chinese green tea every week for at least one year had a reduced risk of this disease (around 50%).[19] It was also suggested that green tea does not affect blood pressure or blood glucose concentrations directly.[31] Residual albuminuria is also another typical nephropathy of diabetes. In a double-blind study of 42 diabetic subjects, when 600 mg green tea polyphenol was given for 12 weeks, it ameliorated albuminuria in diabetic patients, in combination with renin–angiotensin. The mechanism may be due to reduction in podocyte apoptosis by decreasing plasma dickkopf WNT pathway inhibitor 1.[29] A double-blind, randomized, controlled clinical trial was conducted in 35 diabetic subjects. The participants received 1120 mg of green tea extract containing 560 mg of polyphenols per day or cellulose. The results indicated that green tea may contribute to improvement of the bone mineral content in diabetic subjects.[32]

9.2.3 *In Vitro* and *In Vivo* Animal Studies

Tang *et al.*[33] compared the anti-diabetic effects of green and black teas. Streptozotocin-treated mice were given free access to tea extract, dissolved in water (0.1%); and green tea showed better results than black tea regarding the non-fasting glucose level, body weight, and serum insulin. The result was also consistent with most other publications that green tea with higher contents of tea polyphenols exerted better effects than black tea, especially at a low administration dose of 0.01% in water. The green tea extract improved insulin secretion in RINm5F cells treated with iron whose addition decreases insulin secretion. The mechanism may involve the iron chelation and free-radical scavenging abilities of polyphenols.[34] α-Glucosidase was treated as one of the main targets for diabetes, while green tea and tea polyphenols had excellent inhibitory effects against α-glucosidase.[35] Furthermore, green tea also improved uncoupling of endothelial nitric oxide synthase in diabetic rats induced by streptozotocin-induced diabetes. Meanwhile, the tetrahydrobiopterin level also decreased after streptozotocin induction, but green tea could reverse the levels of BH4 and guanosine triphosphate cyclohydrolase I.[36]

As widely published, green tea also has an effect on inhibition of oxidative stress both *in vivo* and *in vitro*. In an animal experiment, the 50 mg kg^{-1} dose of EGCG inhibited the weight gain of high-fat diet induced mice, but the fasting glucose level was not significantly affected compared with the control group. The fasting insulin level as well as insulin sensitivity was decreased significantly after EGCG treatment. We believe that the adopted dose of EGCG is almost the lower limit in many studies; therefore, a more positive effect may be expected with increased doses of EGCG.[37] An interesting result in this rat model fed a high-fat diet, fasting blood glucose was not affected by an EGCG dose of 0.32% for 16 weeks. However, the fasting insulin level, HOMA-IR, and glucose infusion rate were improved.[38] The alcoholic extract of green tea was orally administered to diabetic rats induced by streptozotocin injection which showed a significant decrease of blood glucose levels when the dose was 200 mg extract per kg. If green tea extract was given to normal rats, no change of serum glucose levels was observed.[39] In another study on T2D mice induced by a high-fat diet and streptozotocin injection, EGCG was orally fed to mice at levels of 50, 100, and 200 mg per kg per day. The blood glucose decrease after EGCG treatment followed a typical dose-dependent response. It also decreased insulin resistance and ameliorated insulin sensitivity by protecting β-cells in the pancreas.[40] Liupao tea, a typical type of dark tea, was recognized to possess anti-diabetic activity. Liupao tea extract could significantly alleviate hyperglycaemia, insulin resistance, and plasma lipid profiles in rats induced by streptozotocin and a high-fat, high-sugar diet. The hypoglycaemic effect of high-dose Liupao tea extract was comparable to that of metformin. This beneficial function is likely to be associated with the prebiotic effects on increasing the relative abundance of short chain fatty acid (SCFA)-producing bacteria and correspondingly the level of SCFAs.[41]

As another type of unfermented tea, white tea also exhibited benefits on reversing the abnormal serum lipids of T2D mice; meanwhile, the antioxidant capacities were also improved by white tea (400 mg kg^{-1}).[42] According to previous publications on the anti-inflammatory properties of various types of tea, white tea inhibited pro-inflammatory mediators including tumour necrosis factor-alpha (TNF-α), cyclooxygenase-2, and inducible nitric oxide synthase (iNOS). Recently, there have been many reports about the inhibitory effects of tea catechins or polyphenols on the formation of advanced glycation end products.[43] The prevention and regulation effects of tea on various diabetic complications, such as diabetic nephropathy, CVD, and diabetic retinopathy, have also been observed. A comprehensive review on tea and diabetic complications as well as its health benefits related to diabetes and the mechanisms involved has been published.[44]

Recently, targeting on the glucose producing and transporting pathways, Liu *et al.*[45] compared the hypoglycaemic activities of green, oolong, black, and dark tea extracts. Like previous studies concerning the comparison of various types of tea manufactured by one source of fresh tea leaves, this study also adopted this type of research strategy. Green tea showed the best effect on inhibiting glucose transport. The messenger RNA (mRNA) levels of sodium/glucose cotransporter 1 and glucose transporter 2 (GLUT2) were also down-regulated by green tea, but black tea had the best effect on the relative expression of $Na+/K+$-adenosine triphosphatase.

Although most of the published results on the effects of tea on diabetes are based on tea extract and tea polyphenols, tea polysaccharides also have beneficial effects on T2D. High doses of tea polysaccharides can effectively control body weight, basically approaching the level of the control group. Furthermore, the levels of total, HDL, and LDL cholesterols as well as TAG were also reduced after tea polysaccharide intervention. The mechanism may be due to upregulating the expressions of critical proteins in phosphatidylinositol 3-kinase/protein kinase B signaling pathways, including PI3K, phosphor-protein kinase B, and GLUT4.[46] Wang *et al.*[47] reported that the arabinogalactan of green tea which mainly contained arabinose and galactose at a molar ratio of 1.0 : 0.96 showed increasing effects on insulin secretion. Karadag *et al.*[48] optimized the ultrasonic-assisted extraction condition of green tea polysaccharides. In another related research study conducted by this research group, it was confirmed that the α-glucosidase inhibition rate of green tea polysaccharides was up to 91.86% at 0.5 mg mL^{-1} concentration, which was significantly higher than that of acarbose.[49] In comparison with green tea polysaccharides, black tea polysaccharides showed higher antidiabetic activities on streptozotocin-induced diabetic mice fed with a high-fat diet.[50] Pelvan *et al.*[51] confirmed the antidiabetic properties of black tea polysaccharides with no cytotoxic or mutagenic effects. Therefore, black tea polysaccharides were suggested to be used as a safe potential antidiabetic ingredient.

9.3 Understanding the Antidiabetic Mechanisms of Tea and Tea Polyphenols with Antioxidant Activity

When the ability to generate and eliminate free radicals in the body cannot reach equilibrium, oxidative stress can occur and this leads to the impairment to the normal functions of the macromolecules in cells, thus affecting the occurrence and development of many chronic diseases, including T2D.[52] The natural bioactive components in tea, especially tea polyphenols, have been confirmed to serve as excellent antioxidants in both *in vitro* and *in vivo* studies. The main antioxidant mechanisms of tea may involve improving antioxidant enzymes, inhibiting lipid peroxidation, scavenging of free radicals, and complexation of metal ions.

Chemical assays have been used to assess the *in vitro* antioxidant activity of tea and its bioactive components, including the 2,2′-azinobis(3-ethylbenzothiazoline-6-sulfonate) (ABTS) radical scavenging assay, 2,2-diphenyl-1-picrylhydrazyl (DPPH) radical scavenging assay, ferric reducing antioxidant power (FRAP) assay, total phenolic content (TPC), and oxygen radical absorbance capacity (ORAC) assay as well as cellular antioxidant assay.[53] Tang *et al.*[54] determined the antioxidant capacity of 30 types of tea through various chemical assays, including FRAP [611 to 5375 μmol Fe(II) per g dry weight (dw)], ABTS (326 to 3004 μmol Trolox equivalents per g dw) free radical scavenging assays, and the TPC (37.25 to 254 mg gallic acid equivalents per g dw). The results showed that unfermented and semi-fermented teas, such as green tea, yellow tea, and oolong tea, possessed stronger antioxidant activities, which might indicate an effect of processing on the antioxidant properties and capacity of tea. The main contributors to the antioxidant activity of a fermented tea were determined by component analysis, including catechins, caffeine, theaflavin, gallic acid, chlorogenic acid, ellagic acid, and flavanol-3-*O*-glycosides.[55] Furthermore, a type of white tea polyphenols showed a relatively strong free radical scavenging capacity than vitamin C through DPPH and FRAP radical scavenging assays.[55] As a typical type of dark tea, Fuzhuan brick tea not only possessed ideal ABTS and superoxide radical scavenging activities, but also protected H_2O_2-treated rat pheochromocytoma (PC12) cells from oxidative damage.[56] In addition to polyphenols, tea polysaccharides were also potential antioxidants which were determined through the reducing power, ABTS and DPPH radical scavenging, and ORAC assays, and an improvement of H_2O_2-induced oxidative injury in human umbilical vein endothelial cells induced by tea polysaccharides was also observed.[57]

Furthermore, numerous *in vivo* animal studies have reported determination of the antioxidant activity of tea and its reactive components. For instance, inhalation of green tea aroma could significantly increase the antioxidant capacity of serum in mice and reduce the urinary level of 8-hydroxydeoxyguanosine (8-OHdG), which is a representative marker of oxidative damage, as well as inhibition of the rise of X-ray irradiation induced 8-OHdG levels in urine.[58] In addition, the activities of antioxidant enzymes, such as catalase, glutathione peroxidase, and glutathione reductase, could be improved by green tea extract in carbon tetrachloride (CCl_4)-treated male ICR mice.[59] Tea polyphenols can also increase the levels of rat serum catalase and superoxide dismutase,[60] leading to the prevention of oxidative stress caused by bacterial infection and intestinal damage.[61]

Tea polyphenols have strong antioxidant capacities because of their hydrogen atom donating ability from the hydroxyl group in the catechin skeleton and gallate moiety. They can chelate various metal ions, such as iron, which plays an important role in metal homeostasis and the complications of diabetes. Furthermore, Fe^{2+} and Cu^{+} are involved in the production of reactive oxygen species (ROS). At high concentrations of EGCG, it may also produce ROS *in vitro*. The apoptosis of cancer cells could be induced by the pro-oxidant properties of tea polyphenols. Copper mediated oxidation of EGCG leads to the formation of polymerized polyphenols. It has been indicated that copper oxidized catechins are more efficient prooxidants than their unoxidized forms.

In short, tea and its bioactive components exhibit antioxidant activity both *in vitro* and *in vivo*, among which tea polyphenols (catechins), caffeine, theaflavins, and gallic acid are the major contributors. The mechanisms include scavenging free radicals, enhancing activities of antioxidant enzymes, and chelating metal ions. Furthermore, the antioxidant capacity varies among different types of tea, indicating the potential influence of processing techniques on the antioxidant properties of tea and its bioactive components.

Tea and its bioactive components benefit the prevention and treatment of T2D through several mechanisms, including protecting pancreatic β-cells, improving insulin resistance, regulating the digestion, absorption and metabolism of carbohydrates, as well as regulating the gut microbiota. In addition, diabetic complications, such as diabetic nephropathy and diabetic cardiopathy, could be significantly ameliorated by tea and its bioactive components, of which tea polyphenols, polysaccharides, and polypeptides are the main contributors.[61]

Pancreatic β-cells are responsible for the secretion of the well-known hypoglycaemic hormone insulin within the human body, and thus the normal and healthy function of pancreatic β-cells is of vital importance for the maintenance of the blood glucose level.[62] According to a recent research study, peptide II isolated from black tea extract showed a strong inhibitory activity against dipeptidyl peptidase IV inhibitor, which is a type of effective hypoglycaemic drug, and increased the levels of insulin in streptozotocin-induced diabetic mice. Improved pancreatic β-cell function, as well as inhibition of the proliferation of pancreatic α-cells, was observed.[63] Furthermore, providing green tea and ginseng to streptozotocin-treated diabetic rats decreased blood glucose levels; it also restored and improved the patterns and function of pancreatic β-cells, which might be the result of declined oxidative stress.[64]

Insulin resistance refers to reduced insulin response and subsequent compensation of elevated plasma insulin concentrations which leads to abnormal function in all body tissues. Insulin resistance most frequently results from changes in post-insulin receptor signaling pathways although insulin receptor changes may occur, such as production of receptor antibodies. Therefore, insulin resistance has become a major impetus for streptozotocin T2D. It is genetically associated, but obesity and lack of exercise are also potential causes. Qin *et al.*[65] studied the anti-diabetic effect and mechanism of green tea polyphenols on insulin and lipid metabolism. The main glucose indices were significantly increased after green tea treatment. It lowered the biomarkers of insulin resistance including retinol binding protein 4 and soluble cluster differentiation 36, and improved the expression of GLUT1 and GLUT4; meanwhile it inhibited glycogen synthase kinase-3 beta expression. Decreased blood glucose levels, as well as improved glucose intolerance, could be observed in T2D mice with the intake of green or black tea extract, but green tea extract showed a better effect on reducing blood glucose levels and weight gains in T2D mice compared to the black tea extract. These results provide further evidence about the impact of processing on the antioxidant capacity of tea and its bioactive components. Homeostatic model assessment indicated that the mechanism of the anti-diabetic effect of green tea extract may be associated with insulin resistance.[33] In addition, the levels of fasting blood glucose and insulin levels in high fat diet fed Zucker fatty mice could be decreased by green tea polyphenols. Increased insulin-stimulated glucose uptake in isolated soleus muscle could also be observed, which might be partially associated with the inhibition of protein kinase C-θ and improvement of insulin sensitivity in muscle.[66]

Regulation of carbohydrate hydrolysis, absorption, and metabolism is one of the mechanisms of hypoglycaemic action. As a representative tea polyphenol, EGCG deactivates pancreatic amylase and glucosidase, thus inhibiting the hydrolysis of carbohydrates, and decreases the production of glucose from dietary starch. Green tea also reduces the intestinal sodium/glucose cotransporter 1/GLUT2 ratio, thus decreasing the absorption of glucose from enterocytes. In addition, tea polyphenols also revert the carbohydrate metabolism, through rehabilitating the activities of hexokinase, pyruvate kinase, lactate dehydrogenase, glucose-6-phosphatase, fructose-1,6-bisphosphatase, glucose-6-phosphate dehydrogenase, glycogen synthase, and glycogen phosphorylase in the liver of diabetic rats. Furthermore, it also up-regulates the expression of mRNA of peroxisome proliferator-activated receptor-coactivator-1a, mitochondrial nicotinamide adenine dinucleotide dehydrogenase 5, mitochondrial cytochrome *b*, and mitochondrial cytochrome *c* oxidase III in skeletal muscle, increases the expression of peroxisome proliferator-activated receptor and carnitine palmitoyl transferase-1a in the liver, and decreases hepatic stearoyl-CoA desaturase 1 mRNA transcription. Black tea polyphenols mainly containing high-molecular weight oxidation products could also increase phosphorylation of adenosine 5′-monophosphate (AMP)-activated protein kinase (AMPK) and expression of GLUT4.

The improvement of GLUT4 translocation is associated with glucose uptake in adipocytes and skeletal muscle. Therefore, the main pathway has been recognized to be the AMPK pathway. A group of carbohydrates found in Pu-erh tea polysaccharides were found to inhibit α-glucosidase *in vitro* and exhibit a better hypoglycaemic effect than acarbose, which is a common hypoglycaemic drug in mice.[67] Furthermore, the inhibition rate of polysaccharides or theaflavins of Zijuan tea on α-glucosidase has been suggested to be higher than acarbose and other fractions, while the two fractions could also effectively decrease blood glucose in hyperglycaemic mice.[68] In addition, a related research study confirmed that the dark tea proteins and their hydrolysates had a significant suppressive effect on α-glucosidase, with the half-maximal inhibitory concentration (IC_{50}) values ranging from 0.0103 to 1.3114 mg mL^{-1}.[69]

The gut microbiota is composed of a set of microorganisms that colonize the digestive tract. Humans host thousands of bacteria in the gastrointestinal tract, as much as human cells in their body. The microbiota contains bacteria but also fungi and viruses. The microbiota settles gradually after birth and reaches the adult stage around the age of 3 years and could be modulated by many environmental factors and by the genetics of the host. An imbalanced microbiota, also referred to as dysbiosis, is observed in many human diseases including metabolic diseases like T2D. The microbiota is already considered a potential therapeutic tool and could serve as a diagnostic and prognostic biomarker. Liupao tea extract significantly improved hyperglycaemia and insulin resistance. It also showed a similar effect to that of a high dose of metformin, a common hypoglycaemic drug. The beneficial properties of Liupao tea extract might be associated with its intestinal bacterial regulating function, including up-regulation of the Bacteroidetes/Firmicutes ratio, and increasing the number and variety of SCFA producing bacteria, such as *Prevotella* and *Bacteroides*, in streptozotocin/high-fat and high-sugar diet induced rats.[41] Furthermore, cornstarch–tea (instant tea or matcha) diet significantly reduced blood glucose through regulating the intestinal microbiota, by promoting *Coriobacteriaceae*, *Lactobacillaceae*, *Prevotellaceae*, and *Bifidobacteriaceae*, and reducing *Bacteroidaceae*, *Ruminococcaceae*, *Helicobacteraceae*, and *Enterobacteriaceae*, in alloxan-induced T2D mice.[24]

Tea and its bioactive components could contribute to the alleviation of diabetic complications, such as diabetic nephropathy and diabetic cardiopathy. The apoptosis of podocytes leads to the development of proteinuria and nephropathy in diabetes mellitus. Green tea was found to down-regulate apoptosis and the albumin influx through reducing GSK-3β-protein 53 in high glucose-treated immortalized mouse podocytes. In addition, green tea reduced podocyte apoptosis, foot process effacement, and albuminuria in diabetic spontaneously hypertensive rats.[70] In another study, treatment with Cuiyu tea polypeptides significantly decreased the fasting blood glucose, serum insulin level, total urinary protein, creatinine, and urine nitrogen in high-fat diet/streptozotocin-induced diabetic mice by activating the polyol protein kinase Cζ/c-Jun N-terminal kinase/nuclear factor of kappa light polypeptide gene enhancer in B-cells/TNF-α/iNOS pathways and advanced glycation end products/receptor for advanced glycation end-product/transforming growth factor-β1 pathways, as well as elevating the podocin expression in the glomeruli, and inhibited the release of pro-inflammatory factors.[71] Moreover, EGCG synergistically enhanced autologous adipose-derived stem cells to protect cardiac functions in insulin-dependent diabetes mellitus rats through the expression of survival protein.[72]

9.4 Conclusion

A wide range of studies including epidemiological studies, both *in vitro* and *in vivo*, as well as human intervention studies have been conducted to validate and refine the health benefits of tea and its bioactive components towards T2D. According to the results of most epidemiological studies, tea was capable of reducing the risk of T2D, while a few studies indicated tea drinking was not associated with T2D risk, and fewer suggested it may increase the risk of T2D. In studies with human interventional trials, the positive effects of tea and tea polyphenols on blood glucose and lipids were limited, but those on some antioxidant index and immunological index could be observed. *In vitro* and *in vivo* experiments indicated the beneficial properties of tea and its bioactive components towards T2D and its associated factors. In addition, some studies found the prevention and regulatory effects of tea on various diabetic complications, such as diabetic nephropathy, CVD, and diabetic retinopathy.

Tea and its bioactive components have been proven to be ideal sources of natural antioxidants, of which tea catechins, caffeine, theaflavins, and gallic acid are the major contributors. The main antioxidant mechanisms of tea may involve improvement of antioxidant enzymes, inhibition of lipid peroxidation, scavenging of free radicals, and complexation of metal ions. Tea polyphenols have strong antioxidant capacities due to their electron donating ability from the hydroxyl group in the catechin skeleton and gallate moiety. They can chelate various metal ions, such as iron, which plays an important role in metal homeostasis and the complications of diabetes.

Tea and its bioactive components offer prevention and treatment to T2D through several mechanisms, including protecting pancreatic β-cells, improving insulin resistance, inhibiting the digestion, absorption and metabolism of carbohydrates, as well as regulating the gut microbiota. In addition, diabetic complications, such as diabetic nephropathy and diabetic cardiopathy, could be significantly ameliorated by tea and its bioactive components, of which tea polyphenols and polysaccharides are the main representatives.

References

1. Y. R. Sang, H. Guan, M. L. Zhi, W. K. Cheng, S. Waspadji and P. Palmes, *et al.*, Multi-country study on the prevalence and clinical features of peripheral arterial disease in Asian type 2 diabetes patients at high risk of atherosclerosis, *Diabetes Res. Clin. Pract.*, 2007, **76**, 82–92.

2. B. N. M. Yusof, R. A. K. Talib, N. A. Karim, N. A. K. Chinna and H. Gilbertson, A low-GI diet is associated with a short-term improvement of glycaemic control in Asian patients with type 2 diabetes, *Diabetes, Obes. Metab.*, 2010, **11**, 387–396.
3. L. Zhang, C.-T. Ho, J. Zhou, J. S. Santos, L. Armstrong and D. Granato, Chemistry and biological activities of processed camellia sinensis teas: A comprehensive review, *Compr. Rev. Food Sci. Food Saf.*, 2019, **18**, 1474–1495.
4. P. T. Katzmarzyk, The metabolic syndrome: an introduction, *Appl. Physiol., Nutr., Metab.*, 2007, **32**, 1–3.
5. J. Gupta, Y. H. Siddique, T. Beg, G. Ara and M. Afzal, A Review on the beneficial effects of tea polyphenols on human health, *Int. J. Pharmacol.*, 2008, **4**, 314–338.
6. Y. A. Dound, Health benefits of green tea and its polyphenols: A Review, *Indian Pract.*, 2020, **73**, 33–37.
7. J. V. Higdon and B. Frei, Tea catechins and polyphenols: Health effects, metabolism, and antioxidant functions, *Crit. Rev. Food Sci. Nutr.*, 2003, **43**, 89–143.
8. M. A. Martin, L. Goya and S. Ramos, Protective effects of tea, red wine and cocoa in diabetes: Evidences from human studies, *Food Chem. Toxicol.*, 2017, **109**, 302–314.
9. A. Basu, N. M. Betts, A. Mulugeta, C. Tong, E. Newman and T. J. Lyons, Green tea supplementation increases glutathione and plasma antioxidant capacity in adults with the metabolic syndrome, *Nutr. Res.*, 2013, **33**, 180–187.
10. S. van Dieren, C. S. Uiterwaal, Y. T. van der Schouw, A. D. van der, J. M. Boer and A. Spijkerman, *et al.*, Coffee and tea consumption and risk of type 2 diabetes, *Diabetologia*, 2009, **52**, 2561–2569.
11. S. A. Rebello, C. H. Chen, N. Naidoo, W. Xu, J. Lee and K. S. Chia, *et al.*, Coffee and tea consumption in relation to inflammation and basal glucose metabolism in a multi-ethnic Asian population: a cross-sectional study, *Nutr. J.*, 2011, **10**, 61.
12. S. Oba, C. Nagata, K. Nakamura, K. Fujii, T. Kawachi and N. Takatsuka, *et al.*, Consumption of coffee, green tea, oolong tea, black tea, chocolate snacks and the caffeine content in relation to risk of diabetes in Japanese men and women, *Br. J. Nutr.*, 2010, **103**, 453–459.
13. F. Dai, H. Cai, H. Li, G. Yang, B. T. Ji and W. Zheng, *et al.*, Association of sleep duration and incidence of diabetes modified by tea consumption: a report from the Shanghai men's health study, *Sleep Med.*, 2017, **38**, 135–141.
14. S. R. Vaz, L. M. N. de Amorim, P. V. F. de Nascimento, V. S. P. Veloso, M. S. Nogueira and I. A. Castro, *et al.*, Effects of green tea extract on oxidative stress and renal function in diabetic individuals: A randomized, double-blinded, controlled trial, *J. Funct. Foods*, 2018, **46**, 195–201.
15. C. Y. Liu, C. J. Huang, L. H. Huang, I. J. Chen, J. P. Chiu and C. H. Hsu, Effects of green tea extract on insulin resistance and glucagon-like

peptide 1 in patients with type 2 diabetes and lipid abnormalities: a randomized, double-blinded, and placebo-controlled trial, *PLoS One*, 2014, **10**, e91163.
16. E. Dardashti Pour, F. Yaghobian, F. Dehghan and M. A. Azarbayjani, Forecast of ameliorating effect of dietary flavonol consumption in white tea with or without aerobic training on type 2 diabetes (T2D) in females, *Clin. Nutr. ESPEN*, 2021, **45**, 134–140.
17. A. Essmat and M. S. Hussein, Green tea extract for mild-to-moderate diabetic peripheral neuropathy: A randomized controlled trial, *Complement. Ther. Clin. Pract.*, 2021, **43**, 101317.
18. M. Alves Ferreira, A. P. Oliveira Gomes, A. P. Guimarães de Moraes, M. L. Ferreira Stringhini, J. F. Mota and A. Siqueira Guedes Coelho, *et al.*, Green tea extract outperforms metformin in lipid profile and glycaemic control in overweight women: A double-blind, placebo-controlled, randomized trial, *Clin. Nutr. ESPEN*, 2017, **22**, 1–6.
19. Q. Ma, D. Chen, H. P. Sun, N. Yan, Y. Xu and C. W. Pan, Regular Chinese green tea consumption is protective for diabetic retinopathy: A clinic-based case-control study, *J. Diabetes Res.*, 2015, **2015**, 231570.
20. A. L. Brown, J. Lane, J. Coverly, J. Stocks, S. Jackson and A. Stephen, *et al.*, Effects of dietary supplementation with the green tea polyphenol epigallocatechin-3-gallate on insulin resistance and associated metabolic risk factors: randomized controlled trial, *Br. J. Nutr.*, 2009, **101**, 886–894.
21. R. Xu, Y. Bai, K. Yang and G. Chen, Effects of green tea consumption on glycemic control: a systematic review and meta-analysis of randomized controlled trials, *Nutr. Metab.*, 2020, **17**, 56.
22. N. R. Stendell-Hollis, C. A. Thomson, P. A. Thompson, J. W. Bea, E. C. Cussler and I. A. Hakim, Green tea improves metabolic biomarkers, not weight or body composition: a pilot study in overweight breast cancer survivors, *J. Hum. Nutr. Diet.*, 2010, **23**, 590–600.
23. F. Mahmoud, D. Haines, E. Al-Ozairi and A. Dashti, Effect of black tea consumption on intracellular cytokines, regulatory t cells and metabolic biomarkers in type 2 diabetes patients, *Phytother. Res.*, 2016, **30**, 454–462.
24. H. H. Zhang, J. Liu, Y. J. Lv, Y. L. Jiang, J. X. Pan and Y. J. Zhu, *et al.*, Changes in intestinal microbiota of type 2 diabetes in mice in response to dietary supplementation with instant tea or matcha, *Can. J. Diabetes*, 2020, **44**, 44–52.
25. O. Asbaghi, F. Fouladvand, M. J. Gonzalez, V. Aghamohammadi, R. Choghakhori and A. Abbasnezhad, The effect of green tea on C-reactive protein and biomarkers of oxidative stress in patients with type 2 diabetes mellitus: A systematic review and meta-analysis, *Complementary Ther. Med.*, 2019, **46**, 210–216.
26. O. Asbaghi, F. Fouladvand, S. Moradi, D. Ashtary-Larky, R. Choghakhori and A. Abbasnezhad, Effect of green tea extract on lipid profile in patients with type 2 diabetes mellitus: A systematic review and meta-analysis, *Diabetes Metab. Syndr.*, 2020, **14**, 293–301.

27. A. Mousavi, M. Vafa, T. Neyestani, M. Khamseh and F. Hoseini, The effects of green tea consumption on metabolic and anthropometric indices in patients with type 2 diabetes, *J. Res. Med. Sci.*, 2013, **18**, 1080–1086.
28. H. Mozaffari-Khosravi, Z. Ahadi and K. Barzegar, The effect of green tea and sour tea on blood pressure of patients with type 2 diabetes: A randomized clinical trial, *J. Diet Suppl.*, 2013, **10**, 105–115.
29. C. M. Borges, A. Papadimitriou, D. A. Duarte, J. M. Lopes de Faria and J. B. Lopes de Faria, The use of green tea polyphenols for treating residual albuminuria in diabetic nephropathy: A double-blind randomised clinical trial, *Sci. Rep.*, 2016, **6**, 28282.
30. K. C. Silva, M. A. Rosales, D. E. Hamassaki, K. C. Saito, A. M. Faria and P. A. Ribeiro, *et al.*, Green tea is neuroprotective in diabetic retinopathy, *Invest. Ophthalmol. Vis. Sci.*, 2013, **54**, 1325–1336.
31. P. D. Ribaldo, D. S. Souza, S. K. Biswas, K. Block, J. M. Lopes de Faria and J. B. Lopes de Faria, Green tea (*Camellia sinensis*) attenuates nephropathy by downregulating Nox4 NADPH oxidase in diabetic spontaneously hypertensive rats, *J. Nutr.*, 2009, **139**, 96–100.
32. L. M. N. de Amorim, S. R. Vaz, G. Cesário, A. S. G. Coelho and P. B. Botelho, Effect of green tea extract on bone mass and body composition in individuals with diabetes, *J. Funct. Foods*, 2018, **40**, 589–594.
33. W. P. Tang, S. M. Li, Y. Liu, M. T. Huang and C.-T. Ho, Anti-diabetic activity of chemically profiled green tea and black tea extracts in a type 2 diabetes mice model *via* different mechanisms, *J. Funct. Foods*, 2013, **5**, 1784–1793.
34. P. Koonyosying, C. Uthaipibull, S. Fucharoen, E. V. Koumoutsea, J. B. Porter and S. Srichairatanakool, Decrement in cellular iron and reactive oxygen species, and improvement of insulin secretion in a pancreatic cell line using green tea extract, *Pancreas*, 2019, **48**, 636–643.
35. X. Yang and F. Kong, Evaluation of the *in vitro* α-glucosidase inhibitory activity of green tea polyphenols and different tea types, *J. Sci. Food Agric.*, 2016, **96**, 777–782.
36. A. M. Faria, A. Papadimitriou, K. C. Silva, J. M. Lopes de Faria and J. B. Lopes de Faria, Uncoupling endothelial nitric oxide synthase is ameliorated by green tea in experimental diabetes by re-establishing tetrahydrobiopterin levels, *Diabetes*, 2012, **61**, 1838–1847.
37. H. J. Jang, S. D. Ridgeway and J. A. Kim, Effects of the green tea polyphenol epigallocatechin-3-gallate on high-fat diet-induced insulin resistance and endothelial dysfunction, *Am. J. Physiol. Endoc. M.*, 2013, **305**, E1444–E1451.
38. H. Hou, W. Yang, S. Bao and Y. Cao, Epigallocatechin gallate suppresses inflammatory responses by inhibiting Toll-like receptor 4 signaling and alleviates insulin resistance in the livers of high-fat-diet rats, *J. Oleo Sci.*, 2020, **69**, 479–486.
39. F. Haidari, K. Omidian, H. Rafiei, M. Zarei and M. Mohamad Shahi, Green tea (*Camellia sinensis*) supplementation to diabetic rats improves

serum and hepatic oxidative stress markers, *Iran. J. Pharm. Res.*, 2013, **12**, 109–114.
40. Z. Ren, Z. Yang, Y. Lu, R. Zhang and H. Yang, Anti-glycolipid disorder effect of epigallocatechin-3-gallate on high-fat diet and STZ-induced T2DM in mice, *Mol. Med. Rep.*, 2020, **21**, 2475–2483.
41. Q. Ding, B. Zhang, W. Zheng, X. Chen, J. Zhang and R. Yan, *et al.*, Liupao tea extract alleviates diabetes mellitus and modulates gut microbiota in rats induced by streptozotocin and high-fat, high-sugar diet, *Biomed. Pharmaco Ther.*, 2019, **118**, 109262.
42. X. Xia, X. Wang, H. Wang, Z. Lin, K. Shao, J. Xu and Y. Zhao, Ameliorative effect of white tea from 50-year-old tree of *Camellia sinensis* L. (Theaceae) on kidney damage in diabetic mice via SIRT1/AMPK pathway, *J. Ethnopharmacol.*, 2021, **272**, 113919.
43. Q. Wu, Y. Ouyang, Y. Kong, Y. Min, J. Xiao and S. Li, *et al.*, Catechin inhibits the release of advanced glycation end products during glycated bovine serum albumin digestion and corresponding mechanisms *in vitro*, *J. Agric. Food Chem.*, 2021, **69**, 8807–8818.
44. J. M. Meng, S. Y. Cao, X. L. Wei, R. Y. Gan, Y. F. Wang and S. X. Cai, *et al.*, Effects and mechanisms of tea for the prevention and management of diabetes mellitus and diabetic complications: An updated review, *Antioxidants*, 2019, **8**, 170.
45. S. Liu, Z. Ai, Y. Meng, Y. Chen and D. Ni, Comparative studies on the physicochemical profile and potential hypoglycemic activity of different tea extracts: Effect on sucrase-isomaltase activity and glucose transport in Caco-2 cells, *Food Res. Int.*, 2021, **148**, 110604.
46. S. Li, H. Chen, J. Wang, X. Wang, B. Hu and F. Lv, Involvement of the PI3K/Akt signal pathway in the hypoglycemic effects of tea polysaccharides on diabetic mice, *Int. J. Biol. Macromol.*, 2015, **81**, 967–974.
47. H. Wang, S. Shi, B. Bao, X. Li and S. Wang, Structure characterization of an arabinogalactan from green tea and its anti-diabetic effect, *Carbohydr. Polym.*, 2015, **124**, 98–108.
48. A. Karadag, E. Pelvan, K. Dogan, N. Celik, D. Ozturk and K. Akalın, *et al.*, Optimisation of green tea polysaccharides by ultrasound-assisted extraction and their *in vitro* antidiabetic activities, *Qual. Assur. Saf. Crops Foods*, 2019, **11**, 479–490.
49. A. Karadag, K. Dogan, E. Pelvan, A. Tas, O. A. Eklioglu and S. Aksu, *et al.*, α-Glucosidase inhibitory activities, functional properties, and safety of green tea polysaccharides as a potential source of dietary supplement, *J. Food Bioact.*, 2018, **3**, 124–132.
50. K. Akalın, A. T. Ekiz, F. Karakaya, A. Karadag, E. Pelvan and K. Dogan, *et al.*, *In vivo* antidiabetic activities of green and black tea polysaccharides using streptozotocin-induced diabetic mice fed with a high-fat diet, *J. Food Bioact.*, 2019, **8**, 74–83.
51. E. Pelvan, A. Karadag, K. Dogan, S. Aksu, A. Tas and K. Akalın, *et al.*, *In vitro* antidiabetic activities, chemical compositions, antioxidant

activities, and toxicity of black tea polysaccharides as a potential source of dietary ingredients, *J. Food Bioact.*, 2021, **13**, 93–101.
52. L. Ilaria, R. Gennaro, C. Francesco, B. Giulia, A. Luisa and D. M. David, *et al.*, Oxidative stress, aging, and diseases, *Clin. Interv. Aging.*, 2018, **13**, 757–772.
53. A. Shang, J. Li, D. D. Zhou, R. Y. Gan and H. B. Li, Molecular mechanisms underlying health benefits of tea compounds, *Free Radic. Biol. Med.*, 2021, **172**, 181–200.
54. G. Y. Tang, C. N. Zhao, X. Y. Xu, R. Y. Gan, S. Y. Cao and Q. Liu, *et al.*, Phytochemical composition and antioxidant capacity of 30 Chinese teas, *Antioxidants*, 2019, **8**, 180.
55. Y. Zhou, F. Tan, C. Li, W. Li, W. Liao and Q. Li, *et al.*, White peony (fermented *Camellia sinensis*) polyphenols help prevent alcoholic liver injury *via* antioxidation, *Antioxidants*, 2019, **8**, 524.
56. G. Chen, M. Wang, M. Xie, P. Wan, D. Chen and B. Hu, *et al.*, Evaluation of chemical property, cytotoxicity and antioxidant activity *in vitro* and *in vivo* of polysaccharides from Fuzhuan brick teas, *Int. J. Biol. Macromol.*, 2018, **116**, 120–127.
57. M. Fan, J. Zhu, Y. Qian, W. Yue, Y. Xu and D. Zhang, *et al.*, Effect of purity of tea polysaccharides on its antioxidant and hypoglycemic activities, *J. Food Biochem.*, 2020, **44**, e13277.
58. Y. S. Li, Y. Kawasaki, I. Tomita and K. Kawai, Antioxidant properties of green tea aroma in mice, *J. Clin. Biochem. Nutr.*, 2017, **61**, 14–17.
59. C. F. Tsai, Y. W. Hsu, H. C. Ting, C. F. Huang and C. C. Yen, The *in vivo* antioxidant and antifibrotic properties of green tea (*Camellia sinensis*, Theaceae), *Food Chem.*, 2013, **136**, 1337–1344.
60. N. A. Ahmed, N. M. Radwan, H. S. Aboul Ezz and N. A. Salama, The antioxidant effect of green tea mega EGCG against electromagnetic radiation-induced oxidative stress in the hippocampus and striatum of rats, *Electromagn. Biol. Med.*, 2016, **36**, 63–73.
61. L. Zhang, S. Q. Gui, J. Wang, Q. R. Chen, J. L. Zeng and A. L. Liu, *et al.*, Oral administration of green tea polyphenols (TP) improves ileal injury and intestinal flora disorder in mice with *Salmonella typhimurium* infection via resisting inflammation, enhancing antioxidant action and preserving tight junction, *J. Funct. Foods*, 2020, **64**, 103654.
62. Y. Itoh, Y. Kawamata, M. Harada, M. Kobayashi, R. Fujii and S. Fukusumi, *et al.*, Free fatty acids regulate insulin secretion from pancreatic beta cells through GPR40, *Nature*, 2003, **422**, 173–176.
63. Y. Lu, P. Lu, Y. Wang, X. Fang, J. Wu and X. Wang, A novel dipeptidyl peptidase IV inhibitory tea peptide improves pancreatic β-cell function and reduces α-cell proliferation in streptozotocin-induced diabetic mice, *Int. J. Mol. Sci.*, 2019, **20**, 322.
64. T. Karaca, M. Yoruk, I. H. Yoruk and S. Uslu, Effects of streptozotocin-extract of green tea and ginseng on pancreatic beta cells and levels of serum glucose, insulin, cholesterol and triglycerides in rats with experimentally induced diabetes: A histochemical and immunohistochemical study, *J. Anim. Vet. Adv.*, 2010, **9**, 102–107.

65. B. Qin, M. M. Polansky, D. Harry and R. A. Anderson, Green tea polyphenols improve cardiac muscle mRNA and protein levels of signal pathways related to insulin and lipid metabolism and inflammation in insulin-resistant rats, *Mol. Nutr. Food Res.*, 2010, **54**, S14–S23.
66. J. Cheng, Y. Tan, J. Zhou, L. Xiao, M. Johnson and X. Qu, Green tea polyphenols ameliorate metabolic abnormalities and insulin resistance by enhancing insulin signalling in skeletal muscle of Zucker fatty rats, *Clin. Sci.*, 2020, **134**, 1167–1180.
67. Y. T. Deng, S. Y. Lin-Shiau, L. F. Shyur and J. K. Lin, Pu-erh tea polysaccharides decrease blood sugar by inhibition of alpha-glucosidase activity *in vitro* and in mice, *Food Funct.*, 2015, **6**, 1539–1546.
68. D. Chen, J. Sun, W. Dong, Y. Shen and Z. Xu, Effects of polysaccharides and polyphenolics fractions of Zijuan tea (*Camellia sinensis var. kitamura*) on α-glucosidase activity and blood glucose level and glucose tolerance of hyperglycaemic mice, *Int. J. Food Sci. Tech.*, 2018, **53**, 2335–2341.
69. K. Y. Su, X. L. Mao, L. P. Ai and X. W. Zhang, *In vitro* assessment of anti-diabetic potential of four kinds of dark tea (*Camellia sinensis* L.) protein hydrolysates, *J. Appl. Bot. Food Qual.*, 2019, **92**, 57–63.
70. E. B. Peixoto, A. Papadimitriou, D. A. Teixeira, C. Montemurro, D. A. Duarte and K. C. Silva, *et al.*, Reduced LRP6 expression and increase in the interaction of GSK3β with p53 contribute to podocyte apoptosis in diabetes mellitus and are prevented by green tea, *J. Nutr. Biochem.*, 2015, **26**, 416–430.
71. X. Deng, L. Sun, X. Lai, L. Xiang, Q. Li and W. Zhang, *et al.*, Tea polypeptide ameliorates diabetic nephropathy through RAGE and NF-κB signaling pathway in type 2 diabetes mice, *J. Agric. Food Chem.*, 2018, **66**, 11957–11967.
72. T. S. Chen, S. Y. Liou, C. H. Kuo, L. F. Pan, Y. L. Yeh and J. Liou, *et al.*, Green tea epigallocatechin gallate enhances cardiac function restoration through survival signaling expression in diabetes mellitus rats with autologous adipose tissue-derived stem cells, *J. Appl. Physiol.*, 2017, **123**, 1081–1091.

CHAPTER 10

Brain Health: Cognition, Depression, and Neurodegenerative Diseases

YEN-CHUN KOH,[a] CHI-TANG HO[b] AND MIN-HSIUNG PAN*[a,c,d]

[a] Institute of Food Sciences and Technology, National Taiwan University, Taipei, Taiwan; [b] Department of Food Science, Rutgers University, New Brunswick, NJ, USA; [c] Department of Medical Research, China Medical University Hospital, China Medical University, Taichung, Taiwan; [d] Department of Health and Nutrition Biotechnology, Asia University, Taichung, Taiwan
*Email: mhpan@ntu.edu.tw

10.1 Introduction

The brain is a part of the central nervous system (CNS), and it controls all bodily functions, including memory and thoughts, speech, limb movement, and organ activities. Therefore, brain disorders can lead to severe physical and psychological disabilities. There are several causes of brain disorders, the major categories of which are (1) trauma or injuries, (2) tumours, (3) neurodegeneration, and (4) mental disorders.

Brain injuries can be traumatic or non-traumatic. The former results from an external mechanical force that causes permanent loss of brain function, while the latter may be caused by a non-physical force, such as a virus, a toxic compound, tumour, stroke, or an inadequate supply of brain oxygen.[1] Among these causes, brain tumours (primary tumours such as glioblastoma,

Food Chemistry, Function and Analysis No. 39
Dietary Supplements with Antioxidant Activity: Understanding Mechanisms and Potential Health Benefits
Edited by Cesarettin Alasalvar, Fereidoon Shahidi and Chi-Tang Ho

Published by the Royal Society of Chemistry, www.rsc.org

the most common primary tumour of the CNS) are devastating with poor prognoses and high mortality.[2] Unfortunately, it is impossible to completely remove malignant gliomas by surgical resection because tumour cells infiltrate the brain structures.[3] Unlike brain tumours, neurodegeneration normally occurs at a later age and is normally caused by gradual damage of neuronal cells.[4] As the elderly population continues to grow worldwide, neurodegenerative diseases (NDD) such as Alzheimer's disease, Parkinson's disease, and Huntington's disease will also increase, and they can ultimately lead to cognitive impairment and dementia.[5] Therefore, NDD have become health problems of widespread concern. Besides NDD, depression has also been associated with the incidence of and rapid conversion to dementia from cognitive decline later in life.[6] Although depression can be an age-related disease, it is also a major cause of disability for individuals under the age of 40. Genetics, neuroendocrinology, epigenetic regulation, and psychosocial factors all influence the pathogenicity of depression.[7] In addition to depression, bipolar disorder and schizophrenia are also severe mental disorders that can impair cognition, regulate emotions, and influence individual behaviours.[8]

The blood–brain barrier provides unique protection to the central nervous system, especially from exogenous injuries.[9] Microglia also play an important role, specifically in protecting the immune defence system of the brain, but abnormal activation of microglia can lead to neuroinflammation-mediated neuronal degeneration.[9] Furthermore, neuroinflammation may be linked with the signs of depression as the serum of depressed patients contains elevated levels of inflammatory markers.[10] Besides neuroinflammation, it has been reported that hippocampal cell death induced by oxidative stress is observed in progressive depressed patients. Therefore, promoting autophagy-related cellular survival may be another strategy for fighting depression and offering neuroprotection.[11]

This chapter focuses on the mechanisms that may protect neurons, avoid brain diseases, mainly depression and cognitive disorders, and impart neuroprotection *via* phytochemical intervention.

10.2 Health Benefits of Dietary Supplements on Brain Health

10.2.1 Pathogenesis of Brain Diseases

It is important to identify brain disease indicators. For example, increased serum corticosterone and decreased serum and hippocampal serotonin are considered abnormalities observed during depressed behaviour, and the level of serum serotonin might be a reliable indicator for depression monitoring.[12] Excess hippocampal corticosterone and glutamate have been reported to cause nerve cell death *via* elevation of intracellular reactive oxygen species (ROS).[13] Moreover, the efficacy of brain-derived neurotrophic factor (BDNF), a neurotrophin, may be highly correlated to the pathophysiology of depression.[14] Reduced BDNF was also observed in aging mice, and the decrease was suggested to be related to memory impairment.[15]

Subventricular and subgranular zones are the major sites associated with learning and memory. Cognition deficits may be promoted by neurogenesis impairment, which is detrimental to maintaining synaptic plasticity and hippocampal functionality.[16] Memory deficits are a part of cognitive function decline related to aging and might be due to increased oxidative stress, decreased antioxidant defences, and inhibition of Na^+- or K^+-ATPase activity affected by oxidative modifications.[15] It has also been suggested that inhibition of Na^+- or K^+-ATPase could be mitigated by reducing abnormal increases of reactive species, although the underlying mechanism has not been clarified.

NDD such as Parkinson's disease and Alzheimer's disease are commonly accompanied by pathological features, such as oxidative stress and mitochondrial dysfunction.[17,18] Therefore, strategies focusing on antioxidation and mitochondrial rescue are promising in the prevention of neurodegeneration-related issues.

Accumulation of amyloid beta (Aβ)-peptide is recognized as the major cause of Alzheimer's disease. It results in cytotoxic injury, causing neuron death and consequently memory loss and cognitive decline.[19] In the progression of Alzheimer's disease or Huntington's disease, aggregation of oxidation modified glyceraldehyde-3-phosphate dehydrogenase (GAPDH) has been observed. It has been demonstrated that GAPDH is involved in the pathogenesis of NDD, such as accelerated amyloidogenesis in Alzheimer's disease.[20] Therefore, avoiding GAPDH resulting from oxidation has become a promising target in NDD prevention.

The pathogenesis of Parkinson's disease normally involves the loss and dysfunction of dopaminergic neurons, especially in the striatum. Depigmentation processes of substantia nigra and locus coeruleus are the major hallmarks of Parkinson's disease, accompanied by accumulation of α-synuclein (α-syn) and the presence of Lewy bodies.[21,22] Some phytochemicals can slow the process of amyloidogenesis or accumulation of α-syn, while others ameliorate the degenerative, oxidative, or neuroinflammatory effects during disease progression. In the following sections, the potential benefits of the application of phytochemicals in brain disease progression are discussed.

10.2.2 *In Vitro* Studies

10.2.2.1 Phytochemicals Prevent Neuroinflammation

Over-activation of astrocytes, the glial cells that function to maintain brain homeostasis, has exhibited a close relationship with neuroinflammation (Figure 10.1) and neurodegeneration.[23] In an *in vitro* study, resveratrol mediated glioprotection was found to eliminate the inflammatory response induced by lipopolysaccharides (LPS) *via* activation of adenosine receptors.

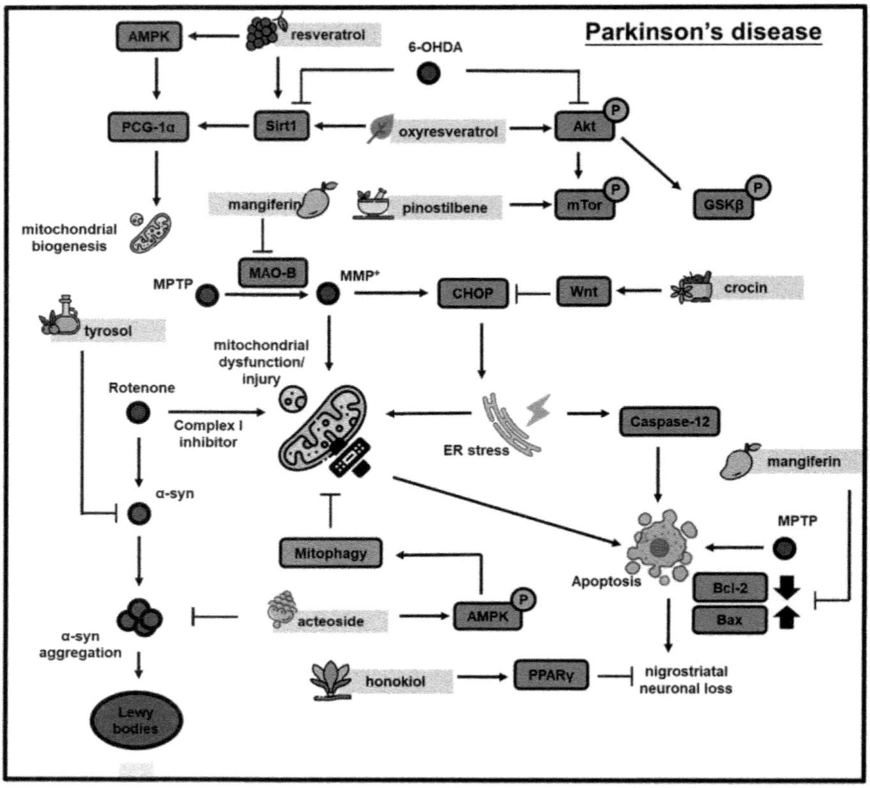

Figure 10.1 Potential mechanisms of phytochemical involvement in Parkinson's disease prevention/amelioration. Abbreviations: Akt, protein kinase B; AMPK, adenosine monophosphate-activated protein kinase; Bax, Bcl-2-associated X; Bcl-2, B-cell lymphoma 2; CHOP, CCAAT/enhancer-binding protein homologous protein; ER, endoplasmic reticulum; GSKβ, glycogen synthase kinase 3 beta; MAO-B, monoamine oxidase B; MMP^+, 1-methyl-4-phenylpyridinium; MPTP, 1-methyl-4-phenyl-1,2,3,6-tetrahydropyridine; mTOR, mammalian target of rapamycin; 6-OHDA, 6-hydroxydopamine; PGC-1α, peroxisome proliferator-activated receptor-γ coactivator-1 alpha; PPARγ, peroxisome proliferator-activated receptor-gamma; Sirt1, sirtuin 1; α-syn, alpha-synuclein; Wnt, wingless-related integration site.

The suppressive effect of resveratrol on the Neuro-2A cell line and C8-B4 microglia was demonstrated to result in neuroprotection.[24] Macasiamenene F is a prenylated stilbenoid from a novel group of stilbenes isolated from *Mammea siamensis*, and its anti-neuroinflammatory benefits have been confirmed.[25] In that study, Macasiamenene F exhibited reduced systemic inflammation of monocytes *via* the inhibitor of NF-κB/nuclear factor-kappa B (IκB/NF-κB) pathway and reduced pro-inflammatory responses in LPS-treated microglia.

Activation of NF-κB has become one of the major molecular targets in the strategy of neuroinflammation prevention. An isolated prenylflavonoid, dehydroglyasperin C, has been identified as a potent inducer of antioxidant enzymes and revealed to inhibit NF-κB activation *via* several pathways.[26] Dehydroglyasperin C inhibited microglial hyperactivation by reducing phosphorylation of mitogen-activated protein kinase p38/extracellular signal-regulated kinase (MAPK p38/ERK) signalling pathways in the BV-2 cell line. As a negative regulator of MAPKs, MAPK phosphatase 1 (MKP-1) was significantly upregulated after being treated with dehydroglyasperin C that subsequently suppressed NF-κB activation.[26] Anti-neuroinflammation is a widely studied strategy in brain disease prevention, and it is discussed in a later section.

10.2.2.2 Anti-neurodegenerative Effects – Oxidative Stress

Oxidative stress is a major inducer of neural cell death. The benefits of antioxidants such as phytochemicals are largely to rescue neural cells from dying (Table 10.1). Accumulation of β-amyloid can trigger neuronal apoptosis induced by mitochondrial dysfunction. Loss of mitochondrial membrane potential can be induced by elevated ROS, resulting in the release of cytochrome C which subsequently leads to apoptosis.[19] Intervention with tetrahydroxystilbene glucoside effectively protected neuronal cell damage by restoring mitochondrial function. In another study of phytochemicals preventing oxidative stress-induced apoptosis, the authors demonstrated glutamate-induced cytotoxicity in murine hippocampal neuronal HT22 cells *via* ROS-related apoptosis. Furthermore, pterostilbene showed an ameliorative effect by enhancing nuclear factor erythroid 2 (NFE2)-related factor 2/heme oxygenase 1/nicotinamide adenine dinucleotide (NADH) quinone oxidoreductase 1 (Nrf2/HO-1/NQO-1) and intracellular levels of the antioxidative enzyme, glutathione (GSH), and superoxide dismutase (SOD).[27]

Another stilbene, pinosylvin, showed significant neuroprotective benefits against oxygen and glucose deprivation/reperfusion (OGD/R) damages on PC12 cells, greater than other stilbenes (pterostilbene, resveratrol, and 4-methoxystilbene). The study revealed that pinosylvin exhibited antioxidative and anti-apoptotic effects and improved mitochondrial dysfunction.[28] In another study, four stilbenes (cajaninstilbene acid, longistylin A, longistylin C, and cajanolactone A) inhibited oxidative stress on the PC12 cell line induced by glutamate and corticosterone by increasing hippocampal SOD and catalase (CAT) levels.[13] Without antioxidants, oxidative stress accumulation may lead to neural cell apoptosis.

10.2.2.3 Anti-neurodegenerative Effects – Apoptosis

As noted above, neurotoxicity could be induced by several causes, depending on the disease pathogenesis. For example, a prion protein fragment (PrP 106–126) has been reported to induce mitochondrial dysfunction by

Table 10.1 Preventive effect of phytochemicals on depression, cognitive decline, and neurodegeneration.

Phytochemical	Model of study	Treatment protocol	Underlying mechanism	Ref.
EGCG	• Depression • Animal model	• Chronic unpredictable mild stress-induced depression-like behaviour in Wistar rats	• Reduced levels of IL-6 and NO • Suppressed apoptosis (mRNA of caspase-3 and caspase-9)	11
Resveratrol	• Depression • Animal model	• Wistar–Kyoto rats (non-induced animal model of depression)	• Elevated hippocampal BDNF level	14
	• Depression	• Chronic restraint stress-induced depression-like behaviour in Wistar rats for 3 weeks	• Upregulated BDNF and p-ERK levels in the brain region • Increased mRNA levels of Bcl-2 and Bax	33
	• Cognition impaired • Animal model	• Chronic unpredictable mild stress-induced rats (5 weeks)	• Normalized serum corticosterone • Upregulated BDNF, p-CREB, and p-ERK levels	49
	• Depression • Animal model	• Repeated corticosterone (40 mg kg^{-1})-induced depression in mice for 21 days	• Ameliorated depressive-like behaviour • Recovered BDNF level in the hippocampus	50
Chrysin	• Cognitive decline • Animal model	• Aged (20 month old) Swiss Albino mice	• Reduced level of reactive species • Increased activities of SOD, CAT, GPx, and Na^+- and K^+-ATPase • Upregulated expression of BDNF in the prefrontal cortex and hippocampus	15
	• Neurodegeneration • Huntington's disease • Animal model	• 3-Nitropropionic acid-induced neurotoxicity in Wistar rats	• Reduced lipid peroxidation, nitrite, and protein carbonyl levels • Induced SOD, catalase, and GSH levels • Upregulated Bcl-2 and reduced Bax and Bad mRNA levels to reduce neuronal apoptosis	42

Dehydroglyasperin C	• Inflammation-induced neurodegeneration • Cell model	• Primary cortical neurons from ICR mice treated with a conditioned medium of BV-2 for 48 h	• Inhibited microglial hyperactivation by increasing MKP-1 expression • Downregulated activation of NF-κB and TNF-α expression	26
Resveratrol and pinostilbene	• Neurodegeneration • Animal model	• Age-related neurodegeneration	• Alleviated motor decline • Promoted dopaminergic neuron survival • Increased ERK1/2 activation	35
4-Hydroxyisophthalic acid	• Neurodegeneration • Animal model	• Paraquat-induced movement disorder and neurodegeneration in the *Drosophila* brain	• Reduced ROS and lipid peroxidation • Restored Cu/Zn SOD, Mn-SOD, GSH, catalase, and acetylcholinesterase activity in the brain	43
Naringin	• Neurodegeneration • Animal model	• 3-Nitropropionic acid (10 mg kg^{-1})-induced neurodegeneration in rats	• Activated Nrf2 and enhanced phase II enzyme • Reduced hydroperoxide, nitrite, and hydroxy radicals • Decreased expression of pro-inflammatory mediators	46
Total glycosides isolated from *Cistanche tubulosa*	• Depression • Animal model	• Chronic unpredictable mild stress-induced ICR mice (4 weeks)	• Alleviated hyperactivation of the HPA axis and neuroinflammation • Restored 5-HT and BDNF in the hippocampus • Mitigated colonic inflammation and intestinal barrier disruption • Reduced expression of IDO1 and suppressed tryptophan–kynurenine metabolism	51

Abbreviations: 5-HT, 5-hydroxytryptamine; Bad, Bcl-2 associated agonist of cell death; Bax, Bcl-2-associated X protein; Bcl-2, B-cell lymphoma 2; BDNF, brain-derived neurotrophic factor; CAT, catalase; CREB, cAMP-response element binding protein; EGCG, epigallocatechin gallate; ERK, extracellular signal-regulated kinase; GPx, glutathione peroxidase; GSH, glutathione; HPA, hypothalamic–pituitary–adrenal axis; IDO1, indoleamine 2,3-dioxygenase 1; IL-6, interleukin-6; MPK-1, MAPK phosphatase 1; NF-κB, nuclear factor-kappa B; NO, nitric oxide; Nrf2, nuclear factor erythroid 2-related factor 2; ROS, reactive oxygen species; SOD, superoxide dismutase; TNF-α, tumour necrosis factor-alpha.

translocation of B-cell lymphoma 2 (Bcl-2)-associated X (Bax) into mitochondria in prion disease. It was suggested that regulating mitochondrial homeostasis and inducing autophagy using resveratrol could protect against neurotoxicity.[29] 6-Hydroxydopamine is commonly used to mimic the development of Parkinson's disease in cell models (Figure 10.2 and Table 10.2). It could lead to neurotoxicity and apoptosis in SH-SY5Y cells *via* phosphorylation of c-Jun N-terminal kinase (JNK) and c-Jun, which could be prevented by pinostilbene and oxyresveratrol.[30,31] Notably, it was revealed that the expression of sirtuin 1 (Sirt1) is induced by oxyresveratrol, and it might play an important role in counteracting the neurotoxicity of 6-hydroxydopamine (6-OHDA).[31] The importance of Sirt1 was also mentioned

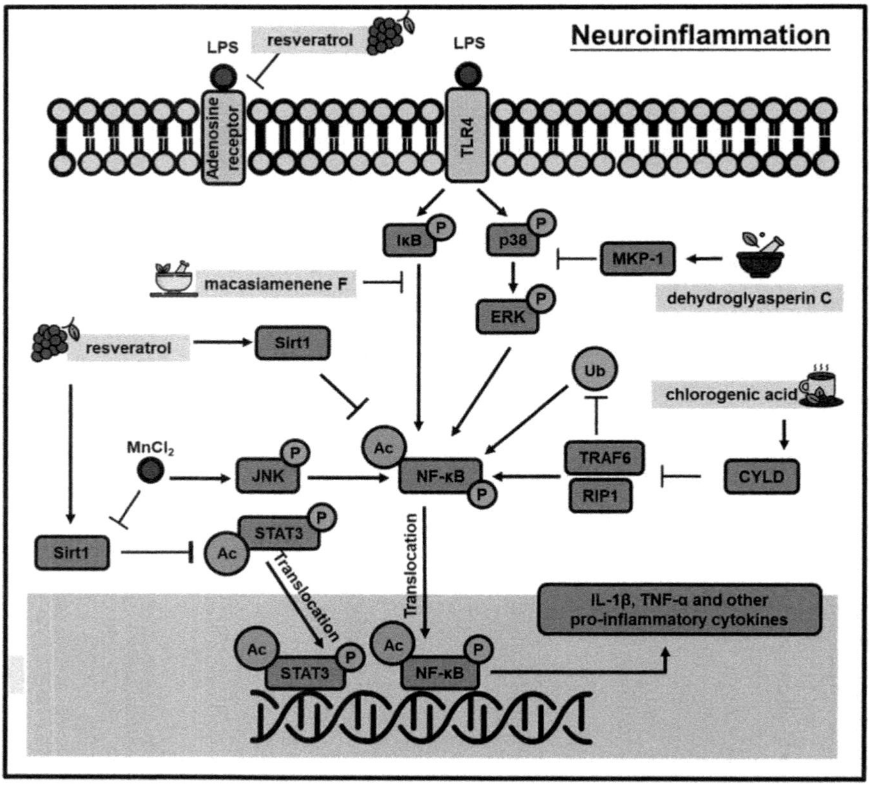

Figure 10.2 Potential mechanisms by which phytochemicals may improve neuroinflammation. Abbreviations: Ac, acetylation; CYLD, cylindromatosis tumour suppressor protein; ERK, extracellular signal-regulated kinase; IκB, inhibitor of NF-κB; IL-1β, interleukin-1beta; JNK, c-Jun N-terminal kinase; LPS, lipopolysaccharide; MPK-1, MAPK phosphatase 1; NF-κB, nuclear factor-kappa B; RIP1, receptor-interacting protein 1; Sirt1, sirtuin 1; STAT3, signal transducer and activator of transcription 3; TLR4, toll-like receptor 4; TNF-α, tumour necrosis factor alpha; TRAF6, TNF receptor-associated factor 6; Ub, ubiquitination.

in a study that demonstrated the capability of piceatannol to prevent GAPDH oxidative modifications.[20]

Some novel strategies have been discussed regarding neural cell apoptosis. Recently, Aimaiti *et al.*[22] suggested that acteoside could enhance mitophagy and the autophagy-lysosome pathway to protect rotenone-induced neuronal cell injury and apoptosis in a Parkinson's disease model. 1-Methyl-4-phenylpyridinium (MMP^+) is a neurotoxin that induces Parkinson's disease-related neuronal injury by impairing mitochondrial respiration and energy metabolism. The phytochemical crocin was found to reduce CCAAT/enhancer-binding protein homologous protein (CHOP) expression induced by MPP^+ in the PC12 cell line, which preserved wingless-related integration site (Wnt) activity.[32] Elevation of CHOP and 78 kDa glucose-regulated protein (GRP78), accompanied by cleavage of caspase-12, suggested MMP^+ induced endoplasmic reticulum (ER) stress. Post-intervention of crocin was found to reverse the ER stress by regulating the CHOP/Wnt pathway, subsequently rescuing mitochondrial dysfunction.[32] Therefore, several strategies are involved in the phytochemical rescue from neural cell death, and these *in vitro* studies may provide ideas applicable in preclinical studies.

10.2.3 *In Vivo* Animal Studies

10.2.3.1 Neuronal Degeneration – Oxidative Stress and Apoptosis

Research on oxidative stress inducing neural cell apoptosis has mostly focused on brain disease prevention/amelioration by phytochemicals in animal studies, as most phytochemicals are recognized for their anti-oxidative advantages. Therefore, in this section, some possible underlying mechanisms involved in neural apoptosis prevention by phytochemicals are discussed.

Research has revealed that apoptosis caused by chronic stress could be avoided by resveratrol through inducing the expression of Bcl-2 and reducing the Bax level in the hippocampus and prefrontal cortex.[33] Another study reported the regulatory effect of mangiferin on Bcl-2 and Bax, which exhibited protective effects on neuronal cells against 1-methyl-4-phenyl-1,2,3,6-tetrahydropyridine (MPTP)-induced neuronal apoptosis.[34] Furthermore, another animal study suggested that resveratrol and pinostilbene improved motor function, and resveratrol and pinostilbene may provide protective effects on distinct brain regions.[35] Focal cerebral ischemic injury caused by an acute ischemic stroke can lead to neuronal death. Behavioural deficits were observed after inducing transient focal cerebral ischemia in rats with middle cerebral artery occlusion (2 h), followed by 24 h reperfusion.[36] Stilbenes including (+)-ampelopsin A, γ-2-viniferin, and *trans*-ε-viniferin identified in *Vitis amurensis* leaves were found to protect against ischemic brain damage by reducing proteins associated with apoptosis.

Table 10.2 Preventive effect of phytochemicals on Parkinson's disease.

Phytochemical	Model of study	Treatment protocol	Underlying mechanism	Ref.
Quercetin	• Parkinson's disease • Animal model	• 6-Hydroxydopamine (300 μg)-induced Sprague-Dawley rats (single intracisternal injection)	• Increased striatal dopamine and antioxidant enzyme levels • Decreased protein carbonyl content • Increased neuronal survivability	21
Acteoside	• Parkinson's disease • Cell model	• Rotenone-induced neuronal injury in PC-12 cells	• Reduced apoptotic rate and ROS • Promoted aggregation of α-syn *via* the autophagy-lysosome pathway	22
Pinostilbene	• Parkinson's disease • Cell model	• 6-Hydroxydopamine-induced neurotoxicity in SH-SY5Y	• Reduced activity of caspase-3 • Attenuated phosphorylation of JNK and c-Jun • Restored phosphorylation of mTOR and GSK-3β	30
Oxyresveratrol	• Parkinson's disease • Cell model	• 6-Hydroxydopamine-induced neurotoxicity in SH-SY5Y	• Reduced activity of caspase-3 • Attenuated phosphorylation of JNK and c-Jun • Increased expression of p-Akt and Sirt1	31
Crocin	• Parkinson's disease • Cell model	• MMP^+-induced PC12 cells	• Preserved membrane potential and ATP synthesis • Inhibited ER stress by regulating the CHOP/Wnt pathway	32
Mangiferin	• Parkinson's disease • Animal model	• MPTP-induced Parkinson's disease in mice	• Reversed behavioural deficits, dopamine depletion, and dopaminergic neuronal degeneration • Attenuated oxidative stress and apoptosis	34
Kaempferol	• Parkinson's disease • Cell model	• Rotenone-mediated acute toxicity in the SY-SY5Y cell line	• Reduced caspase activation, ROS level, and mitochondrial carbonyl level • Increased mitochondrial fission rate and mitochondria-containing autophagosomes • Induced LC3-II	41

Table 10.2 (*Continued*)

Phytochemical	Model of study	Treatment protocol	Underlying mechanism	Ref.
Tyrosol	• Parkinson's disease • Animal model	• *Caenorhabditis elegans* strain • NL5901 (overexpress α-syn) • UA44 (age-dependent neurodegeneration)	• Decreased α-syn aggregation • Modulated chaperone and scavenged ROS	44
Honokiol	• Parkinson's disease • Animal model	• 6-Hydroxydopamine (single injection, 15 μg/3 μL) into the right striatum of mice to induce hemiparkinsonism	• Inhibited oxidative stress and astrogliosis by activating PPARγ • Recovered motor function • Reversed dopaminergic neuronal loss • Increased tyrosine hydroxylase density • Induced dopamine transporter and VMAT2 level	45
Valproic acid	• Parkinson's disease • Animal model	• *Caenorhabditis elegans* expressing α-syn in dopaminergic neurons	• Restored ERK–MAPK signalling for dopaminergic neuron protection	67

Abbreviations: Akt, protein kinase B; CHOP, CCAAT/enhancer-binding protein homologous protein; ER, endoplasmic reticulum; ERK, extracellular signal-regulated kinase; GSK-3β, glycogen synthase kinase-3 beta; JNK, c-Jun N-terminal kinase; LC3, microtubule-associated protein 1A/1B-light chain 3B; MAPK, extracellular signal-regulated kinase; MMP^+, 1-methyl-4-phenylpyridinium; MPTP, 1-methyl-4-phenyl-1,2,3,6-tetrahydropyridine; mTOR, mammalian target of rapamycin; PPARγ, peroxisome proliferator-activated receptor gamma; ROS, reactive oxygen species; Sirt1, sirtuin 1; VMAT2, vesicular monoamine transporter 2; Want, wingless-related integration site; α-syn, alpha-synuclein.

Autophagy and antioxidative enzymes could also be responsible for apoptosis prevention. In 2019, it was revealed that crocin, a carotenoid compound isolated from saffron, exhibited neuroprotective effects in middle cerebral artery occlusion/reperfusion-induced rats by inhibiting excessive autophagy.[37] Specifically, it was suggested that activation of the mammalian target of rapamycin (mTOR) and inactivation of adenosine monophosphate-activated protein kinase (AMPK) by crocin were responsible for the reduction of autophagy. In traumatic brain injury induced by the weight-drop model, brain injury and neural cell death could be effectively ameliorated by sodium aescinate, a triterpene saponin, *via* activation of the Nrf2/antioxidant responsive element (ARE) pathway and elevation of HO-1 and NQO-1.[38] Furthermore, quercetin, a flavonoid with high antioxidative capability, was found to reduce the protein carbonyl content in the striatum. The protein carbonyl level indirectly indicates the damages caused by oxidative stress in the brain if amino acid residues are attacked by ROS.[39] The effect of sesquiterpene farnesol on neuronal apoptosis mediated by ROS has also been investigated.[40] Farnesol could reduce lipid peroxidation, protein carbonyl content, and 8-hydroxydeoxyguanosine induced by LPS injection, indicating stress reduction. Activation of caspase-3 and p53 elevation, accompanied by Bax increase and Bcl-2 reduction, were significantly suppressed by farnesol intervention. It was suggested that apoptosis induced by LPS was reversed by the antioxidative activity of the enzymes involved, including SOD, CAT, GSH, and glutathione *S*-transferases.[40] Impairment of memory may generally occur during aging, and it may be due to increased oxidative stress. Chrysin is a compound naturally found in some edible flowers and mushrooms. Intervention with chrysin prevented memory decline observed in aging, likely by increasing antioxidative enzymes and recovering BDNF loss in the prefrontal cortex and hippocampus.[15] Therefore, antioxidation and anti-apoptosis have been the major strategies of neurodegeneration prevention in previous animal studies instead of autophagy. However, as the benefits of autophagy in rescuing cell death have been widely discussed, they should be studied further.

10.2.3.2 Anti-neurodegenerative Effects – Environmental Toxin-induced Brain Damage

Besides disease-induced neurodegeneration or neuronal damage, environmental toxins can also induce oxidative damage. For example, as an inhibitor of mitochondrial complex I, rotenone can lead to oxidative brain damage in the pathogenesis of Parkinson's disease.[41] It has also been reported that autophagy induced by phytochemicals such as kaempferol could protect against mitochondrial toxins but not apoptotic stimuli. Therefore, phytochemicals may not exhibit neuroprotective effects under specific circumstances. Environmental toxins such as 3-nitropropionic acid can lead to behavioural alteration, probably by mitochondrial dysfunction and brain damage caused by oxidative stress. It was also revealed that chrysin showed an ameliorative effect in 3-nitropropionic acid-induced behavioural despair in rats by improving antioxidant status and preventing cell apoptosis.[42]

Paraquat, which is a widely used herbicide, has been used as an inducer for the symptoms of Parkinson's disease in *Drosophila*, causing locomotor impairment and NDD related to oxidative stress and mitochondrial disorder.[43] Hydroxyisophthalic acid is a natural antioxidant molecule in the roots of *Decalepis hamiltonii* and could ameliorate the symptoms of paraquat exposure in *Drosophila*. Besides inducing antioxidant enzymes and reducing markers of oxidative stress, neuronal death and mitochondrial fragmentation were attenuated.[43]

In the *Caenorhabditis elegans* model of Parkinson's disease, the phenol, tyrosol, isolated from extra-virgin olive oil, alleviated neurodegeneration.[44] Tyrosol inhibited α-syn accumulation, reduced ROS by elevating antioxidant enzymes, and promoted the mRNA level of chaperones (hsp-70, hsp-16.2, and hsp-12.6). The neurotoxin 6-OHDA is widely used to induce Parkinsonism by inducing toxicity in dopaminergic neurons, thereby lowering the dopamine level.[21] Hemiparkinsonian mice induced by 6-OHDA injection at the right striatum were found to have motor dysfunction and progressive dopaminergic damage.[45] Proliferator-activated receptor-gamma (PPARγ) could be activated by honokiol, a bioactive polyphenol isolated from *Magnolia officinalis*, to reverse the adverse effect of 6-OHDA by reducing nigrostriatal neuronal loss, oxidative response (decreased $gp91^{PHOX}$), and activation of astrocytes (glial fibrillary acidic protein/GFAP).[45]

10.2.3.3 *Neuroinflammation*

Anti-neuroinflammation is another important strategy in brain disease prevention. In previous animal studies, inflammation was promoted by seizures in the hippocampus. However, this situation was reversed by resveratrol that subsequently ameliorated the disease.

Environmental toxins may also lead to neuroinflammation that then results in neurodegenerative or cognitive disorders. For instance, neuronal loss in the striatum is a common pathogenic marker of Huntington's disease, and 3-nitropropionic acid is an environmental toxin that can cause selective neuronal degeneration by inhibiting the enzyme succinate dehydrogenase in mitochondrial complex II.[46] Naringin, a flavanone found in citrus fruits, could effectively mitigate neurodegeneration induced by enhancing gene expression of phase II and antioxidant enzymes, resulting in decreased pro-inflammatory cytokines in 3-nitropropionic acid-induced neurodegenerative rats. Environmental exposure to manganese may also result in neurotoxicity and hippocampal apoptosis. Cong *et al.*[47] suggested that exposure to $MnCl_2$ might lead to induction of JNK and activation of NF-κB. The authors proposed that the signal transducer and activator of transcription 3 (STAT3) in neuroinflammation might be a result of Sirt1 reduction. They suggested that this adverse effect could be reversed by resveratrol intervention in hippocampal tissue.

Deep hypothermic circulatory arrest (DHCA) is a surgical method that may cause the inducement of neural oxidative stress and inflammatory effects as side effects. Microglial and NF-κB activation with elevated circulatory cytokines was observed after DHCA.[48] It was suggested that chlorogenic acid could suppress NF-κB activation with ubiquitination *via* elevation of cylindromatosis tumour suppressor protein (CYLD) that inhibited TNF receptor-associated factor 6 (TRAF6) and receptor-interacting protein 1 (RIP1), subsequently regulating ubiquitin ligase activity.

10.2.3.4 *Depression*

Resveratrol is a non-flavonoid phenolic compound found in grapes and grapevines, and it is well-recognized for its capabilities as an antioxidant and in preventing multiple diseases. The anti-depressive and neuroprotective effects of resveratrol have been widely investigated in recent years. For example, resveratrol administration showed a dose-dependent elevation in hippocampal (but not frontal cortical) BDNF.[14] In fact, BDNF elevation could be responsible for the positive effects of resveratrol in depression and neuroprotection.[49,50] Furthermore, resveratrol exhibited an ameliorative effect on chronic, unpredictable, mild stress-induced depression-like behaviour.[12] It was suggested that suppressing neuroinflammation, oxidative stress, and apoptosis could be the mode of anti-depressant-like effect of resveratrol.

Gut microbiota modulation is suggested to act bidirectionally with phytochemicals to alleviate depression. For example, besides mitigating hyperactivation of the hypothalamic–pituitary–adrenal (HPA) axis, neuroinflammation and deficiencies in 5-hydroxytryptamine (5-HT), and BDNF in the hippocampus, total glucosides extracted from the stems of *Cistanche tubulosa* could also modulate the abundance of *Tyzzerella_3*, *Acetatifactor*, and *norank_f_Lachnospiraceae* which were found to be negatively correlated with 5-HT and BDNF.[51]

10.2.3.5 Cognitive Disorders

Cognitive function decline is one of the behavioural indicators of Alzheimer's disease, and up to 95% of Alzheimer's disease (Figure 10.3 and Table 10.3) is age-related.[52] As demonstrated by Chang *et al.*,[52] pterostilbene could demonstrate neuroprotective effects in the SAMP8 mouse model, an accelerated aging mouse model, *via* activation of PPARα and subsequent enhancement of the manganese superoxide dismutase level and inhibited NF-κB. In addition, pterostilbene suppression of cellular stress *via* phosphorylation of JNK/paired helical filaments (PHF) was also observed. Therefore, pterostilbene has been suggested to ameliorate the pathology of Alzheimer's disease, including neuroinflammation and cellular stress, and its administration resulted in improved cognitive function.

As mentioned earlier, neurogenesis impairment may contribute to NDD that manifests cognition deficits. Mirza *et al.*[53] demonstrated that rosmarinic acid and ursolic acid could reverse the deficits induced by Aβ_{1-42} in terms of cognition improvement, promoting hippocampal neurogenesis (Ki67, NeuN, and DCX as neurogenic markers) and stabilizing synaptic regulation (Syn I, II, III, synaptophysin, and PSD-95 as synaptic markers). By using an amyloid precursor protein/presenilin 1 (APP/PS1) double-transgenic mouse model, it was revealed that cognitive impairment could be improved by andrographolide, a labdane diterpene isolated from *Andrographis paniculata*, particularly by inhibiting Aβ deposition and microglial activation (downregulating Itgax, TLR2, CD14, CCL3, CCL4, TLR1, and C3ar1 markers).[54] Presenilin stabilization factor-like protein (PSFL) is the protein stabilizing presenilin, the membrane protein involved in Aβ peptides. It has been targeted in the strategy of Alzheimer's disease prevention, and three phytochemicals, saponin, casuarictin, and enoxolone, were identified as potential inhibitors of PSFL.[55]

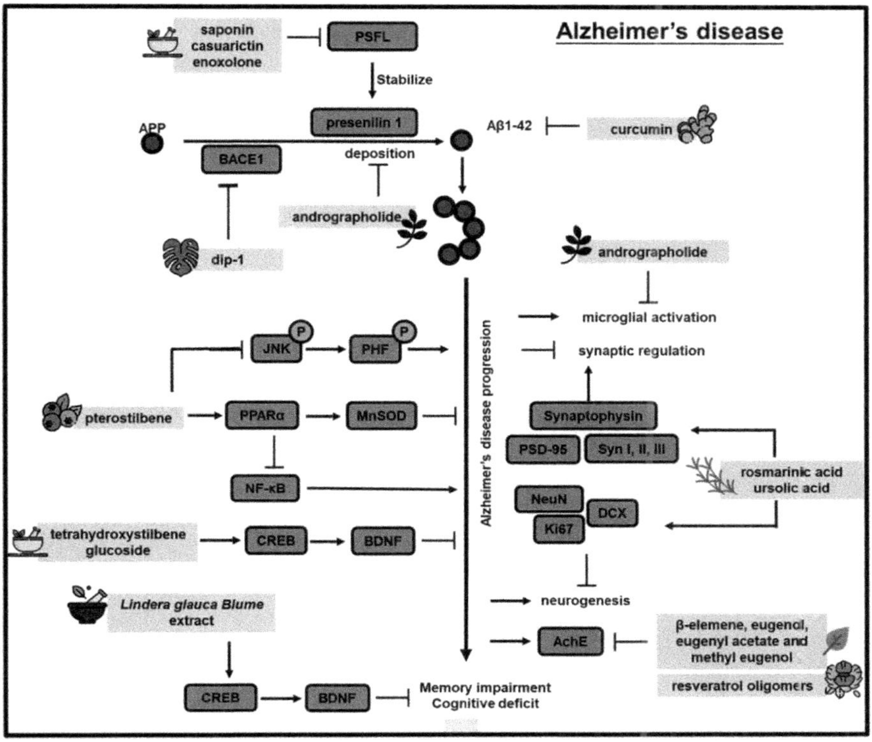

Figure 10.3 Potential mechanisms of phytochemical involvement in Alzheimer's disease prevention/alleviation. Abbreviations: Aβ, amyloid beta; AchE, acetylcholinesterase; BACE1, β-secretase 1; BDNF, brain-derived neurotrophic factor; CREB, cAMP-response element binding protein; DCX, doublecortin; JNK, c-Jun N-terminal kinase; Ki67, MKI67; NF-κB, nuclear factor-kappa B; MnSOD, manganese superoxide dismutase; NeuN, neuronal nuclear protein; PHF, paired helical filaments; PPARα, peroxisome proliferator-activated receptor alpha; PSD-95, postsynaptic density 95; PSFL, presenilin stabilization factor-like protein; Syn, synuclein.

In another platform, combining chronic cerebral hypoperfusion and diabetes mellitus in rats, cognitive impairment was induced, and curcumin alleviated impairments in spatial learning and memory performance.[56] It was further revealed that curcumin played a regulatory role in neuroprotection by inhibiting the triggering receptor expressed on myeloid cell 2/toll-like receptor 4/NF-κB (TREM2/TLR4/NF-κB), apoptosis (Bax/Bcl/caspase pathway), and pyroptosis [NLR family pyrin domain containing 3 (NLRP3)/caspase1/interleukin (IL)-18] in the rat hippocampus and microglial cell line BV2.

Retarding or preventing the progression of brain disease that may lead to cognitive decline is a major recent research strategy. However, the progression of these NDD could also lead to several problematic issues that exacerbate the disease, including oxidative stress and neuroinflammation.

Table 10.3 Preventive effect of phytochemicals on Alzheimer's disease.

Phytochemical	Model of study	Treatment protocol	Underlying mechanism	Ref.
Tetrahydroxystilbene glucoside	• Alzheimer's disease • Cell model	• Murine hippocampal neuronal cells (HT-22) induced with Aβ_{1-42} (10 $\mu g\,mL^{-1}$) for 24 h	• Restoring mitochondrial function *via* the Nrf2–HO-1 pathway • Reduced mitochondrial related cell cytotoxic death	19
Pterostilbene	• Alzheimer's disease • Animal model	• SAMP8 mice	• Increased PPARα expression and reduced phosphorylation of tau and JNK • Improved radial arm water maze function	52
Rosmarinic acid and ursolic acid isolated from *Rosmarinus officinalis*	• Alzheimer's disease • Animal model	• Stereotactic injection of Aβ_{1-42} peptide (1 $\mu g\,\mu L^{-1}$) in the CA1 region of the hippocampus	• Normalized neuronal density • Increased expression of neurogenic proteins (Ki67, NeuN, and DCX) • Increased expression of synaptic markers (Syn I, II, III, synaptophysin, and PSD-95)	53
Andrographolide	• Alzheimer's disease • Animal model	• APP/PS1 mice	• Alleviated cognitive impairment • Reduced Aβ deposition • Inhibited microglial activation and proinflammatory cytokine secretion • Decreased gene expression of Itgax, TLR2, CD14, CCL3, CCL4, TLR1, and C3arl	54
Saponin, casuarictin, and enoxolone	• Alzheimer's disease • Molecular docking	• PFSL protein	• Inhibited PSFL by direct binding	55
Platismatia glauca extract	• Alzheimer's disease • Animal model	• Bilateral injection of the Aβ_{1-42} peptide into the	• Restored neuronal marker (NeuN)	64

Table 10.3 (*Continued*)

Phytochemical	Model of study	Treatment protocol	Underlying mechanism	Ref.
		hippocampus in mice	• Reduced activation of astrocytes and GFAP level • Upregulated phosphorylated CREB and BDNF levels	
Curcumin	• Alzheimer's disease • Cell model	• H_2O_2-induced SH-SY5Y cells	• Suppressed SUMO-1ylation • Inhibited phosphorylation of JNK and tau • Decreased degradation pathways (caspase-3 and LC3B)	70
Dip-1 isolated from *Dipteris wallichii*	• Alzheimer's disease • Molecular docking	• BACE1	• Having high drug likeness and high BBB permeability • BACE1 inhibitor	79

Abbreviations: Aβ, β-amyloid; BACE1, β-secretase 1; BBB, blood–brain barrier; BDNF, brain-derived neurotrophic factor; CA1, *cornu ammonis* area 1; CREB, cAMP-response element binding protein; DCX, doublecortin; GFAP, glial fibrillary acidic protein; HO-1, heme oxygenase 1; JNK, c-Jun N-terminal kinase; LC3B, microtubule-associated protein 1A/1B-light chain 3B; NeuN, neuronal nuclear protein; Nrf2, nuclear factor erythroid 2-related factor 2; PPARα, proliferator-activated receptor-alpha; PSFL, presenilin stabilization factor like protein; SUMO-1, small ubiquitin-related modifier 1.

10.2.4 Human Intervention Studies

Compared to synthetic drugs, plant-derived natural medicinal products have gained significant attention and demonstrated their importance in multiple ways, including cost benefits, tolerable toxicity, and availability. However, evaluating the clinical use of compounds requires assessment of several parameters, including their bioavailability, safety, and effectiveness.[57]

Among the well-studied compounds, curcumin's potential benefits have been the most frequently discussed and reviewed. As curcumin has been found to be beneficial in several animal studies at molecular and behavioural levels, it has been used in clinical trials for disease amelioration, including the cognitive deficit related to Alzheimer's disease. For example, 32 preclinical studies and five clinical studies out of 190 total met inclusion criteria. It was reported that consumption of curcumin could be beneficial *via* a β-amyloid reduction in the brain, a result supported by neuroimaging.[58] The limitations of curcumin clinical studies have been discussed, including the lack of oxidative stress and inflammatory biomarker measurements, lack of translational and methodological consistency in cognitive function evaluation or performance, absence of consistent pathological aspects, and the difficulties in comparing the effectiveness of curcumin among studies due to the differences in formulation and dosage.[58]

In addition to curcumin, resveratrol has also been involved in clinical trials as it has been well studied and exhibited notable advantages in neurodegenerative diseases. Only two cases reviewed here were published in 2018, and no significant effects were observed. Nevertheless, it has been suggested that the effects of resveratrol should focus on the progression of cognitive and functional impairment but not prevention.[59]

10.3 Underlying Mechanisms of Dietary Supplements for Brain Health

As mentioned above, antioxidation, anti-neuroinflammation, and anti-apoptosis are the main strategies in brain disease improvement. However, there are other possible mechanisms proffered in disease research that are not yet widely applied. In this section, we focus on these possible pathways to provide ideas for future studies.

10.3.1 Neuronal Proliferation and Differentiation

Glycogen synthase kinase-3 beta (GSK-3β)/β-catenin is a potential pathway involved in the prevention of brain diseases. Decreased levels of β-catenin and elevated phosphorylated GSK-3β (p-GSK-3β) have been found in depressed subjects.[12] As a transcription factor associated with dendritic growth, brain development and cognitive activity, β-catenin can be a good molecular target to promote cell proliferation and differentiation. Activation of the Wnt signalling pathway by resveratrol has supportively shown the potential of resveratrol in preventing brain diseases. In addition, resveratrol also showed a positive impact on the expression of Bcl-2 and inhibition of NF-κB, suggesting its effect on neuron cell survival and neuroinflammation.

Bisphenol A (BPA) is a component of plastic that may lead to neurogenesis and myelination process impairment which consequently causes cognitive dysfunction. In an *in vivo* study, myelin sheath degeneration and neurobehavioural impairment induced by BPA could be effectively improved by curcumin by increasing Notch pathway genes, including notch homolog 1 (Notch-1), hairy and enhancer of split-1 (Hes-1), and mind bomb1 (Mib-1).[60] Activation of the notch pathway by curcumin could potentially protect oligodendrocyte progenitor cells from demyelination.

10.3.2 Neurite Growth and Synaptogenesis

Inducing nerve regeneration after injury is one strategy to recover the nervous system. Successful nerve regeneration is finalized with axonal regrowth of injured neurons, reconnection with target organs, and functional restoration.[61] Elevation of cyclic adenosine monophosphate (cAMP) can stimulate intrinsic growth capacity which subsequently increases the regrowth of axons *via* the IL-6/Janus kinase (JAK)/STAT3 pathway which finally activates Gap-43. In the progression of Alzheimer's disease, collapsing response mediator protein 2 (CRMP-2) hyperphosphorylation has been reported to be associated with neurofibrillary tangle formation and neural process regression. As a promising drug for Alzheimer's disease, curcumin could inhibit hyperphosphorylation of CRMP-2, which in turn prevented axonal regression and promoted axonal regrowth [evidenced by neurofilament 200 (NF-200), an axon-specific marker], and a reversal effect on spatial learning deficits in mice.[58]

It has been reported that depressive disorder may be associated with the immune system and that anti-neuroinflammation was suggested as a potential anti-depressant strategy. It was revealed that neuroinflammation induced by LPS could reduce dendritic spines and inhibit synaptic markers (PSD95 and snap25), accompanied by the level of synaptophysin.[62] It was suggested that synaptogenesis defects could be prevented *via* anti-neuroinflammation and antioxidative stress.

10.3.3 Anti-neuronal Degeneration

Alteration of the ERK/cAMP-response element binding protein (CREB) pathway has been recommended to improve stress-induced cognitive deficits. This signalling pathway has been associated with several brain functions, including memory, neuroplasticity, cell proliferation and differentiation, and cellular responses to stress, among others.[49] Phosphorylation of ERK1/2 by resveratrol and pinostilbene was suggested to protect the SH-SY5Y cell line from dopamine-induced cell death.[35] In an animal study, resveratrol protected the *cornu ammonis* area 1 (CA1) region of the hippocampus against cerebral ischemia *via* activation of the ERK/CREB signalling pathway.[63] BDNF, discussed in Section 10.2.1, was one of the transcribed molecules of the ERK/CREB pathway that could be activated by resveratrol. In addition, activation of the CREB–BDNF pathway by extracts of the herb *Lindera glauca* Blume was also suggested to be beneficial against $A\beta_{1-42}$-induced memory impairment in mice.[64] In addition to the CREB/BDNF pathway, it was suggested that suppression of the protein kinase B (Akt)/GSK-3 phosphorylation pathway might contribute to neuroprotection, especially in alcohol-induced neurodegeneration, possibly *via* anti-apoptosis, antioxidation, and anti-inflammation.[65] Akt/GSK-3β pathway activation is also responsible for Alzheimer's disease alleviation as GSK-3β activity could be deactivated by phosphorylation, which reduces the level of acetylcholine.[66] It was revealed that curcumin could activate Akt/GSK-3β in a mouse memory retrieval deficit model induced by scopolamine.

The formation of cytoplasmic inclusion bodies, so-called Lewy bodies, from accumulated aggregated α-syn in dopaminergic neurons is a known pathology of Parkinson's disease.[44] Besides inhibiting the expression of α-syn, ERK–MAPK has been suggested as a potential pathway to be explored as the accumulation of the phosphorylated ERK/MAPK pathway in the cytoplasm has been detected in the substantia nigra of patients with Parkinson's disease.[67] The involvement of ERK as a target pathway in NDD therapy has been discussed and reviewed previously.[68] It was demonstrated that a high-fat, high-fructose diet could lead to neurodegeneration with increases in advanced glycation end products (AGE), activation of the receptor for advanced glycation end products (RAGE) level, and hyperphosphorylation of tau in the brain cortex and hippocampus. Furthermore, a reduction in Nrf2, BDNF, phosphatidylinositol-3-kinase (PI3K), and ERK was also observed.[69] PI3K and ERK were suggested to be associated with neuronal death, and activation by phosphorylation could be a potential ameliorative strategy. Colocalization of the tau protein in nuclear bodies is linked to many NDD, and SUMO-1ylation and activation of JNK may be involved in inducing cellular death.[70] Hyperactivation of the small ubiquitin-like modifier protein 1 (SUMO-1)–JNK–Tau axis induced by H_2O_2, which leads to apoptosis and degradation in the SH-SY5Y cell line, could be counteracted by curcumin.

10.3.4 Antioxidative Stress and Mitochondrial Dysfunction Rescue

Oxidative stress injury is one of the major inducers of neuronal cell apoptosis. Activation of Nrf2 could exhibit neuroprotection by elevating the levels of enzymes involved in antioxidation, including HO-1, NQO-1, GSH, and SOD.[27] Specifically, activation of Akt by phosphorylation could lead to translocation of Nrf2 onto the ARE promoter from Keap1 and result in antioxidative enzyme production.[71]

Rescuing mitochondrial dysfunction has been suggested as potentially beneficial in Parkinson's disease treatment. The AMPK/Sirt1/peroxisome proliferator-activated receptor-γ coactivator-1α (PGC-1α) pathway is acknowledged in mitochondrial biogenesis, and most importantly AMPK and Sirt1 are the molecular targets of some phytochemicals, such as resveratrol.[17] Moreover, restoring mitochondrial function by antioxidation may be another method to rescue mitochondrial dysfunction. The Nrf2/HO-1 pathway was activated to antagonize neuronal cell damage induced by β-amyloid.[19] Activation of Sirt1/PGC-1α is known as a transcription coactivator of Nrf2.[72] Activation of Nrf2 can induce the expression of NAD(P)H:quinone oxidoreductase-1, heme oxygenase-1, glutathione *S*-transferase P1, and gamma-glutamylcysteine ligase.[46]

ER stress was suggested to be related to mitochondrial dysfunction induced by MPP^+ when mimicking neuronal injury.[32] It was demonstrated that ER stress could be ameliorated by suppressing (CHOP) expression and increasing Wnt activity and β-catenin translocation.

Regulating apoptosis-related genes can be a potential strategy in neuroprotection. It is suggested that upregulating Bcl-2 expression or reducing the level of pro-apoptotic genes such as Bax and Bcl-2 associated agonist of cell death (Bad) protein could effectively prevent cell apoptosis.[42]

10.3.5 Anti-neuroinflammation

Neuroinflammation can be prevented by inhibiting the activation of NF-κB. Phytochemicals such as resveratrol have shown their ability to inhibit TLR4/p38 MAPK/NF-κB or the new molecular target adenosine receptor/p38 MAPK/NF-κB pathway, which results in decreased transcription of pro-inflammatory cytokines and proteins such as inducible nitric oxide synthase (iNOS).[23] Moreover, activation of Sirt1 by resveratrol had also been reported to inhibit phosphorylation and acetylation of NF-κB and STAT3 which subsequently prevent neuroinflammation.[47]

In an animal study of seizure-induced hippocampal inflammation, it was suggested that inhibition of mTOR *via* AMPK could effectively decrease the expression of inflammatory markers including NF-κB, iNOS, cyclooxygenase-2 (COX-2), and IL-1β, and notably the mechanism might be Sirt1-dependent.[73] Furthermore, inhibition of NF-κB could also be achieved by activation of PPARα and elevation of CYLD to attenuate neuroinflammation.[48,52]

Other than NF-κB, the NLR family pyrin domain containing 3 (NLRP3) inflammasome was also reported to be induced by oxidative stress, and inhibition of NLRP3 could be a possible strategy for neuroinflammation.[74] It was suggested that autophagy and activation of NF-κB might be responsible for the NLRP3/caspase1 pathway, resulting in inhibition of pro-inflammatory cytokine secretion.[75]

10.3.6 Autophagy

Activation of NF-κB and NLRP3 represents an important strategy in neuroprotection and induction of autophagy. AMPK activation and phosphorylated mTOR reduction could induce autophagy *via* phosphorylation of UNC-51-like kinase (ULK) to inhibit neuroinflammation.[75] Autophagy is a strategy suggested in preventing neuronal cell death. As mentioned in Section 10.3.4, the Sirt1/PGC-1α pathway is recognized as a protective mechanism regulating mitochondrial homeostasis and preventing oxidative damage in mitochondria. Sirt1 activation has also been suggested to induce autophagy during mitochondrial dysfunction.[29] Therefore, Sirt1 may exhibit multi-targeting effects in disease amelioration *via* different mechanisms.

Autophagy is commonly recognized as a protective strategy in neuron protection. However, excessive autophagy may lead to autophagy-dependent cell death and apoptosis.[37] Therefore, inhibition of autophagy has become a potential strategy to ameliorate brain diseases such as stroke or ischemia *via* inhibition of the AMPK–Unc-51 like autophagy activating kinase 1 (ULK1) pathway. Furthermore, mTOR could also directly phosphorylate ULK1, complexed with AMPK, which in turn inhibits autophagy and promotes proliferation.[37]

10.3.7 Enzymatic Regulation Related to Neurodegeneration

There are several key enzymes responsible for the pathogenesis of NDD. For example, acetylcholinesterase (AchE), ecto-nucleoside triphosphate diphosphohydrolase (E-NTPase), Na^+/K^+-ATPase, and monoamine oxidase (MAO) are found to be associated with depression, Parkinson's disease, Alzheimer's disease, and neuronal death.[76] Some plant-based compounds such as caffeine or caffeic acid showed regulatory effects on these enzymes, resulting in anti-neurodegenerative potential. Monoamine oxidase B (MAO-B) is the enzyme responsible for converting 1-methyl-4-phenyl-1,2,3,6-tetrahydropyridine (MPTP) into MPP^+, leading to neurotoxicity and neuronal death in a Parkinson's disease mouse model. MAO-B is involved in the metabolism of the oxidative deamination reaction of neurotransmitters (dopamine, adrenaline, and serotonin), and the high expression of MAO-B is usually found in Parkinson's and Alzheimer's disease patients.[34] It was revealed that the active component of *Mangifera indica* (mangiferin) could effectively act as an MAO-B inhibitor in the MPTP-induced mouse model of Parkinson's disease which partially contributes to disease attenuation.

The involvement of AchE in the progression of Alzheimer's disease has been widely studied, and some drugs have been developed to inhibit AchE. Among the various compounds isolated were *Piper divaricatum* (using supercritical CO_2), β-elemene, eugenol, eugenyl acetate, and methyl eugenol, which showed an inhibitory effect on AchE; His477 could be the active site involved in the interaction revealed by molecular docking.[77] Notably, resveratrol oligomers isolated from *Paeonia suffruticosa*, which had been reported to inhibit β-secretase, exhibited an inhibitory effect on AchE and butyrylcholinesterase *in vitro*.[78] β-Secretase (BACE1) is an enzyme involved in producing amyloid precursor protein from Aβ oligomers. A novel phytochemical identified from the pteridophyte *Dipteris wallichii*, named Dip-1, has been recently found to have an inhibitory effect on BACE1 and it could be a potential compound to alleviate Alzheimer's disease progression.[79] Although the evidence we have reviewed suggests that phytochemicals may positively affect NDD *via* enzyme regulation, more evidence from animal studies is needed.

10.4 Conclusion

Brain diseases are modern diseases that are increasing worldwide. Synthetic drugs are costly and often accompanied by unavoidable side effects. Therefore, the interest in developing new medicinal solutions from natural sources is increasing. In this chapter, many phytochemicals with the potential to provide new cures are presented. Most have been widely used as traditional herbs or in nutraceutical products in different regions of the world. However, they all still lack adequate scientific preclinical and clinical evidence for pharmaceutical and medical uses. Furthermore, consistent methodologies for assessing medicinal phytochemicals could be developed internationally to increase the efficiency of evaluation. We urge future studies to investigate how brain diseases could be effectively improved by phytochemical intervention.

Acknowledgements

This study was supported by the Ministry of Science and Technology [109-2320-B-002-012-MY3].

References

1. S. Yousefvand, M.-A.-R. Hadjzadeh, F. Vafaee and H. Dolatshad, The protective effects of prolactin on brain injury, *Life Sci.*, 2020, **263**, 118547.
2. A. Vienne-Jumeau, C. Tafani and D. Ricard, Environmental risk factors of primary brain tumors: A review, *Rev. Neurol.*, 2019, **175**, 664–678.
3. M. C. de Gooijer, E. M. Kemper, L. C. M. Buil, C. H. Çitirikkaya, T. Buckle and J. H. Beijnen, *et al.*, ATP-binding cassette transporters restrict drug delivery and efficacy against brain tumors even when blood-brain barrier integrity is lost, *Cell Rep. Med.*, 2021, **2**, 100184.
4. M. S. Tahir, M. Almezgagi, Y. Zhang, A. Bashir, H. M. Abdullah and M. Gamah, *et al.*, Mechanistic new insights of flavonols on neurodegenerative diseases, *Biomed. Pharmacother.*, 2021, **137**, 111253.
5. H. H. H. Adams, R. F. A. G. de Bruijn, A. Hofman, A. G. Uitterlinden, C. M. van Duijn and M. W. Vernooij, *et al.*, Genetic risk of neurodegenerative diseases is associated with mild cognitive impairment and conversion to dementia, *Alzheimer's Dementia*, 2015, **11**, 1277–1285.
6. R. S. Mackin, P. S. Insel, S. Landau, D. Bickford, R. Morin and E. Rhodes, *et al.*, Late-life depression is associated with reduced cortical amyloid burden: Findings from the Alzheimer's disease neuroimaging initiative depression project, *Biol. Psychiatry*, 2021, **89**, 757–765.
7. H. Gan, Y. Lei, N. Yuan, K. Tang, W. Hao and Q. Ma, *et al.*, Circular RNAs in depression: Biogenesis, function, expression, and therapeutic potential, *Biomed. Pharmacother.*, 2021, **137**, 111244.
8. G. Z. Tuncer and Z. Çetinkaya Duman, An examination of emotional eating behavior in individuals with a severe mental disorder, *Arch. Psychiatr. Nurs.*, 2020, **34**, 531–536.
9. J. Huang, N. Huang, S. Xu, Y. Luo, Y. Li and H. Jin, *et al.*, Signaling mechanisms underlying inhibition of neuroinflammation by resveratrol in neurodegenerative diseases, *J. Nutr. Biochem.*, 2021, **88**, 108552.
10. B. Shal, A. Khan, M. Naveed, N. Ullah Khan, H. Ihsan Ul and S. D AlSharari, *et al.*, Effect of 25-methoxy hispidol A isolated from *Poncirus trifoliate* against bacteria-induced anxiety and depression by targeting neuroinflammation, oxidative stress and apoptosis in mice, *Biomed. Pharmacother.*, 2019, **111**, 209–223.
11. J. Wang, S. Xu, X. Chen, L. Wang, J. Li and G. Li, *et al.*, Antidepressant effect of EGCG through the inhibition of hippocampal neuroinflammation

in chronic unpredictable mild stress-induced depression rat model, *J. Funct. Foods*, 2020, **73**, 104106.
12. A. A. Abd El-Fattah, A. T. Fahim, N. A. H. Sadik and B. M. Ali, Resveratrol and dimethyl fumarate ameliorate depression-like behaviour in a rat model of chronic unpredictable mild stress, *Brain Res.*, 2018, **1701**, 227–236.
13. Y.-M. Liu, S.-N. Shen, F.-B. Xia, Q. Chang, X.-M. Liu and R.-L. Pan, Neuroprotection of stilbenes from leaves of *Cajanus cajan* against oxidative damage induced by corticosterone and glutamate in differentiated PC12 cells, *Chin. Herb. Med.*, 2015, 7, 238–246.
14. L. L. Hurley, L. Akinfiresoye, O. Kalejaiye and Y. Tizabi, Antidepressant effects of resveratrol in an animal model of depression, *Behav. Brain Res.*, 2014, **268**, 1–7.
15. L. C. Souza, M. S. Antunes, C. B. Filho, L. Del Fabbro, M. G. de Gomes and A. T. R. Goes, *et al.*, Flavonoid chrysin prevents age-related cognitive decline *via* attenuation of oxidative stress and modulation of BDNF levels in aged mouse brain, *Pharmacol., Biochem. Behav.*, 2015, **134**, 22–30.
16. A. Marín-Burgin and A. F. Schinder, Requirement of adult-born neurons for hippocampus-dependent learning, *Behav. Brain Res.*, 2012, **227**, 391–399.
17. A. Ferretta, A. Gaballo, P. Tanzarella, C. Piccoli, N. Capitanio and B. Nico, *et al.*, Effect of resveratrol on mitochondrial function: Implications in parkin-associated familiar Parkinson's disease, *Biochim. Biophys. Acta, Mol. Basis Dis.*, 2014, **1842**, 902–915.
18. C. H.-L. Hung, S. S.-Y. Cheng, S. M.-Y. Lee and R. C.-C. Chang, Mitochondria as a therapeutic target for Alzheimer's disease, in *The Functions, Disease-Related Dysfunctions, and Therapeutic Targeting of Neuronal Mitochondria*, ed. V. K. Gribkoff, E. A. Jonas and J. M. Hardwick, Wiley, Oxford, UK, 2015, pp. 322–338.
19. C. Jiao, F. Gao, L. Ou, J. Yu, M. Li and P. Wei, *et al.*, Tetrahydroxy stilbene glycoside (TSG) antagonizes Aβ-induced hippocampal neuron injury by suppressing mitochondrial dysfunction *via* Nrf2-dependent HO-1 pathway, *Biomed. Pharmacother.*, 2017, **96**, 222–228.
20. J. Gerszon, M. Wojtala, S. Michlewska and A. Rodacka, Piceatannol effectively counteracts glyceraldehyde-3-phosphate dehydrogenase aggregation and nuclear translocation in hippocampal cells, *J. Funct. Foods*, 2019, **58**, 180–188.
21. N. Haleagrahara, C. J. Siew, N. K. Mitra and M. Kumari, Neuroprotective effect of bioflavonoid quercetin in 6-hydroxydopamine-induced oxidative stress biomarkers in the rat striatum, *Neurosci. Lett.*, 2011, **500**, 139–143.
22. M. Aimaiti, A. Wumaier, Y. Aisa, Y. Zhang, X. Xirepu and Y. Aibaidula, *et al.*, Acteoside exerts neuroprotection effects in the model of Parkinson's disease *via* inducing autophagy: Network pharmacology and experimental study, *Eur. J. Pharmacol.*, 2021, **903**, 174136.
23. L. D. Bobermin, R. H. A. Roppa and A. Quincozes-Santos, Adenosine receptors as a new target for resveratrol-mediated glioprotection, *Biochim. Biophys. Acta, Mol. Basis Dis.*, 2019, **1865**, 634–647.

24. N. Steiner, R. Balez, N. Karunaweera, J. M. Lind, G. Münch and L. Ooi, Neuroprotection of Neuro2a cells and the cytokine suppressive and anti-inflammatory mode of action of resveratrol in activated RAW264.7 macrophages and C8-B4 microglia, *Neurochem. Int.*, 2016, **95**, 46–54.
25. V. Leláková, S. Béraud-Dufour, J. Hošek, K. Šmejkal, V. Prachyawarakorn and P. Pailee, *et al.*, Therapeutic potential of prenylated stilbenoid macasiamenene F through its anti-inflammatory and cytoprotective effects on LPS-challenged monocytes and microglia, *J. Ethnopharmacol.*, 2020, **263**, 113147.
26. J. Kim, J. Kim, J. Shim, S. Lee, J. Kim and S. S. Lim, *et al.*, Licorice-derived dehydroglyasperin C increases MKP-1 expression and suppresses inflammation-mediated neurodegeneration, *Neurochem. Int.*, 2013, **63**, 732–740.
27. B. Wang, H. Liu, L. Yue, X. Li, L. Zhao and X. Yang, *et al.*, Neuroprotective effects of pterostilbene against oxidative stress injury: Involvement of nuclear factor erythroid 2-related factor 2 pathway, *Brain Res.*, 2016, **1643**, 70–79.
28. H. Xu, R. Deng, E. T. S. Li, J. Shen and M. Wang, Pinosylvin provides neuroprotection against cerebral ischemia and reperfusion injury through enhancing PINK1/Parkin mediated mitophagy and Nrf2 pathway, *J. Funct. Foods*, 2020, **71**, 104019.
29. J.-K. Jeong, M.-H. Moon, B.-C. Bae, Y.-J. Lee, J.-W. Seol and H.-S. Kang, *et al.*, Autophagy induced by resveratrol prevents human prion protein-mediated neurotoxicity, *Neurosci. Res.*, 2012, **73**, 99–105.
30. J. Chao, H. Li, K.-W. Cheng, M.-S. Yu, R. C.-C. Chang and M. Wang, Protective effects of pinostilbene, a resveratrol methylated derivative, against 6-hydroxydopamine-induced neurotoxicity in SH-SY5Y cells, *J. Nutr. Biochem.*, 2010, **21**, 482–489.
31. J. Chao, M.-S. Yu, Y.-S. Ho, M. Wang and R. C.-C. Chang, Dietary oxyresveratrol prevents parkinsonian mimetic 6-hydroxydopamine neurotoxicity, *Free Radic. Biol. Med.*, 2008, **45**, 1019–1026.
32. G.-F. Zhang, Y. Zhang and G. Zhao, Crocin protects PC12 cells against MPP^+-induced injury through inhibition of mitochondrial dysfunction and ER stress, *Neurochem. Int.*, 2015, **89**, 101–110.
33. X. Wang, Y. Xie, T. Zhang, S. Bo, X. Bai and H. Liu, *et al.*, Resveratrol reverses chronic restraint stress-induced depression-like behaviour: Involvement of BDNF level, ERK phosphorylation and expression of Bcl-2 and Bax in rats, *Brain Res. Bull.*, 2016, **125**, 134–143.
34. M. Kavitha, J. Nataraj, M. M. Essa, M. A. Memon and T. Manivasagam, Mangiferin attenuates MPTP induced dopaminergic neurodegeneration and improves motor impairment, redox balance and Bcl-2/Bax expression in experimental Parkinson's disease mice, *Chem. – Biol. Interact.*, 2013, **206**, 239–247.
35. E. N. Allen, S. Potdar, V. Tapias, M. Parmar, C. S. Mizuno and A. Rimando, *et al.*, Resveratrol and pinostilbene confer neuroprotection

against aging-related deficits through an ERK1/2-dependent mechanism, *J. Nutr. Biochem.*, 2018, **54**, 77–86.
36. J. Y. Kim, H. Y. Jeong, H. K. Lee, S. Kim, B. Y. Hwang and K. Bae, *et al.*, Neuroprotection of the leaf and stem of *Vitis amurensis* and their active compounds against ischemic brain damage in rats and excitotoxicity in cultured neurons, *Phytomedicine*, 2012, **19**, 150–159.
37. Z. Huang, J. Xu, X. Huang, G. Sun, R. Jiang and H. Wu, *et al.*, Crocin induces anti-ischemia in middle cerebral artery occlusion rats and inhibits autophagy by regulating the mammalian target of rapamycin, *Eur. J. Pharmacol.*, 2019, **857**, 172424.
38. L. Zhang, M. Fei, H. Wang and Y. Zhu, Sodium aescinate provides neuroprotection in experimental traumatic brain injury *via* the Nrf2-ARE pathway, *Brain Res. Bull.*, 2020, **157**, 26–36.
39. Z. I. Alam, S. E. Daniel, A. J. Lees, D. C. Marsden, P. Jenner and B. Halliwell, A generalised increase in protein carbonyls in the brain in Parkinson's but not incidental lewy body disease, *J. Neurochem.*, 1997, **69**, 1326–1329.
40. R. Santhanasabapathy and G. Sudhandiran, Farnesol attenuates lipopolysaccharide-induced neurodegeneration in Swiss albino mice by regulating intrinsic apoptotic cascade, *Brain Res.*, 2015, **1620**, 42–56.
41. G. Filomeni, I. Graziani, D. De Zio, L. Dini, D. Centonze and G. Rotilio, *et al.*, Neuroprotection of kaempferol by autophagy in models of rotenone-mediated acute toxicity: Possible implications for Parkinson's disease, *Neurobiol. Aging*, 2012, **33**, 767–785.
42. S. Thangarajan, S. Ramachandran and P. Krishnamurthy, Chrysin exerts neuroprotective effects against 3-nitropropionic acid induced behavioral despair-Mitochondrial dysfunction and striatal apoptosis *via* upregulating Bcl-2 gene and downregulating Bax-Bad genes in male wistar rats, *Biomed. Pharmacother.*, 2016, **84**, 514–525.
43. S. Niveditha and T. Shivanandappa, Neuroprotective action of 4-Hydroxyisophthalic acid against paraquat-induced motor impairment involves amelioration of mitochondrial damage and neurodegeneration in Drosophila, *NeuroToxicology*, 2018, **66**, 160–169.
44. J. C. Garcia-Moreno, M. Porta de la Riva, E. Martínez-Lara, E. Siles and A. Cañuelo, Tyrosol, a simple phenol from EVOO, targets multiple pathogenic mechanisms of neurodegeneration in a C. elegans model of Parkinson's disease, *Neurobiol. Aging*, 2019, **82**, 60–68.
45. H.-H. Chen, P.-C. Chang, S.-P. Wey, P.-M. Chen, C. Chen and M.-H. Chan, Therapeutic effects of honokiol on motor impairment in hemiparkinsonian mice are associated with reversing neurodegeneration and targeting PPARγ regulation, *Biomed. Pharmacother.*, 2018, **108**, 254–262.
46. K. Gopinath and G. Sudhandiran, Naringin modulates oxidative stress and inflammation in 3-nitropropionic acid-induced neurodegeneration through the activation of nuclear factor-erythroid 2-related factor-2 signalling pathway, *Neuroscience*, 2012, **227**, 134–143.

47. L. Cong, M.-Y. Lei, Z.-Q. Liu, Z.-F. Liu, Z. Ma and K. Liu, *et al.*, Resveratrol attenuates manganese-induced oxidative stress and neuroinflammation through SIRT1 signaling in mice, *Food Chem. Toxicol.*, 2021, **153**, 112283.
48. Q. Chen, Y.-Q. Lei, J.-F. Liu, Z.-C. Wang and H. Cao, Beneficial effects of chlorogenic acid treatment on neuroinflammation after deep hypothermic circulatory arrest may be mediated through CYLD/NF-κB signaling, *Brain Res.*, 2021, **1767**, 147572.
49. D. Liu, Q. Zhang, J. Gu, X. Wang, K. Xie and X. Xian, *et al.*, Resveratrol prevents impaired cognition induced by chronic unpredictable mild stress in rats, *Prog. Neuropsychopharmacol. Biol. Psychiatry*, 2014, **49**, 21–29.
50. S. H. Ali, R. M. Madhana, A. K. V. E. R. Kasala, L. N. Bodduluru and S. Pitta, *et al.*, Resveratrol ameliorates depressive-like behavior in repeated corticosterone-induced depression in mice, *Steroids*, 2015, **101**, 37–42.
51. L. Fan, Y. Peng, J. Wang, P. Ma, L. Zhao and X. Li, Total glycosides from stems of *Cistanche tubulosa* alleviate depression-like behaviors: bidirectional interaction of the phytochemicals and gut microbiota, *Phytomedicine*, 2021, **83**, 153471.
52. J. Chang, A. Rimando, M. Pallas, A. Camins, D. Porquet and J. Reeves, *et al.*, Low-dose pterostilbene, but not resveratrol, is a potent neuromodulator in aging and Alzheimer's disease, *Neurobiol. Aging*, 2012, **33**, 2062–2071.
53. F. J. Mirza, S. Amber, Sumera, D. Hassan, T. Ahmed and S. Zahid, Rosmarinic acid and ursolic acid alleviate deficits in cognition, synaptic regulation and adult hippocampal neurogenesis in an Aβ1-42-induced mouse model of Alzheimer's disease, *Phytomedicine*, 2021, **83**, 153490.
54. J. Zhang, Y. Zheng, Y. Zhao, Y. Zhang, Y. Liu and F. Ma, *et al.*, Andrographolide ameliorates neuroinflammation in APP/PS1 transgenic mice, *Int. Immunopharmacol.*, 2021, **96**, 107808.
55. R. Kumari, A. Chaudhary and A. Mani, Casuarictin: A new herbal drug molecule for Alzheimer's disease as inhibitor of presenilin stabilization factor like protein, *Heliyon*, 2020, **6**, e05546.
56. Y. Zheng, J. Zhang, Y. Zhao, Y. Zhang, X. Zhang and J. Guan, *et al.*, Curcumin protects against cognitive impairments in a rat model of chronic cerebral hypoperfusion combined with diabetes mellitus by suppressing neuroinflammation, apoptosis, and pyroptosis, *Int. Immunopharmacol.*, 2021, **93**, 107422.
57. S. Kumar, D. Kumar, A. Bhat and A. Kumar, Phytochemicals in clinical studies: Current perspective, in *Functional Food and Human Health*, ed. V. Rani and U. C. S. Yadav, Springer Singapore, Singapore, 2018, pp. 471–511.
58. S. D. Voulgaropoulou, T. A. M. J. van Amelsvoort, J. Prickaerts and C. Vingerhoets, The effect of curcumin on cognition in Alzheimer's disease and healthy aging: A systematic review of pre-clinical and clinical studies, *Brain Res.*, 2019, **1725**, 146476.

59. K. Drygalski, E. Fereniec, K. Koryciński, A. Chomentowski, A. Kiełczewska and C. Odrzygóźdź, *et al.*, Resveratrol and Alzheimer's disease. From molecular pathophysiology to clinical trials, *Exp. Gerontol.*, 2018, **113**, 36–47.
60. A. Tandon, S. J. Singh, M. Gupta, N. Singh, J. Shankar and N. Arjaria, *et al.*, Notch pathway up-regulation *via* curcumin mitigates bisphenol-A (BPA) induced alterations in hippocampal oligodendrogenesis, *J. Hazard. Mater.*, 2020, **392**, 122052.
61. M.-M. Chen, Z.-Q. Yin, L.-Y. Zhang and H. Liao, Quercetin promotes neurite growth through enhancing intracellular cAMP level and GAP-43 expression, *Chin. J. Nat. Med.*, 2015, **13**, 667–672.
62. W. Li, T. Ali, K. He, Z. Liu, F. A. Shah and Q. Ren, *et al.*, Ibrutinib alleviates LPS-induced neuroinflammation and synaptic defects in a mouse model of depression, *Brain, Behav., Immun.*, 2021, **92**, 10–24.
63. Z. Li, F. Fang, Y. Wang and L. Wang, Resveratrol protects CA1 neurons against focal cerebral ischemic reperfusion-induced damage via the ERK-CREB signaling pathway in rats, *Pharmacol., Biochem. Behav.*, 2016, **146–147**, 21–27.
64. Y. Kim and S.-H. Cho, *Lindera glauca* blume ameliorates amyloid-β1-42-induced memory impairment in mice with neuroprotection and activation of the CREB-BDNF pathway, *Neurochem. Int.*, 2021, **147**, 105071.
65. M. Motaghinejad, S. Safari, S. Feizipour and S. Sadr, Crocin may be useful to prevent or treatment of alcohol induced neurodegeneration and neurobehavioral sequels *via* modulation of CREB/BDNF and Akt/GSK signaling pathway, *Med. Hypotheses*, 2019, **124**, 21–25.
66. R. Soukhaklari, L. Moezi, F. Pirsalami, N. Ashjazadeh and M. Moosavi, Curcumin ameliorates scopolamine-induced mice memory retrieval deficit and restores hippocampal p-Akt and p-GSK-3β, *Eur. J. Pharmacol.*, 2018, **841**, 28–32.
67. B. B. Kautu, A. Carrasquilla, M. L. Hicks, K. A. Caldwell and G. A. Caldwell, Valproic acid ameliorates *C. elegans* dopaminergic neurodegeneration with implications for ERK-MAPK signaling, *Neurosci. Lett.*, 2013, **541**, 116–119.
68. M. H. Farzaei, D. Tewari, S. Momtaz, S. Argüelles and S. M. Nabavi, Targeting ERK signaling pathway by polyphenols as novel therapeutic strategy for neurodegeneration, *Food Chem. Toxicol.*, 2018, **120**, 183–195.
69. H. Liao, L.-M. Chou, Y.-W. Chien, C.-H. Wu, J.-S. Chang and C.-I. Lin, *et al.*, Grape powder consumption affects the expression of neurodegeneration-related brain proteins in rats chronically fed a high-fructose–high-fat diet, *J. Nutr. Biochem.*, 2017, **43**, 132–140.
70. L. Buccarello, J. Dragotto, F. Iorio, K. Hassanzadeh, M. Corbo and M. Feligioni, The pivotal role of SUMO-1-JNK-Tau axis *in* an *in vitro* model of oxidative stress counteracted by the protective effect of curcumin, *Biochem. Pharmacol.*, 2020, **178**, 114066.
71. Q. Zhang, G. Su, T. Zhao, B. Sun, L. Zheng and M. Zhao, Neuroprotection of round scad (*Decapterus maruadsi*) hydrolysate in glutamate-damaged

PC12 cells: Possible involved signaling pathways and potential bioactive peptides, *J. Funct. Foods*, 2020, **64**, 103690.
72. A. P. Gureev, E. A. Shaforostova and V. N. Popov, Regulation of mitochondrial biogenesis as a way for active longevity: Interaction between the Nrf2 and PGC-1α signaling pathways, *Front. Genet.*, 2019, **10**, 435.
73. S.-J. Wang, Q.-Y. Bo, X.-H. Zhao, X. Yang, Z.-F. Chi and X.-W. Liu, Resveratrol pre-treatment reduces early inflammatory responses induced by status epilepticus *via* mTOR signaling, *Brain Res.*, 2013, **1492**, 122–129.
74. D. Zhang, Y. Feng, H. Pan, Z. Xuan, S. Yan and Y. Mao, *et al.*, 9-Methylfascaplysin exerts anti-ischemic stroke neuroprotective effects *via* the inhibition of neuroinflammation and oxidative stress in rats, *Int. Immunopharmacol.*, 2021, **97**, 107656.
75. Z. Huang, X. Zhou, X. Zhang, L. Huang, Y. Sun and Z. Cheng, *et al.*, Pien-Tze-Huang, a Chinese patent formula, attenuates NLRP3 inflammasome-related neuroinflammation by enhancing autophagy *via* the AMPK/mTOR/ULK1 signaling pathway, *Biomed. Pharmacother.*, 2021, **141**, 111814.
76. S. F. Akomolafe, A. J. Akinyemi, O. B. Ogunsuyi, S. I. Oyeleye, G. Oboh and O. O. Adeoyo, *et al.*, Effect of caffeine, caffeic acid and their various combinations on enzymes of cholinergic, monoaminergic and purinergic systems critical to neurodegeneration in rat brain-*in vitro*, *NeuroToxicology*, 2017, **62**, 6–13.
77. M. S. de Oliveira, J. N. da Cruz, S. Gomes Silva, W. A. da Costa, S. H. B. de Sousa and F. W. F. Bezerra, *et al.*, Phytochemical profile, antioxidant activity, inhibition of acetylcholinesterase and interaction mechanism of the major components of the *Piper divaricatum* essential oil obtained by supercritical CO2, *J. Supercrit. Fluids*, 2019, **145**, 74–84.
78. S. Liu, Y. Li, F. Yi, Q. Liu, N. Chen and X. He, *et al.*, Resveratrol oligomers from *Paeonia suffruticosa* protect mice against cognitive dysfunction by regulating cholinergic, antioxidant and anti-inflammatory pathways, *J. Ethnopharmacol.*, 2020, **260**, 112983.
79. P. Chetia, M. K. Mazumder, S. Mahanta, B. De and M. Dutta Choudhury, A novel phytochemical from *Dipteris wallichii* inhibits human β-secretase 1: Implications for the treatment of Alzheimer's disease, *Med. Hypotheses*, 2020, **143**, 109839.

CHAPTER 11

Improvement of the Immune System by Dietary Supplements and Natural Products

SUI KIAT CHANG*[a] AND CESARETTIN ALASALVAR[b]

[a] Department of Allied Health Sciences, Faculty of Science, Universiti Tunku Abdul Rahman, Kampar, Perak, Malaysia; [b] TÜBİTAK Marmara Research Center, Life Sciences, Gebze-Kocaeli, Turkey
*Email: suikiatchang@gmail.com.tr

11.1 Introduction

Epidemics and pandemics are always a test for humanity, and COVID-19 is no exception. Since the beginning of the COVID-19 epidemic, people around the world have been under constant stress and are experiencing significant negative emotions, fear, anxiety, and anger. All these immediately affect the state of their immune system and reduces the resistance of the body not only to the SARS-CoV-2 virus but also to other viral and bacterial infections.[1] Therefore, supporting or boosting the immune system has become an important reason by using plant-based foods, beverages, dietary supplements, and herbal extracts, among others.

The immune system can be divided into two categories of innate and adaptive immunity. Innate immunity, the body's first defence against an intruder, includes physical and chemical barriers, such as the skin and stomach acid, and responds quickly to threats. Comparatively, the adaptive immune system responds more slowly and prevents us from future

Food Chemistry, Function and Analysis No. 39
Dietary Supplements with Antioxidant Activity: Understanding Mechanisms and Potential Health Benefits
Edited by Cesarettin Alasalvar, Fereidoon Shahidi and Chi-Tang Ho

Published by the Royal Society of Chemistry, www.rsc.org

sicknesses, as it involves specialized systemic cells that eliminate pathogens the body has previously encountered.[2,3]

It is believed that the only sustainable way to survive in the current pandemic situations (COVID-19) is to boost the immune system. Many traditional foods and herbs have antiviral and immunomodulatory properties.[1,4,5] A balanced diet and dietary intake of nutrients affect the immune system through gene expression, cell activation, and modification of signalling molecules. In addition, various food ingredients are determinants of the gut microbial composition and subsequently form immune response in the body.[1,6] A diet combined with medicinal plants with immunomodulatory, antiviral, and anti-inflammatory effects can significantly enhance this protection.[5] This chapter discusses the improvement of the immune system by natural products and supplements (foods, herbs, essential oils, dietary supplements, and bioactives).

11.2 Sources of Natural Products and Supplements for Immune Health

Sources of natural products and supplements [medicinal plants and plant-based foods, vitamins, minerals, propolis, essential oils, bioactives, omega-3 (ω-3) fatty acids, prebiotics, probiotics, and others] for immune health are reviewed in detail (Figure 11.1).

11.2.1 Medicinal Plants and Plant-based Foods

Medicinal plants have been used for centuries in the form of whole herbal products, extracts, oils, powders, capsules, or lotions to enhance health promotion and disease prevention as well as enriching food products as flavourants and colourants, among others. Essential oils and extracts of some medicinal plants and their bioactive compounds have been on the market as nutraceuticals/dietary supplements/herbal products for decades. Therefore, they play important roles in daily life for numerous health purposes, one of which is the enhancement of the immune system.[5]

Herbs are traditionally used in many therapeutic practices, if not as the main, then as the accompanying therapy in combination with medications, aimed at boosting immunity for disease prevention.[1] In addition, plants from traditional Chinese medicine (TCM) are rich sources of compounds used for development of medicines for a wide range of ailments (from coughs and colds to parasitic infections and inflammations).[7]

Pelvan *et al.*[5] reviewed the *in vitro*, *in vivo*, and clinical evidence supporting the immunomodulatory role of eleven plant-based medicinal herbs (bay laurel, black cumin, clove, fennel, lemon balm, lemongrass, marjoram, peppermint, rosemary, sage, and thyme) and their essential oils. They found that some herbs support a strong immune system, possess important immunomodulatory properties, and are safe to consume at recommended amounts.

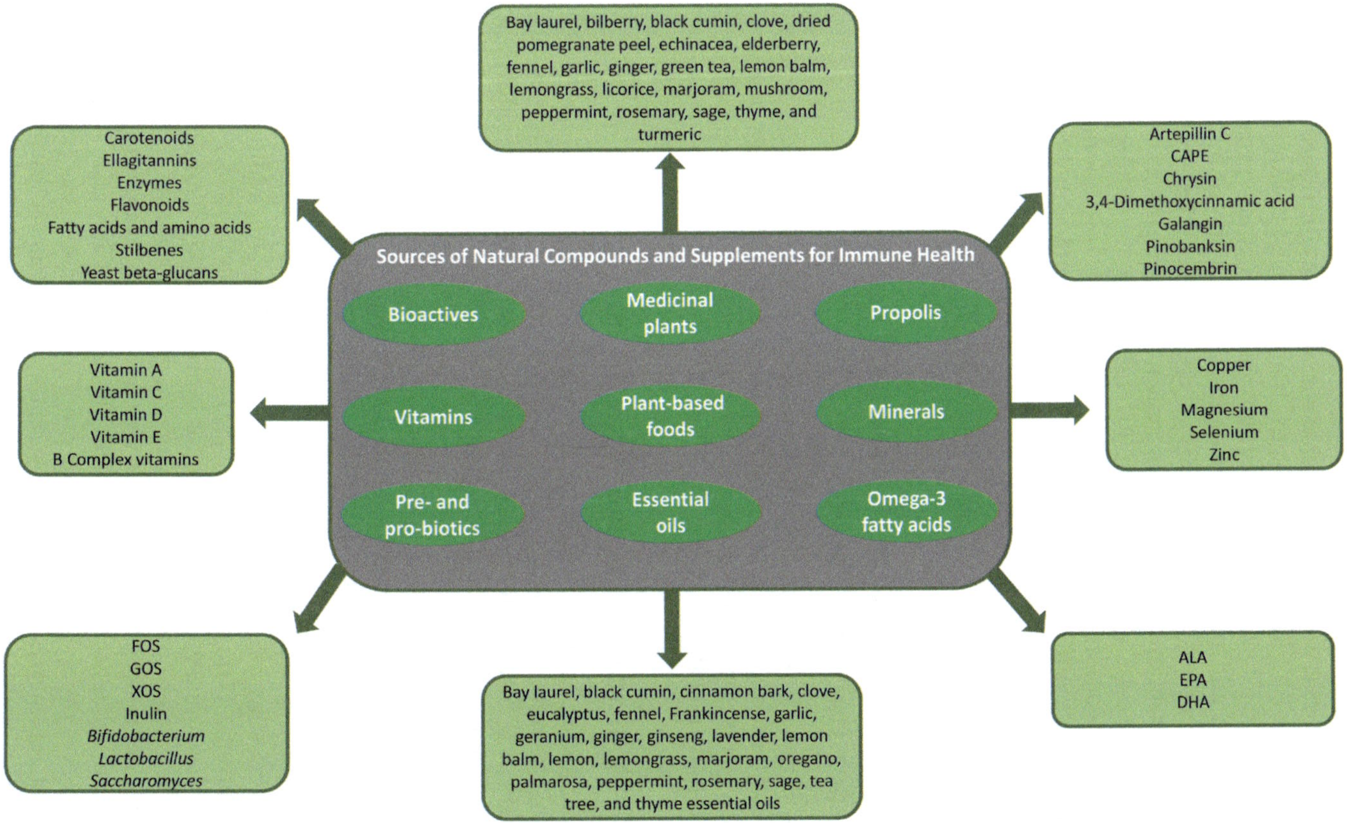

Figure 11.1 Sources of natural compounds and supplements for immune health. Abbreviations: ALA, α-linolenic acid; CAPE, caffeic acid phenethyl ester; DHA, docosahexaenoic acid; EPA, eicosapentaenoic acid; FOS, fructooligosaccharides; GOS, galactooligosaccharides; XOS, xylooligosaccharides.

Additional randomized clinical trials (RCTs) are needed to validate the detailed immunomodulatory effects of these herbs.

Despite the fact that there are several medicinal plants and plant-based foods boosting the immune system, garlic, turmeric, ginger, echinacea, elderberry, and licorice are only a few of the many potent immune-boosting plants.

The benefits of garlic to health have been proclaimed for centuries; however, only recently have *Allium sativum* and its derivatives been proposed as promising candidates for maintaining the homeostasis of the immune system. Garlic appears to enhance the functioning of the immune system by stimulating certain cell types, such as macrophages, lymphocytes, natural killer cells, dendritic cells, and eosinophils, by mechanisms including modulation of cytokine secretion, immunoglobulin production, phagocytosis, and microphage activation. Garlic has been shown to contain a high level of allicin, a critical plant compound that acts as a germicide.[8] Research indicates that turmeric (curcumin is the most potent compound) might also boost the immune system by exerting immunomodulatory effects on several cells and organs of the immune system.[9] Similarly, ginger can combat chronic inflammation while maintaining a healthy immune system. Ginger and its compounds can enhance the immune response and act as a protective measure due to the potent antiviral, anti-inflammatory, and antioxidative effects of ginger in numerous animal models.[10] Moreover, *Echinacea*, also known as the purple coneflower, is one of the best immune-boosting medicinal plants. Several *in vivo* investigations have been conducted on the immunomodulatory and anti-inflammatory properties of *Echinacea* and these suggest that its administration boosts innate immunity and bolsters the ability of the immune system to fight pathogenic infections by activating neutrophils, macrophages, polymorphonuclear leukocytes and natural killer (NK) cells.[11] The roots and the aerial part of *Echinacea* contain caffeic acid derivatives, alkamides, ketoalkenes, polysaccharides, and glycoproteins which are thought to be responsible for its immunostimulatory and anti-inflammatory properties.[12] *In vitro* studies on human cells have also shown its immunomodulatory effects with a notable decrease in proinflammatory cytokines, such as tumour necrosis factor-alpha (TNF-α) and stimulation of NK cells.[13] Moreover, the ripe fruits of elderberry are used for dietary supplements because they contain flavonoids, vitamins, and essential oils. They have anti-inflammatory, antioxidant, antiviral, and immunological effects.[13,14] Finally, the data obtained from experimental and clinical studies show that licorice possesses antiviral, anti-inflammatory, immunomodulatory, and antimicrobial activities.[15] The ethanol extracts of polysaccharides derived from licorice have proven to be able to stimulate the immune system *via* increment of serum immunoglobulin (IgA, IgG, and IgM) levels along with an increase in proliferation of spleen lymphocytes.[16] The surprising fact is that licorice appears in the top 10 main ingredients used in TCM prescriptions for treatment of epidemic diseases.[17]

More recently, Öznur *et al.*[18] developed a chewable tablet formulation with propolis and herbal extracts (green tea, dried pomegranate peel, and bilberry) against Wuhan (B.1.36) and Omicron (BA.1.1) variants of SARS-CoV-2 infection. Results have confirmed that a combination of herbal extracts with propolis at suitable concentrations can effectively be used as a food supplement to prevent against Wuhan and Omicron variants of SARS-CoV-2 virus.

Crawford *et al.*[4] examined a total of 30 select dietary supplement products that online platforms marketed to support or boost the immune system. Supporting or boosting the immune system has become an important reason for using dietary supplements and many consumers are purchasing products through online platforms. The 30 immune health dietary supplement products tested and analysed had claims related to immune support, immune defence, and bolstering or boosting the immune system. Examples of such claims included "all seasons immune support", "immune strengthening ingredients", "a powerhouse immune system booster", and "bolster up your body's immune support". Fifteen products additionally had scientific sounding claims by using terms such as "research based" or "research supported", "clinically studied", "scientifically proven", "supported by...gold-standard clinical studies", and "backed by science". The total number of ingredients listed on the Supplement Facts labels of these products ranged from a single ingredient in one product up to 312 listed ingredients.

The 312 total ingredients listed across these products included herbs and other botanicals [143 (45.8%)], vitamins [74 (23.7%)], minerals [59 (18.9%)], amino acids [6 (1.9%)], and other substances, such as enzymes, organ tissue, and metabolites [12 (3.8%)]. Prebiotics and probiotics [7 (2.2%)] and various mushroom extracts [11 (3.5%)] were also included in some products. The most frequent ingredients listed within these categories across products were *Echinacea* (14 products), elderberry (18 products), vitamin C (24 products), vitamin D (15 products), and zinc (25 products) (Figure 11.2).[4]

11.2.2 Vitamins

A wide range of fat-soluble vitamins (such as A, D, and E) and water-soluble vitamins (such as C and B complex)[19–24] have been shown to display essential immunomodulatory effects.[2,25] Nutrition is undoubtedly an important building block of immunity and the influence of micronutrients on immune function has been widely investigated.[2] Vitamins have long been known to influence immune functions, playing a role in both innate and adaptive immune responses.[22] While the effects of some vitamins, such as vitamins C and E as well as B vitamins, are brought about in a relatively nonspecific manner (*e.g.*, by anti-oxidative action), other vitamins, such as vitamins A and D, influence immune responses in highly specific ways.[22]

Regarding fat-soluble vitamins, vitamin A is a retinoid and a fat-soluble vitamin involved in the development of the immune system and in the regulation of cellular and humoral immune processes.[20] Vitamin D has often

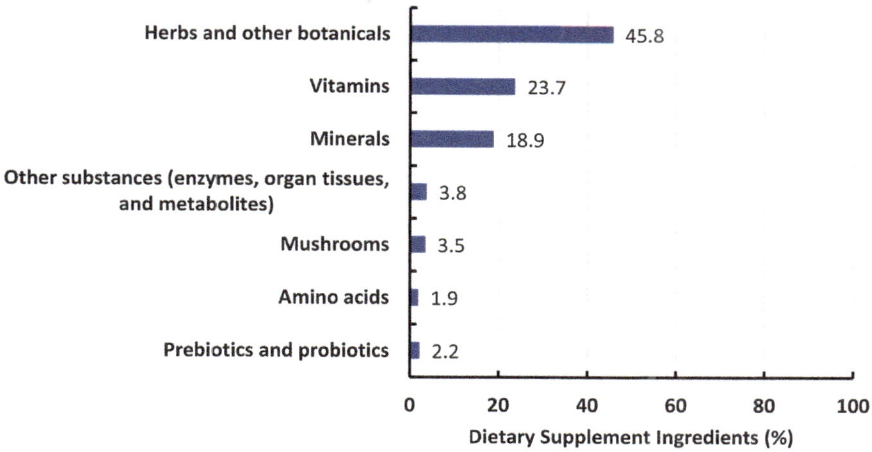

Figure 11.2 Types of dietary supplement ingredients advertised on product labels across 30 dietary supplements containing a total of 312 ingredients marketed for immune health. Reproduced from ref. 4, http://jamanetwork.com/article.aspx?doi=10.1001/jamanetworkopen.2022.26040, under the terms of the CC BY 4.0 License, https://creativecommons.org/licenses/by/4.0/.

been suggested to be involved in decreasing the risk of microbial infection[26] by increasing physical barriers, cellular natural immunity, and adaptive immunity.[23] Vitamin E is a lipid-soluble antioxidant that has been found to have immunomodulatory effects in cell-based, preclinical, and clinical studies.[21]

With regard to water-soluble vitamins, vitamin C (ascorbic acid) is found in many foods, particularly citrus fruits and green vegetables. It is a potent antioxidant with anti-inflammatory and immune-supporting effects. It accumulates at high levels in most immune cells and is involved in the immune defence through its support of various cellular functions of both the innate and adaptive immune systems.[19] B vitamins act as components of coenzymes. They may mediate the interaction with immune cells involved in inflammation and various pathophysiological pathways.[24] Vitamin B complex deficiency may weaken the host immune response.[27]

Figure 11.2 shows the types of dietary supplement ingredients advertised on product labels across 30 dietary supplements containing a total of 312 ingredients for immune health. Vitamins with 74 ingredients represent 23.7% of dietary supplement ingredients after herbs and other botanicals.[4] Vitamins, in general, share a good percentage of the immune health market.

11.2.3 Minerals

Some minerals (such as zinc, selenium, magnesium, iron, and copper)[28–31] have been shown to play essential immunomodulatory effects.[2,25]

Zinc is an essential trace element, whose role in immune function has been known for several decades and has been investigated extensively. Altered zinc homeostasis plays a crucial role in innate and adaptive immunity. The effects of zinc deficiency on immune cells can result in altered host defence, an increased inflammation risk and even death. Chronic zinc deficiency causes an increase in the production of pro-inflammatory cytokines, affecting the outcome of various inflammatory diseases.[28] Selenium offers a wide array of health effects. Dietary selenium levels and selenoprotein expression have been shown to impact the regulation of immune cell functions. Low selenium status has been found to be associated with poor immune functioning, while a higher selenium status promotes antiviral effects.[29] Magnesium has been shown in animal models to play a role in both innate and acquired immune responses. A compromised immune system may be found in elderly people, and a balanced magnesium homeostasis may be important in strengthening resistance to infection in the elderly.[32] Suggestions have also been made that magnesium deficiency could contribute to the immunological alterations observed following strenuous exercise.[33] Moreover, iron and immunity are closely linked. Iron is an essential micronutrient for both humans and pathogenic microbes. Most human pathogens require iron, and the innate immune system is involved in limiting iron availability to invading microbes.[30] Finally, copper plays an important role in the human immune system. It is involved in the functions of T helper cells, B cells, NK cells, and macrophages.[34] Dietary copper deficiency has been shown to affect both innate and adaptive immunity.[31]

11.2.4 Propolis

Propolis plays an important role in immunomodulatory effects.[35,36] Its complex chemical composition includes polyphenols, phenolic aldehydes, sesquiterpenes, quinins, coumarins, amino acids, steroids, and inorganic compounds, among others. Propolis is known to have a broad spectrum of biological properties, including antimicrobial, antioxidant, anti-inflammatory, antiallergic, dermatoprotective, laxative, anti-diabetic, antitumor, and immunomodulatory activities.[36] The immunomodulatory activity of propolis has been well-researched. This activity is attributed to flavonoids and some phenolic acids, mainly caffeic acid phenethyl ester (CAPE) and artepillin C (3,5-diprenyl-4-hydroxycinnamic acid). CAPE is the most bioactive compound in propolis. Propolis and these components exhibit immunomodulatory effects on a wide spectrum of immune cells, including cells of lymphoid or monocytic lineages, mediated by the extracellular signal-regulated kinase 2 and mitogen-activated protein-kinase signalling pathway and by eukaryotic transcription factors: nuclear factor of activated T cells and nuclear factor kappa B.[36]

Pelvan *et al.*[37] identified forty-three natural phenolic compounds: these include 19 phenolic acids, 18 flavonoids, and 6 non-flavonoids. Among quantified 17 compounds, pinocembrin predominated in propolis extract, followed by galangin, chrysin, pinobanksin, CAPE, and 3,4-dimethoxycinnamic acid. They also developed propolis and essential oils containing oral/throat spray formulation against SARS-CoV-2 infection. The spray with 1:640-fold dilution provided the highest efficacy and the cytopathic effect was delayed for 54 h at this dilution, and the antiviral activity rate was 85.3%. A combination of natural products with essential oils at the right concentrations can be used as a supplement for the prevention of SARS-CoV-2 infection.

Due to its richness in phenolics, propolis can suppress the virus replication and modulate the host immune response. It is a common problem that influenza viruses, which cause respiratory tract infections, regularly mutate, thus making it difficult for successful antiviral therapies.[38] They also reported that propolis components (such as catechin, *p*-coumaric acid, and flavonols) display angiotensin-converting enzyme-inhibitory activity. There is plenty of evidence for the interference of propolis and/or its components with viral replication and infectivity, potentially reducing lung inflammation owing to anti-inflammatory properties while promoting immune system fortification. These are beneficial properties that could help diminish the symptoms and harmful effects of COVID-19.

11.2.5 Essential Oils

Essential oils contain compounds with a specific fragrance and biological properties that can be used for medicinal purposes.[39] They can either fortify the immune system or eliminate harmful bacteria and viruses. While all essential oils contain antibacterial, antiviral, and antiseptic properties, to some extent, some essential oils have more of one than the others. The main volatile components of essential oils are terpenes which are aromatic compounds found in plants. Monoterpenes and sesquiterpenes are the most common aromatic components in essential oils.

The biological effects of essential oils on humans are complex; however, research continues to examine the immunomodulatory properties of essential oils and their potential application in infectious diseases. Several studies have demonstrated that certain essential oils have potential antimicrobial, antifungal, anti-inflammatory, and antiviral effects that may be beneficial to the immune system.[5,40]

Pelvan *et al.*[5] reviewed the *in vitro*, *in vivo*, and clinical evidence supporting the immunomodulatory role of eleven plant-based essential oils of medicinal herbs (bay laurel, black cumin, clove, fennel, lemon balm, lemongrass, marjoram, peppermint, rosemary, sage, and thyme). They found that some essential oils support a strong immune system, possess important immunomodulatory properties, and are safe to consume at recommended amounts. Although there are many medicinal herbs rich in essential oils for immune boosting, some of the most important ones are reviewed below.

Black cumin essential oil has recently been investigated in experimental studies for its anti-inflammatory and immunomodulatory effects. Silva *et al.*[41]

carried out various *in vitro* studies in human mammary carcinoma (MCF-7) and melanoma (A375) cell lines treated with 1 mL of 0.002% (w/v) 2,2-diphenyl-1-picrylhydrazyl solution in dimethyl sulfoxide. Clove essential oil volatiles increased the stabilization of protein structures, whereas non-volatile compounds (mainly thymoquinone present in black cumin essential oil) increased antioxidant activity, POTE ankyrin domain family member F and Heat Shock Protein 90-β expression, enzyme inhibition, and cytotoxic activity in cancer cells. The main essential oils in black cumin are *p*-cymene, γ-terpinene, thymoquinone, β-pinene, carvacrol, terpinen-4-ol, and longifolene, among which thymoquinone is the most active compound.[42]

Cinnamon bark essential oil mainly consists of *trans*-cinnamaldehyde, *o*-methoxy-cinnamaldehyde, cinnamyl aldehyde, benzaldehyde, phenylethanol, borneol, eugenol, coumarin, and cinnamic acid,[43] among which *trans*-cinnamaldehyde is the predominant one, contributing 80% to the total essential oil.[44] Chowdhury *et al.*[45] conducted a study to evaluate the effect of three essential oils with different main chemical structures (cinnamon bark oil, clove bud oil, and ajwain seed oil) as an alternative to antibiotic growth promoters (AGPs) on intestinal health, immune response, and antioxidant status in broiler chickens. Supplementation of cinnamon bark oil could be more advantageous than the AGPs in diets due to its combined beneficial effects on immune response, gut health, antioxidant status, and blood cholesterol in broiler.

Clove essential oil is one of the most robust antibacterial essential oils. Its high percentage of eugenol, generally between 70% and 80%, supports strong antimicrobial and antifungal properties that cleanse harmful molecules from the air.[44,46] It is suggested that clove essential oil promotes the regeneration of specific cells and proteins associated with the immune system and inflammation.[47] The main components of clove essential oil are phenylpropanoids, which include eugenol, carvacrol, thymol, and cinnamaldehyde. These and other compounds found in clove essential oil, such as terpenes and flavonoids, have demonstrated immune-supportive capabilities.[48]

Eucalyptus essential oil is rich in 1,8-cineole, generally at 60–90%.[49] With its strong antiseptic and anti-inflammatory properties, it strengthens the immune system and invigorates the body.[50] Mahmoodi *et al.*[51] reported that the sum of eucalyptus (1,8-cineole, α-pinene, and *p*-cymene) constitutes around 60–80% of the essential oil.

Frankincense essential oil showed a strong immunostimulant effect when based on a lymphocyte proliferation assay. Human peripheral venous blood lymphocytes were stimulated with the mitogen phytohaemagglutinin and the mitogenic response induced in the presence of frankincense essential oil (90% lymphocyte proliferation) was comparable to that of known immunostimulants like *Echinacea purpurea* aqueous extract (95%) and levamisole (85%).[40,52] Frankincense oil has several pharmacological properties, of which many allude to its anti-infective potential. The main components in the oil include α-pinene, α-thujene, β-pinene, myrcene, sabinene, limonene, *p*-cymene, and β-caryophyllene.[53]

Lavender essential oil is one of the most widely used across the globe. Linalool and linalyl acetate are principal constituents and range from 11.4% to 46.7% and 7.4% to 44.2%, respectively.[54] It has antiviral, antimicrobial, and antibacterial properties that make it a great addition to our body's natural defence system. Another added benefit of using lavender essential oil is its ability to relieve stress, which is often one of the biggest factors in promoting immune system health.[55] Lavender essential oil in the study of Giovannini *et al.*[56] enhanced the innate immune response by stimulating phagocytosis, but also mitigated a consequent inflammatory response, thus supporting and balancing the overall immune reaction.

Lemon essential oil (rich in D-limonene, around 65%) offers strong antibacterial and antioxidant activities that support the immune system.[44] It is also associated with enhanced production of lymphocytes, the body's main type of immune cells.[40] The major constituents of lemon essential oil are D-limonene, β-pinene, and γ-terpinene.[57]

Rosemary essential oil has potent antibacterial properties. Because of its high percentages of 1,8-cineole (38.5%), α-pinene (12.3%), limonene (6.23%), camphene (6%), and linalool (5.70%), it has an anti-inflammatory effect on airway illnesses.[58] The essential oil has fortifying qualities boosting the immune system because of its psychological benefits. Research shows the feel-good properties of rosemary, reducing cortisol levels and thus increasing the strength of the immune system.[59]

Oregano essential oil has demonstrated immune-supportive capabilities including anti-microbial, antifungal, antiparasitic, antioxidant, anti-inflammatory, and antitumor effects. The primary immune-supportive constituents of oregano essential oil are carvacrol and thymol.[60] Carvacrol represents more than 70% to the total essential oil in oregano.[44] Dietary supplementation of oregano essential oil to sows during lactation led to a greater number of T lymphocytes in the milk but had no effect on immune responses (immunoglobulin concentrations, T lymphocytes, and NK cell activity) in suckling piglets.[40,61]

Peppermint essential oil has medicinal potential as an anti-inflammatory agent. It contains 1–3% essential oil, of which menthol and menthone constitute 50–60% of its essential oil.[62] In addition, Lang *et al.*[44] assessed a peppermint essential oil composed mainly of menthol (32.5%) and (DL)-menthone (26.3%). They found that the inhibitory activity of peppermint essential oil inhibited the lipopolysaccharide-induced phagocytosis of RAW 264.7 cells by 42% at 100 $\mu g\,mL^{-1}$.

Tea tree essential oil fights infections and inhibits harmful bacteria. With potent antibacterial and antifungal qualities due to its high levels of α-sabinene and antiseptic qualities because of terpinen-4-ol, tea tree essential oil is a notable immune booster.[46] Research shows that it significantly suppressed proteins associated with an inflammatory response.[63]

Thyme essential oil is one of the most aromatic and medicinal plant oils used worldwide. The oil, consisting of two main components, thymol and carvacrol, has antioxidant, antimicrobial, antitussive, antibacterial, and anti-inflammatory effects.[64] The sum of thymol, carvacrol, and *p*-cymene constitutes approximately 60–80% of the oil.[65] The effect of thyme essential oil on the symptoms of COVID-19 patients was recently investigated in a randomized controlled trial by Sardari *et al.*[66] The results showed that 5 mL of syrup or thyme essential oil 3 times per day for one week mitigated the symptoms of SARS-CoV-2 while decreasing neutrophil and increasing lymphocyte counts.

Individual or combinations of essential oils can be used for immune support, antiviral activity, and prevention against SARS-CoV-2 infection. Pelvan *et al.*[37] developed propolis extract, cold-pressed black cumin oil, and essential oils (Anatolian sage, oregano, lemon, and geranium) containing oral/throat spray formulation against SARS-CoV-2 infection. The oral/throat spray with 1:640-fold dilution provided the highest efficacy and the cytopathic effect was delayed for 54 h at this dilution, and the antiviral activity rate was 85.3%.

There are numerous other essential oils from different medicinal herbs (*e.g.*, garlic, geranium, ginger, ginseng, palmarosa, and sage, among others) that have also been reported supporting the immune system.[40]

11.2.6 Bioactives

Some bioactives (such as carotenoids, flavonoids, stilbenes, tannins, and ω-3 fatty acids, among others)[67–70] have been shown to exert essential immunomodulatory effects.[2]

Numerous investigations using non-provitamin A carotenoids have been able to show the immunomodulatory action of dietary carotenoids. Non-provitamin A carotenoids were found to be active or even more active than β-carotene in enhancing cellular and humoral immune responses in both experimental animals and humans.[67]

In addition to the well-established antioxidant properties of polyphenols including flavonoids, evidence suggests that they may be capable of modulating immune responses. The antiviral, antimicrobial, anti-inflammatory and cytotoxic capacity of flavonoids has been suggested to benefit the immune system.[68] Quercetin is a flavonol found in fruits and vegetables and has biological properties that may reduce the risk of infections. It has been shown to exhibit anti-inflammatory properties in cells from animals and humans and to ameliorate experimentally induced impairment of immunity function in animal models.[71]

Resveratrol (stilbene) is a naturally occurring polyphenolic phytoalexin, which is present in the skin and seeds of various plants. The bioavailability of orally administered resveratrol is very low. The naturally dimethylated analogue of resveratrol, pterostilbene, may be more effective due to its higher *in vivo* bioavailability.[72] The findings of experimental studies suggest a potential activity of resveratrol in a variety of human diseases, including viral infections.[70]

Ellagitannins are a complex class of bioactive compounds, mainly composed of hydrolyzed tannins, found in pomegranates, strawberries, ground elm, blueberries, raspberries, tea, walnuts, chestnuts, and mulberries.[73] Ellagitannins, ellagic acid, and their metabolites (such as urolithin) have been reported to exhibit numerous beneficial effects on human health including the anti-inflammatory effect and immune health.[74] The ω-3 fatty acids are given in detail in the next section.

11.2.7 ω-3 Fatty Acids

Some dietary components have been found to have immuno-regulatory properties in boosting the immune system, including ω-3 fatty acids. The effects of dietary polyunsaturated fatty acids (PUFAs) on the immune system have been studied for more than a decade, with a special emphasis on ω-3 PUFAs, such as α-linolenic acid (ALA), eicosapentaenoic acid (EPA), and docosahexaenoic acid (DHA). The food sources of ALA are usually plant oils, nuts, and seeds, whereas EPA and DHA are mostly found in fish and algal oils. A comprehensive review of the food sources of ω-3 fatty acids has been conducted by Cholewski *et al.*[75] and Shahidi and Santhiravel.[76]

ω-3 fatty acids regulate immune cell function by presenting several cell-type specific features. Long chain PUFAs, such as arachidonic acid, EPA, and DHA, are the components of the cell membranes of the immune system and act through multiple interaction mechanisms to influence the immune function.[77] Both EPA and DHA are well known to have favourable effects on immunity and inflammation.[78,79] Evidence to date indicates that ω-3 fatty acids incorporate themselves into the cellular membrane of all the immune cells effectively. The reason is that PUFAs possess multiple double bonds in their carbon chain. The incorporation of PUFAs into the plasma membrane enhanced the fluidity of the plasma membrane.[80] Several review articles have reported the effects of ω-3 fatty acids on the immune system by assessing their clinical effects on immune-related diseases.[81,82] Hence, this section discusses the specific effects of ω-3 fatty acids on different types of immune cells according to recent findings in the field.

Generally, ALA, DHA, and EPA inhibited the activation of immune cells from both the innate and the adaptive branch.[69] Interestingly, some specific immune functions are promoted by dietary ω-3 fatty acids in specific immune cell types, for example, phagocytosis by macrophages and neutrophils or differentiation of regulatory T cells, suggesting that ω-3 fatty acids do not act as unspecific immune-repressors. Besides, supplementation with EPA and DHA has been shown to inhibit the activation of basophils, mast cells and T cells.[69] Cho and Park[83] reported that dietary supplementation with ω-3 fatty acids decreases the circulating levels of basophils in healthy humans. In another study, dietary supplementation with plant ω-3 fatty acids decreases basophil production of leukotrienes *ex vivo* in asthmatic patients.[84] This indicates that EPA and DHA play important roles in reducing the severity of allergy and asthma. Both EPA and DHA increase the production of IgM by

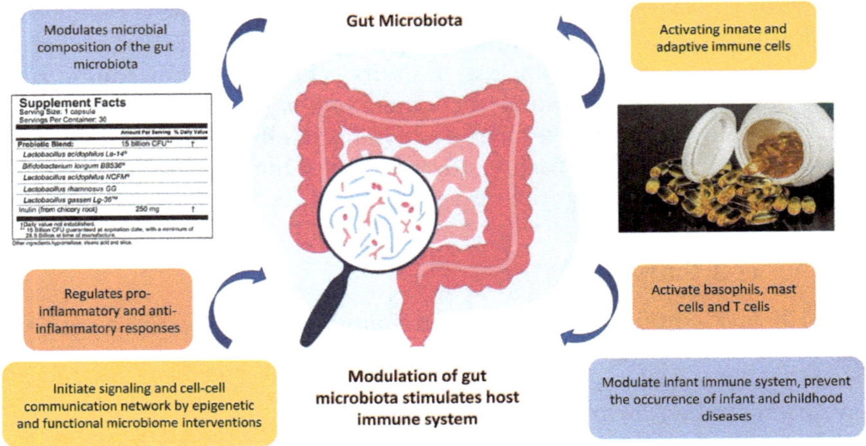

Figure 11.3 Mechanisms of how ω-3 fatty acids, prebiotics, and probiotics modulate and enhance the host immune system.

B cells by increasing the number of antibody-producing cells in mice and humans.[85,86] This effect is quite specific for IgM since ω-3 fatty acids do not affect B cell production of IgA, IgG, or IgD.[87,88]

It has been demonstrated that increased consumption of ω-3 fatty acids during pregnancy may be useful to modulate the infant immune system and prevent the occurrence of infant and childhood allergic diseases.[77] Moreover, increased consumption of ω-3 fatty acids during pregnancy lowers the sensitivity to common food allergens and reduces the risk and severity of atopic dermatitis during the first year of an infant's life.[89] Studies reported that infants supplemented with ω-3 fatty acids demonstrate immune differences compared to the controls. This suggests stronger immune maturation with lower risk of allergic diseases and respiratory illnesses during the first years of life.[89,90] In another scenario, ω-3 fatty acids exert antiviral effects by inhibiting influenza virus replication, indicating its important roles in preventing COVID-19 and influenza infections.[91]

However, the precise mechanisms by which ω-3 fatty acids exert their immunoregulatory effects are still not fully understood. Hence, more well-designed explorative studies are needed to validate this unanswered question. The scientific community and consumers expect more exciting results or more convincing results of the immunomodulatory effects of ω-3 fatty acids in the coming years. Figure 11.3 illustrates the mechanisms of how ω-3 fatty acids, prebiotics, and probiotics modulate and enhance the host immune system.

11.2.8 Prebiotics and Probiotics

There are trillions of microbes residing on the human body that are known to influence and regulate the physiology of the human body. Most microbes are inhabitants in the gastrointestinal tract of humans and are known as gut microbiota. The human gut microbiota comprises four major (*Bacteroidetes*, *Firmicutes*, *Proteobacteria*, and *Actinobacteria*) and two minor phyla (*Verrucomicrobia* and *Fusobacteria*).[92] These commensal bacteria communicate with each other and also with the host gut epithelium to maintain the gut homeostasis and improve the host immunity.[93]

Prebiotics are non-viable substrates that serve as nutrients for beneficial microorganisms harboured by the host, including the administered probiotic strains and the resident microorganisms.[94] Probiotics are live microorganisms that, when administered in adequate amounts, confer a health benefit on the host.[95] The most common types of probiotics belong to the genera *Bifidobacterium*, *Lactobacillus*, and *Saccharomyces*, and the most common seven types of probiotics include *Bifidobacterium animalis* subsp. *animalis*, *Bifidobacterium animalis* subsp. *lactis*, *Bifidobacterium longum*, *Lactobacillus acidophilus*, *Lactobacillus reuteri*, *Lactobacillus casei*, and *Saccharomyces boulardii*.[96]

The gut microbiota has been implicated in immune cell activity and overall immunity based on probiotic and prebiotic supplementations. Probiotics and prebiotics play crucial roles in managing the intestinal microbiota in order to improve host health, even though their influence on other organs is not well studied.[97] Different studies on hosts and cell cultures demonstrated the implication of probiotics and prebiotics in the regulation of pro-inflammatory and anti-inflammatory responses and their effectiveness in controlling inflammation-associated diseases (ulcerative colitis, allergy, dermatitis, and cancer, among others).[98] There is a complex and intricate relationship between the gut, gut associated lymphoid tissue, immune cells, and non-immune cells.[97] This interface can involve but is not limited to cytokines, metabolite production, and microbiota interference. Thus, probiotics and prebiotics may initiate a network of signalling and cell–cell communication leading to systemic inflammatory response control *via* epigenetic and functional microbiome interventions.[98]

Probiotics modulate the immune response directly[99] by changing the microbial population of the gut microbiota[100] and/or by modifying the microbial metabolite levels[101] (Figure 11.3). Supplementation of *Bifidobacterium infantis* 35624 has been shown to modulate the activity of the dendritic cells, resulting in the increase of CD103(+) dendritic cells in the lamina propria. This mechanism, in turn, reduced the severity of dextran sulfate sodium-induced colitis.[102] On the other hand, probiotics have been shown to modulate stimulation of dendritic cells by modulating bacterial populations in the gut microbiota. A study demonstrated that administration of *Lactobacillus plantarum* modulated the population of segmented filamentous bacteria in the intestine of immunocompromised mice, which, in turn, changed the Th17 activation.[103]

Prebiotics are usually indigestible fibre components that pass through the upper gastrointestinal tract by stimulating the growth of probiotics,[104] for example, fructooligosaccharides (FOS), galactooligosaccharides (GOS), inulin, and xylooligosaccharides (XOS). Besides, other food components, such as proteins or peptides, unsaturated fatty acids, phenolic compounds, micronutrients, and macronutrients, have also been shown to stimulate the growth of probiotics.[104] A study conducted by Vulevic *et al.*[105] investigated the immune stimulating efficacy of prebiotics in a human study for 10 weeks by supplementing some elderly individuals daily with GOS. Results demonstrated that the NK cells were more proactive whereby their phagocytic activities increased.[105]

Arslanoglu *et al.*[106] reported that supplementing healthy term infants who are at atopy risk with prebiotic-supplemented hypoallergenic formula for 6 months demonstrated a significant reduction in the prevalence of allergies even after 5 years of the study period. In another similar study, infants fed with long-chain FOS and GOS formula demonstrated reduced wheezing, atopic dermatitis and urticaria conducted in a double-blind, randomized, placebo-controlled trial as compared to infants fed with non-prebiotic formula.[107] Finally, prebiotics have important function in the metabolic processes associated with immunomodulation. When prebiotics reaches the gut microbiota, fermentation of prebiotics releases short-chain fatty acids, such as acetate, propionate, and butyrate, which, in turn, enhances the integrity of the gut barrier that is associated with immunity.[108]

11.2.9 Others

Consumers have a growing awareness of the ability for yeast β-glucans to help immune support health. It is notable that not all β-glucans can provide the same immune support advantages. The strain, as well as the method of β-glucan extraction, can impact its benefits for the immune system.[25] De Macro Castro *et al.*[109] reviewed the role of β-glucans in immune health and highlighted their benefits for groups of healthy populations that have increased risk of respiratory infections. The review focused on their mechanism of action as immunomodulators, exploring the theory they "train" the body's immune cells, triggering changes that help fight off pathogens.

Besides yeast β-glucans, amino acids, mushrooms, and other substances (such as enzymes, organ tissues, and metabolites) have also been reported as dietary supplements marketed for immune health (Figure 11.1).[4]

11.3 Future Perspectives

The spread of misinformation is serious during the coronavirus pandemic on issues, such as the marketing of possible cures, treatments, and prevention strategies.[110] One of the most important strategies suggested to fight against the coronavirus infection is by stimulating the immune system in relation to numerous nutritional products and practices. Our immune system is complex, and how various nutrients affect the immune system remains not completely understood. We need to be cautious that overly active immune systems could lead to autoimmune diseases. A content analysis of

Google search results conducted by Rachul *et al.*[110] demonstrated the seriousness of conflictive messaging available on various online platforms which presents serious challenges for the general public seeking scientifically accurate information and advice during and after the COVID-19 pandemic.

A recent case series analysis reported by Crawford *et al.*[4] demonstrated that most dietary supplements related to immune boosting advertised and sold through online platforms had incorrect labels, and the claims made on those labels may mislead consumers into purchasing the incorrect products. In addition, some health claims made on most of these immune boosting products are inconsistent with any of the categories of claims defined by Food and Drug Administration regulations.[4] The results indicated that these products may potentially not contain what is stated on the label. Hence, the general public should be aware of this. In view of this, further validated research and evidence-based educational resources are needed to assist the general public in making the correct decisions when purchasing dietary supplements.

The nutritional compositions of many foods can be retrieved from nutrient databases, such as the websites of the United States Department of Agriculture and the nutrient databases of most countries. However, to the best of our knowledge, information on dietary supplements is still scarce. Hence, data on the nutritional compositions and health claims of dietary supplements should be documented in these national databases. These data serve as useful references for epidemiological studies to determine the association between the consumption of dietary supplements and human health. Besides, long-term prospective epidemiological studies should also be carried out on populations to determine the health outcomes after consuming these immune boosting dietary supplements. The scientific community is expecting more exciting results on dietary prenylated phenolics in the coming years.

11.4 Conclusion

A balanced diet containing diverse food bioactives, essential oils from medicinal herbs, macronutrients, micronutrients, and natural products, among others, may have supportive effects on immune health and COVID-19. Essential oils and extracts of some medicinal herbs and their bioactive compounds have been marketed as nutraceuticals/dietary supplements/herbal products for decades. Key dietary components, such as vitamins A, C, D, and E, as well as B-complex vitamins, certain minerals, and ω-3 PUFAs, have immune health effects. Therefore, these products play important roles to enhance the immune system due to their immunomodulatory properties and are safe to consume at recommended amounts. In addition, ω-3 PUFAs, propolis, prebiotics, and probiotics demonstrated excellent immunomodulatory properties, thus enhancing the immune system. Probiotics and prebiotics play crucial roles to improve host immunity by stimulating the gut microbiota, even though their influence on other organs is not completely understood.

Additional RCTs are needed to validate the detailed immunomodulatory effects of these products, supplements, and bioactives highlighted in this chapter.

References

1. O. Babich, S. Sukhikh, A. Prosekov, L. Asyakina and S. Ivanova, Medicinal plants to strengthen immunity during pandemic, *Pharmaceuticals*, 2020, **13**, 313.
2. K. W. Lange and Y. Nakamura, Food bioactives, micronutrients, immune function and COVID-19, *J. Food Bioact.*, 2020, **10**, 1–8.
3. J. S. Marshall, R. Warrington, W. Watson and H. L. Kim, An introduction to immunology and immunopathology, *Allergy, Asthma, Clin. Immunol.*, 2018, **14**(Suppl. 2), 49.
4. C. Crawford, B. Avula, A. T. Lindsey, A. Walter, K. Katragunta and I. A. Khan, *et al.*, Analysis of selected dietary supplement products marketed to support or boost the immune system, *JAMA Network Open*, 2022, **5**, e2226040.
5. E. Pelvan, Ö. Karaoğlu, E. Ö. Fırat, K. B. Kalyon, E. Ros and C. Alasalvar, Immunomodulatory effects of selected medicinal herbs and their essential oils: A comprehensive review, *J. Funt. Foods*, 2022, **94**, 105108.
6. F. Aman and S. Masood, How nutrient can help fight against COVID-19 pademic?, *Pak. J. Med. Sci.*, 2020, **36**, COVID19-S4.
7. H. Luo, Q. L. Tang, Y. X. Shang, S.-B. Liang, M. Yang and N. Robinson, *et al.*, Can Chinese medicine be used for prevention of Corona Virus Disease 2019 (COVID-19)? A review of historical classics, research evidence and current prevention programs, *Chin. J. Integr. Med.*, 2020, **26**, 243–250.
8. R. Arreola, S. Quintero-Fabián, R. I. López-Roa, E. O. Flores-Gutiérrez, J. P. Reyes-Grajeda and L. Carrera-Quintanar, Immunomodulatory and anti-inflammatory effects of garlic compounds, *J. Immunol. Res.*, 2015, **2015**, 401630.
9. A. A. Momtazi-Borojeni, S. M. Haftcheshmeh, S.-A. Esmaeili, T. P. Johnston, E. Abdollahi and A. Sahebkar, Curcumin: A natural modulator of immune cells in systemic lupus erythematosus, *Autoimmun. Rev.*, 2018, **17**, 125–135.
10. A. Jafarzadeh, S. Jafarzadeh and M. Nemati, Therapeutic potential of ginger against COVID-19: Is there enough evidence?, *J. Tradit. Chin. Med. Sci.*, 2021, **8**, 267–279.
11. A. A. Khalaf, S. Hussein, A. F. Tohamy, S. Marouf, H. D. Yassa and A. R. Zaki, *et al.*, Protective effect of *Echinacea purpurea* (Immulant) against cisplatin-induced immunotoxicity in rats, *DARU, J. Pharm. Sci.*, 2019, **27**, 233–241.
12. C. Burlou-Nagy, F. Bănică, T. Jurca, L. G. Vicas, E. Marian and M. E. Muresan, *Echinacea purpurea* (L.) Moench: Biological and pharmacological properties – A Review, *Plants*, 2022, **11**, 1244.

13. M. Catanzaro, E. Corsini, M. Rosini, M. Racchi and C. Lanni, Immunomodulators inspired by nature: A review on curcumin and echinacea, *Molecules*, 2018, **23**, 2778.
14. H. Kim and A. I. Calderón, Rational and safe use of the top two botanical dietary supplements to enhance the immune system, *Comb. Chem. High Throughput Screening*, 2022, **25**, 1129–1130.
15. Y. J. Ming Yang and Y. Li-Ping, A systematic summary of natural compounds in radix Glycyrrhizae, *Tradit. Med. Res.*, 2018, **3**, 82–94.
16. Y. K. Hong, H. T. Wu, T. Ma, W. J. Liu and X. J. He, Effects of *Glycyrrhiza glabra* polysaccharides on immune and antioxidant activities in high-fat mice, *Int. J. Biol. Macromol.*, 2009, **45**, 61–64.
17. S. L. Ng, K.-Y. Khaw, Y. S. Ong, H. P. Goh, N. Kifli and S. P. Teh, *et al.*, Licorice: A potential herb in overcoming SRAS-CoV-2 infections, *J. Evidence-Based Integr. Med.*, 2021, **26**, 1–8.
18. Ö. Karaoğlu, M. Serhatli, E. Pelvan, B. Karadeniz, İ. Demirtas and H. Sipahi, *et al.*, Chewable tablet with herbal extracts and propolis arrests Wuhan and Omicron variants of SARS-CoV-2 virus, *J. Funct. Foods*, 2023, accepted.
19. A. Ang, J. M. Pullar, M. J. Currie and M. C. M. Vissers, Vitamin C and immune cell function in inflammation and cancer, *Biochem. Soc. Trans.*, 2018, **46**, 1147–1159.
20. Z. Huang, Y. Liu, G. Qi, D. Brand and S. G. Zheng, Role of vitamin A in the immune system, *J. Clin. Med.*, 2018, 7, 258.
21. E. D. Lewis, S. N. Meydani and D. Wu, Regulatory role of vitamin E in the immune system and inflammation, *IUBMB Life*, 2019, **71**, 487–494.
22. J. R. Mora, M. Iwata and U. H. von Andrian, Vitamin effects on the immune system: vitamins A and D take centre stage, *Nat. Rev. Immunol.*, 2008, **8**, 685–698.
23. B. Prietl, G. Treiber, T. R. Pieber and K. Amrein, Vitamin D and immune function, *Nutrients*, 2013, **5**, 2502–2521.
24. E. Spinas, A. Saggini, S. K. Kritas, G. Cerulli, A. Caraffa and P. Antinolfi, Crosstalk between vitamin B and immunity, *J. Biol. Regul. Homeostatic Agents*, 2015, **29**, 283–288.
25. L. M. Ohr, Focus on immune health, *Food Technol.*, 2020, **74**, 65–69.
26. W. B. Grant, H. Lahore, S. L. McDonnell, C. A. Baggerly, C. B. French and J. L. Aliano, *et al.*, Evidence that vitamin D supplementation could reduce risk of inuenza and COVID-19 infections and deaths, *Nutrients*, 2020, **12**, 988.
27. K. Yoshii, K. Hosomi, K. Sawane and J. Kunisawa, Metabolism of dietary and microbial vitamin B family in the regulation of host immunity, *Front. Nutr.*, 2019, **6**, 48.
28. P. Bonaventura, G. Benedetti, F. Albarede and P. Miossec, Zinc and its role in immunity and inammation, *Autoimmun. Rev.*, 2014, **14**, 277–285.
29. C. S. Broome, F. McArdle, J. A. Kyle, F. Andrews, N. M. Lowe and C. A. Hart, *et al.*, An increase in selenium intake improves immune

function and poliovirus handling in adults with marginal selenium status, *Am. J. Clin. Nutr.*, 2004, **80**, 154–162.
30. J. E. Cassat and E. P. Skaar, Iron in infection and immunity, *Cell Host Microbe*, 2013, **13**, 509–519.
31. C. Muñoz, E. Rios, J. Olivos, O. Brunser and M. Olivares, Iron, copper and immunocompetence, *Br. J. Nutr.*, 2007, **98**(Suppl. 1), S24–S28.
32. M. Tam, S. Gómez, M. González-Gross and A. Marcos, Possible roles of magnesium on the immune system, *Eur. J. Clin. Nutr.*, 2003, **57**, 1193–1197.
33. M. J. Laires and C. Monteiro, Exercise, magnesium and immune function, *Magnesium Res.*, 2008, **21**, 92–96.
34. S. S. Percival, Copper and immunity, *Am. J. Clin. Nutr.*, 1998, **67**, 1064S–1068S.
35. G. C. F. Chan, K. W. Cheung and S. M. Y. Daniel, The immunomodulatory and anticancer properties of propolis, *Clin. Rev. Allergy Immunol.*, 2013, **44**, 262–273.
36. K. Wolska, A. Gorska, K. Antosik and K. Lugowska, Immunomodulatory effects of propolis and its components on basic immune cell functions, *Indian J. Pharm. Sci.*, 2019, **81**, 575–588.
37. E. Pelvan, M. Serhatlı, Ö. Karaoğlu, B. Karadeniz, C. P. Kodolbaş and N. A. Öncü, *et al.*, Development of propolis and essential oils containing oral/throat spray formulation against SARS-CoV-2 infection, *J. Funt. Foods*, 2022, **97**, 105225.
38. S. M. Osés, P. Marcos, P. Azofra, A. de Pablo, M. Á. Fernández-Muíño and M. T. Sancho, Phenolic profile, antioxidant capacities and enzymatic inhibitory activities of propolis from different geographical areas: Needs for analytical harmonisation, *Antioxidants*, 2020, **9**, 75.
39. D. A. A. K. Kalemba and A. Kunicka, Antibacterial and antifungal properties of essential oils, *Curr. Med. Chem.*, 2003, **10**, 813–829.
40. A. Peterfalvi, E. Miko, T. Nagy, B. Reger, D. Simon and A. Miseta, *et al.*, Much more than a pleasant scent: A review on essential oils supporting the immune system, *Molecules*, 2019, **24**, 4530.
41. A. F. C. Silva, P. I. Haris, M. L. Serralheiro and R. Pacheco, Mechanism of action and the biological activities of *Nigella sativa* oil components, *Food Biosci.*, 2020, **38**, 100783.
42. S. Bourgou, A. Pichette, B. Marzouk and J. Legault, Bioactivities of black cumin essential oil and its main terpenes from Tunisia, *S. Afr. J. Bot.*, 2010, **76**, 210–216.
43. R. Wang, R. Wang and B. Yang, Extraction of essential oils from five cinnamon leaves and identification of their volatile compound compositions, *Innovative Food Sci. Emerging Technol.*, 2009, **10**, 289–292.
44. M. Lang, P. J. Ferron, J. Bursztyka, A. Montjarret, E. Duteil and A. Bazire, Evaluation of immunomodulatory activities of essential oils by high content analysis, *J. Biotechnol.*, 2019, **303**, 65–71.
45. S. Chowdhury, G. P. Mandal, A. K. Patra, P. Kumar and I. Samanta, *et al.*, Different essential oils in diets of broiler chickens: 2. Gut

microbes and morphology, immune response, and some blood profile and antioxidant enzymes, *Anim. Feed Sci. Technol.*, 2018, **236**, 39–47.

46. G. Sandner, M. Heckmann and J. Weghuber, Immunomodulatory activities of selected essential oils, *Biomolecules*, 2020, **10**, 1139.
47. X. Han and T. L. Parker, Anti-inflammatory activity of clove (*Eugenia caryophyllata*) essential oil in human dermal fibroblasts, *Pharm. Biol.*, 2017, **55**, 1619–1622.
48. K. Chaieb, H. Hajlaoui, T. Zmantar, A. B. Kahla-Nakbi, M. Rouabhia and K. Mahdouani, *et al.*, The chemical composition and biological activity of clove essential oil, *Eugenia caryophyllata* (*Syzigium aromaticum* L. Myrtaceae): a short review, *Phytother. Res.*, 2007, **21**, 501–506.
49. A. Serafino, P. S. Vallebona, F. Andreola, M. Zonfrillo, L. Mercuri and M. Federici, *et al.*, Stimulatory effect of eucalyptus essential oil on innate cell-mediated immune response, *BMC Immunol.*, 2008, **9**, 17.
50. S. Tadtong, N. Kamkaen, R. Watthanachaiyingcharoen and N. Ruangrungsi, Chemical components of four essential oils in aromatherapy recipe, *Nat. Prod. Commun.*, 2015, **10**, 1934578X1501000673.
51. A. Mahmoodi, M. Soltani, A. Akhondzadeh Basti, L. Roomiani, A. Kamali and S. Taheri, Chemical composition and antibacterial activity of essential oils and extracts from *Rosmarinus officinalis*, *Zataria multiflora*, *Anethum graveolens* and *Eucalyptus globulus*, *Global Vet.*, 2012, **9**, 73–79.
52. B. R. Mikhaeil, G. T. Maatooq, F. A. Badria and M. M. A. Amer, Chemistry and immunomodulatory activity of frankincense oil, *Z. Naturforsch., C: J. Biosci.*, 2003, **58**, 230–238.
53. S. F. Van Vuuren, G. P. P. Kamatou and A. M. Viljoen, Volatile composition and antimicrobial activity of twenty commercial frankincense essential oil samples, *S. Afr. J. Bot.*, 2010, **76**, 686–691.
54. K. Pokajewicz, M. Białon, L. Svydenko, R. Fedin and N. Hudz, Chemical composition of the essential oil of the new cultivars of *Lavandula angustifolia* Mill. Bred in Ukraine, *Molecules*, 2021, **26**, 5681.
55. K. Wińska, W. Mączka, J. Łyczko, M. Grabarczyk, A. Czubaszek and A. Szumny, Essential oils as antimicrobial agents – myth or real alternative?, *Molecules*, 2019, **24**, 2130.
56. D. Giovannini, A. Gismondi, A. Basso, L. Canuti, R. Braglia and A. Canini, *et al.*, *Lavandula angustifolia* Mill. essential oil exerts antibacterial and anti-inflammatory effect in macrophage mediated immune response to *Staphylococcus aureus*, *Immunol. Invest.*, 2016, **45**, 11–28.
57. N. S. Dosoky and W. N. Setzer, Biological activities and safety of Citrus spp. essential oils, *Int. J. Mol. Sci.*, 2018, **19**, 1966.
58. N. R. F. Nascimento, R. M. D. C. Refosco, E. C. F. Vasconcelos, M. R. Kerntopf, C. F. Santos and F. J. A. Batista, *et al.*, 1,8-Cineole induces relaxation in rat and guinea-pig airway smooth muscle, *J. Phar. Pharmacol.*, 2009, **61**, 361–366.

59. H. Rahimi, M. Nakhaei, N. Mehrpooya, S. M. Hatami and S. A. Vagharseyyedin, The effect of inhaling the aroma of rosemary essential oil on the pre-hospital emergency personnel stress and anxiety: A Quasi-Experimental Study, *Mod. Care J.*, 2019, **16**, e95082.
60. A. Lombrea, D. Antal, F. Ardelean, S. Avram, I. Z. Pavel and L. Vlaia, *et al.*, A recent insight regarding the phytochemistry and bioactivity of *Origanum vulgare* L. essential oil, *Int. J. Mol. Sci.*, 2020, **21**, 9653.
61. C. Ariza-Nieto, M. Bandrick, S. K. Baidoo, L. Anil, T. W. Molitor and M. R. Hathaway, Effect of dietary supplementation of oregano essential oils to sows on colostrum and milk composition, growth pattern and immune status of suckling pigs, *J. Anim. Sci.*, 2011, **89**, 1079–1089.
62. N. Khodambashi Emami, A. Samie, H. R. Rahmani and C. A. Ruiz-Feria, The effect of peppermint essential oil and fructooligosaccharides, as alternatives to virginiamycin, on growth performance, digestibility, gut morphology and immune response of male broilers, *Anim. Feed Sci. Technol.*, 2012, **175**, 57–64.
63. M. N. M. Nogueira, S. G. Aquino, C. R. Junior and D. M. P. Spolidório, Terpinen-4-ol and alpha-terpineol (tea tree oil components) inhibit the production of IL-1β, IL-6 and IL-10 on human macrophages, *Inflammation Res.*, 2014, **63**, 769–778.
64. F. C. Fachini-Queiroz, R. Kummer, C. F. Estevao-Silva, M. D. B. Carvalho, J. M. Cunha and R. Grespan, *et al.*, Effects of thymol and carvacrol, constituents of *Thymus vulgaris* L. essential oil on the inflammatory response, *J. Evidence-Based Complementary Altern. Med.*, 2012, 657026.
65. S. J. Lee, K. Umano, T. Shibamoto and K. G. Lee, Identification of volatile components in basil (*Ocimum basilicum* L.) and thyme leaves (*Thymus vulgaris* L.) and their antioxidant properties, *Food Chem.*, 2005, **91**, 131–137.
66. S. Sardari, A. Mobaien, L. Ghassemifard, K. Kamali and N. Khavasi, Therapeutic effect of thyme (*Thymus vulgaris*) essential oil on patients with COVID19: A randomized clinical trial, *J. Adv. Med. Biomed. Res.*, 2021, **29**, 83–91.
67. B. P. Chew and J. S. Park, Carotenoid action on the immune response, *J. Nutr.*, 2004, **134**, 257S–261S.
68. J. González-Gallego, M. V. García-Mediavilla, S. Sánchez-Campos and M. J. Tuñón, Fruit polyphenols, immunity and inflammation, *Br. J. Nutr.*, 2010, **104**(Suppl. 3), S15–S27.
69. S. Gutiérrez, S. L. Svahn and M. E. Johansson, Effects of omega-3 fatty acids on immune cells, *Int. J. Mol. Sci.*, 2019, **20**, 5028.
70. L. Malaguarnera, Inuence of resveratrol on the immune response, *Nutrients*, 2019, **11**, 946.
71. Y. Li, J. Yao, C. Han, J. Yang, M. T. Chaudhry and S. Wang, *et al.*, Quercetin, inflammation and immunity, *Nutrients*, 2016, **8**, 167.
72. T. Walle, F. Hsieh, M. H. DeLegge, J. E. Oatis Jr and U. K. Walle, High absorption but very low bioavailability of oral resveratrol in humans, *Drug Metab. Dispos.*, 2004, **32**, 1377–1382.

73. B. Cerdá, F. A. Tomás-Barberán and J. C. Espín, Metabolism of antioxidant and chemopreventive ellagitannins from strawberries, raspberries, walnuts, and oak-aged wine in humans: identification of biomarkers and individual variability, *J. Agric. Food Chem.*, 2005, **53**, 227–235.
74. P. Chen, Z. Guo, F. Chen, Y. Wu and B. Zhou, Recent advances and perspectives on the health benefits of urolithin B, A bioactive natural product derived from ellagitannins, *Front. Pharmacol.*, 2022, **13**, 917266.
75. M. Cholewski, M. Tomczykowa and M. A. Tomczyk, Comprehensive review of chemistry, sources and bioavailability of omega-3 fatty acids, *Nutrients*, 2018, **10**, 1662.
76. F. Shahidi and S. Santhiravel, Novel marine bioactives: Application in functional foods, nutraceuticals, and pharmaceuticals, *J. Food Bioact.*, 2022, **19**, 4–96.
77. E. A. Miles, C. E. Childs and P. C. Calder, Long-chain polyunsaturated fatty acids (LCPUFAs) and the developing immune system: a narrative review, *Nutrients*, 2021, **13**, 247.
78. L. S. Sorensen, O. Thorlacius-Ussing, H. H. Rasmussen, S. Lundbye-Christensen, P. C. Calder and K. Lindorff-Larsen, *et al.*, Effects of perioperative supplementation with omega-3 fatty acids on leukotriene B(4) and leukotriene B(5) production by stimulated neutrophils in patients with colorectal cancer: A randomized, placebo-controlled intervention trial, *Nutrients*, 2014, **6**, 4043–4057.
79. E. A. Gurzell, H. Teague, M. Harris, J. Clinthorne, S. R. Shaikh and J. I. Fenton, DHA-enriched fish oil targets B cell lipid microdomains and enhances *ex vivo* and *in vivo* B cell function, *J. Leukocyte Biol.*, 2013, **93**, 463–470.
80. M. Hashimoto and S. Hossain, Fatty acids: from membrane ingredients to signaling molecules, in *Biochemistry and Health Benefits of Fatty Acids*, ed. V. Waisundara, IntechOpen Limited, London, UK, 2018, pp. 1–20.
81. M. O. Husson, D. Ley, C. Portal, M. Gottrand, T. Hueso and J. L. Desseyn, *et al.*, Modulation of host defence against bacterial and viral infections by omega-3 polyunsaturated fatty acids, *J. Infect.*, 2016, **73**, 523–535.
82. C. M. Yates, P. C. Calder and G. Ed Rainger, Pharmacology and therapeutics of omega-3 polyunsaturated fatty acids in chronic inflammatory disease, *Pharmacol. Ther.*, 2014, **141**, 272–282.
83. E. Cho and Y. Park, Association between serum fatty acid composition and innate immune markers in healthy adults, *Nutr. Res. Pract.*, 2016, **10**, 182–187.
84. J. P. Arm, J. A. Boyce, L. Wang, H. Chhay, M. Zahid and V. Patil, *et al.*, Impact of botanical oils on polyunsaturated fatty acid metabolism and leukotriene generation in mild asthmatics, *Lipids Health Dis.*, 2013, **12**, 141.

85. H. Teague, C. J. Fhaner, M. Harris, D. M. Duriancik, G. E. Reid and S. R. Shaikh, n-3 PUFAs enhance the frequency of murine B-cell subsets and restore the impairment of antibody production to a T-independent antigen in obesity, *J. Lipid Res.*, 2013, **54**, 3130–3138.
86. S. Ramon, F. Gao, C. N. Serhan and R. P. Phipps, Specialized pro-resolving mediators enhance human B cell differentiation to antibody-secreting cells, *J. Immunol.*, 2012, **189**, 1036–1042.
87. H. Teague, M. Harris, J. Fenton, P. Lallemand, B. M. Shewchuk and S. R. Shaikh, Eicosapentaenoic and docosahexaenoic acid ethyl esters differentially enhance B-cell activity in murine obesity, *J. Lipid Res.*, 2014, **55**, 1420–1433.
88. C. Weise, K. Hilt, M. Milovanovic, D. Ernst, R. Ruhl and M. Worm, Inhibition of IgE production by docosahexaenoic acid is mediated by direct interference with STAT6 and NFkappaB pathway in human B cells, *J. Nutr. Biochem.*, 2011, **22**, 269–275.
89. C. Furuhjelm, K. Warstedt, M. Fagerås, K. Fälth-Magnusson, J. Larsson and M. Fredriksson, *et al.*, Allergic disease in infants up to 2 years of age in relation to plasma omega-3 fatty acids and maternal fish oil supplementation in pregnancy and lactation, *Pediatr. Allergy Immunol.*, 2011, **22**, 505–514.
90. C. Furuhjelm, K. Warstedt, J. Larsson, M. Fredriksson, M. F. Böttcher and K. Fälth-Magnusson, *et al.*, Fish oil supplementation in pregnancy and lactation may decrease the risk of infant allergy, *Acta Paediatr.*, 2009, **98**, 1461–1467.
91. H. Shakoor, J. Feehan, A. S. Al Dhaheri, H. I. Ali, C. Platat and L. C. Ismail, *et al.*, Immune-boosting role of vitamins D, C, E, zinc, selenium and omega-3 fatty acids: Could they help against COVID-19?, *Maturitas*, 2021, **143**, 1–9.
92. P. B. Eckburg, E. M. Bik, C. N. Bernstein, E. Purdom, L. Dethlefsen and M. Sargent, *et al.*, Diversity of the human intestinal microbial flora, *Science*, 2005, **308**, 1635–1638.
93. C. Gubert, G. Kong, T. Renoir and A. J. Hannan, Exercise, diet and stress as modulators of gut microbiota: Implications for neurodegenerative diseases, *Neurobiol. Dis.*, 2020, **134**, 104621.
94. G. R. Gibson, R. Hutkins, M. E. Sanders, S. L. Prescott, R. A. Reimer and S. J. Salminen, *et al.*, Expert consensus document: The International Scientific Association for Probiotics and Prebiotics (ISAPP) consensus statement on the definition and scope of prebiotics, *Nat. Rev. Gastroenterol. Hepatol.*, 2017, **14**, 491–502.
95. C. Hill, F. Guarner, G. Reid, G. R. Gibson, D. J. Merenstein and B. Pot, *et al.*, Expert consensus document: The International Scientific Association for Probiotics and Prebiotics consensus statement on the scope and appropriate use of the term probiotic, *Nat. Rev. Gastroenterol. Hepatol.*, 2014, **11**, 506–514.
96. J. M. Khaled, Probiotics, prebiotics, and COVID-19 infection: A review article, *Saudi J. Biol. Sci.*, 2021, **28**, 865–869.

97. T. J. Ashaolu, Immune boosting functional foods and their mechanisms: A critical evaluation of probiotics and prebiotics, *Biomed. Pharmacother.*, 2020, **130**, 110625.
98. N. Yahfoufi, J. F. Mallet, E. Graham and C. Matar, Role of probiotics and prebiotics in immunomodulation, *Curr. Opin. Food Sci.*, 2018, **20**, 82–91.
99. D. M. Martirosyan and J. Singh, A new definition of functional food by FFC: what makes a new definition unique?, *Funct. Foods Health Dis.*, 2015, **5**, 209–223.
100. M. K. Tripathi and S. K. Giri, Probiotic functional foods: survival of probiotics during processing and storage, *J. Funct. Foods*, 2014, **9**, 225–241.
101. S. H. Al-Sheraji, A. Ismail, M. Y. Manap, S. Mustafa, R. M. Yusof and F. A. Hassan, Prebiotics as functional foods: a review, *J. Funct. Foods*, 2013, **5**, 1542–1553.
102. P. Konieczna, R. Ferstl, M. Ziegler, R. Frei, D. Nehrbass and R. P. Lauener, *et al.*, Immunomodulation by *Bifidobacterium infantis* 35624 in the murine lamina propria requires retinoic acid-dependent and independent mechanisms, *PLoS One*, 2013, **8**, 1–21.
103. Y. Goto, C. Panea, G. Nakato, A. Cebula, C. Lee and M. G. Diez, *et al.*, Segmented filamentous bacteria antigens presented by intestinal dendritic cells drive mucosal Th17 cell differentiation, *Immunity*, 2014, **40**, 594–607.
104. I. A. Neri-Numa and G. M. Pastore, Novel insights into prebiotic properties on human health: a review, *Food Res. Int.*, 2020, **131**, 108973.
105. J. Vulevic, A. Juric, G. E. Walton, S. P. Claus, G. Tzortzis and R. E. Toward, *et al.*, Influence of galacto-oligosaccharide mixture (B-GOS) on gut microbiota, immune parameters and metabonomics in elderly persons, *Br. J. Nutr.*, 2015, **114**, 586–595.
106. S. Arslanoglu, G. E. Moro, G. Boehm, F. Wienz, B. Stahl and E. Bertino, Early neutral prebiotic oligosaccharide supplementation reduces the incidence of some allergic manifestations in the first 5 years of life, *J. Biol. Regul. Homeostatic Agents*, 2012, **26**, 49–59.
107. O. S. Ivakhnenko and S. L. Nyankovskyy, Effect of the specific infant formula mixture of oligosaccharides on local immunity and development of allergic and infectious disease in young children: randomized study, *Pediatr. Pol.*, 2013, **88**, 398–404.
108. J. A. Krumbeck, H. E. Rasmussen, R. W. Hutkins, J. Clarke, K. Shawron and A. Keshavarzian, Probiotic Bifidobacterium strains and galacto-oligosaccharides improve intestinal barrier function in obese adults but show no synergism when used together as synbiotics, *Microbiome*, 2018, **6**, 121.
109. E. De Marco Castro, P. C. Calder and H. M. Roche, β-1,3/1,6-glucans and immunity: state of the art and future directions, *Mol. Nutr. Food Res.*, 2020, **65**, 1901071.
110. C. Rachul, A. R. Marcon, B. Collins and T. Caulfield, COVID-19 and 'immune boosting' on the internet: a content analysis of Google search results, *BMJ Open*, 2020, **10**, e040989.

CHAPTER 12

Potential of Naturally-occurring Compounds for the Development of Dietary Supplements with Antiviral Activity

FEREIDOON SHAHIDI* AND RENAN DANIELSKI

Department of Biochemistry, Memorial University of Newfoundland, St. John's, NL, Canada
*Email: fshahidi@mun.ca

12.1 Introduction

The coronavirus disease (COVID-19) pandemic caused by the severe acute respiratory syndrome coronavirus 2 (SARS-CoV-2) virus has reminded the global population of the catastrophic and long-lasting consequences that can stem from widespread viral infections. In order to curb the transmission of the virus and prevent the collapse of the health care system, lockdowns were imposed in all parts of the world, causing a major strain on the global economy and the everyday life of citizens. The psychological long-term impact of the COVID-19 pandemic still needs to be assessed, while the total number of deaths accounts for over six million as of August 2022.[1] With the availability of COVID-19 vaccines, the level of protection against more severe

Food Chemistry, Function and Analysis No. 39
Dietary Supplements with Antioxidant Activity: Understanding Mechanisms and Potential Health Benefits
Edited by Cesarettin Alasalvar, Fereidoon Shahidi and Chi-Tang Ho

Published by the Royal Society of Chemistry, www.rsc.org

forms of the disease has significantly increased, allowing the economy to gradually reopen in many parts of the world. Nevertheless, vaccine supply is still far from ideal and treatment options are scarce.

As seen with the coronavirus pandemic, virus outbreaks are common from time to time with varying degrees of severity. This is potentialized by various aspects of modern life, such as migration, global travel, and urbanization, facilitating the spread of viruses among the population and leading to major public health crises. As an aggravating factor, vaccines are not always an option to prevent viral infections, such as the case of the human immunodeficiency virus (HIV), and even when they are, viral mutation and the emergence of new strains may significantly reduce vaccine efficacy.[2]

Therefore, the procurement of antiviral treatments has also been a priority. Viruses hijack the cell mechanisms of the host in order to survive and replicate, which may prompt disease. The main challenge of this course of treatment consists in finding drugs capable of targeting and destroying the virus, while causing no harm to the host. Antiviral drugs generally act on viral enzymes related to the acquisition of the virus shape or replication cycle. However, this type of medicine presents considerable drawbacks, including various side effects and the possibility of emerging drug-resistant strains.[3]

In order to overcome the hurdles associated with traditional antiviral drugs, research has been focused on finding natural compounds with antiviral activity that could serve as an alternative treatment option. Naturally occurring molecules, such as alkaloids, polyphenols, and terpenoids, have shown antiviral potential and represent the target molecules serving as the basis for the future development of dietary supplements focused on treating viral infections. As such, this chapter discusses aspects involving the underlying mechanisms of viral infections, as well as recent scientific data on promising naturally-occurring molecules with antiviral potential.

12.2 Viral Infections and Pathogenesis

Viruses are submicroscopic (20–300 nm) agents that use the reproductive system of living cells in order to replicate. Their genome (RNA or DNA) can be enveloped by proteins (capsids) or lipids, although non-enveloped viruses also exist. Viral infections can lead to disease, which will be determined by the ability of a virus to thrive. In most cases, viruses are arrested by the antibodies before the manifestation of symptoms. This process is known as subclinical infection, and it is essential for conferring immunity to the population. Nevertheless, if the virus is able to surpass the inhibitory effects imposed by physical barriers, host defences, and differing cellular susceptibilities to infection, disease will occur. In addition, within the virus population, highly virulent strains may become dominant as a consequence of specific selection pressures. However, in the majority of cases, less virulent and mild strains are predominant since severe disease or even the death of the host would impair the replication and transmission of the virus.[3] The main factors defining the occurrence of viral pathogenesis are as follows:

(a) virus access to the tissue,
(b) susceptibility of cells to virus multiplication, and
(c) vulnerability of the virus to host defences.

The pathogenic mechanism of viruses involves the following steps:[4]

(a) Virus entry: viruses can attach to cell-surface receptors (proteins, lipids, and carbohydrates) and fuse with (enveloped viruses) or penetrate through (non-enveloped viruses) cellular membranes. Complete virus particles (virions) can enter the host cells through multiple routes (*e.g.*, respiratory, skin, genital, and gastrointestinal) and deliver their genome to the cell cytoplasm.
(b) Spread from the portal of entry: the virus can reach the systemic circulation and enter the target organs. Alternatively, it can disseminate *via* the nervous system, such as rabies and herpes viruses.
(c) Incubation: during this period, the host is still asymptomatic. Meanwhile, local multiplication of the virus occurs, with subsequent spreading.
(d) Virus shedding: viruses are mainly shed through the respiratory, alimentary, and urogenital tracts, as well as blood, subsequently spreading in the environment.

As viruses cannot synthesize their own nucleic acids and proteins, they hijack the host cells for this purpose. Therefore, in a situation where the body defences are not able to cope with the viral infection, the energy and macromolecules of the cells are impaired, leading to cell death and disease. The mechanisms of cellular pathogenesis subvert the normal functions of living cells, placing a higher priority toward the synthetic needs of the virus.[2,3]

Factors such as cell susceptibility, physical barriers, pH, temperature, and level of host defences can also determine the affinity of viruses for specific tissue, which is known as tropism. Some examples include polioviruses, which display selectivity toward nerve cells; rhinoviruses, specific for the upper respiratory tract since they are mostly adapted to low temperature and pH, as well as high oxygen tension; and enteroviruses, which predominantly infect the intestine due to their resistance to digestive enzymes' inactivation, bile, and acid.[2,3]

12.2.1 Herpes Simplex Virus (HSV)

The HSV belongs to the *Herpesviridae* family, being HSV-1 and HSV-2 their most common members. Oral and facial infections, as well as encephalitis, are caused by HSV-1, while HSV-2 usually causes genital infection.

The significance of HSV lies in the fact that this is a widespread virus present in approximately 80% of the world population. Within infected people, around 40% will develop at least one infection in their lifetime. Close contact with infected people is the main transmission route, and the infection is usually asymptomatic. Moreover, HSV infection also increases the risk of HIV infection.[5]

HSV is classified as a non-eradicable virus, meaning that it enters the nerve cells causing lesions. Some factors may trigger its reactivation and recurrence, namely fever, stress, and immune suppression. With no vaccines available, the administration of drugs is the main course of treatment for the herpes virus. These include virucidal agents, immunomodulators, and chemotherapeutic agents. Nevertheless, side effects and the emergence of drug-resistant strains are associated with their continuous use, as it is also common for other antiviral drugs.[2,5]

12.2.2 Human Immunodeficiency Virus (HIV)

According to the World Health Organization (WHO),[6] there are 75 million people living with HIV in the world. This single-stranded enveloped RNA virus is included in the *Retroviridae* family and is characterized by the ability to merge a DNA copy of its genome with the DNA of the host cells. After entering the cell, the RNA genome of HIV is reversely transcribed into double-stranded DNA by the viral enzyme reverse transcriptase. Then, another viral enzyme, integrase, is responsible for incorporating the viral DNA into the cellular DNA. After this process, the virus becomes latent, and the immune system of the host is not capable of detecting it. The incubation period of HIV can be up to 10 years following primary infection, and during this time symptoms are not manifested by the host.[7]

The transmission occurs through the exchange of bodily fluids, such as blood, semen, rectal and vaginal fluids, and breast milk. Contaminated blood transfusion, hypodermic needles, mother-to-fetus contamination, and breastfeeding are also among the common routes of transmission. If untreated, the infection can lead to acquired immune deficiency syndrome (AIDS), where a destruction of the T cells over time causes an increased risk of opportunistic diseases caused by bacteria and viruses, as well as cancer. Notwithstanding, since the year 2000, the mortality rate of HIV-positive patients has had a 50% decline. Although effective HIV vaccines are not available, drug treatment has evolved throughout the years, significantly improving the quality of life of seropositive subjects.[2,7]

Antiretroviral therapy (ART) is the standard treatment against HIV, which uses a combination of drugs with multiple targets, such as reverse transcriptase, protease, and integrase. ART is extremely effective in preventing AIDS and keeping the virus load undetectable. However, considerable side effects are associated with ART, such as allergic reactions, bleeding, bone loss, high blood sugar, and acidosis, among others. Besides, multidrug resistance may also occur. Therefore, natural products with anti-HIV properties have been procured.[7]

12.2.3 Influenza

The influenza virus belongs to the *Orthomyxoviridae* family and it is characterized as a single-stranded RNA virus presenting a fragmented genome enveloped in capsids that can be transmitted to humans and birds. Cold and dry climates favour the spread of this seasonal virus. Different types of influenza exist and the severity of their infection varies according to the strain and the immunological vulnerability of each patient. In terms of the infection rate, type A influenza overcomes type B. Besides, avian influenza is highly virulent to humans and its infection presents a high mortality rate.[8]

The influenza virus usually sheds after one day of infection, and its pathology includes fever, muscle pain, headache, and tiredness, with a peak of these symptoms around two to three days after the infection. Mild infections are typically resolved within 1–2 weeks. Risk groups that can manifest more severe forms of the disease include infants, seniors, pregnant women, and immunocompromised people. Vaccines can be administered in order to prevent acute disease. However, the emergence of new strains represents a challenge for vaccine effectiveness, requiring constant updates.[8]

The influenza virus is highly mutable, which may cause the evasion of immunity and generate pandemic strains. The 1918 Spanish flu pandemic was caused by the influenza virus (H1N1 strain), leaving an estimated death toll of 17 to 50 million people worldwide.[9] On that account, novel compounds and dietary supplements are procured in order to curb the influenza virus. Several bioactive classes have shown antiviral activity against influenza, including flavonols (*e.g.*, catechin and luteolin), other flavonoids, proanthocyanidins, and phenolic acids (*e.g.*, caffeic and ellagic acids), among others.

12.2.4 Hepatitis C

The hepatitis C virus is an enveloped single-stranded RNA virus, belonging to the *Flaviviridae* family. Approximately 185 million people have been infected with this virus, with a mortality rate around 33%. Acute liver inflammation is the main characteristic of hepatitis C pathology, with the possibility of the inflammation becoming chronic. This factor increases the risk of the development of cirrhosis and hepatocellular carcinoma. The major routes of transmission are through contaminated syringes (mostly among drug users) and infected medical equipment.[10]

With no vaccines available as a prevention tool, hepatitis C treatment usually involves the co-administration of ribavirin and pegylated interferon-α for 24–48 weeks. However, the treatment is not exempted from side effects, with teratogens (substances that can lead to fetal malformation) representing the most concerning one. Additionally, drug resistance is also associated with hepatitis C treatment.[10] As an alternative to the traditional treatment, flavonoids and phenolic acids have been studied for their inhibitory activity against hepatitis C virus.[2]

12.2.5 SARS-CoV-2

The first cases of transmission of the SARS-CoV-2 virus were recorded in Wuhan, Hubei Province, China in December 2019, being quickly disseminated to the rest of the world and culminating on the WHO declaring COVID-19, the disease caused by SARS-CoV-2, a global pandemic on March 11, 2020. More than two years later, the pandemic is still ongoing.[11]

Coronaviruses are positive-stranded RNA viruses. Their envelope formed by spike glycoproteins gives them a crown appearance when observed under a microscope. SARS-CoV-2 belongs to the same subgenus as the SARS-CoV and the Middle East respiratory syndrome coronavirus, which have caused epidemics in the past decades.[12] Most people infected by SARS-CoV-2 will develop symptoms within 11.5 days after infection, and the clinical manifestation of COVID-19 can include fever, cough, shortness of breath, sore throat, and headache. While it is estimated that 17.9% to 33.3% of infected people do not develop symptoms, the most severe forms of the disease can lead to acute respiratory failure, septic shock, and/or multiple organ dysfunction.[11] COVID-19 vaccines have been developed and more than 60% of the world population has been immunized. In addition, less severe strains of the virus have become dominant, decreasing the burden on the health care system seen in the early months of the pandemic. Nevertheless, seniors and immunocompromised people are still at high risk of hospitalization and death from COVID-19 even when fully vaccinated. Therefore, the quest for effective treatment against the disease continues. Curcumin, other polyphenols, essential oils (terpenoids), and alkaloids have been some of the active phytoconstituents studied for the formulation of dietary supplements to treat COVID-19.

12.3 Natural Compounds for the Development of Antiviral Dietary Supplements

Polyphenols, alkaloids, and terpenoids are some of the most studied naturally-occurring bioactive molecules targeting the inhibition of viral infections, showing promising outcomes for the future development of antiviral dietary supplements. A compilation of some of these studies is shown in Table 12.1.

12.3.1 Phenolic Compounds

Phenolic compounds are a large group of plant phytochemicals with over 10 000 compounds identified to this day. They are produced by plants as a

Table 12.1 Antiviral activity of bioactive compounds found in natural sources.

Antiviral molecule	Virus	Dose	Mechanism of action	Type of study	Ref.
Phenolic compounds					
Vitis vinifera leaf extract (rich in quercetin derivatives, luteolin, kaempferol, apigenin, isorhamnetin, myricetin, chrysoeriol, biochanin, isookanin, and scutellarein)	HSV-1 and SARS-CoV-2	10 $\mu g\,mL^{-1}$	Inhibition of virus replication	Cell line model (Vero, HCT-116, Hacat, HepG2, and MCF7 cell lines)	13
Moquiniastrum floribundum extract (rich in flavones and chlorogenic acid derivatives)	HIV	0.24 $mmol\,L^{-1}$	Inhibition of HIV-1 reverse transcriptase	*In vitro*	14
Hydrolysable tannins (rich in punicalagin, punicalin, and geraniin)	Hepatitis B	1–30 μM	Prevention of covalently closed circular DNA formation	*In vitro*	15
Resveratrol derivatives	Influenza virus (H1N1)	3.56–186 μM	Blocking of virus adsorption to the host cell Progeny virus release inhibition	*In vitro*	16
Alkaloids					
Debromohymenialdisine Hymenialdisine	HIV-1	3.1 μM 13 μM	Inhibition of virus replication	*In vitro*	17
Isoquinoline alkaloids	Hepatitis B	0.1–100 000 nM	Not determined	Cell line model (human hepatoma HepG2.2.15)	18
Alkaloid-rich *Peganum harmala* seed extract	Influenza A	10–60 $\mu g\,mL^{-1}$	Inhibition of viral RNA replication and polymerase activity	Cell line model (MDCK cells)	19
Terpenoids					
β-pinene Limonene	Herpes simplex virus	1–50 $\mu g\,mL^{-1}$	Inhibition of viral multiplication	Cell line model (RC-37 African green monkey kidney cells)	20
Damsin Canrenone	HIV-1	0.0001–10 000 $\mu g\,mL^{-1}$	Inhibition of HIV-1 transcription	Cell line model (MT-2 cells)	21

Abbreviations: HSV, herpes simplex virus; HIV, human immunodeficiency virus; SARS-CoV-2, severe acute respiratory syndrome coronavirus 2.

defense mechanism against biotic and abiotic stresses, being synthesized from phenylalanine, and to a lesser extent tyrosine. The basic structure of a phenolic compound comprises at least one phenoxyl ring to which one (monophenols) or more (polyphenols) hydroxyl groups are attached. Structural differences based on the degrees of methylation, hydroxylation, and polymerization serve to categorize phenolic compounds into several subgroups, namely flavonoids (15-carbon skeleton consisting of two phenoxyl rings and a heterocyclic ring), phenolic acids (hydroxycinnamic and hydroxybenzoic acid derivatives), lignans (backbone formed by two phenylpropanoid units), stilbenes (two phenoxyl rings connected by a 2-carbon methylene bridge), proanthocyanidins (polymers of catechin and epicatechin), and hydrolysable tannins (esters of gallic and ellagic acids). As the subclass with the highest number of compounds, flavonoids can be further divided into flavonols, flavan-3-ols, flavones, isoflavones, flavanones, and anthocyanidins.[22]

Although phenolic compounds are generally associated with antioxidant capacity, displayed by their ability to scavenge free radicals and chelate transition metals (phenols bearing a galloyl moiety),[22] they have also demonstrated a myriad of other biological activities, including the antiviral effect. Polyphenols, particularly tannins (proanthocyanidins and hydrolysable tannins), are able to form hydrogen and ionic bonds with the proteins composing the viral envelope due to their multiple hydroxyl groups. This reduces protein activity, hindering the interaction between the virus and the host cells. Nevertheless, phenolic compounds also show antiviral activity against non-enveloped viruses, such as noroviruses, although their mechanism of action is still not entirely understood.[23] The chemical structures of several phenolic compounds with antiviral activity are presented in Figure 12.1.

Phenolic compounds not only act on the attachment of viruses to the host cell, but they have also demonstrated inhibition of reverse transcriptase. Resveratrol, a well-known stilbene found in grapes and wine, could suppress the replication of HIV molecular clone NL4-3 bearing the mutant M184V reverse transcriptase at a concentration of 5 μM. This effect was owed to the hindering of DNA synthesis during the reverse transcription step of HIV.[24]

Özçelik *et al.*[25] studied the antiviral activity of flavonoids (quercetin, apigenin, genistein, naringin, silymarin, and silibinin) and phenolic acids (gallic, caffeic, chlorogenic, and quinic acids) against HSV-1 using Madin–Darby bovine kidney cell lines. Flavonoids have shown significant inhibitory efficiency against the herpes virus, with apigenin and naringin providing the best anti-HSV-1 effect at concentrations varying from 0.4 to 1.6 $\mu g\, mL^{-1}$. The same has been observed for phenolic acids, especially gallic and quinic acids, which are capable of inhibiting HSV-1 at doses as low as 0.05 $\mu g\, mL^{-1}$. Such results are comparable to the effect displayed by acyclovir, an antiviral medication used in herpes treatment.

In another cell line study,[26] the antiviral activity against HSV-1 and HSV-2 of methanolic and acetone phenolic-rich extracts from apple pomace was tested in Vero cells. At a concentration of 1000 $\mu g\, mL^{-1}$, the acetone extracts completely prevented the replication of HSV-1 and HSV-2 in infected cells.

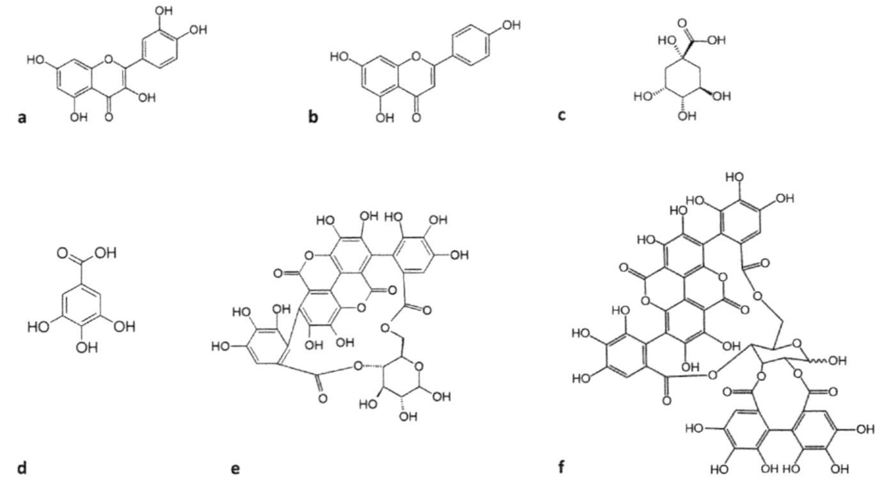

Figure 12.1 Chemical structures of polyphenols with antiviral activity: (a) quercetin, (b) apigenin, (c) quinic acid, (d) gallic acid, (e) punicalin, and (f) punicalagin.

The same effect was observed for the methanolic extract at a 1200 $\mu g\,mL^{-1}$ dose. Neither concentration displayed cytotoxicity, presenting more than 80% cell viability. According to the authors, the antiviral activity of apple pomace extracts could be due to the presence of flavonols at high amounts, such as hyperin, reynoutrin, and quercitrin, which are associated with the inhibition of herpes virus.

Sanna *et al.*[27] extracted polyphenols from different parts of pomegranate and tested the capacity of individual phenolic compounds present in the extracts to inhibit HIV enzymes. Extracts from the leaves were rich in ellagic acid and flavones, while bark and peel extracts showed a high abundance of hydrolysable tannins (punicalins and punicalagins) as well as ellagic acid. The hydrolysable tannins and ellagic acid exhibited high inhibitory power against HIV-1 reverse transcriptase-associated ribonuclease H (RNase H) and integrase, with half-maximal inhibitory concentrations (IC_{50}) ranging from 0.12 to 1.4 μM and 0.065 to 0.09 μM, respectively. Luteolin and apigenin were less efficient, with IC_{50} varying between 3.7 and 22 μM for both enzymes. RNase H was the first successful enzymatic target for HIV therapy, and drugs that inhibit this enzyme are currently in use.

A study conducted by Song *et al.*[28] investigated the effect of green tea catechins, namely epigallocatechin gallate (EGCG), epicatechin gallate (ECG), and epigallocatechin (EGC), on the inhibition of influenza virus replication. The study was conducted on Madin–Darby bovine kidney cell culture and demonstrated that EGCG and ECG exerted inhibitory activity toward A/H1N1, A/H3N2, and B strains. Moreover, high doses of these catechins were also able to suppress viral RNA synthesis. According to the authors, the 3-galloyl moiety of the catechin skeleton plays a crucial role in the antiviral activity reported.

The potential of complex polyphenols for the treatment of COVID-19 has also been studied. In a recent study, Lu *et al.*[29] targeted NSP13 helicase, a non-structural protein important for SARS-CoV-2 replication and transcription. Punicalagin, a hydrolysable tannin, functioned as an allosteric inhibitor for NSP13 helicase by preventing its binding to the DNA substrate. A molecular docking analysis showed that upon the binding of punicalagin, NSP13 remains in an open conformation, which allosterically induces conformational changes in the nucleic acid binding area. Additionally, the polyphenol could also inhibit viral replication in Vero cells, with a half-maximal effective concentration (EC_{50}) of 68.56 nM.

The use of *in silico* approaches, in particular molecular docking, has been extensively used for the investigation of novel SARS-CoV-2 enzyme inhibitors. This technique helps clarify the binding mechanism between the enzyme and the inhibitor molecule besides calculating the binding affinity between them. Apigenin and its analogues (apigenin 7-glucoside-4′-*p*-coumarate, apigenin 7-glucoside-4′-*trans*-caffeate, and apigenin 7-*O*-β-D-glucoside) were docked with SARS-CoV-2 protease, with apigenin 7-glucoside-4′-*p*-coumarate showing the lowest binding energy (-8.8 kcal mol^{-1}). Therefore, the highest affinity for the viral main protease, even surpassing the affinity displayed by boceprevir, a standard protease inhibitor used to treat hepatitis. Binding between apigenin 7-glucoside-4′-*p*-coumarate and protease occurred at the active site, involving nine hydrogen bonds.[30] A similar outcome was reported for the anthocyanin cyanidin 3-*O*-glucosyl-rutinoside, which showed good binding affinity for SARS-CoV-2 helicase, spike protein, RNA-dependent RNA polymerase, and main protease, antiviral targets for SARS-CoV-2.[31]

Although the antiviral properties of phenolic compounds are promising, their biological fate should not be ignored, as this directly impacts their activity. Once absorbed in the small intestine, phenolic compounds can go through phase I and phase II metabolism, commonly undergoing methylation, glucuronidation, and sulfation, significantly altering their original conformation. Undigested phenolic compounds can also suffer a series of transformations when they reach the colon and are fermented by the colonic bacteria. There are a great number of possible phenolic metabolites, with some of them improving their bioactivities compared to the original compounds and others having their bioactivities reduced or even lost. Therefore, as suggested by Lu *et al.*,[29] the antiviral effects of phenolic metabolites should also be examined. Besides, drug delivery strategies to prevent the gastric degradation of phenolic compounds also need to be considered to avoid loss of antiviral efficiency.

12.3.2 Alkaloids

Similar to phenolic compounds, alkaloids are also a group of plant secondary metabolites with a variety of biological activities (*e.g.*, anticarcinogenic, antidepressant, and anti-inflammatory activities). They are present in plant tissues either as organic acid salts or as weak bases. Some alkaloids may occur as glycosides, amides, and esters. Alkaloids fall into Lipinski's criteria for

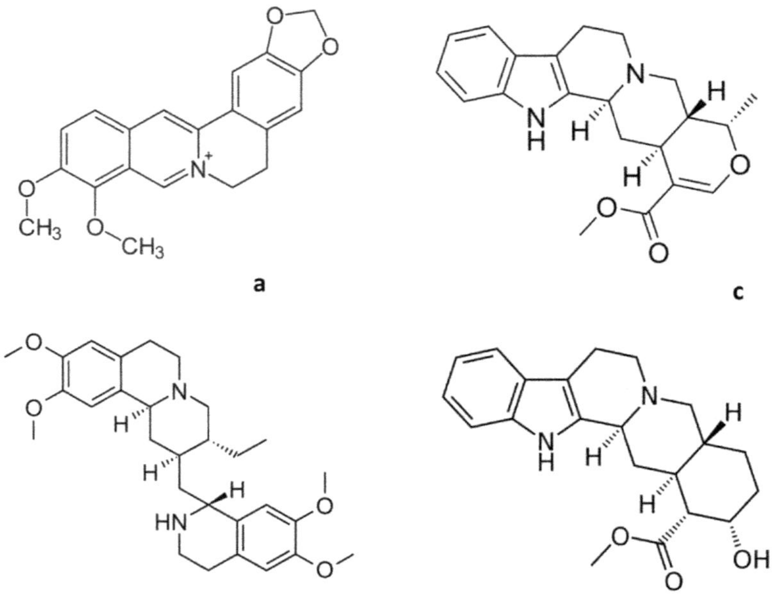

a b c d

Figure 12.2 Chemical structures of alkaloids with antiviral activity: (a) berberine, (b) emetine, (c) ajmalicine, and (d) yohimbine.

drug-like potential molecules. They have a low to moderate molecular weight and are water-soluble under acidic/neutral pH and lipid-soluble under basic pH, which gives them the ability to be transported in the protonated form and penetrate cell membranes in the neutral form. As such, these molecules are the basis of some chemotherapeutic drugs, such as camptothecin, a topoisomerase I inhibitor, and vinblastine, an antitumor medicine. Besides, research has been conducted on the capacity of alkaloids to act on several targets for the antiviral effect.[32] The structures of selected antiviral alkaloids can be seen in Figure 12.2.

Alkaloids usually display a dose-dependent antiviral effect. In most studies, concentrations lower than 10 μM are used with no toxicity effect observed. Several mechanisms can be used by alkaloids in order to suppress viral infections, including:[32]

(a) Interference with virus attachment and cell entry: due to their amphipathic nature (lipophilic skeleton with a polar functional group) with a large hydrophobic portion, some alkaloids can penetrate cell membranes and interact with their components (lipids, proteins, receptors, and ion channels). Consequently, the lipid bilayer is altered, and, in some cases, alkaloids can merge with the membrane, binding the glycoproteins of the viral envelope, which hinders the interaction between the viruses and the host cells.

(b) Suppression of virus replication by inhibiting DNA and RNA synthesis: alkaloids are also able to attack the viral genome and enzymes present in the viral life cycle, which interrupts their replication. Alkaloids may take part in nucleic acid intercalation (molecule insertion between the planar bases of DNA), a process that interrupts DNA-associated pathways, negatively impacting DNA structure and processing. They have been shown to intercalate with DNA polymerase I and reverse transcriptase. Some alkaloids can also affect RNA polymerases and DNA topoisomerase, interfering with base-pairing and obstructing DNA replication, repair, and transcription.

(c) Inhibition of protein synthesis: alkaloids can inhibit viral protein synthesis, which is believed to happen mainly through blocking the ribosomal protein, leading to viral translation inhibition.

The alkaloid berberine has demonstrated inhibitory activity against H1N1 influenza A strains PR/8/34 and WS/33 on RAW 264.7 macrophage-like cells, with an IC_{50} of 0.01 μM. A deeper look into the antiviral mechanism of berberine suggested that this alkaloid may hinder protein trafficking by promoting post-translational modifications, inhibiting virus growth.[33]

Emetine, the main alkaloid found in ipecac, exhibits inhibition against HIV reverse transcriptase.[34] Molecular docking simulations showed that this alkaloid is able to bind reverse transcriptase mainly by hydrophobic interaction. When emetine was applied to cellular models at a concentration of 0.1 mM, a drastic reduction (up to 80%) in H1V-1 replication was observed with no cytotoxic effects whatsoever. According to the study, emetine is able to penetrate HIV and bind reverse transcriptase, blocking the reverse transcription process.

Alkaloids are also potential candidates for the inhibition of the novel coronavirus. Mohseni *et al.*[35] conducted a molecular docking screening of over 700 natural alkaloids that could serve as inhibitors of SARS-CoV-19. From this pool, a total of nine were selected based on the interaction with the active pocket of the virus. Ajmalicine and yohimbine, which best performed on the screening, were chosen for molecular dynamics simulations. The binding of both compounds to the active site was mainly through hydrophobic interaction and hydrogen bonding. The complexes showed high stability, and the alkaloids showed the ability to deeply penetrate the hydrophobic pocket of the virus, making them promising SARS-CoV-2 blockers.

12.3.3 Terpenoids

Terpenoids are a large class of naturally occurring molecules with over 50 000 compounds reported to date. They are mostly found in plant materials, particularly in the essential oils of medicinal plants. This structurally diverse group is derived from mevalonic acid and its compounds can be sub-divided according to their number of isoprene units, which include monoterpenes

(C10), sesquiterpenes (C15), diterpenes (C20), triterpenes (C30), tetraterpenes (C40), and polyterpenes ($C>40$). In their natural sources, terpenoids can also be present as glycosides, alcohols, aldehydes, carboxylic acids, ketones, and esters.[36]

Several terpenoids have commercial importance, such as menthol, used as a raw material for spices, flavouring, and cosmetic products, as well as limonoids, which are part of the formulation of many pesticides. The biological activity of this group is widely studied, with anti-inflammatory, antitumor, antibacterial, antioxidant, and neuroprotective effects, and the ability to lower blood sugar being reported. In addition, terpenoids have also been procured as antiviral agents.[36] Terpenoid-rich essential oils have presented strong antiviral action against a myriad of DNA and RNA viruses, including herpes simplex virus types 1 and 2, poliovirus, dengue virus type 2, and influenza virusadeno type 3. It is important to note that although terpenoids represent the majority of low-molecular weight compounds present in essential oils, phenolic compounds and straight-chain compounds without any side chain are usually also part of the composition. Therefore, the activity presented by essential oils may stem from the synergism between all of these components.[37,38] The structures of some terpenoids found in essential oils with antiviral activity are shown in Figure 12.3.

Astani *et al.*[39] compared the anti-HSV-1 activity of eucalyptus, tea tree, and thyme essential oils with the activity provided by their isolated monoterpenes (α-terpinene, γ-terpinene, α-pinene, *p*-cymene, terpinene-4-ol, α-terpineol, thymol, citral, and 1,8-cineole). While the essential oils were able to inactivate over 96% of viral particles, the individual monoterpenes reached a viral inactivity of about 80%, indicating a possible synergistic effect between the monoterpenes and the other essential oil components. The antiviral mechanism of action was mainly through direct inactivation of virus particles, interfering with the structures present in the virion envelope, which are essential for virus adsorption to the host cells.

The protein *trans*-activator of transcription (Tat) is fundamental for HIV transcription, which makes it an attractive target for the inhibition of HIV viral infection. Feriotto *et al.*[40] investigated the capacity of essential oils (*Thymus vulgaris*, *Cananga odorata*, *Cymbopogon citratus*, and *Rosmarinus officinalis*) to interfere with the functions of Tat protein. According to gas chromatography/mass spectrometry analysis, α-pinene, eucalyptol, β-pinene, *cis*-citrol, thymol, and linalool were among the major terpenoids composing the essential oils studied. The administration of essential oils sharply decreased HIV-1 LTR transcription promoted by Tat protein. This is believed to occur through the interaction of essential oil components with Tat, which prevents it from participating in the transcription process.

The terpenoid saikosaponin b2 occurs naturally in the roots of *Bupleurum kaoi* and it has demonstrated antiviral activity against hepatitis C virus. The compound proved to effectively act on the early stages of the viral life cycle by preventing virus attachment and entry to hepatoma cells at a non-cytotoxic dose of 50 μM.[41]

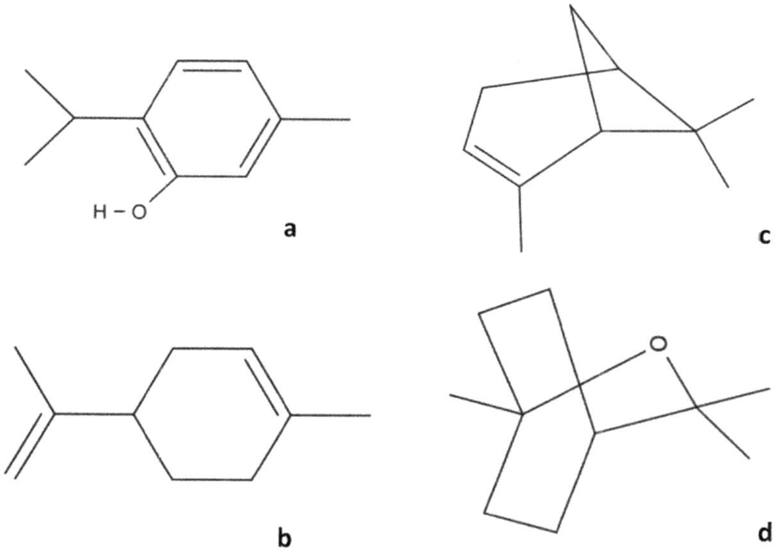

Figure 12.3 Chemical structures of terpenoids with antiviral activity: (a) thymol, (b) limonene, (c) α-pinene, and (d) eucalyptol.

The essential oil extracted from the peels of *Citrus x aurantium* is mainly composed of limonene and it has demonstrated antiviral effect against influenza A (H1N1) on MDCK cells, according to Fadilah *et al.*[42] A virus reduction of 99% was observed, and the authors suggested its use as a household disinfectant against air-borne viruses. Nevertheless, limonoid's antiviral mechanism against H1N1 remains unclear.[42]

An *in silico* molecular docking approach has been used in order to assess the inhibitory potential of terpenoids against SARS-CoV-2. Several terpenoids were docked with the S1 receptor binding domain of the spike glycoprotein of SARS-CoV-2, a target for hindering virus attachment to the host cells. Thymol, camphene, pulegone, ocimene, and menthol exhibit good binding affinity for the glycoprotein, and the complexes formed are stabilized by hydrophobic interaction.[43]

The angiotensin-converting enzyme 2 (ACE2) receptor located in the host cells plays a crucial role in the SARS-CoV-2 virus entry process. This is due to the strong affinity between the spike protein located in the virus envelope and ACE2. Therefore, the blocking of this receptor has been investigated as a valid strategy to suppress the viral infection caused by the novel coronavirus. Kumar *et al.*[44] tested the ability of geranium and lemon essential oils to serve as blockers of the ACE2 receptor. Geranium essential oil was mainly composed of citronellol, geraniol, and neryl acetate, while limonene was the primary compound found in lemon essential oil. Both oils were able to inhibit ACE2 in epithelial cells in a dose-dependent manner (12.5–50 $\mu g\,mL^{-1}$ for geranium oil and 6.25–25 $\mu g\,mL^{-1}$ for lemon oil). They also promoted the downregulation of ACE2 expression in human colorectal adenocarcinoma cells.

12.3.4 Other Compounds

Several other bioactive compounds have displayed antiviral activity and could be considered as potential active ingredients for the development of natural supplements focused on the suppression of viral infections.

Curcumin, the most abundant polyphenol in turmeric, can interfere with viral transcription regulation, replication cycle, genome replication, and virus entry and attachment. These effects have been detected against a myriad of viruses, including HIV, herpes simplex virus, Ebola virus, and influenza type A virus.[45] Molecular docking simulations have also demonstrated the ability of curcumin and its derivatives to complex with the target proteins of SARS-CoV-2, such as papain-like protease, with strong binding affinity, which signals the pharmaceutical potential of these molecules for COVID-19 treatment.[46,47]

Vitamin D has also been shown to act as a natural antiviral agent, with promising results in what comes to anti-hepatitis C virus inhibition. The administration of vitamin D was able to inhibit the production of the hepatitis C virus in hepatoma cells. The effect was owed to calcitriol, a hormonally active metabolite of vitamin D that can upregulate the production of interferon-β, the immediate cellular response to viral infection.[48]

Mushrooms also present antiviral activity, which has been partly attributed to their high content of polysaccharides, especially heteropolysaccharides and glucans. Such polysaccharides demonstrated actions against the influenza virus by promoting immunomodulatory effects, herpes simplex virus by hampering viral replication, HIV through the suppression of protease activity, and hepatitis C through the inhibition of RNA-dependent RNA-polymerase.[49]

12.4 Antiviral Dietary Supplements

Nutraceuticals represent the intersection between nutrition and pharmaceuticals. Research in this area focuses on the key components of food that display health-promoting benefits and can serve as alternatives to synthetic drugs. Nowadays, with the emergence of a healthy lifestyle trend, nutraceuticals have become more popular than ever.[50]

Several dietary supplements using isolated bioactive compounds from natural sources as raw materials are available in the market. These include quercetin, resveratrol, green tea extract, vitamin C, vitamin E, B-complex vitamins, zinc, and copper supplements, among others. They usually come in the form of capsules with varying dosages and are marketed for their most well-known properties, such as antioxidant activity and immune system enhancement. Nevertheless, antiviral properties are seldom mentioned among their benefits, which could be due to the lack of quality data from *in vivo* studies, especially clinical trials, supporting this type of evidence.[51]

These nutraceuticals mostly act by modulating the inflammatory response of the body to viral infections, inhibiting the formation of pro-inflammatory cytokines and hindering viral replication. For instance, resveratrol can suppress virus protein production, which has been observed for the influenza virus, hepatitis C virus, HSV, HIV, and others. On the other hand, quercetin is effective against both DNA and RNA viruses. This phenolic compound is able to hinder DNA replication and affect post-viral healing by interacting with signaling pathways linked to post-transcriptional modulators. Another popular phenolic compound used in nutraceuticals is curcumin, which has been observed to bind receptors associated with viral infection mechanisms. Curcumin can also block the ACE2 cell receptor, which binds the spike glycoprotein of SARS-CoV-2, besides reducing the formation of pro-inflammatory cytokines.[51]

Therefore, it is important to note that the claim of dietary supplements boosting the immune system can also be related to their antiviral efficacy. The increase in adaptive antibody response, along with T-cell-mediated immune response, positively affects the ability of the body to fight viral infections.[51] In order to market dietary supplements as antivirals, bridging the gap between *in vitro* and *in vivo* studies is of the utmost importance, particularly in what comes to obtaining the approval of regulatory agencies for making such claims.

12.5 Conclusion

Polyphenols, alkaloids, terpenoids, and other bioactive compounds are naturally found in plant materials and present a range of biological activities, including antiviral effects. These molecules are usually able to inhibit viral infections by interacting with the components of the virus envelope, mostly proteins, hindering their attachment and entry to the host cells. Inhibition of key enzymes for the life cycle of the virus and blocking of genome synthesis are also among the strategies used by natural compounds to decrease virus multiplication. Therefore, they stand out as possible candidates for the synthesis of novel dietary supplements targeting viral infections. Notwithstanding, it is important to note that data supporting the antiviral activity of these molecules mainly come from cell line and other *in vitro* studies. Therefore, *in vivo* experiments, preferentially using human subjects, when possible, should follow in order to properly investigate if the same outcome will be observed, as digestion and metabolization can greatly influence the biological effects of phytochemicals. The synergism between different bioactive compounds should also be taken into consideration, as this may work to increase their antiviral activity.

References

1. World health organization (WHO), Coronavirus Disease (COVID-19) Pandemic, Published online at: https://www.who.int/emergencies/diseases/novel-coronavirus-2019 (accessed August 15, 2022).

2. M. Denaro, A. Smeriglio, D. Barreca, C. De Francesco, C. Occhiuto and G. Milano, *et al.*, Antiviral activity of plants and their isolated bioactive compounds: An update, *Phytother. Res.*, 2020, **34**, 742–768.
3. S. Baron, M. Fons and T. Albrecht, Viral pathogenesis, in *Medical Microbiology*, ed. S. Baron, University of Texas Medical Branch at Galveston, Galveston, TX, 4th edn, 1996, pp. 1–12.
4. D. S. Dimitrov, Virus entry: molecular mechanisms and biomedical applications, *Nat. Rev. Microbiol.*, 2004, **2**, 109–122.
5. S. T. Hassan, R. Masarčíková and K. Berchová, Bioactive natural products with anti-herpes simplex virus properties, *J. Pharm. Pharmacol.*, 2015, **67**, 1325–1336.
6. World Health Organization (WHO), HIV/AIDS. Published online at: https://www.who.int/health-topics/hiv-aids#tab=tab_1 (accessed August 18, 2022).
7. J. V. Lazarus, K. Safreed-Harmon, S. E. Barton, D. Costagliola, N. Dedes and J. del Amo Valero, *et al.*, Beyond viral suppression of HIV - the new quality of life frontier, *BMC Med.*, 2016, **14**, 1–5.
8. F. Krammer, Emerging influenza viruses and the prospect of a universal influenza virus vaccine, *Biotechnol. J.*, 2015, **10**, 690–701.
9. M. Martini, V. Gazzaniga, N. L. Bragazzi and I. Barberis, The Spanish Influenza Pandemic: a lesson from history 100 years after 1918, *J. Prev. Med. Hyg.*, 2019, **60**, e64.
10. J. Gao and C. Ju, Research progress on the direct antiviral drugs for hepatitis C virus, *BioSci. Trends*, 2017, 41–45.
11. R. Bhattacharya, K. Dev and A. Sourirajan, Antiviral activity of bioactive phytocompounds against coronavirus: An update, *J. Virol. Methods*, 2021, **290**, 114070.
12. R. Goyal, R. Bala, R. K. Sindhu, M. Zehravi, R. Madaan and S. Ramproshad, *et al.*, Bioactive based nanocarriers for the treatment of viral infections and SARS-CoV-2, *Nanomaterials*, 2022, **12**, 1530.
13. C. Zannella, R. Giugliano, A. Chianese, C. Buonocore, G. A. Vitale and G. Sanna, *et al.*, Antiviral activity of *Vitis vinifera* leaf extract against SARS-CoV-2 and HSV-1, *Viruses*, 2021, **13**, 1263.
14. C. I. Tamayose, P. B. Torres, N. Roque and M. J. P. Ferreira, HIV-1 reverse transcriptase inhibitory activity of flavones and chlorogenic acid derivatives from *Moquiniastrum floribundum* (Asteraceae), *S. Afr. J. Bot.*, 2019, **123**, 142–146.
15. C. Liu, D. Cai, L. Zhang, W. Tang, R. Yan and H. Guo, *et al.*, Identification of hydrolyzable tannins (punicalagin, punicalin and geraniin) as novel inhibitors of hepatitis B virus covalently closed circular DNA, *Antiviral Ther.*, 2016, **134**, 97–107.
16. C. Li, J. S. Fang, W. W. Lian, X. C. Pang, A. L. Liu and G. H. Du, *In vitro* antiviral effects and 3D QSAR study of resveratrol derivatives as potent inhibitors of influenza H1N1 neuraminidase, *Chem. Biol. Drug Des.*, 2015, **85**, 427–438.

17. A. O'Rourke, S. Kremb, T. M. Bader, M. Helfer, P. Schmitt-Kopplin and W. H. Gerwick, *et al.*, Alkaloids from the sponge *Stylissa carteri* present prospective scaffolds for the inhibition of human immunodeficiency virus 1 (HIV-1), *Mar. Drugs*, 2016, **14**, 28.
18. M. Aljofan, H. J. Netter, A. N. Aljarbou, T. B. Hadda, I. E. Orhan and B. Sener, *et al.*, Anti-hepatitis B activity of isoquinoline alkaloids of plant origin, *Arch. Virol.*, 2014, **159**, 1119–1128.
19. M. T. Moradi, A. Karimi, M. Rafieian-Kopaei and F. Fotouhi, *In vitro* antiviral effects of *Peganum harmala* seed extract and its total alkaloids against influenza virus, *Microb. Pathog.*, 2017, **110**, 42–49.
20. A. Astani and P. Schnitzler, Antiviral activity of monoterpenes beta-pinene and limonene against herpes simplex virus *in vitro*, *Iran, J. Microbiol.*, 2014, **6**, 149.
21. L. A. Ticona, P. Bermejo, J. A. Guerra, M. J. Abad, M. Beltran and R. M. Lázaro, *et al.*, Ethanolic extract of *Artemisia campestris* subsp. glutinosa (Besser) Batt. inhibits HIV-1 replication *in vitro* through the activity of terpenes and flavonoids on viral entry and NF-κB pathway, *J. Ethnopharmacol.*, 2020, **263**, 113163.
22. F. Shahidi and P. Ambigaipalan, Phenolics and polyphenolics in foods, beverages and spices: Antioxidant activity and health effects – A review, *J. Funct. Foods*, 2015, **18**, 820–897.
23. N. N. Besednova, B. G. Andryukov, T. S. Zaporozhets, S. P. Kryzhanovsky, L. N. Fedyanina and T. A. Kuznetsova, *et al.*, Antiviral effects of polyphenols from marine algae, *Biomedicines*, 2021, **9**, 200.
24. A. Heredia, C. Davis, M. N. Amin, N. M. Le, M. A. Wainberg and M. Oliveira, *et al.*, Targeting host nucleotide biosynthesis with resveratrol inhibits emtricitabine (FTC)-resistant HIV-1, *AIDS*, 2014, **28**, 317.
25. B. Özçelik, M. Kartal and I. Orhan, Cytotoxicity, antiviral and antimicrobial activities of alkaloids, flavonoids, and phenolic acids, *Pharm. Biol.*, 2011, **49**, 396–402.
26. B. Suárez, Á. L. Álvarez, Y. D. García, G. del Barrio, A. P. Lobo and F. Parra, Phenolic profiles, antioxidant activity and *in vitro* antiviral properties of apple pomace, *Food Chem.*, 2010, **120**, 339–342.
27. C. Sanna, A. Marengo, S. Acquadro, A. Caredda, R. Lai and A. Corona, *et al.*, *In vitro* anti-HIV-1 reverse transcriptase and integrase properties of *Punica granatum* L. leaves, bark, and peel extracts and their main compounds, *Plants*, 2021, **10**, 2124.
28. J.-M. Song, K.-H. Lee and B.-L. Seong, Antiviral effect of catechins in green tea on influenza virus, *Antiviral Res.*, 2005, **68**, 66–74.
29. L. Lu, Y. Peng, H. Yao, Y. Wang, J. Li and Y. Yang, *et al.*, Punicalagin as an allosteric NSP13 helicase inhibitor potently suppresses SARS-CoV-2 replication *in vitro*, *Antiviral Res.*, 2022, 105389.
30. A. Farhat, H. Ben Hlima, B. Khemakhem, Y. Ben Halima, P. Michaud and S. Abdelkafi, *et al.*, Apigenin analogues as SARS-CoV-2 main protease inhibitors: *In-silico* screening approach, *Bioengineered*, 2022, **13**, 3350–3361.

31. P. A. Akinnusi, S. O. Olubode and W. A. Salaudeen, Molecular binding studies of anthocyanins with multiple antiviral activities against SARS-CoV-2, *Bull. Natl. Res. Cent.*, 2022, **46**, 1–9.
32. F. L. Abookleesh, B. S. Al-Anzi and A. Ullah, Potential antiviral action of alkaloids, *Molecules*, 2022, **27**, 903.
33. C. E. Cecil, J. Davis, N. B. Cech and S. M. Laster, Inhibition of H1N1 influenza A virus growth and induction of inflammatory mediators by the isoquinoline alkaloid berberine and extracts of goldenseal (*Hydrastis canadensis*), *Int. Immunopharmacol.*, 2011, **11**, 1706–1714.
34. A. L. C. Valadão, C. M. Abreu, J. Z. Dias, P. Arantes, H. Verli and A. Tanuri, *et al.*, Natural plant alkaloid (emetine) inhibits HIV-1 replication by interfering with reverse transcriptase activity, *Molecules*, 2015, **20**, 11474–11489.
35. M. Mohseni, H. Bahrami, B. Farajmand, F. S. Hosseini, M. Amanlou and H. Salehabadi, Indole alkaloids as potential candidates against COVID-19: an *in-silico* study, *J. Mol. Model.*, 2022, **28**, 1–13.
36. W. Yang, X. Chen, Y. Li, S. Guo, Z. Wang and X. Yu, Advances in pharmacological activities of terpenoids, *Nat. Prod. Commun.*, 2020, **15**, 1934578X20903555.
37. A. R. Wani, K. Yadav, A. Khursheed and M. A. Rather, An updated and comprehensive review of the antiviral potential of essential oils and their chemical constituents with special focus on their mechanism of action against various influenza and coronaviruses, *Microb. Pathog.*, 2021, **152**, 104620.
38. A. Astani and P. Schnitzler, Antiviral activity of monoterpenes beta-pinene and limonene against herpes simplex virus *in vitro*, *Iran, J. Microbiol.*, 2014, **6**, 149.
39. A. Astani, J. Reichling and P. Schnitzler, Comparative study on the antiviral activity of selected monoterpenes derived from essential oils, *Phytother. Res.*, 2010, **24**, 673–679.
40. G. Feriotto, N. Marchetti, V. Costa, S. Beninati, F. Tagliati and C. Mischiati, Chemical composition of essential oils from *Thymus vulgaris*, *Cymbopogon citratus*, and *Rosmarinus officinalis*, and their effects on the HIV-1 Tat protein function, *Chem. Biodiversity*, 2018, **15**, e1700436.
41. L. T. Lin, C. Y. Chung, W. C. Hsu, S. P. Chang, T. C. Hung and J. Shields, *et al.*, Saikosaponin b2 is a naturally occurring terpenoid that efficiently inhibits hepatitis C virus entry, *J. Hepatol.*, 2015, **62**, 541–548.
42. N. Q. Fadilah, A. Jittmittraphap, P. Leaungwutiwong, P. Pripdeevech, D. Dhanushka and C. Mahidol, *et al.*, Virucidal activity of essential oils from Citrus x aurantium L. against influenza A virus H1N1: Limonene as a potential household disinfectant against virus, *Nat. Prod. Commun.*, 2022, **17**, 1934578X211072713.
43. S. A. Kulkarni, S. K. Nagarajan, V. Ramesh, V. Palaniyandi, S. P. Selvam and T. Madhavan, Computational evaluation of major components from plant essential oils as potent inhibitors of SARS-CoV-2 spike protein, *J. Mol. Struct.*, 2020, **1221**, 128823.

44. K. J. S. Kumar, M. G. Vani, C. S. Wang, C. C. Chen, Y. C. Chen and L. P. Lu, *et al.*, Geranium and lemon essential oils and their active compounds downregulate angiotensin-converting enzyme 2 (ACE2), a SARS-CoV-2 spike receptor-binding domain, in epithelial cells, *Plants*, 2020, **9**, 770.
45. D. Mathew and W. L. Hsu, Antiviral potential of curcumin, *J. Funct. Foods*, 2018, **40**, 692–699.
46. H. Alici, H. Tahtaci and K. Demir, Design and various *in silico* studies of the novel curcumin derivatives as potential candidates against COVID-19-associated main enzymes, *Comput. Biol. Chem.*, 2022, **98**, 107657.
47. E. Pitsillou, J. Liang, K. Ververis, A. Hung and T. C. Karagiannis, Interaction of small molecules with the SARS-CoV-2 papain-like protease: *In silico* studies and *in vitro* validation of protease activity inhibition using an enzymatic inhibition assay, *J. Mol. Graphics Modell.*, 2021, **104**, 107851.
48. M. Gal-Tanamy, L. Bachmetov, A. Ravid, R. Koren, A. Erman and R. Tur-Kaspa, *et al.*, Vitamin D: an innate antiviral agent suppressing hepatitis C virus in human hepatocytes, *Hepatology*, 2011, **54**, 1570–1579.
49. S. Chun, J. Gopal and M. Muthu, Antioxidant activity of mushroom extracts/polysaccharides – their antiviral properties and plausible antiCOVID-19 properties, *Antioxidants*, 2021, **10**, 1899.
50. V. Sachdeva, A. Roy and N. Bharadvaja, Current prospects of nutraceuticals: A review, *Curr. Pharm. Biotechnol.*, 2020, **21**, 884–896.
51. S. Singh, P. Kola, D. Kaur, G. Singla, V. Mishra and P. S. Panesar, *et al.*, Therapeutic potential of nutraceuticals and dietary supplements in the prevention of viral diseases: a review, *Front. Nutr.*, 2021, **8**, 679312.

CHAPTER 13

Gut Microbiota and Health

TERESA GERVASI,[a,†] TERRI GRASSBY,*[b,†] DAVIDE BARRECA[c] AND GIUSEPPINA MANDALARI*[c]

[a] Department of Biomedical and Dental Sciences and Morphofunctional Imaging, University of Messina, Messina, Italy; [b] Department of Nutritional Sciences, University of Surrey, Guildford, Surrey, UK; [c] Department of Chemical, Biological, Pharmaceutical and Environmental Science, University of Messina, Messina, Italy
*Emails: t.grassby@surrey.ac.uk; gmandalari@unime.it

13.1 Introduction

Antioxidants are natural or synthetic compounds able to counteract the negative effects of reactive species that are produced during normal physiological processes or at the onset of a stress condition. These reactive species are, mainly, centred on oxygen or nitrogen, creating reactive oxygen species (ROS) or reactive nitrogen species (RNS). These chemical compounds have a dual role inside the organism; they can damage biological macromolecules (such as proteins, DNA, and phospholipids), cells, and tissues (being highly reactive compounds), but they are also molecular signals, which activate inherent stress responses.[1] They start to create problems when there is an imbalance in the ratio of oxidants to antioxidants; excess oxidants initiate a series of uncontrolled radical reactions. Oxidative stress is commonly associated with various physiological malfunctions and diseases, including Parkinson's and Alzheimer's diseases,[2,3] aging,[4] impaired immunomodulation,[5] cancer,[6] and

[†]These authors equally contributed to this work.

Food Chemistry, Function and Analysis No. 39
Dietary Supplements with Antioxidant Activity: Understanding Mechanisms and Potential Health Benefits
Edited by Cesarettin Alasalvar, Fereidoon Shahidi and Chi-Tang Ho

Published by the Royal Society of Chemistry, www.rsc.org

programmed cell death.[7] Antioxidants are able to block the initiation and/or progression of the process by different mechanisms, which are linked to their chemical structure.[8] The term "antioxidants" represents compounds with very different chemical characteristics, such as phenolic and polyphenol compounds, amino acids and their derivatives, carotenoids, tocopherols, terpenoids, vitamins, and minerals, among others (Figure 13.1).

Living organisms can produce some antioxidants ("endogenous" antioxidants), but these are often insufficient to counteract all the oxidative stress, so "exogenous" antioxidants, supplied by certain plant-based foods rich in these compounds, are needed.

The gut microbiota is an intricate and heterogeneous mix of microorganisms, which colonises the human gastrointestinal tract (GIT) and consists of bacteria, archaea, and eukarya. They have co-evolved with the host, over thousands of years of evolution, to create a series of mutually beneficial relationships that influence the general condition of the host during homeostasis and disease.[9–11] The interactions and mutual exchanges between the gut microbiota and the host (the tissues of the GIT and other peripheral tissues in particular), as well as the components present in the diet, are fundamental to the appropriate development of the organism and its wellness.[9–11] The development of advanced analytical methods (shotgun sequencing, mass spectrometry, and nuclear-magnetic-resonance-based metabolomics) to identify and characterize the complex ecosystems being developed, coupled with a reduction in sequencing costs and innovations in bioinformatics, has highlighted the role of the gut microbiota in host wellness. This includes a system for harvesting energy, new compounds produced by fermentation and metabolic processes, protection against pathogens, strengthening of the host immunity system, and protection of the intestinal epithelium.[12–15] Its importance is evident during dysbiosis, which may contribute to the pathogenesis of type 2 diabetes, obesity, cardiometabolic diseases, nonalcoholic fatty liver disease, and malnutrition.[16,17]

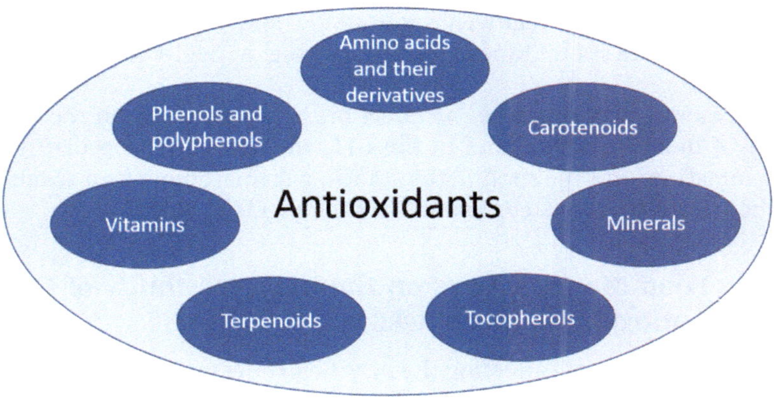

Figure 13.1 The different classes of antioxidants.

The gut microbiota is dominated by commensal or mutualist organisms and the proportions of the different species are influenced by several factors, including the mode of birth (vaginal or caesarean), infant nutrition (breast milk or formula), lifestyle, intake of pharmaceuticals (type, timing, and frequency), and the genetics of the host. It is responsible for producing an abundance of bioactive metabolites during the digestion and metabolism of food and drugs, modifying their action. It also regulates gut endocrine functions and neurological signalling, trains the host immune system, and transforms toxins so they can be eliminated.[17] In this chapter, we focus on the role played by dietary antioxidants in the gut microbiota and their effects on the metabolism of the healthy host, providing perspectives for future basic and translational studies in this nascent and promising research field.

13.2 Potential Beneficial Effects of Dietary Antioxidants on Gut Health

Diet is considered as one of the main drivers in shaping the gut microbiota. In an average lifetime, around 60 tonnes of food pass through the human GIT (Figure 13.2), along with an abundance of microorganisms from the environment, which may pose a threat to gut integrity.

Many of the ingested dietary antioxidants reach the large bowel without degradation in the upper GIT. In the colon, they are substrates for the indigenous microbiota, which are known to include species and genera able to exert varying effects on the host health.[18,19] While the majority of dietary phytochemicals are hydrolysed by the colonic microbiota, small percentages are taken up during transit through the small intestine, and in some cases it is believed that the presence of a glucose moiety may enhance absorption.[20,21] Simple flavonoid glucosides are generally taken up by epithelial cells *via* sodium-dependent glucose transporter 1 (SGLT-1) and other hexose transporters, and the aglycones can then be readily absorbed through passive diffusion by epithelial cells in the small intestine. However, polyphenol glycosides are relatively hydrophilic and do not passively diffuse across biological membranes, thus requiring a deglycosylation step for absorption.

Here, we will now examine the effect of the food matrix on the bioaccessibility of dietary antioxidants in the GIT, the absorption occurring in the small intestine, and the modulation of the gut microbiota as a result of the metabolism of bioactive compounds by the resident flora.

13.2.1 Food Matrix Effects on the Bioaccessibility of Dietary Antioxidants in the GIT

One of the main factors affecting the beneficial effects of polyphenols is their bioaccessibility and absorption in the upper GIT, followed by their metabolism by the gut microbiota.[22]

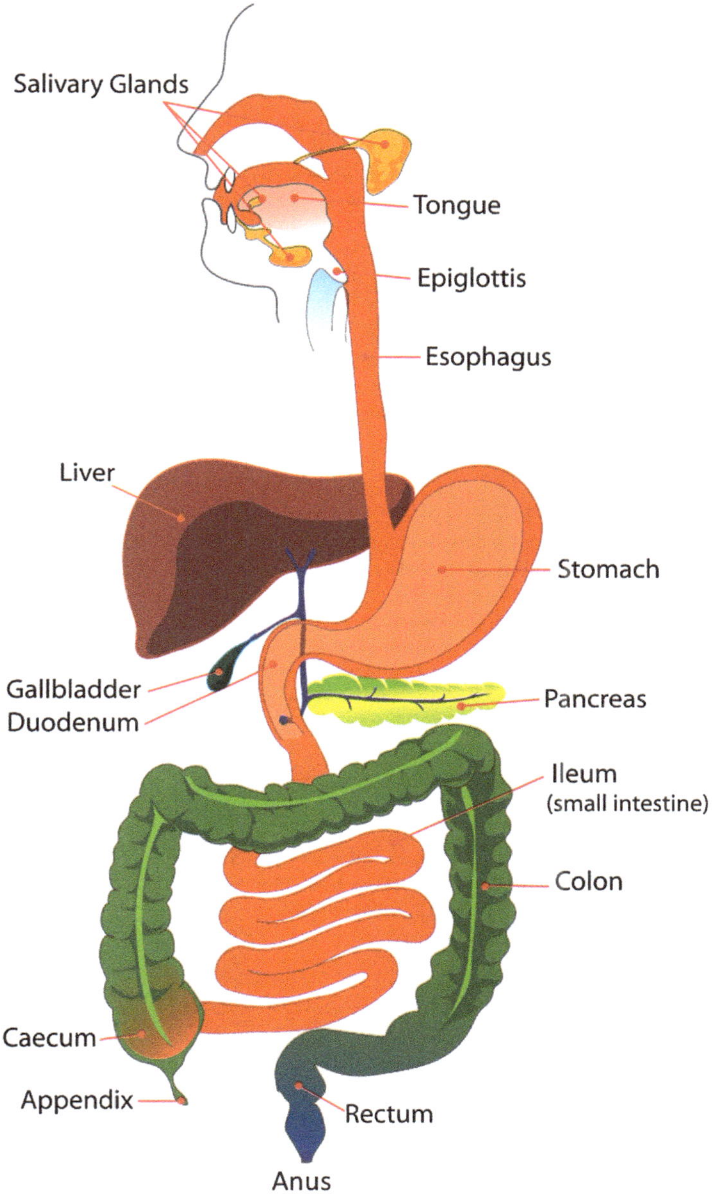

Figure 13.2 A simplified diagram of the human GIT.

Bioaccessibility refers to the proportion of a nutrient or phytochemical compound "released" from a complex food matrix during digestion and, therefore, becoming potentially available for absorption in the GIT.[23] This process is affected by the physico-chemical properties of the food matrix and how it changes during digestion. Several studies have established that polyphenol release in the gut is highly dependent on the food matrix.[24,25]

In a previous investigation, we have reported that incorporating natural and roasted salted pistachios in a food matrix (muffin) decreased the bioaccessibility of certain bioactive compounds, such as protocatechuic acid and luteolin, during simulated human digestion.[26] Furthermore, different food matrices had a significant impact on the bioaccessibility of polyphenols from almond skin using a dynamic gastric model: full fat milk lowered the recovery of all classes of polyphenols and affected the free total phenols and the antioxidant status, suggesting the possibility that phenolic compounds could bind protein within the matrix.[27] On the other hand, results from a clinical trial showed that milk did not affect absorption of (−)-epicatechin from cocoa powder in healthy humans.[28] The same authors later reported differences in metabolite excretion profiles, mainly glucuronide and sulphate conjugates, detected in the urine of volunteers after the intake of two cocoa beverages.[29] Therefore, the interaction of polyphenols with other compounds, such as protein, needs to be considered when determining the overall bioactivity.[30] For example, epigallocatechin-3-*O*-gallate possesses a high affinity for blood proteins, which could potentially extend its half-life in the blood.[31] Binding to albumin could also influence the delivery of polyphenols and their metabolites to cells and tissues. Finally, the interactions with other phenolic compounds with similar mechanisms of absorption could influence the bioavailability of polyphenols.

Digestion conditions, including pH, temperature, bile salts, gastric, and pancreatic enzymes, may also result in altered bioaccessibility of polyphenols. An *in vitro* gastrointestinal digestion study of pomegranate juice indicated that anthocyanins were largely transformed into non-red forms due to pH and could, therefore, be partially degraded.[32]

Taken together, these data indicate that the matrix in which polyphenols are consumed could affect their metabolism and excretion, and this may have implications for their biological activity. Research focused on the bioaccessibility of polyphenols from food matrices is extremely important to evaluate the effect of these different metabolite profiles on the body. Novel technological strategies, including micro-encapsulation, could be explored to enhance bioavailability.

13.2.2 Hydrolysis in the Mouth

Polyphenols are a huge and heterogeneous group of compounds characterised by complex structures and polymerisation. Their chemical structure and any substituents largely influence their absorption, defining whether polyphenols will be absorbed in the small intestine, or enter the colon, where they could be metabolised by the colonic microbiota.[33]

The hydrolysis of some flavonoid glycosides might initially take place in the oral cavity after contact with glycosidase enzymes of the oral microflora. Kamonpatana *et al.*[34] proved that degradation of chokeberry anthocyanins in the saliva depends on their structure and is mostly performed by the oral microbiota. This study revealed that in both natural and artificial saliva the glycosides of delphinidin and petunidin were degraded more easily than those of cyanidin, pelargonidin, peonidin, and malvidin and that the stability of di- and tri-saccharide conjugates of anthocyanidins was higher than that of monosaccharides. The authors concluded that the degradation of anthocyanin in the mouth is both structure-dependent and mostly mediated by the oral microbiota.

13.2.3 Absorption in the Small Intestine

There are different and specific kinetics for the absorption of the various classes of polyphenols. Isoflavones and phenolic acids are certainly the best absorbed polyphenols, followed by catechins, flavanones, and quercetin glucosides. The proanthocyanidins, galloylated tea catechins, and anthocyanins, because of their large molecular weight, are the least absorbed polyphenols.[35]

In the small intestine only about 5–10% of the total polyphenol intake may be absorbed. This is mostly made up of low-molecular-weight polyphenols, characterised by monomeric and dimeric structures.[36] It is also known that polyphenols are usually glycosylated with glucose or rhamnose as well as with galactose, arabinose, xylose, glucuronic acid, or other sugars,[37] and their metabolism generally starts with the removal of the sugar by glycosidases (Figure 13.3).

Some glycoside phenolic compounds are hydroxylated into their corresponding aglycones inside the small intestine, due to the membrane-bound lactase phlorizin hydrolase, which is located on the brush-border of the small intestine epithelial cells. The aglycone product, due to its increased lipophilicity and proximity to the cellular membrane,[38] can then pass into the epithelial cells by passive diffusion. Some polyphenols can be transported as glycosides by sugar transporters through the epithelium. In the epithelial cells, these glycosides are hydrolysed by cytosolic β-glucosidase, forming new aglycones.[39]

Monomeric flavan-3-ols, such as (−)-epicatechin, which are often acylated by gallic acid, but never glycosylated, are generally absorbed at the enterocyte level with no prior chemical modification, such as deconjugation or hydrolysis.[40] However, deconjugation is more often catalysed by faecal microbiota enzymes,[39] as reported below.

13.2.4 Bio-transformation in the Large Bowel

About 90–95% of the ingested polyphenols are not absorbed in the small intestine and reach the large bowel intact, where they are metabolised by the resident microbial community. The final derivatives or aglycones, which are absorbed in the enterocytes of the small intestine or the colon, are metabolised by sulfotransferases, uridine-5′-diphosphate glucuronosyltransferases,

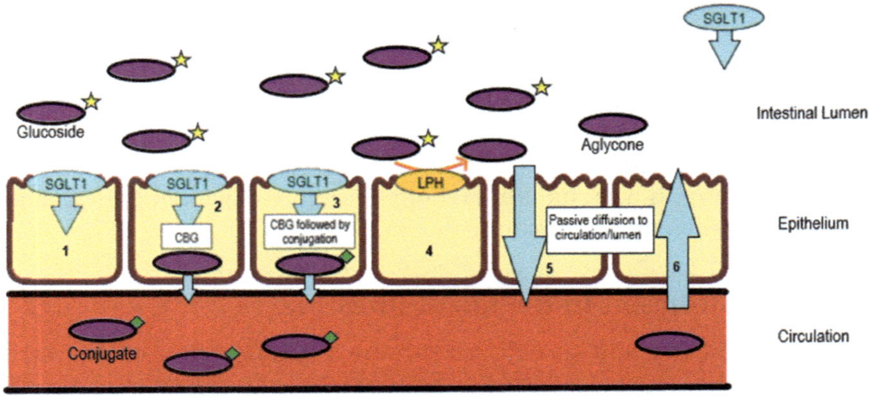

Figure 13.3 Specifics of absorption and enteric metabolism of polyphenolic glucosides in the small intestine: (1) glucosides (*e.g.*, quercetin-3-glucoside) transported through the cellular membrane by the active sugar transporter (SGLT-1); (2) hydrolysis of glucosides to aglycones within the endothelial cells by CBG, followed by passive diffusion into the circulation; (3) conjugation and transport into the circulation; (4) hydrolysis by LPH in the brush border of the endothelial cells to form the aglycone; (5) passive diffusion of the aglycone through the cellular membrane due to increased lipophilicity; and (6) excretion of molecules from the circulation into the gut lumen. Drawn from the text in Murota *et al.*[48] Abbreviations: CBG, cytosolic β-glucosidase; LPH, lactase phlorizin hydrolase; SGLT-1, sodium-dependent glucose transporter-1.

and catechol-*O*-methyltransferases forming sulphate, glucuronide, and/or methylated metabolites. Having entered the bloodstream, these metabolites reach the liver *via* the portal vein. In the liver, they are subjected to more phase II metabolism, and these molecules are then transported to the bloodstream again prior to urinary excretion.[40] These structural changes may increase or decrease the bioactivity of the initial compound.

A portion of these conjugates is excreted by the liver as a component of the bile into the small intestine, therefore reaching the colon in a different chemical form than the original polyphenols. Here, the microbial enzymes deconjugate compounds before being reabsorbed. The unabsorbed metabolites are eliminated *via* the faeces.[40]

The pathway followed is common to drug metabolism, and much of the information available on the metabolism of polyphenols derives from comparisons with it. There are significant differences between prescribed drug doses and usual dietary polyphenol consumption. Unlike drugs, polyphenols in food would not be expected to saturate the metabolic pathways.[41]

13.2.5 Biotransformation by the Gut Microbiota

The human gut is the most important bacterial ecosystem in the human body. The microbial content of the GIT changes along its length, starting with narrow diversity, and low numbers of microbes in the stomach and duodenum, around 10^1–10^3 bacteria per g, there is wider diversity and higher numbers, almost 10^4–10^7 bacteria per g, in the jejunum and ileum, with the highest numbers (10^{11-12} bacteria per g) in the large intestine.[42] The diversity of a healthy adult gut ecosystem is generally defined to comprise thousands of species-level phylogenetic types.[42] About 90–99% of this variety is represented by the bacterial phyla *Firmicutes*, the dominant *Firmicutes* (50–80%), mainly composed of bacteria belonging to the *Clostridium* clusters XIVa and IV, and *Bacteroidetes*. Other phyla present in the human gut are *Actinobacteria* (3–15%), *Proteobacteria* (1–20%), *Verrucomicrobia* (0.1%), *Fusobacteria*, *Cyanobacteria*, *Spyrochaetes*, and *Lentisphaerae.*[43]

These microorganisms play a key role in polyphenol biotransformation, facilitating their breakdown into low-molecular-weight phenolic metabolites, which could be absorbed or excreted by faeces. This could result in the inactivation and/or degradation of phytochemicals or in the production of compounds with enhanced biological activity or bioavailability.[36]

In vitro studies have highlighted how polyphenol catabolism is strictly related to the gut microorganisms and is influenced by the different capacities of the bacteria constituting the gut microbiota, which metabolise the polyphenols and produce different derivatives.[44]

The microorganisms in the large intestine are responsible for the cleavage of glycosidic linkages by enzymes such as rhamnosidase, β-glucosidase, and β-glucuronidase, and the breakdown of the heterocyclic backbone.[36] In particular, the bacterial biotransformation mostly consists of the following reactions: hydroxylation, dehydroxylation, *O*-methylation, *O*-demethylation, glycosylation, deglycosylation, dehydrogenation, hydrogenation, C ring cleavage of the benzo-γ-pyrone system, cyclisation, and carbonyl reduction.[45] Several pathways of the bacterial metabolism of the most important groups of polyphenolic compounds are demonstrated below.

13.2.5.1 Isoflavones

Given the different number and/or position of hydroxyl and methoxy substituents of A and B rings which can be responsible for specific reaction types, isoflavone biotransformation may result in a large variety of metabolites depending on the structural differences of the parent aglycones.[46]

The biotransformation of isoflavones usually starts with aglycone release by hydrolysis of the glycosidic bond. After ingestion, small intestinal brush-border membrane enzymes and microbial β-glucosidases remove the glycosyl group, activating these phytochemicals. Potentially, bacterial enzymes of the small intestine are able to hydrolyse all glycoconjugates; however, the enzymes of colonic bacteria only exert their activity on those molecules that escape hydrolysis and absorption in the small intestine.[46] The aglycones are readily absorbed and become bioactive.

The isoflavone aglycones are further transformed. The main change, which is usually typical to all classes of isoflavones, is hydrogenation of the double bond between C2 and C3 in ring C. For example, the most common isoflavones consumed in food, daidzein and genistein become dihydrodaidzein and dihydrogenistein, respectively.

These dihydroisoflavones are then metabolised by the gut microbiota and may go through two different pathways of bacterial transformation: a reductive metabolism, which promotes the production of equol or equol-related molecules or the cleavage of the C ring with the fission of the molecule into two moieties.[47] In the latter, the C-ring cleavage of dihydroisoflavones always starts with the breakage of the bond between the oxygen in the pyrone ring and the C-2, yielding *O*-desmethylangolensin (*O*-DMA). If this compound does not have a hydroxyl group in position 6′, no further reactions occur on the C-ring. Instead, if the compound related to *O*-DMA has a hydroxyl group in position 6′, thanks to the bond breakage between positions 1 and 1′ promoted by the intestinal bacteria, the molecule is cleaved into two parts.[46]

One of the most studied isoflavones is daidzein because equol, its major metabolite, is known to have a potent oestrogenic effect. It has been reported that after the daidzeinglycosidic linkage is broken by bacterial and human β-glucosidase, daidzein, its free form, may be directly taken up into the blood stream or can be metabolised by the microbiota into various reduction products, producing dihydrodaidzein, which can undergo a C-ring cleavage, ending up as *O*-DMA, or to a reductive reaction producing equol. In addition to *O*-DMA and equol, microbial metabolites of daidzein have been observed, such as dihydrodaidzein, tetrahydrodaidzein, 3′-hydroxydaidzein, 6-hydroxydaidzein, 8-hydroxydaidzein, 3-(4-hydroxyphenyl)benzopyran-4,7-diol, and 2-dehydro-*O*-DMA.

Many different bacteria, including *Escherichia coli*, *Bacteroides ovatus*, *Ruminococcus productus*, *Streptococcus intermedius*, *Eggerthella* spp. strain YY7918, *Adlercreutzia equolifaciens*, SNU-Julong 732 (AY310748), *Enterococcus faecium* EPI1, *Lactobacillus mucosae* EPI2, *Finegoldia magna* EPI3, *Veillonella* spp. EP, *Clostridium* spp. HGHA136, and *Eubacterium ramulus*, are involved in the metabolism of daidzein and are responsible for the formation of the final products.[39,48] However, equol may be produced by some bacteria from both daidzein and its glycoside daidzin, but not all microorganisms can produce equol, unless other species of bacteria are present and metabolise daidzin to the aglycone daidzin and then dihydrodaidzein or tetrahydrodaidzein (*cis* and *trans*) and dehydroequol, or other derivatives.[48] Equol has been found in only 30–50% of the population, even if *O*-DMA has been found in 80–90% of the population. The differences in the gut microbial community of individuals are responsible for the ability to produce equol.[48]

13.2.5.2 Flavanones

As previously reported, flavanones seem to be more available for absorption in comparison with other flavonoids, such as flavonols or flavanones. This is probably because these compounds are metabolised less than other flavonoids by colonic microorganisms and consequently are more available for absorption, even in the distal part of the intestine.[39] These compounds are usually in a glycoside form, generally as rutinosides (6-*O*-α-D-rhamnosyl-D-glucosides) and neohesperidosides (2-*O*-α-L-rhamnosyl-D-glucosides) attached at the seventh-carbon position. Their metabolism consists in deglycosylation, followed by C-ring cleavage, which takes place between the 1- and 2-positions or the 4- and 10-positions.[39] The most common flavanones in food, especially in citrus, are hesperetin and naringin.

Borges *et al.*[49] showed, using urinary data, that almost 30% of hesperetin absorption occurs in the small intestine confirming previous data, which reported the plasma pharmacokinetic profiles of hesperetin-*O*-glucuronides obtained with healthy volunteers.[50] However, hesperetin and naringin mainly reach the colon intact due to being resistant to enzymes in the stomach and small intestine.

In vivo and *in vitro* studies showed that, in the colon, the aglycones hesperetin and naringenin released by the microbial α-rhamnosidases and α-glucosidases can then be absorbed through the intestinal epithelium through passive diffusion and proton-coupled active transport. They can be further metabolised by C-ring cleavage, demethylation and dehydroxylation due to bacterial enzymes into phenolic acids and simple phenolics, such as dihydrocaffeic acid, isoferulic acid, 4-hydroxyphenylacetic acid, dihydroferulic acid, ferulic acid, resorcinol, phloroglucinol, 2,4-dihydroxyphenylacetic acid, 4-hydroxybenzoic acid, phloretic acid, phloroglucinic acid, hydroxycinnamic acid, 3-(3′-hydroxyphenyl)propionic acid, protocatechuic acid, and hippuric acid.[51]

13.2.5.3 Flavones

Dietary flavones are mostly ingested as glycosides. The biotransformation of flavone *O*-glycosides, as many other flavonoid glycosides, consists of their hydrolysis or degradation by digestive enzymes or by gut bacteria into aglycones, which are metabolised by phase II enzymes to form *O*-glucuronides and *O*-sulphates in both the small intestine and liver. These compounds are then transferred back to the intestinal lumen and the bloodstream.[52]

There is evidence that the small intestine is a major site of hydrolysis and metabolism of flavone *O*-glycosides and that the membrane bound enzyme, lactase-phlorizin hydrolase, is the most likely enzyme to act on flavone glycosides in this site.[52]

Although several dietary flavonoids occur in the *O*-glycosidic form, numerous studies have investigated the metabolism of *C*-glycosylated flavones, which are widespread in a large variety of plants. For this class of compounds, two different metabolic pathways have been described for the flavone *C*-multiglycosides and the flavone *C*-monoglucosides, respectively. The flavone *C*-multiglycosides are absorbed in the intestine and, without extensive modification in the liver, then return to the gut *via* the enterohepatic circulation. Conversely, the flavone *C*-monoglucosides, such as orientin, vitexin, homoorientin, and isovitexin, are poorly absorbed so they reach the colon where they are deglycosylated and degraded by human intestinal bacteria to smaller metabolites such as phloroglucinol, hydroxycaffeic acid, and phloretic acid, yielding few metabolites in the urine and blood.[53]

13.2.5.4 Flavonols

Flavanol biotransformation is largely influenced by the initial glycosylation pattern type and the hydroxyl group position. The metabolism of a *C*-glycosidic bond has been reported to be slower than the hydrolysis of an *O*-glycosidic bond and the same results have been reported for the flavonoids without hydroxyl groups at the C5, C7, and C4 positions. In addition, the metabolism of flavonol di- and trisaccharides is slower than that of monosaccharides.[40] Some species responsible for this hydrolysis are *Bacteroides distasonis*, *Bacteroides uniformis*, *Bacteroides ovatus*, *Enterococcus casseliflavus*, and *Eubacterium ramulus*. The remaining step in their metabolism is a C-ring fission. Some species involved in the C-ring cleavage are *Eubacterium oxidoreducens*, *Eubacterium ramulus*, *Enterococcus casseliflavus*, *Clostridium orbiscidens*, and others belonging to the *Butyrivibrio* genus. Then the dehydroxylation at the two remaining free phenolic rings follows and it produces different types of phenolic compounds, which are absorbed at the colon level.[40] Considering quercetin and myricetin, the gut metabolites are 2-(3,4-dihydroxyphenyl)acetic acid (from A-ring) and protocatechuic acid (from B-ring) for quercetin, and 2-(3,5-dihydroxyphenyl)acetic acid (from A-ring) and gallic acid (from B-ring) for myricetin or 2-(3-hydroxyphenyl)acetic acid from both metabolites.[40]

13.2.5.5 Flavanols

Flavanols are a heterogeneous class of flavonoids consisting of simple flavanols (catechin, epicatechin, gallocatechin, epigallocatechin, and the corresponding gallate esters) and their oligomeric and polymeric forms (tannins and proanthocyanidins). Their bioavailability depends on their structure, such as the degree of polymerisation and galloylation. Monomers are readily absorbed in the small intestine, whereas oligomers with a degree of polymerisation >3 and polymers need to be biotransformed by the colonic microbiota.[54]

The small intestinal and hepatic metabolism of flavanols produces conjugated derivatives of monomeric flavanols, such as glucuronic acid or sulphate esters, methyl ethers, or their combined forms, which are usually eliminated in the bile.[54] For the metabolism of flavanols, particularly proanthocyanidins, the colon plays a very important role.

In particular, the galloylated monomeric flavan-3-ols, such as (−)-epigallocatechin gallate and (−)-epicatechin gallate, are metabolised by cleavage of the gallic acid ester moiety by microbial esterases, producing gallic acid, which is further decarboxylated into pyrogallol (Figure 13.4). After these steps, the C-ring is opened, producing diphenylpropan-2-ol, which is converted into 5-(3′,4′-dihydroxyphenyl)-γ-valerolactone or 5-(3′,4′,5′-trihydroxyphenyl)-γ-valerolactone [in the case of (epi)catechins) and (epi)gallocatechins], respectively. The valerolactone ring later breaks and 5-(3′,4′-dihydroxyphenyl) valeric acid and/or 4-hydroxy-5-(3′,4′-dihydroxyphenyl) valeric acid are obtained. Further biotransformation of these valeric acids results in hydroxyphenylpropionic and hydroxybenzoic acids. The last step of the microbial catabolism involves the dehydroxylation of 3,4-dihydroxylated phenolic acids generally at C-4′, but also at C-3′, producing 3- and 4-monohydroxylated phenolic acids, respectively. With respect to (epi)gallocatechins, prior to this dehydroxylation step there is another at C-5.

In the case of hydroxyphenylvalerolactones, the dehydroxylation of 5-(3′,4′,5′-trihydroxyphenyl)-γ-valerolactone produced a 3,5-dihydroxylated derivative, showing that the dehydroxylation at C-4′ occurs.[54]

The microbial metabolites may be metabolised in the liver, after absorption, by phase II enzymes and the conjugated derivatives are eliminated in urine or as non-conjugated microbial metabolites in faeces.[54]

13.2.5.6 Anthocyanins

As previously reported, anthocyanins may be partially degraded in saliva. In the stomach, anthocyanins are rapidly absorbed and may be detected in blood and urine in their original form or methylated, glucuronidated, and sulfoconjugated.[55]

However, the majority of dietary anthocyanins, which remain unabsorbed, are likely to be metabolised by the colonic microbiota. In fact, only a small proportion of dietary anthocyanins are absorbed in the small intestine. Previous studies have shown that less than 1% of anthocyanin parent compounds are identified in the systemic circulation and urine, in particular delphinidin 3-*O*-rutinoside, cyanidin 3-*O*-rutinoside, delphinidin 3-*O*-glucoside, and cyanidin 3-*O*-glucoside, which were found to be directly absorbed and excreted as the glycosylated forms.[53]

The microbial degradation of anthocyanins, including monoglucosides, diglucosides, and acylated anthocyanins, involves the cleavage of the sugar moiety, due to the colon bacterial enzymes α,L-rhamnosidase, β,D-glucosidase, and β,D-glucuronidase. Then, the breakdown of the anthocyanidin heterocycle occurs and various metabolites are produced.[55]

Eker *et al.*[56] reported the microbial metabolites obtained by the gut metabolism of different anthocyanins. In particular, cyanidin derivatives generally produce protocatechuic acid, vanillic acid, *p*-coumaric acid, 2,4,6-trihydroxybenzaldehyde, gallic acid, syringic acid, 2,4,6-trihydroxybenzaldehyde, 2,3-dihydroxybenzoic acid, 3-hydroxycinnamic acid, and 3-hydroxyphenylpropionic

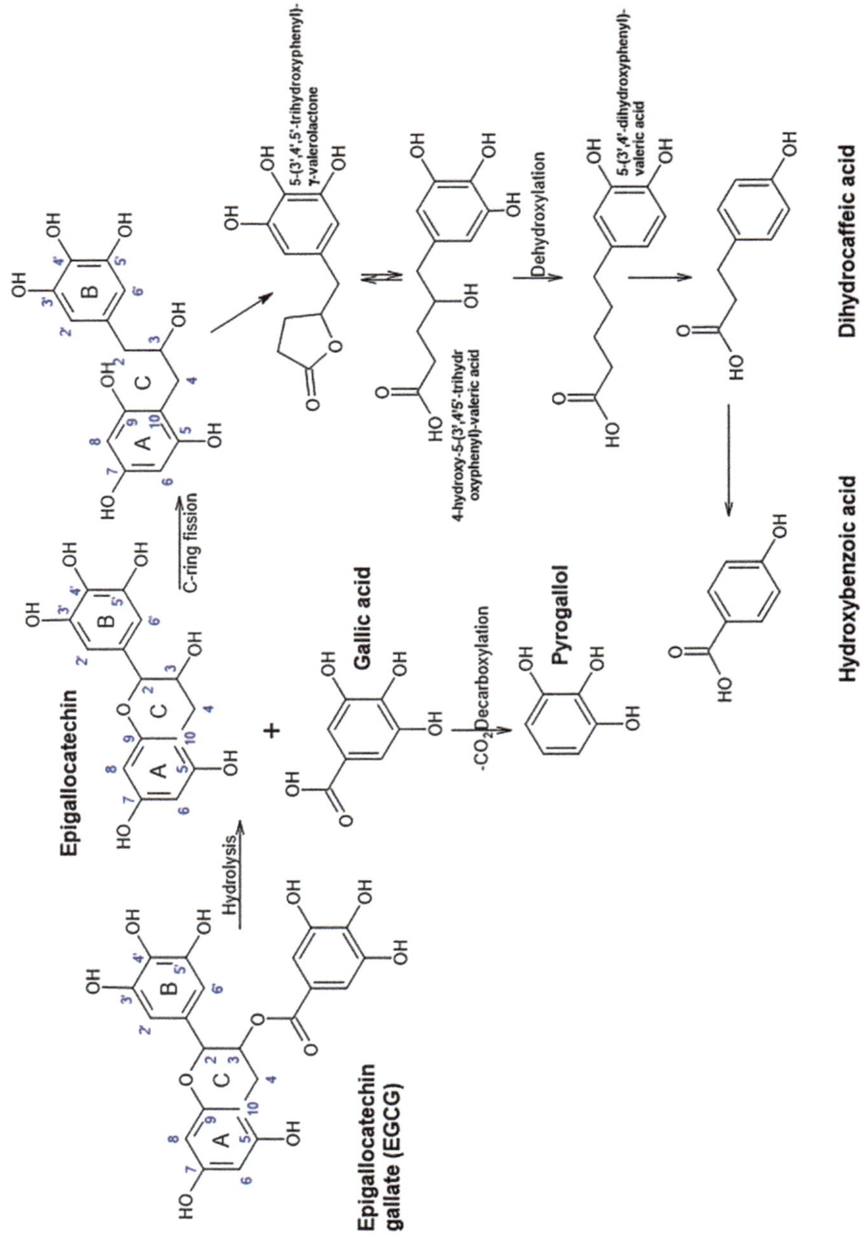

Figure 13.4 Microbial metabolism pathway of epigallocatechin gallate.

acid. In addition, considering the gut metabolism of malvidin derivatives, syringic acid, *p*-coumaric acid, 4-hydroxybenzoic acid, and homovanillic acid are obtained, and pelargonidin-3-glucoside produces tyrosol, hydroxyphenylpropionic acid, hydroxyphenylacetic acid, and *p*-hydroxybenzoic acid.

13.2.5.7 Phenolic Acids

Phenolic acids can originate from the food matrix or from the biotransformation of other polyphenols. Once ingested in the free form, these molecules are rapidly absorbed by the small intestine and conjugated, initially in the small intestine and later in the liver.

The conjugation steps of phenolic acids and the quantity of metabolites formed by the gut microflora in the colon depend on the chemical structures of the starting compounds and the ingested dose.[41] The conjugation reactions are catalysed by various enzymes. Catechol-*O*-methyltransferase catalyses the transfer of a methyl group from *S*-adenosyl-L-methionine to polyphenols that present an *O*-diphenolic moiety. These reactions usually occur in the liver and kidneys. Sulfotransferases promote the transfer of a sulphate moiety from 3′-phosphoadenosine-5′-phosphosulfate to a hydroxyl group on the polyphenols. This reaction takes place in the liver. Uridine 5′-diphospho-glucuronosyltransferase (UDP) catalyses the transfer of a glucuronic acid to polyphenols from UDP-glucuronic acid. This glucuronidation usually occurs in the enterocytes before conjugation in the liver.[57] The esterification of phenolic acid has been proven to highly reduce its absorption relative to the free form. In fact, the absorption of chlorogenic acid, the ester of caffeic acid and quinic acid, takes place primarily in the colon, thanks to hydrolysis by microbial esterases, given that in human tissue there are no esterases able to release caffeic acid from chlorogenic acid.[57]

After the microbial hydrolysation of these molecules, the colonic microorganisms can degrade the aglycones, releasing aromatic acids and reducing the efficiency of their absorption.[57]

Previous *in vitro* studies have shown the catabolism of phenolic acids by the gut microbiota. The biotransformation of caffeic acid and its esters, chlorogenic acid and caftaric acid, has been studied by using an *in vitro* fermentation model with human faecal microbiota. After 2 hours of fermentation, caffeic, quinic, and tartaric acids were no longer detected, but the microbial metabolites, such as 3-hydroxyphenylpropionic and benzoic acids, were detected.[58]

13.2.6 Modulation of the Gut Microbiota

It has been shown that polyphenols may be absorbed directly or may undergo transformation by intestinal bacteria. The biological properties of the obtained products could be substantially different from those of the original compounds. Thus, the biotransformation of polyphenols may result in the production of inactive compounds or in products with an enhanced activity. It is known that polyphenol biotransformation is strictly influenced by the gut microbiota, but these molecules may be capable of influencing the large-intestinal bacterial population.

In order to understand the modulation of gut microbiota by polyphenols or their metabolites, various experimental approaches have been used, from *in vitro* tests to assess the effect of polyphenols on the growth of isolated intestinal bacteria to fermentation experiments or compositional analysis of animal and human faecal samples.

The inhibition of specific microorganisms, including commensal and pathogenic bacteria, by flavonoids has been shown by many studies. Furthermore, while prenyl groups (3-methylbut-2-en-1-yl), alkylamino chains, alkyl chains, and heterocyclic moieties containing nitrogen or oxygen and hydroxyl groups on aromatic rings improve antimicrobial function, methylation decreases such activity. At the same time, flavonoids can increase the presence of specific bacteria within the gut community.[59]

Several studies show the positive modulation of the gut microbiota promoted by polyphenols, highlighting that a polyphenol-based diet promotes beneficial prebiotic effects resulting in the enhancement of the growth and establishment of *Lactobacillus*, *Bifidobacterium*, *Roseburia*, *Akkermansia*, *Faecalibacterium*, and *Bacteroides* and a significant decrease of *Enterococcus faecalis*, *Lachnospiraceae*, and the ratio of *Firmicutes* to *Bacteroides*.[60] This modulation promotes an increase in the composition of short chain fatty acids and a decrease in obesity, inflammation, and the metabolism of adipogenesis and lipogenesis.[61] However, it has to be mentioned that Kemperman *et al.*[62] using an *in vitro* gut microbial ecosystem reported that polyphenols from black tea increased the pathogenic bacteria such as *Klebsiella* and decreased the *Bifidobacteria* in the gut. Considering the influence of phenolic compounds on gut microbiota several studies have reported the inhibitory or stimulatory effect of phenolic compounds on beneficial or pathogenic bacteria, and their ratio in the gut.[53]

Clavel *et al.*[63] showed that isoflavone supplementation (100 mg per day) led to marked changes in the intestinal ecosystem of postmenopausal women. These biomolecules stimulated dominant microorganisms of the *Clostridium coccoides–Eubacterium rectale* cluster, *Lactobacillus–Enterococcus* group, *Faecalibacterium prausnitzii* subgroup, and *Bifidobacterium* genus. In addition, the authors suggested that the stimulation of the *Clostridium coccoides–Eubacterium rectale* cluster depended on the capacity of women to excrete equol and was transient, with the exception of a prolonged bifidogenic effect.

Several *in vitro*, animal, and human studies are in line with the finding that flavanone supplementation, mostly from citrus, inhibits *Enterobacteriaceae*.[51] One study showed that drinking a pasteurised orange juice (with an unknown flavanone content) for two months resulted in a significant increase in *Lactobacillus* spp. and total anaerobes.[64] A second study on orange juice consumption, from two different cultivars of *Citrus sinensis*, resulted in an increase of *Clostridia* operational taxonomic units from *Mogibacteriaceae*, *Tissierellaceae*, *Veillonellaceae*, *Odoribacteraceae*, and *Ruminococcaceae* families *versus* the control drink.[65]

An *in vitro* study, simulating the distal colon, highlighted the different influences of (−)-epicatechin or (+)-catechin on the gut microbiota. While (+)-catechin promotes a significant increase in the *Clostridium coccoides–Eubacterium rectale* group, *Bifidobacterium* spp., and *Escherichia coli* and an inhibition of the *Clostridium histolyticum* group, (−)-epicatechin promotes just an increase of the *Clostridium coccoides–Eubacterium rectale* group.[66] In addition, *in vivo* tests proved that in healthy volunteers the intake of proanthocyanidin-rich extract from grape seeds significantly increased the number of *Bifidobacterium* and decreased the number of putrefactive bacteria and enterobacteria.[67] Another study confirmed these results by showing that the consumption of a high-flavan-3-ol cocoa drink promoted a significant increase in the bifidobacterial and lactobacillus populations, and a decrease in the clostridia count, when compared with a low-flavan-3-ol cocoa drink.[68]

As reported earlier, a large proportion of dietary anthocyanins can reach the colon, where they are metabolised by the microbiota. Thanks to the bacterial metabolism, the anthocyanins are converted into smaller phenolic compounds, such as gallic acid, syringic acid, *p*-coumaric acid, and pyrogallol. Anthocyanins and their metabolites may represent an important tool for a positive modulation of the gut microbiota. These molecules have been proven to significantly enhance the growth of *Bifidobacterium* spp., *Lactobacillus* spp., and *Enterococcus* spp.[55] These bacterial strains have positive effects in the large intestine by promoting the competitive exclusion of potentially detrimental bacteria, such as *Clostridium histolyticum*, and by the production of short-chain fatty acids that have an antimicrobial effect on pathogenic bacteria.[55]

Phenolic acids, together with other aromatic acids, are the prevalent molecules found in faecal water; some of these acids may originate directly from the diet, but the majority result from the microbial metabolism of other dietary phenolic compounds, such as flavanols, flavonols, flavones, and anthocyanins.[69] Their modulation of the gut microorganisms has been widely investigated. Cueva *et al.*[69] showed their antimicrobial potential against several microorganisms such as *Escherichia coli*, *Lactobacillus* spp., *Staphylococcus aureus*, *Pseudomonas aeruginosa*, and *Candida albicans*. Although it was strain-dependent, it was also highly reliant on the number and position of substitutions on the benzene ring of phenolic acids and the saturated side-chain length.

Another *in vitro* test proved that gallic acid incubated in a faecal slurry reduced a group of potentially harmful bacteria such as *Clostridium histolyticum* without any negative effects on beneficial bacteria. In addition, it significantly reduced *Bacteroides* spp. growth and enhanced both the total bacterial number and the abundance of *Atopobium* spp.[70]

Many other studies have described the modulation of the gut microbiota promoted by flavonoid-rich food. Lee *et al.*[71] showed that the tea phenolic compounds, including epicatechin, catechin, 3-*O*-methylgallic acid, gallic acid, and caffeic acid, and the aromatic metabolites obtained from these compounds by the gut microbiota repress the growth of *Clostridium perfringens*, *Clostridium difficile*, and *Bacteroides* spp., whereas they slightly affected commensal anaerobes such as *Clostridium* spp. and *Bifidobacterium* spp., and probiotics such as *Lactobacillus* spp.

Mandalari *et al.*[72] observed a significant increase in *Bifidobacteria* and the *Clostridium coccoides–Eubacterium rectale* group and a decrease in the *Clostridium histolyticum* group using natural and blanched almond skins in a fermentation model with faecal microbiota. The authors related the prebiotic effect not only to a high amount of dietary fibre, but also to some phenolic compounds such as ferulic acid, flavanols, and flavonols present in the almond skins.

It is clear that polyphenols, as single compounds or in mixed formulations, affect the growth of single microorganisms or the entire gut microbiota. However, *in vivo* studies showed that the numerous variations in the gut microbiota among individuals may result in a different response to the same polyphenol. In addition, the interaction of polyphenols with other dietary components could interfere with the interaction between polyphenols and bacteria.[44] One example reported by Mansoorian and Combet[73] showed that fermentable fibres (inulin, pectin, and ispaghula) generally inhibited total phenolic acid production from rutin in batch culture with human faecal bacteria. It has been shown that polyphenols and the gut microbiota have a bi-directional relationship where polyphenols may modulate the gut microbiota composition, and the gut bacteria biotransform polyphenols into bioactive compounds which may produce clinical benefits.[73]

13.3 Conclusion

Plant-based foods provide an exogenous source of antioxidants to the body. In order to be beneficial, they need to be bioavailable, which can be affected by the structure of the food and interactions with other components. The focus of this chapter was polyphenols which are structurally heterogeneous, but many have antioxidant properties. Their antioxidant capacity can be enhanced, reduced, or completely nullified by human and bacterial metabolic transformations, especially through hydroxylation and dehydroxylation reactions. Variations in the bacteria present, and their associated enzymes, as well as the other dietary components present, result in different metabolite profiles, which would be expected to have different antioxidant capacities. Polyphenols can also promote and inhibit the proliferation of certain species of bacteria, adding another layer of complexity.

References

1. S. Di Meo, T. T. Reed, P. Venditti and V. M. Victor, Role of ROS and RNS sources in physiological and pathological conditions, *Oxid. Med. Cell. Longevity*, 2016, **2016**, 1245049.

2. N. Bhattacharjee, R. Paul, A. Giri and A. Borah, Chronic exposure of homocysteine in mice contributes to dopamine loss by enhancing oxidative stress in nigrostriatum and produces behavioral phenotypes of Parkinson's disease, *Biochem. Biophys. Rep.*, 2016, **6**, 47–53.
3. S. Hamulakova, P. Poprac, K. Jomova, V. Brezova, P. Lauro and L. Drostinova, *et al.*, Targeting copper(II)-induced oxidative stress and the acetylcholinesterase system in Alzheimer's disease using multifunctional tacrine-coumarin hybrid molecules, *J. Inorg. Biochem.*, 2016, **161**, 52–62.
4. X. Li, Y. Chen, S. Shao, Q. Tang, W. Chen and Y. Chen, *et al.*, Oxidative stress induces the decline of brain EPO expression in aging rats, *Exp. Gerontol.*, 2016, **83**, 89–93.
5. J. Park, J. Min, B. Kim, U. Chae, J. W. Yun and M. Choi, *et al.*, Mitochondrial ROS govern the LPS-induced pro-inflammatory response in microglia cells by regulating MAPK and NF-κB pathways, *Neurosci. Lett.*, 2015, **584**, 191–196.
6. C. Glorieux, J. M. Sandoval, A. Fattaccioli, N. Dejeans, J. C. Garbe and M. Dieu, *et al.*, Chromatin remodeling regulates catalase expression during cancer cells adaptation to chronic oxidative stress, *Free Radical Biol. Med.*, 2016, **99**, 436–450.
7. P. Formichi, E. Radi, C. Branca, C. Battisti, J. Brunetti and P. Da Pozzo, *et al.*, Oxidative stress-induced apoptosis in peripheral blood lymphocytes from patients with POLG-related disorders, *J. Neurol. Sci.*, 2016, **368**, 359–368.
8. M. Carocho and I. C. F. R. Ferreira, A review on antioxidants, prooxidants and related controversy: Natural and synthetic compounds, screening and analysis methodologies and future perspectives, *Food Chem. Toxicol.*, 2013, **51**, 15–25.
9. E. Thursby and N. Juge, Introduction to the human gut microbiota, *Biochem. J.*, 2017, **474**, 1823–1836.
10. F. Backhed, Host-bacterial mutualism in the human intestine, *Science*, 2015, **307**, 1915–1920.
11. A. S. Neish, Microbes in gastrointestinal health and disease, *Gastroenterology*, 2009, **136**, 65–80.
12. J. M. M. Natividad and E. F. Verdu, Modulation of intestinal barrier by intestinal microbiota: Pathological and therapeutic implications, *Pharmacol. Res.*, 2013, **69**, 42–51.
13. G. den Besten, K. van Eunen, A. K. Groen, K. Venema, D.-J. Reijngoud and B. M. Bakker, The role of short-chain fatty acids in the interplay between diet, gut microbiota, and host energy metabolism, *J. Lipid Res.*, 2013, **54**, 2325–2340.
14. A. J. Bäumler and V. Sperandio, Interactions between the microbiota and pathogenic bacteria in the gut, *Nature*, 2016, **535**, 85–93.
15. T. Gensollen, S. S. Lyer, D. L. Kasper and R. S. Blumberg, How colonization by microbiota in early life shapes the immune system, *Science*, 2016, **352**, 539–544.

16. C. Chang and H. Lin, Dysbiosis in gastrointestinal disorders, *Best Pract. Res., Clin. Gastroenterol.*, 2016, **30**, 3–15.
17. Y. Fan and O. Pedersen, Gut microbiota in human metabolic health and disease, *Nat. Rev. Microbiol.*, 2021, **19**, 55–71.
18. A. M. Aura, K. A. O'Leary, G. Williamson, M. Ojala, M. Bailey and R. Puupponen-Pimia, *et al.*, Quercetin derivatives are deconjugated and converted to hydroxyphenylacetic acids but not methylated by human fecal flora *in vitro*, *J. Agric. Food Chem.*, 2002, **50**, 1725–1730.
19. A. M. Aura, P. Martin-Lopez, K. A O'Leary, G. Williamson, K. M. Oksman-Caldentey and K. Poutanen, *et al.*, *In vitro* metabolism of anthocyanins by human gut microflora, *Eur. J. Nutr.*, 2005, **44**, 133–142.
20. P. C. H. Hollman, M. N. C. P. Busyman, Y. van Gameren, P. J. Cnossen, J. H. M. de Vries and M. B. Katan, The sugar moiety is a major determinant of the absorption of dietary flavonoid glycosides in man, *Free Radical Res.*, 1999, **31**, 569–573.
21. P. C. H. Hollman and I. C. W. Arts, Flavonols, flavones and flavonols-nature, occurrence and dietary burden, *J. Sci. Food Agric.*, 2000, **80**, 1081–1093.
22. J. F. Stevens and C. S. Maier, The chemistry of gut microbial metabolism of polyphenols, *Phytochem. Rev.*, 2016, **15**, 425–444.
23. P. R. Ellis, C. W. Kendall, Y. Ren, C. Parker, J. F. Pacy and K. W. Waldron, *et al.*, Role of cell walls in the bioaccessibility of lipids in almond seeds, *Am. J. Clin. Nutr.*, 2004, **80**, 604–613.
24. D. Tagliazucchi, E. Verzelloni, D. Bertolini and A. Conte, *In vitro* bioaccessibility and antioxidant activity of grape polyphenols, *Food Chem.*, 2010, **120**, 599–606.
25. N. Stanisavljevic, J. Samardzic, T. Jankovic, K. Šavikin, M. Mojsin and V. Topalovic, Antioxidant and antiproliferative activity of chokeberry juice phenolics during *in vitro* simulated digestion in the presence of food matrix, *Food Chem.*, 2015, **175**, 516–522.
26. G. Mandalari, C. Bisignano, A. Filocamo, S. Chessa, M. Saro and G. Torre, *et al.*, Bioaccessibility of pistachio polyphenols, xanthophylls, and tocopherols during simulated human digestion, *Nutrition*, 2013, **29**, 338–344.
27. G. Mandalari, M. Vardakou, R. M. Faulks, C. Bisignano, M. Martorana and A. Smeriglio, *et al.*, Food matrix effects of polyphenol bioaccessibility from almond skin during simulated human digestion, *Nutrients*, 2016, **8**, 568.
28. E. Roura, C. Andres-Lacueva, R. Estruch, M. L. Mata-Bilbao, M. Izquierdo-Pulido and A. L. Waterhouse, *et al.*, Milk does not affect the bioavailability of cocoa powder flavonoid in healthy human, *Ann. Nutr. Metab.*, 2007, **51**, 493–498.
29. E. Roura, C. Andres-Lacueva, R. Estruch, M. L. Mata-Bilbao, M. Izquierdo-Pulido and R. M. Lamuela-Raventos, The effects of milk as a food matrix for polyphenols on the excretion profile of cocoa (−)-epicatechin metabolites in healthy human subjects, *Br. J. Nutr.*, 2008, **100**, 846–851.

30. M. D'Archivio, C. Filesi, R. Varì, B. Scazzocchio and R. Masella, Bioavailability of the polyphenols: Status and controversies, *J. Mol. Sci.*, 2010, **11**, 1321–1342.
31. M. Sazuka, T. Itoi, Y. Suzuki, S. Odani, T. Koide and M. Isemura, Evidence for the interaction between (−)-epigallocatechin gallate and human plasma proteins fibronectin, fibrinogen, and histidine-rich glycoprotein, *Biosci., Biotechnol., Biochem.*, 1996, **60**, 1317–1319.
32. A. Perez-Vicente, A. Gil-Izquierdo and C. Garcia-Viguera, *In vitro* gastrointestinal digestion study of pomegranate juice phenolic compounds, anthocyanins, and vitamin C, *J. Agric. Food Chem.*, 2002, **50**, 2308–2312.
33. G. Williamson and M. N. Clifford, Role of the small intestine, colon and microbiota in determining the metabolic fate of polyphenols, *Biochem. Pharmacol.*, 2017, **139**, 24–39.
34. K. Kamonpatana, M. M. Giusti, C. Chitchumroonchokchai, M. Moreno Cruz, K. M. Riedl and P. Kumar, *et al.*, Susceptibility of anthocyanins to *ex vivo* degradation in human saliva, *Food Chem.*, 2012, **135**, 738–747.
35. X. Han, T. Shen and H. Lou, Dietary polyphenols and their biological significance, *Int. J. Mol. Sci.*, 2007, **8**, 950–988.
36. F. Cardona, C. Andrés-Lacueva, S. Tulipani, F. J. Tinahones and M. I. Queipo-Ortuño, Benefits of polyphenols on gut microbiota and implications in human health, *J. Nutr. Biochem.*, 2013, **24**, 1415–1422.
37. *The Flavonoids – Advances in Research since 1986*, ed. J. B. Harborne, Chapman & Hall, London, UK, 1993.
38. A. J. Day, F. J. Cañada, J. C. Díaz, P. A. Kroon, R. McLauchlan and C. B. Faulds, *et al.*, Intestinal uptake of quercetin-3-glucoside in rats involves hydrolysis by lactase phlorizin hydrolase, *FEBS Lett.*, 2000, **468**, 166–170.
39. M. V. Selma, J. C. Espín and F. A. Tomás-Barberán, Interaction between phenolics and gut microbiota: role in human health, *J. Agric. Food Chem.*, 2009, **57**, 6485–6501.
40. L. Marín, E. M. Miguélez, C. J. Villar and F. Lombó, Bioavailability of dietary polyphenols and gut microbiota metabolism: antimicrobial properties, *BioMed Res. Int.*, 2015, **2015**, 905215.
41. A. Scalbert and G. Williamson, Dietary intake and bioavailability of polyphenols, *J. Nutr.*, 2000, **130**(Suppl. 8), 2073S–2085S.
42. M. A. Mahowald, F. E. Rey, H. Seedorf, P. J. Turnbaugh, R. S. Fulton and A. Wollam, *et al.*, Characterizing a model human gut microbiota composed of members of its two dominant bacterial phyla, *Proc. Natl. Acad. Sci. U. S. A.*, 2009, **106**, 5859–5864.
43. E. Biagi, L. Nylund, M. Candela, R. Ostan, L. Bucci and E. Pini, *et al.*, Through ageing, and beyond: gut microbiota and inflammatory status in seniors and centenarians, *PLoS One*, 2010, **5**, e10667.
44. M. L. Y. Wan, V. A. Co and H. El-Nezami, Dietary polyphenol impact on gut health and microbiota, *Crit. Rev. Food Sci. Nutr.*, 2021, **61**, 690–711.

45. H. Cao, X. Chen, A. R. Jassbi and J. Xiao, Microbial biotransformation of bioactive flavonoids, *Biotechnol. Adv.*, 2015, **33**, 214–223.
46. M. Rossi, A. Amaretti, L. Roncaglia, A. Leonardi and S. Raimondi, Dietary isoflavones and intestinal microbiota: metabolism and transformation into bioactive compounds, in *Isoflavones: Biosynthesis, Occurrence and Health Effects*, ed. M. J. Thompson, Nova Science Publishers, Inc., Hauppauge, New York, NY, 2010, pp. 137–162.
47. M. Makarewicz, I. Drożdż, T. Tarko and A. Duda-Chodak, The interactions between polyphenols and microorganisms, especially gut microbiota, *Antioxidants*, 2021, **10**, 188.
48. K. Murota, Y. Nakamura and M. Uehara, Flavonoid metabolism: the interaction of metabolites and gut microbiota, *Biosci., Biotechnol., Biochem.*, 2018, **82**, 600–610.
49. G. Borges, M. E. Lean, S. A. Roberts and A. Crozier, Bioavailability of dietary (poly)phenols: a study with ileostomists to discriminate between absorption in small and large intestine, *Food Funct.*, 2013, **4**, 754–762.
50. G. Borges, W. Mullen, A. Mullan, M. E. Lean, S. A. Roberts and A. Crozier, Bioavailability of multiple components following acute ingestion of a polyphenol-rich juice drink, *Mol. Nutr. Food Res.*, 2010, **54**(Suppl. 2), S268–S277.
51. Y. Stevens, E. V. Rymenant, C. Grootaert, J. V. Camp, S. Possemiers, A. Masclee and D. Jonkers, The intestinal fate of citrus flavanones and their effects on gastrointestinal health, *Nutrients*, 2019, **11**, 1464.
52. G. L. Hostetler, R. A. Ralston and S. J. Schwartz, Flavones: food sources, bioavailability, metabolism, and bioactivity, *Adv. Nutr.*, 2017, **8**, 423–435.
53. T. Ozdal, D. A. Sela, J. Xiao, D. Boyacioglu, F. Chen and E. Capanoglu, The reciprocal interactions between polyphenols and gut microbiota and effects on bioaccessibility, *Nutrients*, 2016, **8**, 78.
54. M. Monagas, M. Urpi-Sarda, F. Sánchez-Patán, R. Llorach, I. Garrido and C. Gómez-Cordovés, *et al.*, Insights into the metabolism and microbial biotransformation of dietary flavan-3-ols and the bioactivity of their metabolites, *Food Funct.*, 2010, **1**, 233–253.
55. L. Tian, Y. Tan, G. Chen, G. Wang, J. Sun and S. Ou, *et al.*, Metabolism of anthocyanins and consequent effects on the gut microbiota, *Crit. Rev. Food Sci. Nutr.*, 2019, **59**, 982–991.
56. M. E. Eker, K. Aaby, I. Budic-Leto, S. R. Brnčić, S. N. El and S. Karakaya, *et al.*, A review of factors affecting anthocyanin bioavailability: possible implications for the inter-individual variability, *Foods*, 2019, **9**, 2.
57. S. A. Heleno, A. Martins, M. J. Queiroz and I. C. Ferreira, Bioactivity of phenolic acids: metabolites versus parent compounds: a review, *Food Chem.*, 2015, **173**, 501–513.
58. M. P. Gonthier, C. Remesy, A. Scalbert, V. Cheynier, J. M. Souquet and K. Poutanen, *et al.*, Microbial metabolism of caffeic acid and its esters

chlorogenic and caftaric acids by human faecal microbiota *in vitro*, *Biomed. Pharmacother.*, 2006, **60**, 536–540.
59. P. I. Oteiza, C. G. Fraga, D. A. Mills and D. H. Taft, Flavonoids and the gastrointestinal tract: Local and systemic effects, *Mol. Asp. Med.*, 2018, **61**, 41–49.
60. S. Mithul Aravind, S. Wichienchot, R. Tsao, S. Ramakrishnan and S. Chakkaravarthi, Role of dietary polyphenols on gut microbiota, their metabolites and health benefits, *Food Res. Int.*, 2021, **142**, 110189.
61. V. Gowd, N. Karim, M. R. I. Shishir, L. Xie and W. Chen, Dietary polyphenols to combat the metabolic diseases *via* altering gut microbiota, *Trends Food Sci. Technol.*, 2019, **93**, 81–93.
62. R. A. Kemperman, S. Bolca, L. C. Roger and E. E. Vaughan, Novel approaches for analysing gut microbes and dietary polyphenols: challenges and opportunities, *Microbiology*, 2010, **156**, 3224–3231.
63. T. Clavel, M. Fallani, P. Lepage, F. Levenez, J. Mathey and V. Rochet, *et al.*, Isoflavones and functional foods alter the dominant intestinal microbiota in postmenopausal women, *J. Nutr.*, 2005, **135**, 2786–2792.
64. A. C. D. Lima, C. Cecatti, M. P. Fidélix, M. A. T. Adorno, I. K. Sakamoto and T. B. Cesar, *et al.*, Effect of daily consumption of orange juice on the levels of blood glucose, lipids, and gut microbiota metabolites: controlled clinical trials, *J. Med. Food*, 2019, **22**, 202–210.
65. E. Brasili, N. M. A. Hassimotto, F. Del Chierico, F. Marini, A. Quagliariello and F. Sciubba, *et al.*, Daily consumption of orange juice from *Citrus sinensis* L. Osbeck cv. Cara cara and cv. Bahia differently affects gut microbiota profiling as unveiled by an integrated meta-omics approach, *J. Agric. Food Chem.*, 2019, **67**, 1381–1391.
66. X. Tzounis, J. Vulevic, G. G. Kuhnle, T. George, J. Leonczak and G. R. Gibson, *et al.*, Flavanol monomer-induced changes to the human faecal microflora, *Br. J. Nutr.*, 2008, **99**, 782–792.
67. J. Yamakoshi, S. Kataoka, T. Koga and T. Ariga, Proanthocyanidin-rich extract from grape seeds attenuates the development of aortic atherosclerosis in cholesterol-fed rabbits, *Atherosclerosis*, 1999, **142**, 139–149.
68. X. Tzounis, A. Rodriguez-Mateos, J. Vulevic, G. R. Gibson, C. Kwik-Uribe and J. P. Spencer, Prebiotic evaluation of cocoa-derived flavanols in healthy humans by using a randomized, controlled, double-blind, crossover intervention study, *Am. J. Clin. Nutr.*, 2011, **93**, 62–72.
69. C. Cueva, M. V. Moreno-Arribas, P. J. Martín-Álvarez, G. Bills, M. F. Vicente and A. Basilio, *et al.*, Antimicrobial activity of phenolic acids against commensal, probiotic and pathogenic bacteria, *Res. Microbiol.*, 2010, **161**, 372–382.
70. M. Hidalgo, M. J. Oruna-Concha, S. Kolida, G. E. Walton, S. Kallithraka and J. P. Spencer, *et al.*, Metabolism of anthocyanins by human gut microflora and their influence on gut bacterial growth, *J. Agric. Food Chem.*, 2012, **60**, 3882–3890.

71. H. C. Lee, A. M. Jenner, C. S. Low and Y. K. Lee, Effect of tea phenolics and their aromatic fecal bacterial metabolites on intestinal microbiota, *Res. Microbiol.*, 2006, **157**, 876–884.
72. G. Mandalari, R. M. Faulks, C. Bisignano, K. W. Waldron, A. Narbad and M. S. Wickham, *In vitro* evaluation of the prebiotic properties of almond skins (*Amygdalus communis* L.), *FEMS Microbiol. Lett.*, 2010, **304**, 116–122.
73. B. Mansoorian and E. Combet, Impact of fermentable fibres on the colonic microbiota metabolism of dietary polyphenols rutin and quercetin, *Int. J. Environ. Res. Public Health*, 2019, **16**, 292.

CHAPTER 14

Digestive Health and Gastrointestinal Protection

SHENG-YI CHEN AND GOW-CHIN YEN*

Department of Food Science and Biotechnology, National Chung Hsing University, Taichung, Taiwan
*Email: gcyen@nchu.edu.tw

14.1 Introduction

The digestive system is directly related to physical health. The American Gastroenterological Association defines good digestive health as a digestive system that fits key features including proper nutrient absorption, gastrointestinal (GI) motility, immune function, and communities of the microbiome.[1] Moreover, the GI mucosa plays a critical role in digestion and absorption of dietary nutrients and is the first protective barrier that defends noxious dietary substances and foreign pathogens.[2] Any abnormalities of the GI mucosa may cause disorders with the decline of organ functions.

GI disorders are common in all regions of the world. Recently, the statistical surveys indicated approximately 40.3% of individuals being with GI disorders globally, such as irritable bowel syndrome (IBS), functional dyspepsia (FD), and rumination syndrome according to Rome IV criteria.[3,4] The two most common GI disorders are IBS and FD, the worldwide prevalence of which was estimated at 32.5% and 38%, respectively.[5,6] IBS includes coeliac disease, inflammatory bowel disease (IBD), colorectal cancer, chronic GI infections, microscopic colitis, and bile acid diarrhea.[7] Accordingly, the annual economic cost of IBS on healthcare systems and maintaining the quality of individual

Food Chemistry, Function and Analysis No. 39
Dietary Supplements with Antioxidant Activity: Understanding Mechanisms and Potential Health Benefits
Edited by Cesarettin Alasalvar, Fereidoon Shahidi and Chi-Tang Ho

Published by the Royal Society of Chemistry, www.rsc.org

life are assessed to be up to £200 million in the UK,[8] €4 billion in Germany,[9] ¥123 billion in China,[10] and $10 billion in the US,[11] indicating GI disorders cause a huge financial burden on individuals and the society.

In addition, considerable scientific experiments and clinical data demonstrate that GI disorders involve broken epithelial barrier integrity, enteroendocrine signalling, gut microbiota, and GI immune system.[12] Many aggressive factors have been linked to GI disorders, such as endogenous factors (genetics, gastric acid, and pepsin) and exogenous stimulations (stress, alcohol consumption, *Helicobacter pylori* infection, and some anti-inflammatory drugs).[13]

Nonsteroidal anti-inflammatory drugs (NSAIDs) include aspirin, ibuprofen, fenbufen, indomethacin, diclofenac sodium, celecoxib, etoricoxib, and meloxicam, among others.[14] More than 30 million people take NSAIDs daily, which are the most common cause of GI injury *via* cyclooxygenase-1 (COX-1) and cyclooxygenase-2 (COX-2) inhibition and subsequent increase in the intestinal permeability and inflammation response.[15] Fortunately, growing evidence supporting dietary interventions play a dominant role in the maintenance of GI barrier integrity and alleviation of the processes of GI disorders.[16] In this chapter, the underlying mechanisms for GI protective effects by consumption of camellia oil and other phytochemicals through antioxidant and anti-inflammatory activities, mucosal healing, and gut microbiota modulation are summarized.

14.2 Health Benefits of Dietary Supplements with Antioxidant Activity for Digestive Health and Gastrointestinal Protection

14.2.1 Epidemiological Studies

Epidemiological studies have suggested that certain dietary substances can improve health outcomes and prevent disease through gut microbiota modulation.[17,18] Mediterranean diet (MedDiet) is the most recognized dietary pattern in the world due to its health benefits *via* antioxidant and anti-inflammatory properties through microbiome communications.[19,20] MedDiet is characterized by high consumption of extra virgin olive oil (EVOO), legumes, nuts, unrefined cereals, fish, fruits, and vegetables; moderate consumption of cheese, yogurt, and wine; and low consumption of meat and meat products in lifestyle habits.[21] In addition, a survey from Southern Italy shows that higher prevalence of IBS and FD is found in participants aged 17–34 years with lower adherence scores of MedDiet (0.57 ± 0.23, $P < 0.001$ and 0.56 ± 0.24, $P < 0.05$, respectively) compared to healthy volunteers (0.62 ± 0.21), indicating that low adherence to MedDiet may enhance GI disorders in younger subjects.[22] Likewise, approximately 184 out of 835 participants (aged 6–18 years) were identified with GI disorders (22%), accompanied by a low MedDiet quality index (KIDMED) score in contrast to the health of adolescents.[23]

Besides, IBDs include ulcerative colitis (UC) and Crohn's disease (CD), which are characterized by the dysregulation of the immune response and gut microbiota in genetically susceptible hosts.[24] Recently, a cohort study comprising 83 147 participants (aged 45–79 years) in Sweden suggested that adherence to MedDiet significantly reduced the risk of later-onset CD.[25] Likewise, the beneficial effects of MedDiet were also found in patients with UC after pouch surgery through a decrease of diagnostic markers such as faecal calprotectin during the trial.[26]

MedDiet is typical with high consumption of fruits and vegetables. The relationship between polyphenol consumption and IBD risk has not been proven by epidemiological approaches.[27] However, high consumption of fruits and vegetables is associated with a lower risk of UC and CD development,[28] indicating the beneficial effects of polyphenols for digestive health and GI protection.

14.2.2 *In Vitro* and *In Vivo* Animal Studies

EVOO is the major component of MedDiet and plays an important role in the amelioration of the symptoms in GI disorders through gut microbiota modulation.[29,30] Previous evidence has shown that pre-treatment with EVOO polyphenols counteracts the oxysterol-induced oxidative stress and inflammation response in intestinal cells.[31] The GI protective effect of EVOO administration is also observed *in vivo* through gut microbiota regulation.[32]

Camellia oil (*Camellia oleifera* Abel. and *Camellia brevistyla*) is one of the most widely consumed edible oils in Asia. It is rich in monounsaturated fatty acids (MUFA), polyphenols, phytosterols (stigmasterol, β-sitosterol, and avenasterol), fat-soluble vitamins (α-tocopherol, β-tocopherol, γ-tocopherol, and δ-tocopherol), catechin, epicatechin, squalene, and other functional substances.[33] Interestingly, camellia oil is known as oriental olive oil because its physicochemical capacity and fatty acid compositions are similar to EVOO, particularly in the content of oleic acid.[33] In addition, information about the role of camellia oil in the alleviation of GI disorder progression is well-established compared to that of EVOO administration. Previous studies have also shown the anti-ulcer properties of camellia oil through its antioxidant and anti-inflammatory activities.[34]

Animal models of GI disorders such as NSAID, ethanol, and acetic acid treatment are established to validate the protective effect of camellia oil. The anti-inflammatory, antioxidative, and microbiota-modulating effects of camellia oil (*Camellia oleifera* Abel. and *Camellia brevistyla*) administration were demonstrated in acetic acid-induced ulcerative colitis and also in GI mucosal injury rats and BALB/c mice *via* NSAID (ketoprofen and indomethacin) induction.[35–38] Additionally, camellia oil (*Camellia oleifera* Abel.) was effective against ethanol-induced acute gastric mucosal damage by reducing the oxidative status in mice.[39] The GI protective capacity of camellia oil is outlined in Table 14.1.

Table 14.1 Effects of camellia oil on GI protection.

Treatment (dosage)	Models (stimulants)	Effects/key findings	Ref.
2 mL per kg bw per day camellia oil (*Camellia brevistyla*)	4% acetic acid (5 mL per kg bw)-induced colitis in SD rats	Mediation of the composition of the gut microbiota through an increase of the *Firmicutes*/*Bacteroidetes* ratio, abundance of *Bifidobacterium*, and a decrease of *Prevotella* which consequently suppress oxidative stress and inflammatory responses.	36
2 mL per kg bw per day camellia oil (*Camellia brevistyla*)	Indomethacin (40 mg per kg bw)-induced gastrointestinal mucosal damage in male BALB/c mice	Down-regulation of IL-6, TNF-α, and COX-2 secretion and up-regulation of the production of HO-1 and VEGF.	35
2 mL per kg bw per day camellia oil (*Camellia oleifera* Abel.)	4% acetic acid (5 mL per kg bw)-induced colitis in SD rats	Raising the diversity of the gut microbiota and promoting the antioxidant system by increasing the levels of GSH and SOD while decreasing the levels of MDA and MPO.	37
2 mL per kg bw per day camellia oil (*Camellia oleifera* Abel.)	Absolute ethanol (5 mL per kg bw)-induced acute gastric mucosal injury in male BALB/c mice	Enhancement of antioxidant enzyme activities and heat shock protein and PGE2 production, as well as the suppression of lipid peroxidation, apoptosis-related proteins, pro-inflammatory cytokines, and NO production.	39
2 mL per kg bw per day camellia oil (*Camellia oleifera* Abel.)	Ketoprofen (50 mg per kg bw)-induced gastrointestinal mucosal damage in male SD rats	Decreasing oxidative damage by inhibition of COX-2, IL-6, and NO expressions and upregulation of HO-1 and VEGF in the gastrointestinal mucosa subsequently reversed the oxidative injury induced by ketoprofen treatment.	38

Abbreviations: bw, body weight; COX-2, cyclooxygenase-2; GI, gastrointestinal; GSH, glutathione; IL-6, interleukin-6; HO-1, heme oxygenase-1; MPO, myeloperoxidase; NO, nitric oxide; PGE2, prostaglandin E2; SOD, superoxide dismutase; TNF-α, tumor necrosis factor-alpha; VEGF, vascular endothelial growth factor.

Furthermore, polyphenolic compounds have been extensively studied in UC and CD prevention through their antioxidant and anti-inflammatory activities. Previous literature reported that resveratrol has important effects on the modulation of intestinal microflora and antioxidant ability in animal models with IBD.[40] Moreover, curcumin administration significantly decreases the expression of inflammatory cytokines and increases antioxidant enzyme activities [catalase, superoxide dismutase-2 (SOD-2), and myeloperoxidase (MPO)] in rats with dextran sulfate sodium (DSS) induced colitis.[41] Likewise, ellagic acid inhibits tumor necrosis factor-alpha (TNF-α)-mediated chronic intestinal inflammation through the reduction of interleukin-6 (IL-6), interleukin-8 (IL-8), and intercellular adhesion molecule-1 (ICAM-1) production and up-regulation of antioxidant pathways in the Caco-2 cell monolayer.[42] In addition, it was found that polyphenols from grape seed, dried blueberries, cranberry extract, dealcoholized muscadine wine, Brazilian berry extract, ellagitannins, and epigallocatechin gallate mitigate IBD *via* prevention of the colon shortening and suppression of inflammatory cytokines and reactive oxygen species (ROS) accumulation by modulation of the gut microbiota in various animal models.[43]

14.2.3 Human Intervention Studies

In addition to *in vitro* and *in vivo* animal studies, a large number of naturally occurring, plant-based dietary polyphenols have abundant antioxidant components for GI protection through gut microbiota modulation in humans.[44] For example, polyphenols extracted from strawberry, green tea, red wine, and other sources have been shown to enhance digestive health and prevent GI disorders by modifying gut microbiota composition.[45]

Approximately 90–95% of unabsorbed polyphenols may accumulate in the colon and metabolized by the resident microbiota.[46] Short-chain fatty acids (SCFA) are produced from the metabolism of dietary polyphenols *via* the gut microbiota and play an important role in maintaining gut microbiota homeostasis.[47] A recent report demonstrated that SCFA (acetic, propionic, and butyric acids) are the major energy source of human colonocytes for maintaining functions such as cell proliferation, differentiation, and metabolism,[48] indicating the crucial role of dietary polyphenols in human digestive health and GI protection. It was shown that berry polyphenols (resveratrol and quercetin) and tea polyphenols (ellagitannins, ellagic acid, and epigallocatechin gallate) exert free radical scavenging and GI tract beneficial effects.[49,50] Moreover, accumulating evidence indicates that dietary polyphenol consumption significantly decreased colitis and colitis-associated colorectal cancer (CAC) through reducing inflammatory activities and elevating butyrate production by *Lactobacillus* and *Bifidobacterium* in humans.[51] Furthermore, food-derived polyphenols (stilbenes, flavonoids, hydroxybenzoic acids, lignans, and coumarins, among others) have been shown to render synergistic effects between the health benefits of their metabolites and modification of the gut microbiota,[52] implying their alternative and complementary roles in the treatment of colitis.

Based on the above information, it is suggested that consumption of camellia oil and plant polyphenols can significantly reduce GI disorders by their antioxidant capacity and the benefits of microbiota metabolism. However, the underlying mechanisms of camellia oil and plant polyphenols for GI protection through antioxidant and anti-inflammatory activities, mucosal healing, and gut microbiota modulation are not integrated.

14.3 Understanding the Mechanisms of Dietary Supplements with Antioxidant Activity for Digestive Health and Gastrointestinal Protection

Oxidative stress is considered the most important etiologic factor involved in several mechanisms for the pathogenesis of GI disorders.[53] Recently published reports indicate that dietary products contain numerous polyphenols which have the potential for prevention of GI disorders *via* their antioxidative ability and gut microbiota modulation.[44,45] In the following section, the effects of camellia oil and some plant polyphenols (resveratrol, curcumin, ellagitannins, ellagic acid, epigallocatechin gallate, quercetin, lycopene, rutin, phloretin, and rosmarinic acid) are summarized and their specific aspects for GI protection through various underlying mechanisms are illustrated in Figure 14.1.

14.3.1 Camellia Oil

As mentioned earlier, camellia oil contains various bioactive compounds involved in colitis alleviation and GI protection. Pre-treatment with camellia oil (2 mL per kg per day, *Camellia oleifera* Abel.) successfully prohibited COX-2, IL-6, and nitric oxide (NO) protein expressions from ketoprofen-induced acute GI ulcers in rats.[38] Likewise, the protective effect of camellia oil (*Camellia brevistyla*) was observed by down-regulation of IL-6, TNF-α, and COX-2 secretion and up-regulation of the production of heme oxygenase-1 (HO-1) and vascular endothelial growth factor (VEGF) in indomethacin-induced GI mucosal damage *in vitro* and *in vivo*.[35] Furthermore, administration of camellia oils (*Camellia oleifera* Abel. and *Camellia brevistyla*) can mediate the composition of the gut microbiota through an increase of the *Firmicutes*/*Bacteroidetes* ratio, the abundance of *Bifidobacterium*, and a decrease of *Prevotella* which consequently suppress oxidative stress and inflammatory responses in acetic acid-induced colitis in rats.[36,37] Therefore, camellia oil is able to reduce oxidative damage and GI disorders by triggering the antioxidative and microbiota modulation capacity.

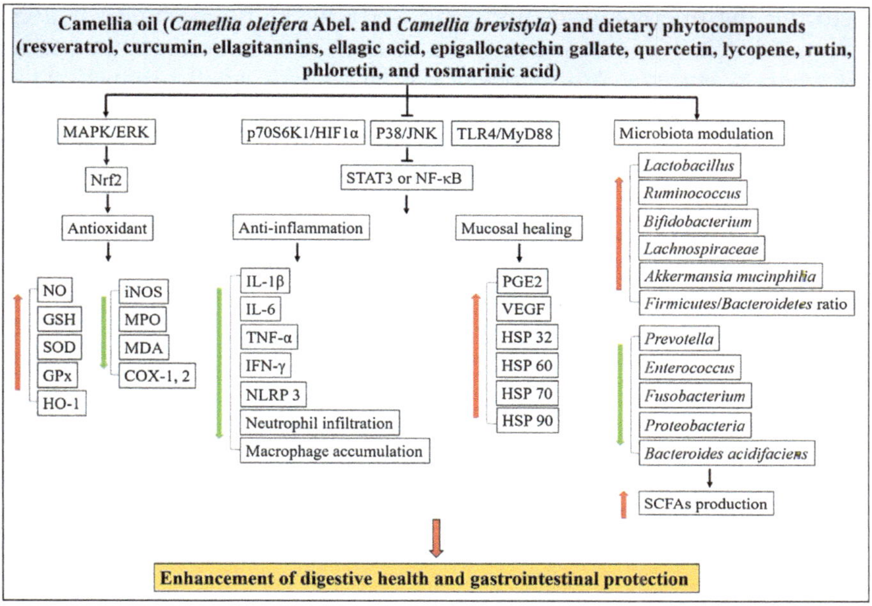

Figure 14.1 The gastrointestinal protective effects of camellia oil and phytocompounds through the underlying mechanisms. Camellia oil and phytochemicals (resveratrol, curcumin, ellagitannins, ellagic acid, epigallocatechin gallate, quercetin, lycopene, rutin, phloretin, and rosmarinic acid) against IBD through multiple biological functions, such as antioxidant, anti-inflammatory, mucosal healing, and microbiota modulation. Abbreviations: COX-1, 2, cyclooxygenase-1, 2; ERK, extracellular signal-regulated kinase; GPx, glutathione peroxidase; GSH, glutathione; HIF-1α, hypoxia-inducible factor-1alpha; HO-1, heme oxygenase-1; HSP, heat shock protein; IBD, inflammatory bowel disease; IFN-γ, interferon-gamma; IL-1β, interleukin-1 beta; IL-6, interleukin-6; iNOS, inducible nitric oxide synthase; JNK, Jun N-terminal kinases; MAPK, mitogen-activated protein kinase; MDA, malondialdehyde; MPO, myeloperoxidase; MyD88, myeloid differentiation factor 88; NF-κB, nuclear factor kappa-light-chain-enhancer of activated B; NLRP3, NLR family pyrin domain containing 3; NO, nitric oxide; Nrf2, nuclear factor erythroid 2-related factor 2; p70S6K1, phosphorribosomal protein S6 kinase beta-1; PGE2, prostaglandin E2; SCFAs, short chain fatty acids; SOD, superoxide dismutase; STAT3, activator of transcription 3; TLR4, toll-like receptor 4; TNF-α, tumor necrosis factor-alpha; VEGF, vascular endothelial growth factor.

14.3.2 Resveratrol

Resveratrol (3,4′,5-trihydroxystilbene) is a natural stilbenoid polyphenol that is synthesized by knotweeds, pine trees, grapevines, and berries.[54] Resveratrol exhibits multiple health benefits and plays important roles in the protection of the GI tract or digestive system. Mechanistically, administration of resveratrol attenuates colonic inflammation and clinical symptoms through restoring the microbiota composition (lowering the abundance of *Bacteroides acidifaciens* and increasing the species of *Ruminococcus gnavus* and *Akkermansia muciniphila*), increasing SCFA (butyric acid) production, and induction of regulatory T cells (Tregs) in the murine model of 2,4,6-trinitrobenzenesulfonic acid (TNBS)-induced colitis.[55] In addition, the results of the clinical trial with patients revealed that 500 mg per day resveratrol supplementation for 6 weeks effectively ameliorated the UC progress and quality of life *via* reduction of the serum level of malondialdehyde (MDA), induction of the level of SOD, and total antioxidant capacity compared that to the placebo group.[56] These results suggest that resveratrol has preventive and therapeutic effects on GI disorders through antioxidative and anti-inflammatory activities, and maintenance of the homeostatic microbiota.

14.3.3 Curcumin

Curcumin is a polyphenolic compound isolated from the rhizome of the plants of the Zingiberaceae family, especially turmeric (*Curcuma longa*).[57] A previous report found that curcumin has good safety and is suitable for various clinical trial applications.[58] However, the major problem with curcumin is its extremely poor bioavailability due to low solubility, stability, and rapid metabolism.[59] The GI effects enhanced by curcumin treatment are observed through the anti-inflammatory, antioxidative, microbiota regulation, and intestinal permeability protective properties.[59] Current hypotheses indicate that curcumin may directly regulate the GI function and gut microbiota as it is detected at a high concentration in the GI tract after oral administration.[60] In addition, oral administration of curcumin at 200 mg per kg bw for 7 days significantly decreased the oxidative stress and inflammatory response by suppression of TNF-α production and neutrophil infiltration in indomethacin-induced gastropathy in rats.[61] Moreover, curcumin mitigates IBD *via* inhibition of free radicals, MPO, COX-1, COX-2, lipoxygenase (LOX), inducible nitric oxide synthase (iNOS), TNF-α, interferon-gamma (IFN-γ), and the transcription factor of nuclear factor kappa-light-chain-enhancer of activated B (NF-κB). It also evocates antioxidants by multiple signalling pathways, especially the mitogen-activated protein kinase (MAPK)/extracellular signal-regulated kinase (ERK) axis in experimental animals and clinical studies.[62]

14.3.4 Ellagitannins

Ellagitannins are hydrolyzable tannins belonging to polyphenols and present in some fruits and nuts, such as pomegranate, strawberries, black raspberries, grapes, pecans, almonds, and walnuts.[63] Articles merely mention that the high antioxidant and anti-inflammation capacity of ellagitannin prevents colitis processes.[64] Earlier evidence demonstrated that ellagitannin mitigated inflammatory response and ulceration through the microRNA 145 (miR-145)/phosphor-ribosomal protein S6 kinase beta-1 (p70S6K1)/hypoxia-inducible factor 1α (HIF1α) axis in rats with DSS-induced colitis and lipopolysaccharide (LPS)-treated human myofibroblasts.[65] In addition, ellagitannin-rich red raspberry showed high antioxidative ability in the inhibition of xanthine oxidase (XO) and increase of catalase contents *via* the activation of AMP-activated protein kinase (AMPK) in DSS-induced acute colitis.[66] Noteworthily, ellagitannin-rich pomegranate byproducts have been shown to suppress the growth of pathogenic bacteria (*Clostridia, Bacteroides fragilis*, and *Staphylococcus aureus*) which may alleviate Crohn's disease and ulcerative colitis development.[67] Furthermore, a double-blind placebo-controlled trial provided valuable information for reducing faecal calprotectin levels and relapse risk in IBD patients by daily supplementation of pomegranate juice (125 mL) for 12 weeks.[68]

14.3.5 Ellagic Acid

Ellagic acid (EA) is a natural dietary polyphenol that exists abundantly in numerous fruits and vegetables, in particular berries, pomegranate, persimmon, tea, and wine.[69] Several reports indicate that EA has high anti-inflammatory and antioxidative abilities and is effective against DSS- or TNBS-induced colitis *via* suppression of COX-2, iNOS, and their related p38, c-Jun N-terminal kinase (JNK), pERK1/2, MAPK, signal transducer and activator of transcription 3 (STAT3), and NF-κB signalling pathways.[70,71] In addition, EA and its metabolites (Urolithin A) have been shown to alleviate or prevent colitis and Crohn's disease *via* protecting the intestinal barrier permeabilization and improving the gut microbiota through the nuclear factor erythroid 2-related factor 2 (Nrf2) signalling pathway.[42,72] Moreover, a study found that EA attenuates DSS-induced colitis and increases the relative abundance of butyrate-producing bacteria (*Ruminococcaceae*) by modulating the miR-145/p70S6K/HIF1α axis in rats.[65] These results imply a protective effect of EA on GI disorders.

14.3.6 Epigallocatechin Gallate

Epigallocatechin gallate (EGCG) is known as the most abundant polyphenolic catechin in green tea.[73] It has high anti-inflammatory activity which significantly reduces intestinal inflammatory response and regulates T helper-1 (Th-1)/Th-2 polarization *via* the toll-like receptor 4 (TLR4)/myeloid differentiation factor 88 (MyD88)/NF-κB signalling pathway in DSS-induced ulcerative colitis in rats.[74] Furthermore, the strong antioxidative potential of EGCG is observed by reduction of histological damage, MDA, MPO, and neutrophils accumulation in colon tissue, while inducing glutathione peroxidase (GPx) and SOD activities in a murine model of colitis.[75] Additionally, supplementation of EGCG enhanced the relative abundance of probiotics such as *Lactobacillus*, *Bifidobacterium*, and *Ruminococcus*, and thus consequently diminished colitis development and prevented colitis-associated colorectal cancer.[51] Besides, EGCG is widely used in clinical studies. For example, an epidemiological study indicated that EGCG is the potent ROS scavenger, which protects patients against oxidative stress and inflammatory cascade related IBD *via* inhibition of the NF-κB signalling pathway.[76]

14.3.7 Quercetin

Quercetin is the most abundant dietary polyphenolic flavonoid and widely distributed in fruits and vegetables such as cranberries, grapes, apples, cilantro, asparagus, and onions, among others.[77] A previous report demonstrated that quercetin inhibits macrophage infiltration and iNOS expression through down-regulation of the NF-κB pathway in rats with DSS-induced colitis.[78] Quercetin is not only an anti-inflammatory agent but also a powerful antioxidant. For instance, the antioxidative effect of quercetin on the alleviation of H_2O_2-evoked intestinal oxidative damage and DSS-induced colitis *via* upregulation of NADPH oxidase (NOX) and glutathione (GSH) and inhibition of aquaporin-3 (AQP3) *in vitro* and *in vivo* has been reported.[79] Besides, quercetin supplementation enhanced the relative abundance of *Bacteroides*, *Bifidobacterium*, *Lactobacillus*, and *Clostridia* and suppressed the relative abundance of *Fusobacterium* and *Enterococcus* in *Citrobacter rodentium*-induced colitis in mice.[80]

14.3.8 Lycopene

Lycopene is a potent antioxidant produced by red vegetables and fruits, such as tomatoes. Previous evidence has shown that lycopene mitigates sulfamethoxazole-induced oxidative stress, inflammation, apoptosis, and tight junction destruction in the intestinal mucosa of grass carps through the AMPK/protein kinase B (ATK)/mammalian target of rapamycin (mTOR) pathway.[81] Moreover, lycopene (100 mg per kg body weight, bw) administration was found to significantly reduce MDA, MPO, and DNA damage in indomethacin-induced gastric ulcers in rats.[82] Additionally, lycopene (50 mg per kg bw) supplementation for 40 days effectively prevented DSS-induced colitis and anxiety-like disorders by increasing the relative abundance of *Bifidobacterium* and *Lactobacillus* and SCFA production in mice.[83] These findings suggest that lycopene plays protective roles in regulating antioxidant, anti-inflammatory, and gut microbiota modulation properties in the gastric mucosa.

14.3.9 Rutin

Rutin is a common dietary flavonoid that is present in various fruits and vegetables such as buckwheat, asparagus, figs, grapes, apples, berries, onions, and citrus, among others.[84] Much evidence has shown that rutin has various bioactivities, including antioxidant, anti-inflammatory, anticancer, antibacterial, cardiovascular protective, and neuroprotective properties.[85] Besides, studies have confirmed the antioxidant activity of rutin in digestive health and gastrointestinal protection. For example, rutin treatment significantly reduced 5-fluorouracil-induced intestinal mucositis through suppression of MDA, MPO, inflammatory response, and COX-2 expression and enhancement of GSH levels in mice.[86] In addition, rutin supplementation significantly improved high-fat diet-induced colitis *via* enhancement of the relative abundance of *Lachnospiraceae* and inhibits the growth of *Proteobacteria*.[87] Moreover, supplementation of a diet containing 0.1% rutin significantly reduced IL-1β, IL-6, and COX-2 expression, consequently improving DSS-induced colitis in mice.[88] The above evidence indicates that rutin has antioxidant and anti-inflammatory capacities for digestive health and gastrointestinal protection.

14.3.10 Phloretin

Phloretin is a dihydrochalcone flavonoid abundant in apples. Growing evidence has demonstrated the protective effects of phloretin against colitis *via* the antioxidant and anti-inflammatory characteristics.[89,90] Mechanically, phloretin (25, 50, and 100 mg per kg bw) pre-administration for 14 days dramatically reduced the colon length shortening and colonic pathological damage through regulating the NF-κB, TLR4, and peroxisome proliferator-activated receptor γ (PPARγ) pathways in mice with DSS-induced colitis.[89] In addition, oral gavage of phloretin (60 mg per kg bw) alleviated UC symptoms, colon inflammation, oxidative stress, and inflammatory cytokines (TNF-α, IL-6, and IL-1β) through inhibiting NF-κB and NLR family pyrin domain containing 3 (NLRP3) activation in DSS-induced colitis in mice.[90] Finally, both faecal microbiota transplantation from phloretin-treated UC mice and phloretin treatment significantly increased the abundance of *Bacteroidetes*, *Alistipes*, and *Lactobacillus*,[90] indicating the potential of phloretin for colitis prevention.

14.3.11 Rosmarinic Acid

Rosmarinic acid is found in several culinary herbs such as rosemary (*Rosmarinus officinalis* Linn.), basil (*Ocimum basilicum*), common thyme (*Thymus vulgaris*), and sage (*Salvia officinalis*). Many beneficial and health-promoting effects of rosmarinic acid have been evidenced, including protection of DNA damage, protection against ultraviolet B (UVB) light, neuroprotective effects, cancer prevention, amelioration of liver fibrosis, and treatment of rheumatoid arthritis.[91] Besides, oral administration of rosmarinic acid significantly increased Nrf2 and HO-1 protein levels and decreased MPO, TNF-α, IL-1β, and NLRP3 expressions in mice with DSS-induced acute colitis.[92] Moreover, rosmarinic acid-rich *Trichodesma khasianum* clarke promotes gastric mucosal healing through down-regulation of anti-inflammatory cytokine expression (IL-1, IL-6, and TNF-α levels) and apoptosis-related protein levels (cytochrome *c* and cleaved caspase-3) and upregulation of heat shock protein (HSP) expression

(HSP32, HSP60, HSP70, and HSP90), antioxidant enzyme activities (such as superoxide dismutase, glutathione peroxidase, and catalase), gut microbiota diversity, and SCFA generation ability in BALB/c mice with ethanol-induced gastric mucosal injury.[93] The above evidence suggests that rosmarinic acid can prevent gastrointestinal disorders through antioxidant, anti-inflammatory, and microbiota modulation capacities.

14.4 Conclusion

The GI tract plays a role in maintaining the digestion and absorption functions, and subsequently sustaining human health. Alterations in GI tract physiology are highly related to the systemic effects of disease development. Dietary bioactive compounds act as powerful protectors that directly prevent disease development *via* GI tract function enhancement and subsequently contribute to overall health by their antioxidative and anti-inflammatory properties. The clear underlying mechanisms of these phytochemicals in GI protection may allow appropriate dietary phytocompounds to be selected for certain applications in the discovery of novel preventive or therapeutic strategies that could be modulated through the antioxidative, anti-inflammatory, and microbiota regulation capacities.

References

1. J. M. Hess, S. S. Jonnalagadda and J. L. Slavin, Dairy foods: Current evidence of their effects on bone, cardiometabolic, cognitive, and digestive health, *Compr. Rev. Food Sci. Food Saf.*, 2016, **15**, 251–268.
2. T. Oshima and H. Miwa, Gastrointestinal mucosal barrier function and diseases, *J. Gastroenterol.*, 2016, **51**, 768–778.
3. P. Oka, H. Parr, B. Barberio, C. J. Black, E. V. Savarino and A. C. Ford, Global prevalence of irritable bowel syndrome according to Rome III or IV criteria: A systematic review and meta-analysis, *Lancet Gastroenterol. Hepatol.*, 2020, **5**, 908–917.
4. A. D. Sperber, S. I. Bangdiwala, D. A. Drossman, U. C. Ghoshal, M. Simren and J. Tack, *et al.*, Worldwide prevalence and burden of functional gastrointestinal disorders, results of rome foundation global study, *Gastroenterology*, 2021, **160**, 99–114.
5. K. M. Fairbrass, S. J. Costantino, D. J. Gracie and A. C. Ford, Prevalence of irritable bowel syndrome-type symptoms in patients with inflammatory bowel disease in remission: A systematic review and meta-analysis, *Lancet Gastroenterol. Hepatol.*, 2020, **5**, 1053–1062.
6. A. C. Ford, S. Mahadeva, M. F. Carbone, B. E. Lacy and N. J. Talley, Functional dyspepsia, *Lancet*, 2020, **396**, 1689–1702.
7. I. Aziz and M. Simren, The overlap between irritable bowel syndrome and organic gastrointestinal diseases, *Lancet Gastroenterol. Hepatol.*, 2021, **6**, 139–148.

8. C. Canavan, J. West and T. Card, Review article: The economic impact of the irritable bowel syndrome, *Aliment. Pharmacol. Ther.*, 2014, **40**, 1023–1034.
9. S. A. Muller-Lissner and O. Pirk, Irritable bowel syndrome in Germany. A cost of illness study, *Eur. J. Gastroenterol. Hepatol.*, 2002, **14**, 1325–1329.
10. F. Zhang, W. Xiang, C. Y. Li and S. C. Li, Economic burden of irritable bowel syndrome in China, *World J. Gastroenterol.*, 2016, **22**, 10450–10460.
11. A. F. Peery, S. D. Crockett, C. C. Murphy, J. L. Lund, E. S. Dellon and J. L. Williams, *et al.*, Burden and cost of gastrointestinal, liver, and pancreatic diseases in the United States: Update 2018, *Gastroenterology*, 2019, **156**, 254–272.
12. N. J. Talley, What causes functional gastrointestinal disorders? A proposed disease model, *Am. J. Gastroenterol.*, 2020, **115**, 41–48.
13. Y. T. Cheng, C. C. Lu and G. C. Yen, Phytochemicals enhance antioxidant enzyme expression to protect against NSAID-induced oxidative damage of the gastrointestinal mucosa, *Mol. Nutr. Food Res.*, 2017, **61**, 1600659.
14. S. Bindu, S. Mazumder and U. Bandyopadhyay, Non-steroidal anti-inflammatory drugs (NSAIDs) and organ damage: A current perspective, *Biochem. Pharmacol.*, 2020, **180**, 114147.
15. I. Bjarnason, C. Scarpignato, E. Holmgren, M. Olszewski, K. D. Rainsford and A. Lanas, Mechanisms of damage to the gastrointestinal tract from nonsteroidal anti-inflammatory drugs, *Gastroenterology*, 2018, **154**, 500–514.
16. S. K. Gill, M. Rossi, B. Bajka and K. Whelan, Dietary fibre in gastrointestinal health and disease, *Nat. Rev. Gastroenterol. Hepatol.*, 2021, **18**, 101–116.
17. A. M. Valdes, J. Walter, E. Segal and T. D. Spector, Role of the gut microbiota in nutrition and health, *Br. Med. J.*, 2018, **361**, k2179.
18. O. Manor, C. L. Dai, S. A. Kornilov, B. Smith, N. D. Price and J. C. Lovejoy, *et al.*, Health and disease markers correlate with gut microbiome composition across thousands of people, *Nat. Commun.*, 2020, **11**, 5206.
19. P. D. Cani and M. Van Hul, Mediterranean diet, gut microbiota and health: when age and calories do not add up!, *Gut*, 2020, **69**, 1167–1168.
20. L. Barrea, G. Muscogiuri, E. Frias-Toral, D. Laudisio, G. Pugliese and B. Castellucci, *et al.*, Nutrition and immune system: From the Mediterranean diet to dietary supplementary through the microbiota, *Crit. Rev. Food Sci. Nutr.*, 2021, **61**, 3066–3069.
21. L. Serra-Majem, B. Roman-Vinas, A. Sanchez-Villegas, M. Guasch-Ferre, D. Corella and C. La Vecchia, Benefits of the Mediterranean diet: Epidemiological and molecular aspects, *Mol. Aspects Med.*, 2019, **67**, 1–55.
22. F. P. Zito, B. Polese, L. Vozzella, A. Gala, D. Genovese and V. Verlezza, *et al.*, Good adherence to Mediterranean diet can prevent gastrointestinal symptoms: A survey from Southern Italy, *World J. Gastrointest. Pharmacol. Ther.*, 2016, **7**, 564–571.

23. C. Agakidis, E. Kotzakioulafi, D. Petridis, K. Apostolidou and T. Karagiozoglou-Lampoudi, Mediterranean diet adherence is associated with lower prevalence of functional gastrointestinal disorders in children and adolescents, *Nutrients*, 2019, **11**, 1283.
24. V. Jairath and B. G. Feagan, Global burden of inflammatory bowel disease, *Lancet Gastroenterol. Hepatol.*, 2020, **5**, 2–3.
25. H. Khalili, N. Hakansson, S. S. Chan, Y. Chen, P. Lochhead and J. F. Ludvigsson, *et al.*, Adherence to a Mediterranean diet is associated with a lower risk of later-onset Crohn's disease: Results from two large prospective cohort studies, *Gut*, 2020, **69**, 1637–1644.
26. L. Godny, L. Reshef, T. Pfeffer-Gik, I. Goren, H. Yanai and H. Tulchinsky, *et al.*, Adherence to the Mediterranean diet is associated with decreased fecal calprotectin in patients with ulcerative colitis after pouch surgery, *Eur. J. Nutr.*, 2020, **59**, 3183–3190.
27. D. A. Martin and B. W. Bolling, A review of the efficacy of dietary polyphenols in experimental models of inflammatory bowel diseases, *Food Funct.*, 2015, **6**, 1773–1786.
28. J. K. Hou, B. Abraham and H. El-Serag, Dietary intake and risk of developing inflammatory bowel disease: A systematic review of the literature, *Am. J. Gastroenterol.*, 2011, **106**, 563–573.
29. R. Estruch, E. Ros, J. Salas-Salvado, M. I. Covas, D. Corella and F. Aros, *et al.*, Primary prevention of cardiovascular disease with a Mediterranean diet supplemented with extra-virgin olive oil or nuts, *N. Engl. J. Med.*, 2018, **378**, e34.
30. J. Lozano-Castellon, A. Lopez-Yerena, J. F. Rinaldi de Alvarenga, J. Romero Del Castillo-Alba, A. Vallverdu-Queralt and E. Escribano-Ferrer, *et al.*, Health-promoting properties of oleocanthal and oleacein: Two secoiridoids from extra-virgin olive oil, *Crit. Rev. Food Sci. Nutr.*, 2020, **60**, 2532–2548.
31. G. Serra, A. Incani, G. Serreli, L. Porru, M. P. Melis and C. I. G. Tuberoso, *et al.*, Olive oil polyphenols reduce oxysterols-induced redox imbalance and pro-inflammatory response in intestinal cells, *Redox Biol.*, 2018, **17**, 348–354.
32. M. Gavahian, A. M. Khaneghah, J. M. Lorenzo, P. E. S. Munekata, I. Garcia-Mantrana and M. C. Collado, *et al.*, Health benefits of olive oil and its components: Impacts on gut microbiota antioxidant activities, and prevention of noncommunicable diseases, *Trends Food Sci. Technol.*, 2019, **88**, 220–227.
33. T. Shi, G. C. Wu, Q. Z. Jin and X. G. Wang, Camellia oil authentication: A comparative analysis and recent analytical techniques developed for its assessment. A review, *Trends Food Sci. Technol.*, 2020, **97**, 88–99.
34. Y. T. Cheng, C. C. Lu and G. C. Yen, Beneficial effects of camellia oil (*Camellia oleifera* Abel.) on hepatoprotective and gastroprotective activities, *J. Nutr. Sci. Vitaminol.*, 2015, **61**(Suppl.), S100–S102.
35. R. Y. Wang, Y. T. Tung, S. Y. Chen, Y. L. Lee and G. C. Yen, Protective effects of camellia oil (*Camellia brevistyla*) against indomethacin-induced

gastrointestinal mucosal damage *in vitro* and *in vivo*, *J. Funct. Foods*, 2019, **62**, 103539.
36. C. C. Wu, Y. T. Tung, S. Y. Chen, W. T. Lee, H. T. Lin and G. C. Yen, Anti-inflammatory, antioxidant, and microbiota-modulating effects of camellia oil from *Camellia brevistyla* on acetic acid-induced colitis in rats, *Antioxidants*, 2020, **9**, 58.
37. W. T. Lee, Y. T. Tung, C. C. Wu, P. S. Tu and G. C. Yen, Camellia oil (*Camellia oleifera* Abel.) modifies the composition of gut microbiota and alleviates acetic acid-induced colitis in rats, *J. Agric. Food Chem.*, 2018, **66**, 7384–7392.
38. Y. T. Cheng, S. L. Wu, C. Y. Ho, S. M. Huang, C. L. Cheng and G. C. Yen, Beneficial effects of camellia oil (*Camellia oleifera* Abel.) on ketoprofen-induced gastrointestinal mucosal damage through upregulation of HO-1 and VEGF, *J. Agric. Food Chem.*, 2014, **62**, 642–650.
39. P. S. Tu, Y. T. Tung, W. T. Lee and G. C. Yen, Protective effect of camellia oil (*Camellia oleifera* Abel.) against ethanol-induced acute oxidative injury of the gastric mucosa in mice, *J. Agric. Food Chem.*, 2017, **65**, 4932–4941.
40. S. Nunes, F. Danesi, D. Del Rio and P. Silva, Resveratrol and inflammatory bowel disease: The evidence so far, *Nutr. Res. Rev.*, 2018, **31**, 85–97.
41. D. Rotrekl, P. Salamunova, L. Parakova, O. Bado, I. Salon and F. Stepanek, *et al.*, Composites of yeast glucan particles and curcumin lead to improvement of dextran sulfate sodium-induced acute bowel inflammation in rats, *Carbohydr. Polym.*, 2021, **252**, 117142.
42. D. E. Iglesias, E. Cremonini, C. G. Fraga and P. I. Oteiza, Ellagic acid protects Caco-2 cell monolayers against inflammation-induced permeabilization, *Free Radical Biol. Med.*, 2020, **152**, 776–786.
43. H. Li, L. M. Christman, R. Li and L. Gu, Synergic interactions between polyphenols and gut microbiota in mitigating inflammatory bowel diseases, *Food Funct.*, 2020, **11**, 4878–4891.
44. M. L. Y. Wan, V. A. Co and H. El-Nezami, Dietary polyphenol impact on gut health and microbiota, *Crit. Rev. Food Sci. Nutr.*, 2021, **61**, 690–711.
45. J. Perez-Jimenez, V. Neveu, F. Vos and A. Scalbert, Identification of the 100 richest dietary sources of polyphenols: an application of the Phenol-Explorer database, *Eur. J. Clin. Nutr.*, 2010, **64**(Suppl. 3), S112–S120.
46. M. L. Y. Wan, V. A. Co and H. El-Nezami, Dietary polyphenol impact on gut health and microbiota, *Crit. Rev. Food Sci. Nutr.*, 2021, **61**, 690–711.
47. C. L. Gentile and T. L. Weir, The gut microbiota at the intersection of diet and human health, *Science*, 2018, **362**, 776–780.
48. S. Zheng, H. Zhang, R. Liu, C. L. Huang, H. Li and Z. Y. Deng, *et al.*, Do short chain fatty acids and phenolic metabolites of the gut have synergistic anti-inflammatory effects? – New insights from a TNF-alpha-induced Caco-2 cell model, *Food Res. Int.*, 2021, **139**, 109833.
49. L. Lavefve, L. R. Howard and F. Carbonero, Berry polyphenols metabolism and impact on human gut microbiota and health, *Food Funct.*, 2020, **11**, 45–65.

50. T. Chen and C. S. Yang, Biological fates of tea polyphenols and their interactions with microbiota in the gastrointestinal tract: Implications on health effects, *Crit. Rev. Food Sci. Nutr.*, 2020, **60**, 2691–2709.
51. Y. Zhao and Q. Jiang, Roles of the polyphenol-gut microbiota interaction in alleviating colitis and preventing colitis-associated colorectal cancer, *Adv. Nutr.*, 2021, **12**, 546–565.
52. Y. T. Loo, K. Howell, M. Chan, P. Zhang and K. Ng, Modulation of the human gut microbiota by phenolics and phenolic fiber-rich foods, *Compr. Rev. Food Sci. Food Saf.*, 2020, **19**, 1268–1298.
53. T. Mousavi, N. Hadizadeh, S. Nikfar and M. Abdollahi, Drug discovery strategies for modulating oxidative stress in gastrointestinal disorders, *Expert Opin. Drug Discovery*, 2020, **15**, 1309–1341.
54. A. Y. Berman, R. A. Motechin, M. Y. Wiesenfeld and M. K. Holz, The therapeutic potential of resveratrol: A review of clinical trials, *npj Precis. Oncol.*, 2017, **1**, 35.
55. H. R. Alrafas, P. B. Busbee, M. Nagarkatti and P. S. Nagarkatti, Resveratrol modulates the gut microbiota to prevent murine colitis development through induction of Tregs and suppression of Th17 cells, *J. Leukocyte Biol.*, 2019, **106**, 467–480.
56. M. Samsamikor, N. E. Daryani, P. R. Asl and A. Hekmatdoost, Resveratrol supplementation and oxidative/anti-oxidative status in patients with ulcerative colitis: A randomized, double-blind, placebo-controlled pilot study, *Arch. Med. Res.*, 2016, **47**, 304–309.
57. L. Shen and H. F. Ji, Bidirectional interactions between dietary curcumin and gut microbiota, *Crit. Rev. Food Sci. Nutr.*, 2019, **59**, 2896–2902.
58. M. Cruz-Correa, L. M. Hylind, J. H. Marrero, M. L. Zahurak, T. Murray-Stewart and R. A. Casero, *et al.*, Efficacy and safety of curcumin in treatment of intestinal adenomas in patients with familial adenomatous polyposis, *Gastroenterology*, 2018, **155**, 668–673.
59. A. L. Lopresti, The problem of curcumin and its bioavailability: Could its gastrointestinal influence contribute to its overall health-enhancing effects?, *Adv. Nutr.*, 2018, **9**, 41–50.
60. B. Scazzocchio, L. Minghetti and M. D'Archivio, Interaction between gut microbiota and curcumin: A new key of understanding for the health effects of curcumin, *Nutrients*, 2020, **12**, 2499.
61. D. Thong-Ngam, S. Choochuai, S. Patumraj, M. Chayanupatkul and N. Klaikeaw, Curcumin prevents indomethacin-induced gastropathy in rats, *World J. Gastroenterol.*, 2012, **18**, 1479–1484.
62. S. S. Patel, A. Acharya, R. S. Ray, R. Agrawal, R. Raghuwanshi and P. Jain, Cellular and molecular mechanisms of curcumin in prevention and treatment of disease, *Crit. Rev. Food Sci. Nutr.*, 2020, **60**, 887–939.
63. E. Bakkalbasi, O. Mentes and N. Artik, Food ellagitannins-occurrence, effects of processing and storage, *Crit. Rev. Food Sci. Nutr.*, 2009, **49**, 283–298.
64. R. Zhao, X. Long, J. Yang, L. Du, X. Zhang and J. Li, *et al.*, Pomegranate peel polyphenols reduce chronic low-grade inflammatory responses by

modulating gut microbiota and decreasing colonic tissue damage in rats fed a high-fat diet, *Food Funct.*, 2019, **10**, 8273–8285.

65. H. Kim, N. Banerjee, M. A. Sirven, Y. Minamoto, M. E. Markel and J. S. Suchodolski, *et al.*, Pomegranate polyphenolics reduce inflammation and ulceration in intestinal colitis-involvement of the miR-145/p70S6K1/HIF1alpha axis *in vivo* and *in vitro*, *J. Nutr. Biochem.*, 2017, **43**, 107–115.
66. S. Bibi, Y. Kang, M. Du and M. J. Zhu, Dietary red raspberries attenuate dextran sulfate sodium-induced acute colitis, *J. Nutr. Biochem.*, 2018, **51**, 40–46.
67. D. Bialonska, S. G. Kasimsetty, K. K. Schrader and D. Ferreira, The effect of pomegranate (*Punica granatum* L.) byproducts and ellagitannins on the growth of human gut bacteria, *J. Agric. Food Chem.*, 2009, **57**, 8344–8349.
68. E. Scaioli, A. Belluzzi, L. Ricciardiello, D. Del Rio, E. Rotondo and P. Mena, *et al.*, Pomegranate juice to reduce fecal calprotectin levels in inflammatory bowel disease patients with a high risk of clinical relapse: Study protocol for a randomized controlled trial, *Trials*, 2019, **20**, 327.
69. S. Alfei, F. Turrini, S. Catena, P. Zunin, M. Grilli and A. M. Pittaluga, *et al.*, Ellagic acid a multi-target bioactive compound for drug discovery in CNS? A narrative review, *Eur. J. Med. Chem.*, 2019, **183**, 111724.
70. M. A. Rosillo, M. Sanchez-Hidalgo, A. Cardeno, M. Aparicio-Soto, S. Sanchez-Fidalgo and I. Villegas, *et al.*, Dietary supplementation of an ellagic acid-enriched pomegranate extract attenuates chronic colonic inflammation in rats, *Pharmacol. Res.*, 2012, **66**, 235–242.
71. M. Marin, R. Maria Giner, J. L. Rios and M. C. Recio, Intestinal anti-inflammatory activity of ellagic acid in the acute and chronic dextrane sulfate sodium models of mice colitis, *J. Ethnopharmacol.*, 2013, **150**, 925–934.
72. R. Singh, S. Chandrashekharappa, S. R. Bodduluri, B. V. Baby, B. Hegde and N. G. Kotla, *et al.*, Enhancement of the gut barrier integrity by a microbial metabolite through the Nrf2 pathway, *Nat. Commun.*, 2019, **10**, 89.
73. Y. Q. Xu, Y. Gao and D. Granato, Effects of epigallocatechin gallate, epigallocatechin and epicatechin gallate on the chemical and cell-based antioxidant activity, sensory properties, and cytotoxicity of a catechin-free model beverage, *Food Chem.*, 2021, **339**, 128060.
74. X. Bing, L. Xuelei, D. Wanwei, L. Linlang and C. Keyan, EGCG maintains Th1/Th2 balance and mitigates ulcerative colitis induced by dextran sulfate sodium through TLR4/MyD88/NF-kappaB signaling pathway in rats, *Can. J. Gastroenterol. Hepatol.*, 2017, **2017**, 3057268.
75. M. Bruckner, S. Westphal, W. Domschke, T. Kucharzik and A. Lugering, Green tea polyphenol epigallocatechin-3-gallate shows therapeutic antioxidative effects in a murine model of colitis, *J. Crohns Colitis*, 2012, **6**, 226–235.

76. G. W. Dryden, Jr., I. Deaciuc, G. Arteel and C. J. McClain, Clinical implications of oxidative stress and antioxidant therapy, *Curr. Gastroenterol. Rep.*, 2005, **7**, 308–316.
77. S. Andres, S. Pevny, R. Ziegenhagen, N. Bakhiya, B. Schafer and K. I. Hirsch-Ernst, *et al.*, Safety aspects of the use of quercetin as a dietary supplement, *Mol. Nutr. Food Res.*, 2018, **62**, 1700447.
78. M. Comalada, D. Camuesco, S. Sierra, I. Ballester, J. Xaus and J. Galvez, *et al.*, *In vivo* quercitrin anti-inflammatory effect involves release of quercetin, which inhibits inflammation through down-regulation of the NF-kappaB pathway, *Eur. J. Immunol.*, 2005, **35**, 584–592.
79. Y. Dong, Q. Hou, J. Lei, P. G. Wolf, H. Ayansola and B. Zhang, Quercetin alleviates intestinal oxidative damage induced by H_2O_2 via modulation of GSH: *In vitro* screening and *in vivo* evaluation in a colitis model of mice, *ACS Omega*, 2020, **5**, 8334–8346.
80. R. Lin, M. Piao and Y. Song, Dietary quercetin increases colonic microbial diversity and attenuates colitis severity in citrobacter rodentium-infected mice, *Front. Microbiol.*, 2019, **10**, 1092.
81. Y. Wang, H. Zhao, Y. Liu, J. Li, X. Nie and P. Huang, *et al.*, Environmentally relevant concentration of sulfamethoxazole-induced oxidative stress-cascaded damages in the intestine of grass carp and the therapeutic application of exogenous lycopene, *Environ. Pollut.*, 2021, **274**, 116597.
82. M. Boyacioglu, C. Kum, S. Sekkin, H. S. Yalinkilinc, H. Avci and E. T. Epikmen, *et al.*, The effects of lycopene on DNA damage and oxidative stress on indomethacin-induced gastric ulcer in rats, *Clin. Nutr.*, 2016, **35**, 428–435.
83. B. Zhao, J. Wu, J. Li, Y. Bai, Y. Luo and B. Ji, *et al.*, Lycopene alleviates DSS-induced colitis and behavioral disorders *via* mediating microbes-gut-brain axis balance, *J. Agric. Food Chem.*, 2020, **68**, 3963–3975.
84. D. Liu, Y. Mao, L. Ding and X. A. Zeng, Dihydromyricetin: A review on identification and quantification methods, biological activities, chemical stability, metabolism and approaches to enhance its bioavailability, *Trends Food Sci. Technol.*, 2019, **91**, 586–597.
85. S. V. Luca, I. Macovei, A. Bujor, A. Miron, K. Skalicka-Wozniak and A. C. Aprotosoaie, *et al.*, Bioactivity of dietary polyphenols: The role of metabolites, *Crit. Rev. Food Sci. Nutr.*, 2020, **60**, 626–659.
86. L. S. Fideles, J. A. L. de Miranda, C. D. S. Martins, M. L. L. Barbosa, H. B. Pimenta and P. V. S. Pimentel, *et al.*, Role of rutin in 5-fluorouracil-induced intestinal mucositis: prevention of histological damage and reduction of inflammation and oxidative stress, *Molecules*, 2020, **25**, 2786.
87. X. Guo, R. Tang, S. Yang, Y. Lu, J. Luo and Z. Liu, Rutin and its combination with inulin attenuate gut dysbiosis, the inflammatory status and endoplasmic reticulum stress in paneth cells of obese mice induced by high-fat diet, *Front. Microbiol.*, 2018, **9**, 2651.

88. K. H. Kwon, A. Murakami, T. Tanaka and H. Ohigashi, Dietary rutin, but not its aglycone quercetin, ameliorates dextran sulfate sodium-induced experimental colitis in mice: Attenuation of pro-inflammatory gene expression, *Biochem. Pharmacol.*, 2005, **69**, 395–406.
89. Z. Zhang, S. Li, H. Cao, P. Shen, J. Liu and Y. Fu, *et al.*, The protective role of phloretin against dextran sulfate sodium-induced ulcerative colitis in mice, *Food Funct.*, 2019, **10**, 422–431.
90. M. Wu, P. Li, Y. An, J. Ren, D. Yan and J. Cui, *et al.*, Phloretin ameliorates dextran sulfate sodium-induced ulcerative colitis in mice by regulating the gut microbiota, *Pharmacol. Res.*, 2019, **150**, 104489.
91. A. Khojasteh, M. H. Mirjalili, M. A. Alcalde, R. M. Cusido, R. Eibl and J. Palazon, Powerful plant antioxidants: A new biosustainable approach to the production of rosmarinic acid, *Antioxidants*, 2020, **9**, 1273.
92. S. Marinho, M. Illanes, J. Avila-Roman, V. Motilva and E. Talero, Anti-inflammatory effects of rosmarinic acid-loaded nanovesicles in acute colitis through modulation of NLRP3 inflammasome, *Biomolecules*, 2021, **11**, 162.
93. G. Y. Wang, S. Y. Chen, Y. Y. Chen, C. J. Hong, Y. H. Hsu and G. C. Yen, Protective effect of rosmarinic acid-rich *Trichodesma khasianum* clarke leaves against ethanol-induced gastric mucosal injury *in vitro* and *in vivo*, *Phytomedicine*, 2021, **80**, 153382.

CHAPTER 15

Natural Dietary Antioxidants and Bone Health

SNIGDHA MISRA,* SANGEETHA SHYAM AND TAN SEOK SHIN

Division of Nutrition and Dietetics, School of Health Sciences, International Medical University, Kuala Lumpur, Malaysia
*Email: snigdha_misra@imu.edu.my

15.1 Introduction

Natural antioxidants are mainly from plant sources. Most plants contain compounds that possess antioxidant activity. They are mostly polyphenolics that occur in all parts of a plant: woods, barks, stems, leaves, fruits, roots, flowers, pollens, and seeds, among others. Foods rich in dietary antioxidants have been associated with a reduced risk of a host of diseases including bone health. Bone is a dynamic organ that provides structural support, facilitates mobility, protects internal organs, and serves as a reservoir for essential minerals.[1] During one's lifetime, bone is constantly renewed through a process of remodelling and modelling involving bone resorption by osteoclasts and bone formation by osteoblasts.[2] Unhealthy bones do not adequately support these functions, may lead to debilitating fractures and cause a loss in quality of life. Various factors affect bone health and these include genetic abnormalities, nutritional deficiencies, hormonal disorders, sedentary lifestyles and inactivity, immobilisation, and smoking.[2] This chapter provides a summary of the emerging evidences of foods or food-derived bioactive compounds with antioxidant properties and their potential molecular mechanisms of action on bone health.

Food Chemistry, Function and Analysis No. 39
Dietary Supplements with Antioxidant Activity: Understanding Mechanisms and Potential Health Benefits
Edited by Cesarettin Alasalvar, Fereidoon Shahidi and Chi-Tang Ho

Published by the Royal Society of Chemistry, www.rsc.org

15.2 Processes Involved in Bone Health

Bone remodelling is a normal physiological process that helps sustain bone health. Bone remodelling maintains skeletal integrity after skeletal development by removing small foci of damaged or effete bone from bone surfaces and replacing them with new bone.[3–6] By this mechanism, the skeleton is continuously renewed throughout life. The cells responsible for bone resorption are shown in Figure 15.1.

Osteoporosis, the most common bone disease, is commonly seen in older adults and considered a multifactorial and chronic systemic disease of the skeleton, closely related to aging.[2,7,8] It occurs due to increased bone loss and resorption and decreased bone formative activities. This imbalance between formation and resorption leads to reduced bone mass, bone mineral density, and consequent structural abnormalities and fragility commonly seen in elderly. Thus, osteoporosis is degenerative in nature and is characterised by a reduced bone strength, which may lead to fractures.[9] More than 200 million individuals worldwide are currently affected by osteoporosis.[9] The World Health Organization estimates that the annual number of hip fractures would increase from 1.7 million in 1990 to

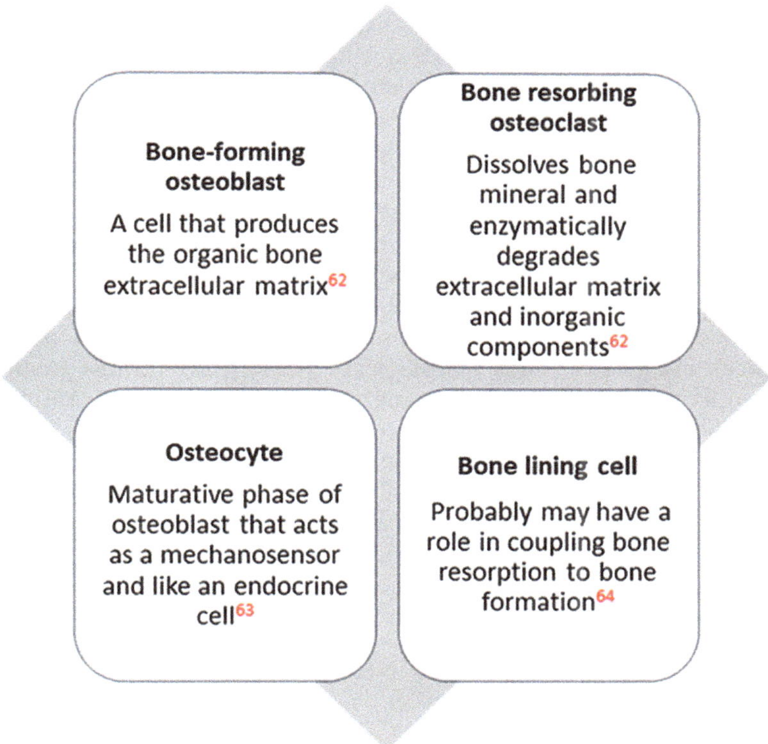

Figure 15.1 Cells responsible for bone resorption.[62–64]

6.3 million in 2050.[10] According to Caroli *et al.*,[10] it is possible that the incidence of osteoporotic fractures would increase further. This increase in incidence is due to the increased life expectancy and other important risk factors such as industrialisation and reduced physical activity. Strategies to avoid osteoporosis include decreasing bone loss induced by acute postmenopausal oestrogen deficiency.[3,4]

15.3 Osteoporosis: An Oxidative-stress Induced Disease

Similar to other chronic diseases, inflammation and related oxidative stress are increasingly recognised as a part of the aetiology of osteoporosis.[3,4] In general, oxidative stress is described as the state of asymmetry between pro-oxidants and antioxidants. Oxidative stress regulates and enhances bone resorption, differentiation, and activity of osteoclast cells, so it has a substantial influence on the incidence of osteoporosis. Therefore, it is hypothesised that there is a correlation between reactive oxygen species (ROS) and the pathogenesis of osteoporosis.[3,4,11–13] ROS are involved in bone resorption through a direct contribution of bone degradation by osteoclast-generated superoxide or an increase in differentiation and function of osteoclasts.[13–16] Oxidative stress (excessive ROS) has been reported to inhibit osteoblastic differentiation[17] *via* extracellular signal-regulated kinases (ERK) and ERK-dependent nuclear factor-kappa B (NF-κB) signalling pathways.[18] However, osteoblasts can produce antioxidants, such as glutathione peroxidase, to protect against ROS.[19] Both human[11,12] and animal studies[13,16] have shown a strong positive correlation between excessive ROS and bone loss in the development of osteoporosis. Excessive ROS can contribute to bone loss. Hence, it is important to elucidate the potential role of natural antioxidants, including diets with adequate amounts of antioxidant-rich fruits, in mitigating bone loss resulting in the development of osteoporosis.

In this chapter, evidence of natural dietary antioxidants in promoting bone health is collated. First, radiological evidence from various populations and the potential mechanisms through which natural dietary antioxidants improve bone health are discussed. In the subsequent sections, the experimental evidence of foods or food-derived bioactive compounds with antioxidant properties and their potential molecular mechanisms of action are presented. It is believed that this broad approach is useful in improving the bone health through both population-based approaches such as development of dietary guidelines and nutraceutical strategies.

15.4 Antioxidant and Anti-inflammatory Mechanisms

The role of a healthy eating strategy with increased consumption of plant-based foods in the prevention of chronic diseases, such as heart disease, cancer, stroke, diabetes, Alzheimer's disease, cataracts, and age-related function decline, is well established.[1,20,21] Evidence for their role in promoting bone health and in the prevention of diseases such as osteoporosis is accruing.

Some of the positive effects of adequate fruit and vegetable intake are ascribed to their antioxidant and anti-inflammatory constituents such as polyphenols, vitamin C, and carotenoids.[22,23] A summary of the role of bioactive compounds in bone health is presented in Figure 15.2. Meanwhile the protective action of nutrients on bone health is shown in Figure 15.3. Polyphenols are phytochemicals found in the plant kingdom and commonly isolated and characterised in fruits and vegetables.[24] Polyphenols can be split into different groups: phenolic acids, flavonoids, stilbenes, tannins, coumarins, and lignans.[25] Currently, there are almost 8000 recognised polyphenols, 500 of which are biologically active.[26] These compounds have several documented biological effects and have been reported to be protective against chronic diseases, including neurodegenerative and cardiovascular pathologies, cancer, and bone resorption pathologies.[24] Recent articles based on the molecular characteristics of dietary polyphenols, specifically flavonoids, have outlined their advantage in the prevention and management of bone resorption diseases.[25,27]

Polyphenols can preserve bone health by five possible mechanisms: (i) decreasing bone loss through antioxidant activity, (ii) diminishing bone loss through anti-inflammatory action, (iii) improving osteoblastogenesis, (iv) reducing osteoclastogenesis, and (v) osteoimmunological activity.[28–32] Polyphenols reduce osteoclast differentiation and activity through a process of downregulation of osteoclast differentiation acting on primary bone marrow cells. In addition, they mediate through the suppression of nuclear factor-activated T-cells, cytoplasmic 1 (NFATC1), which regulates the osteoclast differentiation process.[29–32]

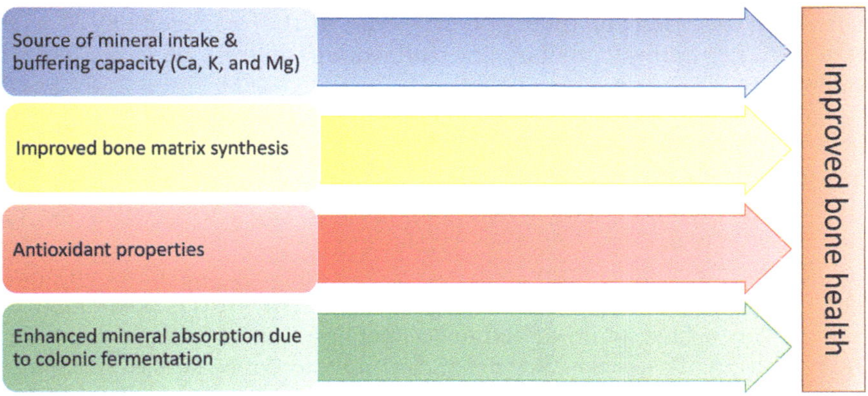

Figure 15.2 Role of bioactive compounds in bone health.

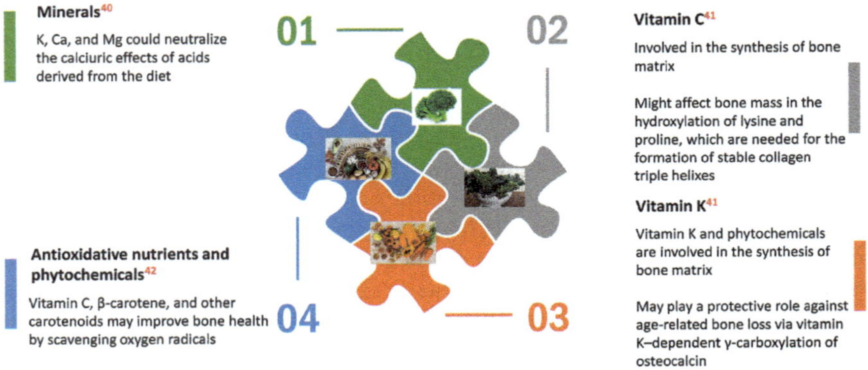

Figure 15.3 Protective action of nutrients on bone health.[40–42]

Recently, molecular targets in cell signalling pathways that influence the bone structure have also been identified. Polyphenols are shown to display bone anabolic action, which has important implications beyond the inhibition of bone resorption through suppressing osteoclast activation. These beneficial actions of polyphenols could be majorly attributed to their antioxidant characteristics, as they can act as scavengers of ROS. Considering these properties, polyphenols could affect bone metabolism through the impairment of inflammatory mediators,[29] such as cytokines, primarily involved in supporting osteoclast differentiation and resorption,[30] consequently enabling a decrease in bone resorption. The suggested mechanisms also involve the downregulation of NF-κB, receptor activator of nuclear factor kappa-B ligand (RANKL), and nitric oxide (NO) production of ROS and inflammatory cytokine synthesis.[31,32]

It is feasible to consume adequate amounts of polyphenols from natural sources including affordable commercial food and beverage products.[27] Plant-based therapies (*e.g.*, polyphenols) are thus safe options as the common treatment of osteoporosis. In this context, various plant components have been assessed for their potential role in the management of osteoporosis.[33] For example, dietary soy isoflavones suppress bone depletion in rodents and post-menopausal women.[27]

Curcumin has also been shown to enhance several aspects of bone health among subjects with osteoporosis by acting on different steps in the recruitment and activation of osteoclast cells, increasing mineral density and mechanical strength. It downregulates the NF-κB, RANKL, and NO production of ROS and inflammatory cytokine synthesis.[31,32]

A review of evidence from 2000 to 2020 showed that intake of vitamin C had a positive correlation with bone mineral density. It was postulated that vitamin C can have multi-directional effects on bone health, ranging from the inhibition of osteoclast activity and stimulation of osteoblast maturation by increasing collagen synthesis to the regulation of gene transcription, DNA, and histone methylation. However, the review also concluded that much of this evidence came from observational studies and there was a need for randomised clinical trials to validate the causal relationship.[34]

Evidence on the benefits of consuming carotenoids, another kind of antioxidant commonly found in red and orange fruits and vegetables, on bone health showed that vegetables rich in β-cryptoxanthin and β-carotene might provide benefits to bone health in postmenopausal Japanese females.[35] Lycopene, another potent carotenoid antioxidant, decreases bone turnover markers by decreasing oxidative stress parameters, suggesting its role in reducing the risk of osteoporosis in women.[36] A preliminary review of the literature suggests that an anthocyanin-rich diet may also have potential positive effects on bone health though the cellular mechanisms of action may need further research.[37]

Dietary fibres, from fruits and vegetables that have cation exchange capacity from unmethylated galacturonic acid residues and phytic acid from cereal fibres, attenuate mineral absorption and retention.[38] This raises concerns for bone health during critical periods of the life span, such as children and elderly. However, these effects have been poorly evidenced and are only observed at high intake levels, and, even if they exist, may be transitory.[39] Furthermore, pectin, various gums, resistant starches, cellulose, certain oligosaccharides such as soy and fructo-oligosaccharides, inulin, lactulose, and related sugars from fruits and vegetables are highly fermentable in the colon.[38–41] Colonic fermentation by the gut microbiota and the subsequent production of short chain fatty acids (predominantly acetate, propionate, and butyrate) reduce the luminal pH and facilitate the solubilisation of calcium, magnesium, and iron, making them more bioavailable.[38,42] This, in turn, makes the luminal environment conducive to diffusive absorption *via* the paracellular route.[38] Thus, fermentable fibres improve metabolic absorption of calcium, magnesium, and iron, in the colon, even in the presence of phytate and are likely to play an essential role in enhanced mineral absorption,[38] thereby improving bone mineralisation. However, it should also be noted that many trials that showed the effect of colonic fermentation of fibres on bone mineralisation used prebiotic supplements[40,41] rather than improving fruit and vegetable intake.

Apart from antioxidants in fruits and vegetables, the functionality of their fermentable fibres and calcium content act in unison to improve bone health. Fruits and vegetables differ in their antioxidant profile and polyphenol containing fibres; therefore, their ability to enhance bone health may vary.

15.5 Antioxidant Activity in Bone Metabolism

15.5.1 Bioactive Compound Extracts

Table 15.1 shows the selected bioactive compounds (curcumin, resveratrol, polyphenols, taxifolin, and catechins) and their suggested mechanism of action on bone health. The antioxidant properties of various nutrients and

Table 15.1 Selected bioactive compounds and the suggested mechanism of action on bone health.

Bioactive compounds	Suggested mechanism of action
Curcumin	Enhances bone health in subjects with osteoporosis by acting on different steps in the recruitment and activation of osteoclast cells, increasing mineral density, and mechanical strength.[31,32]
	Downregulates NF-κB, RANKL, and NO production of ROS and inflammatory cytokine synthesis.[31,32]
Resveratrol	Increases bone mineral density and bone ALP in osteoporotic obese men.[65]
	Eliminates oxidative stress and prevents bone loss in ovariectomised and old rats.[66]
	Acts as an antioxidant, promoting NO production.[43,44]
	The PGF2α effect on OPG synthesis was suppressed by resveratrol through the inhibition of the MAPK pathways in osteoblasts.[43,44]
	Downregulates BMP-4 and stimulates VEGF synthesis through the inhibition of the p70 S6 kinase in osteoblasts *via* SIRT1 activation.[43,44]
Polyphenols	Decrease bone loss through antioxidant activity.[28–32]
	Diminish bone loss through anti-inflammatory action.[28–32]
	Improve osteoblastogenesis.[28–32]
	Reduce osteoclastogenesis and through osteoimmunological activity.[28–32]
	Reduce osteoclast differentiation and activity through downregulation of osteoclast differentiation acting on primary bone marrow cells.[29,32]
	Mediate through the suppression of NFATC1, which regulates the osteoclast differentiation process.[29,32]
Taxifolin	Suppresses the activation of NF-κB and MAPK.[45]
	Decreases osteoclast-specific gene expression, including TRAP, MMP-9, cathepsin K, NFATC1, and RANK.[45]
	Prevents ROS production following RANKL stimulation.[45]
	Alleviates ovariectomy-induced bone loss by repressing osteoclast activity and decreasing the serum levels of TNF-α, IL-1β, IL-6, and RANKL *in vivo*.[45]
Catechins	Counteract the deleterious effects of the imbalance of osteoblastogenesis and osteoclastogenesis.[47]
	Exert effects directly on pre-osteoclasts/osteoclasts or indirectly *via* the modulation of MSC/stromal cell regulation of pre-osteoclasts through activation of the NF-κB (RANK)/RANKL/OPG system.[47]
	Enhance osteoblastogenesis by enhancing osteogenic differentiation of MSCs and increasing osteoblastic survival, proliferation, differentiation, and mineralisation.[47]

Abbreviations: ALP, alkaline phosphatase; BMP-4, bone morphogenic protein-4; IL, interleukin; MAPK, mitogen-activated protein kinase; MMP-9, matrix metalloproteinase-9; MSCs, mesenchymal stem cells; NFATC1, nuclear factor-activated T-cells, cytoplasmic 1; NF-κB, nuclear factor kappa-B; NO, nitric oxide; OPG, osteoprotegerin; PGF2α, prostaglandin F2α; RANK, receptor activator of nuclear factor-kappa B; RANKL, receptor activator of nuclear factor kappa-B ligand; ROS, reactive oxygen species; SIRT1, sirtuin-1; TNF-α, tumour necrosis factor-alpha; TRAP, tartrate-resistant acid phosphatase; VEGF, vascular endothelial growth factor.

their influence on bone metabolism have been reported in recent studies. This is, in particular, for the anti-resorptive therapies on the treatment and prevention of bone loss.[3] Experimental studies both *in vitro* and *in vivo* have supported the roles of polyphenols in the prevention of cardiovascular disease (CVD), cancer, neurodegenerative diseases (NDD), diabetes, and osteoporosis.

Resveratrol, a natural antioxidant, has been shown to increase bone mineral density and bone alkaline phosphatase in osteoporotic obese men.[43] In addition, resveratrol has also demonstrated its effectiveness in eliminating oxidative stress as well as in preventing bone loss in ovariectomised and older rats.[44] It acts as an antioxidant and promotes NO production. The effect of prostaglandin F2α (PGF2α) on osteoprotegerin (OPG) synthesis was suppressed by resveratrol through the inhibition of the mitogen-activated protein kinase (MAPK) pathways in osteoblasts. In addition, resveratrol downregulates bone morphogenic protein-4 (BMP-4) and stimulates vascular endothelial growth factor (VEGF) synthesis through the inhibition of the p70 S6 kinase in osteoblasts *via* sirtuin-1 (SIRT1) activation.[43,44]

Taxifolin is a natural flavonoid with antioxidant and anti-inflammatory properties. It may be considered as a potential alternative therapeutic agent for treating osteoclast-related diseases based on the findings from the *in vitro* and *in vivo* studies conducted. It suppresses the activation of NF-κB and mitogen-activated protein kinase (MAPK) and decreases the expression of osteoclast-specific genes, including tartrate-resistant acid phosphatase (TRAP), matrix metalloproteinase-9 (MMP-9), cathepsin K, NFATC1, and receptor activator of nuclear factor kappa-B (RANK). Besides, taxifolin prevents ROS production following RANKL stimulation. In addition, it alleviates ovariectomy-induced bone loss by repressing osteoclast activity and decreasing the serum levels of tumour necrosis factor-alpha (TNF-α), interleukin (IL)-1β, interleukin-6, and receptor activator of RANKL, *in vivo*. In general, taxifolin inhibits osteoclastogenesis by regulating the modulation of several RANKL signalling pathways.[45]

Catechins, polyphenols found abundantly in green tea with potent antioxidant and anti-inflammatory properties, can counteract the deleterious effects of the imbalance of osteoblastogenesis and osteoclastogenesis caused by osteoporosis. Catechin effects can be directly exerted on pre-osteoclasts/osteoclasts or indirectly exerted *via* the modulation of mesenchymal stem cell (MSC)/stromal cell regulation of pre-osteoclasts through activation of the RANK/RANKL/OPG system. Catechins can also enhance osteoblastogenesis by enhancing osteogenic differentiation of MSCs and increasing osteoblastic survival, proliferation, differentiation, and mineralisation. The *in vitro* effects of catechins on osteogenesis have been confirmed in several animal models, as well as in epidemiological observational studies on human subjects.[46]

15.5.2 Natural Foods

Bone health of selected natural foods and the suggested mechanism of action are presented in Table 15.2. Green tea polyphenols could mitigate bone loss among the female rats with chronic inflammation.[47] Dietary supplementation of *Hypericum perforatum* can induce beneficial effects, in particular, during oxidative stress and active bone resorption caused by excessive physical activity.[48] Berry food and drinks could diminish the effects of age-related bone loss and decrease the risk of osteoporosis in humans.

Table 15.2 Selected natural foods and their suggested mechanisms of action on bone health.

Natural foods	Suggested mechanism of action
Green tea	Downregulates TNF-α modulating cancellous and endocortical bone compartments.[47,51]
	Enhances the ALP activity in osteoblastic cells and bone mineralisation.[47,51]
Berry food and drinks	Activate canonical Wnt signalling in young rats; blueberry enriched diets increase bone density mass through suppression of RANKL in stromal cells.[49,52]
Polygonatum sibiricum polysaccharide	Promotes osteoblastic differentiation and mineralisation *in vitro* through the ERK/GSK-3β/β-catenin signalling pathways; the effects were elicited through phosphorylation of ERK/GSK-3β and, consequently, activation of β-catenin.[58]
Honey	Inhibits activation of the MAPK pathway and NF-κB signalling and subsequently prevents the formation of osteoclasts.[59]
	Favours bone formation by osteoblasts, consequently preserving bone density.[59]
Amaranthus mangostanus L.	Suppresses RANKL-induced osteoclast differentiation and expression of osteoclast-specific genes, TRAP, cathepsin K, NFATC1, and Dc-stamp in RAW 264.7 cells.[60]
	Inhibits ERK, Akt, and NF-κB signalling pathways in RAW 264.7 cells.[60]
Lepidium sativum	Enhances the serum bone formation biomarker LDH activity and osteocalcin levels.[61]
	Inhibits exhaustion of SOD as well as GPx activities and accumulation of lipid peroxides in bone tissues.[61]
	Ameliorates the rise in the markers of bone resorption CTXI and TRAP and modulates RANKL/osteoprotegerin expression.[61]

Abbreviations: Akt, protein kinase B; ALP, alkaline phosphatase; CTXI, carboxyterminal telopeptide, type I; ERK, extracellular signal-regulated kinases; GPx, glutathione peroxidase; GSK-3β, glycogen synthase kinase 3 beta; LDH, lactate dehydrogenase; MAPK, mitogen-activated protein kinase; NFATC1, nuclear factor-activated T-cells, cytoplasmic 1; NF-κB, nuclear factor-kappa B; RANKL, receptor activator of nuclear factor kappa-B ligand; SOD, superoxide dismutase; TNF-α, tumour necrosis factor-alpha; TRAP, tartrate-resistant acid phosphatase; Wnt, wingless-related integration.

Anthocyanins in berry food and drinks can impair oxidative stress as anthocyanin chalcones and quinoidal bases with a double bond linked to the keto group are antioxidants in scavenging free radicals.[29,49–51] Blueberry phenolic components promote bone growth by activating canonical wingless-related integration (Wnt) signalling in young rats, while blueberry enriched diets increase bone density mass through suppression of RANKL in stromal cells.[49,52] In conclusion, dietary supplementation with various types of polyphenols from green tea or *Hypericum perforatum* or blueberry has been shown to diminish trabecular bone loss, inhibit loss of collagen in the bone matrix, constrain senescence pathways in osteoblastic cells, and prevent osteoporosis in ovariectomised rats.[50,51,53,54]

In addition, tea drinking is also reported to demonstrate beneficial effects in maintaining bone density in both menopausal and old women.[55,56] Flavanols in green tea might be the phytomolecules that promote osteogenesis and inhibit osteoclast bone resorption by reducing oxidative stress.[57] Green tea polyphenol inlet reduces degeneration of bone microarchitecture in rats with chronic inflammation by downregulation of TNF-α modulating cancellous and endocortical bone compartments. The bone formation is affected by epigallocatechin 3-gallate through the enhancement of the alkaline phosphatase activity in osteoblastic cells. Bone mineralisation is related to the suppression of osteoclast cell differentiation and formation of oxidative stress-induced calcium stones in rats.[67,68] One of the most studied bioactive compounds in recent years is icariin, a prenylated flavonol glycoside of Chinese herbal tea, which has been reported to have intriguing osteogenic properties both *in vitro* and *in vivo*.[14–17]

Polygonatum sibiricum polysaccharide (PSP), an important natural organic compound in the form of sugar glucoside or free molecules, is one of the medicinal food homologies of Chinese herbal medicine. A study found that PSP promotes osteoblastic differentiation and mineralisation *in vitro* through the extracellular signal-regulated kinase (ERK)/glycogen synthase kinase 3 beta (GSK-3β)/β-catenin signalling pathways, and the effects were elicited through phosphorylation of ERK/GSK-3β and, consequently, activation of β-catenin. Nevertheless, the effects of PSP were not fully representative *in vivo*. Thus, PSP might be useful in the development of a therapeutic approach for the enhancement of bone health and prevention of osteoporosis.[58]

Honey is a potential natural product with a prophylactic effect to prevent bone loss *via* its antioxidant and anti-inflammatory properties, particularly through its polyphenol content that acts upon several signalling pathways, leading to bone anabolic and antiresorptive effects. Menopause and ageing are the main contributing factors to oxidative stress and inflammation, subsequently leading to bone loss. Quercetin, kaempferol, gallic acid, and ascorbic acid are among the phytochemical components found in honey which exert antioxidant and anti-inflammatory actions by inhibiting activation of the MAPK pathway and NF-κB signalling, respectively. This action can prevent the formation of osteoclasts and favour bone formation by osteoblasts, consequently preserving bone density.[59]

Amaranthus mangostanus L. (AM), also known as "bireum" in Korea, is a plant with a powerful antioxidant activity which is widely cultivated in Asia. *In vitro* studies suggested that AM significantly suppressed RANKL-induced osteoclast differentiation and expression of osteoclast-specific genes, TRAP, cathepsin K, NFATC1, as well as Dc-stamp in RAW 264.7 cells. In addition, AM significantly inhibited ERK, protein kinase B (Akt), and NF-κB signalling pathways in RAW 264.7 cells. Also, AM preserved ovariectomy-induced bone loss in mice. In conclusion, experimental results suggest that AM might be a potential candidate for the treatment of postmenopausal osteoporosis.[60]

Lepidium sativum seeds are traditionally used to accelerate healing of bone fracture apart from its culinary uses. *Lepidium sativum* extract exhibited significant anti-osteoporotic actions by mitigating the decrease in relative bone weight, concurrent with improved longitudinal and perpendicular femur compression strength *in vivo*. The extract also enhanced the serum bone formation biomarker lactate dehydrogenase activity and osteocalcin levels, inhibited exhaustion of superoxide dismutase as well as glutathione peroxidase activities and accumulation of lipid peroxides in bone tissues. Moreover, it can ameliorate the rise in the markers of bone resorption carboxyterminal telopeptide, type I (CTXI) and TRAP and modulate RANKL/OPG expression. Glucosinolates, lignans, coumarins, phenolic acids, and alkaloids are among the metabolites that mediate these anti-osteoporotic effects in a synergistic manner.[61]

15.6 Conclusion

Dietary polyphenols from natural foods have been associated with bone health, which may be, in part, attributable to their antioxidant capacity. Such antioxidant-rich foods may be a strategy for slowing down age-related bone loss and improving bone remodelling. Evidence suggests an association of greater fruit and vegetable intake with decreased fracture risk, greater bone mass density, and decreased bone turnover. However, the mechanisms by which these polyphenols directly modify bone remodelling are not fully understood.

References

1. B. L. Clarke and S. Khosla, Physiology of bone loss, *Radiol. Clin. North Am.*, 2010, **48**, 483–495.
2. Office of the Surgeon General (US), *Bone Health and Osteoporosis: A Report of the Surgeon General*, Office of the Surgeon General, Rockville, MD, 2004.
3. V. Domazetovic, G. Marcucci, T. Iantomasi, M. L. Brandi and M. T. Vincenzini, Oxidative stress in bone remodeling: role of antioxidants, *Clin. Cases Miner. Bone Metab.*, 2017, **14**, 209–216.
4. T. J. de Vries, I. el Bakkali, T. Kamradt, G. Schett, I. D. C. Jansen and P. D'Amelio, What are the peripheral blood determinants for increased osteoclast formation in the various inflammatory diseases associated with bone loss?, *Front. Immunol.*, 2019, **10**, 505.
5. B. F. Boyce, J. Li, L. Xing and Z. Yao, Bone remodeling and the role of TRAF3 in osteoclastic bone resorption, *Front. Immunol.*, 2018, **9**, 2263.
6. T. J. Martin and N. A. Sims, RANKL/OPG; Critical role in bone physiology, *Rev. Endocr. Metab. Disord.*, 2015, **16**, 131–139.
7. D. Black and C. J. Rosen, Postmenopausal osteoporosis, *N. Engl. J. Med.*, 2016, **374**, 254–262.
8. T. D. Rachner, S. Khosla and L. C. Hofbauer, New horizons in osteoporosis, *Lancet*, 2011, **377**, 1276–1287.
9. J. A. Cauley, Public health impact of osteoporosis, *J. Gerontol., Ser. A*, 2013, **68**, 1243–1251.

10. A. Caroli, A. Poli, D. Ricotta, G. Banfi and D. Cocchi, Invited review: Dairy intake and bone health: A viewpoint from the state of the art, *J. Dairy Sci.*, 2011, **94**, 5249–5262.
11. S. Yalin, S. Bagis, G. Polat, N. Dogruer, S. Cenk Aksit and R. Hatungil, *et al.*, Is there a role of free oxygen radicals in primary male osteoporosis?, *Clin. Exp. Rheumatol.*, 2005, **23**, 689–692.
12. S. Basu, K. Michaëlsson, H. Olofsson, S. Johansson and H. Melhus, Association between oxidative stress and bone mineral density, *Biochem. Biophys. Res. Commun.*, 2001, **288**, 275–279.
13. G. Banfi, E. L. Iorio and M. M. Corsi, Oxidative stress, free radicals and bone remodeling, *Clin. Chem. Lab. Med.*, 2008, **46**, 1550–1555.
14. S. Yang, P. Madyastha, S. Bingel, W. Ries and L. Key, A new superoxide-generating oxidase in murine osteoclasts, *J. Biol. Chem.*, 2001, **276**, 5452–5458.
15. A. N. Sontakke and R. S. Tare, A duality in the roles of reactive oxygen species with respect to bone metabolism, *Clin. Chim. Acta*, 2002, **318**, 145–148.
16. I. R. Garrett, B. F. Boyce, R. O. Oreffo, L. Bonewald, J. Poser and G. R. Mundy, Oxygen-derived free radicals stimulate osteoclastic bone resorption in rodent bone *in vitro* and *in vivo*, *J. Clin. Invest.*, 1990, **85**, 632–639.
17. N. Mody, F. Parhami, T. A. Sarafian and L. L. Demer, Oxidative stress modulates osteoblastic differentiation of vascular and bone cells, *J. Free Radicals Biol. Med.*, 2001, **31**, 509–519.
18. X. Bai, D. Lu, J. Bai, H. Zheng, Z. Ke and X. Li, *et al.*, Oxidative stress inhibits osteoblastic differentiation of bone cells by ERK and NF-κB, *Biochem. Biophys. Res. Commun.*, 2004, **314**, 197–207.
19. I. Dreher, N. Schütze, A. Baur, K. Hesse, D. Schneider and J. Köhrle, *et al.*, Selenoproteins are expressed in fetal human osteoblast-like cells, *Biochem. Biophys. Res. Commun.*, 1998, **245**, 101–107.
20. R. H. Liu, Potential synergy of phytochemicals in cancer prevention: mechanism of action, *J. Nutr.*, 2004, **134**, 3479S–3485S.
21. W. C. Willett, Balancing life-style and genomics research for disease prevention, *Science*, 2002, **296**, 695–698.
22. H. Ahmadieh and A. Arabi, Vitamins and bone health: beyond calcium and vitamin D, *Nutr. Rev.*, 2011, **69**, 584–598.
23. I. F. F. Benzie, Evolution of dietary antioxidants, *Comp. Biochem. Physiol., Part A: Mol. Integr. Physiol.*, 2003, **136**, 113–126.
24. B. Clara and A. Anna, Antioxidant effects of natural bioactive compounds, *Curr. Pharm. Des.*, 2009, **15**, 3063–3073.
25. J.-J. Li, J.-J. Li, Z. Huang, Z. Huang, R.-Q. Wang and X.-M. Ma, *et al.*, Fruit and vegetable intake and bone mass in Chinese adolescents, young and postmenopausal women, *Public Health Nutr.*, 2013, **16**, 78–86.
26. S. A. New, C. Bolton-Smith, D. A. Grubb and D. M. Reid, Nutritional influences on bone mineral density: a cross-sectional study in premenopausal women, *Am. J. Clin. Nutr.*, 1997, **65**, 1831–1839.

27. M. Messina, A brief historical overview of the past two decades of soy and isoflavone research, *J. Nutr.*, 2010, **140**, 1350S–1354S.
28. K. Austermann, N. Baecker, P. Stehle and M. Heer, Putative effects of nutritive polyphenols on bone metabolism in vivo – evidence from human studies, *Nutrients*, 2019, **11**, 871–884.
29. C. Bodet, F. Chandad and D. Grenier, Cranberry components inhibit interleukin-6, interleukin-8, and prostaglandin E2 production by lipopolysaccharide-activated gingival fibroblasts, *Eur. J. Oral Sci.*, 2007, **115**, 64–70.
30. Z. Yao, L. Xing, C. Qin, E. M. Schwarz and B. F. Boyce, Osteoclast precursor interaction with bone matrix induces osteoclast formation directly by an interleukin-1-mediated autocrine mechanism, *J. Biol. Chem.*, 2008, **283**, 9917–9924.
31. K. Ozaki, Y. Kawata, S. Amano and S. Hanazawa, Stimulatory effect of curcumin on osteoclast apoptosis, *Biochem. Pharmacol.*, 2000, **59**, 1577–1581.
32. A. C. Bharti, Y. Takada and B. B. Aggarwal, Curcumin (diferuloylmethane) inhibits receptor activator of NF-κB ligand-induced NF-κB activation in osteoclast precursors and suppresses osteoclastogenesis, *J. Immunol.*, 2004, **172**, 5940–5947.
33. A. Scalbert, C. Manach, C. Morand, C. Rémésy and L. Jiménez, Dietary polyphenols and the prevention of diseases, *Crit. Rev. Food Sci. Nutr.*, 2005, **45**, 287–306.
34. O. Brzezińska, Z. Łukasik, J. Makowska and K. Walczak, Role of vitamin C in osteoporosis development and treatment – a literature review, *Nutrients*, 2020, **12**, 2394–2416.
35. S. Minoru, Antioxidant vitamins and carotenoids associated with bone mineral density in postmenopause female subjects: Japanese perspectives, in *Nutrition and Health: Nutrition and Diet in Menopause*, ed. C. Hollins Martin, R. Watson and V. Preedy, Humana Press, Totowa, NJ, 2013, pp. 101–115.
36. E. Shea Mackinnon, *The role of the carotenoid lycopene as an antioxidant to decrease osteoporosis risk in women: clinical and in vitro studies*, PhD thesis, University of Toronto, Canada, 2010.
37. C. J. Macdonald-Clarke and H. M. MacDonald, in *Dietary anthocyanidins and bone health, Nutritional Influences on Bone Health*, ed. P. Burckhardt, B. Dawson-Hughes and C. Weaver, Springer, London, UK, 2013, pp. 177–187.
38. S. Ötles and S. Ozgoz, Health effects of dietary fiber, *Acta Sci. Pol., Technol. Aliment.*, 2014, **13**, 191–202.
39. D. Mudgil and S. Barak, Composition, properties and health benefits of indigestible carbohydrate polymers as dietary fiber: A review, *Int. J. Biol. Macromol.*, 2013, **61**, 1–6.
40. S. A. Abrams, I. J. Griffin, K. M. Hawthorne and K. J. Ellis, Effect of prebiotic supplementation and calcium intake on body mass index, *J. Pediatr.*, 2007, **151**, 293–298.

41. S. A. Abrams, I. J. Griffin and K. M. Hawthorne, Young adolescents who respond to an inulin-type fructan substantially increase total absorbed calcium and daily calcium accretion to the skeleton, *J. Nutr.*, 2007, **137**, 2524S–2526S.
42. T. J. Ashaolu, J. O. Ashaolu and S. A. O. Adeyeye, Fermentation of prebiotics by human colonic microbiota *in vitro* and short-chain fatty acids production: a critical review, *J. Appl. Microbiol.*, 2021, **130**, 677–687.
43. M. J. Ornstrup, T. Harsløf, T. N. Kjær, B. L. Langdahl and S. B. Pedersen, Resveratrol increases bone mineral density and bone alkaline phosphatase in obese men: a randomized placebo-controlled trial, *J. Clin. Endocrinol. Metab.*, 2014, **99**, 4720–4729.
44. J. C. Tou, Evaluating resveratrol as a therapeutic bone agent: preclinical evidence from rat models of osteoporosis, *Ann. N. Y. Acad. Sci.*, 2015, **1348**, 75–85.
45. C. Cai, C. Liu, L. Zhao, H. Liu, W. Li and H. Guan, *et al.*, Effects of taxifolin on osteoclastogenesis *in vitro* and *in vivo*, *Front. Pharmacol.*, 2018, **9**, 1286–1296.
46. H.-T. Huang, T.-L. Cheng, S.-Y. Lin, C.-J. Ho, J. Y. Chyu and R.-S. Yang, *et al.*, Osteoprotective roles of green tea catechins, *Antioxidants*, 2020, **9**, 1136–1161.
47. C.-L. Shen, J. K. Yeh, J. J. Cao, O. L. Tatum, R. Y. Dagda and J.-S. Wang, Green tea polyphenols mitigate bone loss of female rats in a chronic inflammation-induced bone loss model, *J. Nutr. Biochem.*, 2010, **21**, 968–974.
48. N. Seferos, L. Petrokokkinos, A. Kotsiou, G. Rallis and C. Tesseromatis, *Hypericum perforatum* L. treatment restored bone mass, *Stomatologija*, 2016, **18**, 9–13.
49. J. Zhang, O. P. Lazarenko, J. Kang, M. L. Blackburn, M. J. J. Ronis and T. M. Badger, *et al.*, Feeding blueberry diets to young rats dose-dependently inhibits bone resorption through suppression of RANKL in stromal cells, *PLoS One*, 2013, **8**, e70438.
50. L. Devareddy, S. Hooshmand, J. K. Collins, E. A. Lucas, S. C. Chai and B. H. Arjmandi, Blueberry prevents bone loss in ovariectomized rat model of postmenopausal osteoporosis, *J. Nutr. Biochem.*, 2008, **19**, 694–699.
51. C.-L. Shen, J. K. Yeh, J. J. Cao, M.-C. Chyu and J.-S. Wang, Green tea and bone health: Evidence from laboratory studies, *Pharmacol. Res.*, 2011, **64**, 155–161.
52. J.-R. Chen, O. P. Lazarenko, X. Wu, J. Kang, M. L. Blackburn and K. Shankar, *et al.*, Dietary-induced serum phenolic acids promote bone growth *via* p38 MAPK/β-catenin canonical Wnt signaling, *J. Bone Miner. Res.*, 2010, **25**, 2399–2411.
53. M. You, D.-W. Kim, K.-S. Jeong, M.-A. Bang, H.-S. Kim and J. Rhuy, *et al.*, St. John's Wort (*Hypericum perforatum*) stimulates human osteoblastic MG-63 cell proliferation and attenuates trabecular bone loss induced by ovariectomy, *Nutr. Res. Pract.*, 2015, **9**, 459–465.
54. J. Zhang, O. P. Lazarenko, M. L. Blackburn, T. M. Badger, M. J. J. Ronis and J.-R. Chen, Blueberry consumption prevents loss of collagen in bone

matrix and inhibits senescence pathways in osteoblastic cells, *AGE*, 2013, **35**, 807–820.

55. A. Devine, J. M. Hodgson, I. M. Dick and R. L. Prince, Tea drinking is associated with benefits on bone density in older women, *Am. J. Clin. Nutr.*, 2007, **86**, 1243–1247.
56. C.-L. Shen, M.-C. Chyu and J.-S. Wang, Tea and bone health: steps forward in translational nutrition, *Am. J. Clin. Nutr.*, 2013, **98**, 1694S–1699S.
57. X. Zeng, J. Tian, K. Cai, X. Wu, Y. Wang and Y. Zheng, *et al.*, Promoting osteoblast differentiation by the flavanes from Huangshan Maofeng tea is linked to a reduction of oxidative stress, *Phytomedicine*, 2014, **21**, 217–224.
58. X. Peng, J. He, J. Zhao, Y. Wu, X. Shi and L. Du, *et al.*, *Polygonatum sibiricum* polysaccharide promotes osteoblastic differentiation through the ERK/GSK-3β/β-catenin signaling pathway *in vitro*, *Rejuvenation Res.*, 2018, **21**, 44–52.
59. M. A. Kamaruzzaman, K.-Y. Chin and E. S. Mohd, Ramli, A review of potential beneficial effects of honey on bone health, *J. Evidence-Based Complementary Altern. Med.*, 2019, **2019**, e8543618.
60. Y.-H. Jeong, H. J. Hur, A. S. Lee, S. H. Lee and M. J. Sung, *Amaranthus mangostanus* inhibits the differentiation of osteoclasts and prevents ovariectomy-induced bone loss, *J. Evidence-Based Complementary Altern. Med.*, 2020, **2020**, e1927017.
61. H. M. Abdallah, M. A. Farag, M. M. Algandaby, M. Z. Nasrullah, A. B. Abdel-Naim and B. G. Eid, *et al.*, Osteoprotective activity and metabolite fingerprint *via* UPLC/MS and GC/MS of *Lepidium sativum* in ovariectomized rats, *Nutrients*, 2020, **12**, 2075–2095.
62. V. Everts, J. M. Delaissé, W. Korper, D. C. Jansen, W. Tigchelaar-Gutter and P. Saftig, *et al.*, The bone lining cell: its role in cleaning howship's lacunae and initiating bone formation, *J. Bone Miner. Res.*, 2002, **17**, 77–90.
63. L. F. Bonewald and M. L. Johnson, Osteocytes, mechanosensing and Wnt signaling, *Bone*, 2008, **42**, 606–615.
64. S. L. Teitelbaum, Osteoclasts: What do they do and how do they do it?, *Am. J. Pathol.*, 2007, **170**, 427–435.
65. M. J. Ornstrup, T. Harslof, T. N. Kjær, B. L. Langdahl and S. B. Pedersen, Resveratrol increases bone mineral density and bone alkaline phosphatase in obese men: a randomized placebo-controlled trial, *J. Clin. Endocrinol. Metab.*, 2014, **99**, 4720–4729.
66. J. C. Tou, Evaluating resveratrol as a therapeutic bone agent: preclinical evidence from rat models of osteoporosis, *Ann. N. Y. Acad. Sci.*, 2015, **1348**, 75–85.
67. C.-L. Shen, J. K. Yeh, J. J. Cao, O. L. Tatum, R. Y. Dagda and J.-S. Wang, Green tea polyphenols mitigate bone loss of female rats in a chronic inflammation-induced bone loss model, *J. Nutr. Biochem.*, 2010, **21**, 968–974.
68. C.-L. Shen, J. K. Yeh, J. J. Cao, M.-C. Chyu and J.-S. Wang, Green tea and bone health: Evidence from laboratory studies, *Pharmacol. Res.*, 2011, **64**, 155–161.

CHAPTER 16

Antioxidants and Antiaging

MURAT TÜRKOĞLU,*[a] ERKİN PEKMEZCİ[b] AND HAKAN SEVİNÇ[a]

[a] Biota Laboratories, R&D Center, Sancaktepe, Istanbul, Turkey;
[b] Department of Dermatology, Istanbul Medipol University, Beykoz, Istanbul, Turkey
*Email: murat.trkgl@gmail.com

16.1 Introduction

The human body consists of multi-billion cells and the cellular activity during energy production in mitochondria produces a large amount of free radicals with a potential to harm other cells and tissues. However, there are internal antioxidants such as superoxide dismutase, catalase, glutathione peroxidase, and glutathione *S*-transferase to fight against oxidative stress. Therefore, a healthy person can manage the fight against oxidative stress internally. In addition, a well-balanced daily nutrition can provide extra antioxidants such as flavonoids, vitamin C, vitamin E, and minerals (copper, zinc, and selenium). These antioxidants can protect the cells and tissues against further degradation and DNA damage. Fruits and vegetables, including their flowers, seeds, and roots, contain many diverse bioactive molecules such as saccharides, sterols, vitamins, flavonoids, amino acids, alkaloids, and minerals, among others, with potential health benefits. Based on daily consumption, the active molecules from these food sources would interact with human cells to form a response as the upregulation or downregulation of the gene expression in the cell. In return, increased or decreased protein synthesis may occur. The net result of these interactions

Food Chemistry, Function and Analysis No. 39
Dietary Supplements with Antioxidant Activity: Understanding Mechanisms and Potential Health Benefits
Edited by Cesarettin Alasalvar, Fereidoon Shahidi and Chi-Tang Ho

Published by the Royal Society of Chemistry, www.rsc.org

could be an antiaging effect for the skin or prevention of a chronic disease. Fighting against aging has been a very old aim of mankind.[1] How can we protect our bodily functions for a longer time? According to an ancient saying in Ayurveda,[2] it depends on the diet: "*When diet is wrong, medicine is of no use; when diet is correct, medicine is of no need.*" When we scientifically investigate the issue of aging and its relationship with antioxidants, phenolics, and specifically flavonoids, we are considering external dietary factors. However, the human body has a genetic code and specific DNA repair mechanisms which differ among individuals. Therefore, the genetic inheritance, geographical location of a person's settlement, cultural daily activities, and availability of rich and diverse polyphenolic food sources can determine the overall outcome of aging or of a chronic disease. For instance, in cold regions where at least six months are winter-like, higher cancer rates were observed than the warmer and sunny regions of the world.[3,4] In some countries, a high alcohol and tobacco consumption is common and they accelerate aging, cause chronic diseases, and lower the life expectancy.[5] In the Mediterranean region, life expectancy is higher and chronic diseases are lower when compared to Northern Europe and North America.[6] The traditional Mediterranean diet is very rich in flavonoids. Daily consumption of extra virgin olive oil, whole grains, olives, legumes, fresh vegetables and spices, nuts, especially lemon, citrus fruits, fish and other seafood, and a low amount of red meat, cheese, and yoghurt are some of the basic components of this diet.[6–8] In some countries, such as Japan, a high daily consumption of black or green tea also contributes to antiaging and to a longer life expectancy.[9] Chronic collagen loss is one of the most significant indicators of aging, affecting bones, skin, blood vessels, eyes, gums, and joints. The collagen content of the bones decreases from a healthy 30% to a low 15% by the age of 70.[10] This chapter intends to contribute to a better understanding of the effects of antioxidant-rich nutrition on aging mechanisms.

16.2 Mechanism of Aging

16.2.1 Free Radicals – Oxidative Stress and Aging

Free radicals and reactive oxygen species (ROS) are molecules which form when oxygen reacts with other compounds creating unpaired electrons in their external shell.[11,12] An oxygen molecule with paired electrons is stable but when it has an unpaired electron, it is extremely reactive because it seeks electrons from vital components of the cell, thus causing impairment in the structure and function of macromolecules.[12] Although ROS are produced by all aerobic cells and involved in the extraction of energy from organic molecules, in immune defence, and in the signalling processes, they have an inductive role in aging as well as in age-related diseases.[13,14] ROS may be generated by endogenous or exogenous sources. Endogenous sources include nicotinamide adenine dinucleotide phosphate oxidase (NOX), myeloperoxidase, and lipoxygenase (LOX).[15] The main exogenous sources of ROS are environmental pollution, smoking, alcohol consumption, exposure to heavy metals, certain drugs, industrial solvents, and radiation. In addition, smoked meat, over-used oil, and saturated fat all produce copious amounts of free radicals in the body.[16] Regardless of being endogenous or exogenous, ROS may invoke oxidative alterations of the major cellular macromolecules such as carbohydrates, lipids, proteins, and DNA.[15] Antioxidant defence mechanisms protect biological systems from oxidative damage, and this process involves both endogenous and exogenous molecules. Endogenous antioxidants include enzymatic and non-enzymatic ones. The primary antioxidant enzymes are superoxide dismutase, catalase, glutathione peroxidase, and thioredoxin.[17,18] The enzymatic antioxidants have more potent protective effects against oxidative attacks due to their ability to decompose ROS.[18] Non-enzymatic antioxidants are molecules that interact with ROS and terminate the free radical chain reactions. Bilirubin, α-tocopherol, β-carotene, albumin, uric acid, melatonin, and ubiquinone (CoQ10) are the main antioxidants in the blood and solid tissues.[19,20] Exogenous antioxidants include L-ascorbic acid, α-tocopherol, selenium, zinc, and polyphenols.[21] Oxidative stress develops when there is a disproportionation between the generation and the clearance of ROS, compromised neutralization processes, and a failure in repairing the resulting damage.[15] Aging can be described as the progressive loss of tissue and organ function over time.[22] As part of the natural aging process, the production of ROS increases, while the defence mechanism weakens.[12] The oxidative stress theory of aging is established on the structural damage hypothesis that age-associated functional losses are due to the accumulation of oxidative harms caused by ROS.[23] Although the accurate mechanism of oxidative stress-induced aging is still not apparent, it is likely that increased ROS levels lead to cellular senescence. Senescent cells acquire an irreversible senescence-associated secretory phenotype (SASP) causing secretion of soluble factors such as interleukin (IL), chemokines, growth factors, degradative enzymes like matrix metalloproteinases (MMPs), and insoluble proteins/extracellular matrix components. All of these compounds can alter the tissue microenvironment and affect neighbouring epithelial cells in a paracrine fashion to promote tumour progression.[11] The age-associated changes of immune functions have an oxidative and inflammatory stress situation as their basis, and competences of immune cells are reliable markers of biological age. The chronic oxidative stress that appears with age affects all cells and especially those of the nervous, endocrine, and immune systems and the communication between each other. Regarding the central role of the immune system in oxidation–inflammation–aging, the administration of adequate amounts of antioxidants in the diet may improve the immune function, decreasing their oxidative stress, and subsequently extend the longevity of individuals.[24,25]

16.3 Aging Related Changes

Developmental changes that occur during fetal growth and postnatal development continue in a genetic and predetermined manner. After the age of 30, our physical parameters reach a plateau and our bodily functions start to decline slowly. This is called "aging".[26] Many people experience aging at different rates. The greying of hair and hair loss, skin wrinkles, changes in nails and the musculoskeletal system, especially affecting bones and joints, hearing loss, vision problems, and changes in the cardiovascular system are among the most important manifestations of aging.[27]

16.3.1 Changes in Hair

With today's increasing life expectancy, the desire to look young plays a bigger role than ever. The hair care industry has realized this and has become more capable of providing active products to meet the consumer demands. The discovery of pharmacological targets and the development of safe and effective medicines also illustrate the strategies of the pharmaceutical industry for maintaining a healthy and beautiful hair. Aging of hair actually means the aging of the hair follicle and a cumulative damage on the hair strands. Hair aging also manifests itself as a decrease in melanocyte function and, in the case of androgenetic alopecia, the miniaturization of hair follicles. Intrinsic factors of hair aging are associated with inter-individual variations, individual genetics, and epigenetic mechanisms.[28]

16.3.2 Chronic Hair Loss

Androgenetic alopecia (AGA), also known as male pattern baldness, is the most common type of progressive hair loss. AGA is a polygenetic condition that changes among individuals depending on the severity, the age of onset, and the location of hair loss on the scalp. Hair loss in men typically occurs on the frontal region and the vertex without affecting the temporal or occipital regions.[29] Androgen-dependent processes are predominantly due to the effect of dihydrotestosterone on the hair follicle.[30] All researchers agree that the hair follicles gradually get smaller (miniaturization), but whether or not there is a true follicle loss is controversial since aging-induced alopecia and male pattern baldness are similar processes.[31]

16.3.3 Greying of Hair

One of the most prominent signs of aging is the greying of the hair. Premature hair greying is having grey hair before the age of 20 in the Caucasian race and before the age of 30 in the African-American population. This process is the same as the aging related hair greying and this hair disorder is mainly caused by oxidative stress.[32,33] Changes in the sympathetic nervous system with age are more likely to trigger some degenerative conditions such as greying of the hair.[34] It was stated that the prevalence and degree of hair greying as a sign of aging are related to cardiovascular risk factors regardless of the established age.[35] This raises the suspicion that hair greying may indeed serve as an important indicator of pathology associated with general systemic aging.[36] Iron, vitamin D, folate, vitamin B_{12}, selenium, vitamin C, zinc, and riboflavin are some important vitamins and minerals, some of which are strong antioxidants that can be used in the treatment of androgenetic alopecia, alopecia areata, and for the treatment of pigment loss in the hair.[37] Table 16.1 summarizes the active ingredients, their benefits, and suggested daily doses.

16.3.4 Aging of Bones and Joints

Age-related bone loss is largely a result of aging mechanisms that affect the bone cell count and function.[45] Some other factors include hormone deficiencies, reduced physical activity, nutritional deficiencies, glucocorticoid excess, and alcohol consumption. Intrinsic mechanisms include skeletal progenitor cell regeneration and differentiation, more general mechanisms of cell aging such as cellular damage and changes in repair mechanisms that affect the bone matrix. These internal mechanisms are of great importance in terms of age-related changes in bone cell functions and skeletal integrity. By aging, slowing down the body movements is an important factor to keep the motion under control.[46,47] Therefore, it was observed that with aging there was a decrease in the control characteristics of the motor functions and in muscle strength.[48] Osteoporosis is a common aging related metabolic bone disease characterized by increased bone fragility and a decrease in bone strength and mass.[49] It may affect all age groups and genders. However, the most vulnerable groups are the white race, the elderly, and women.[50] It is estimated that at least 53 million people in the US are affected by osteoporosis.[51] Adequate nutrition plays an important role in preventing and treating osteoporosis. The most important micronutrients are calcium and vitamin D. However, the role of collagen which contributes to the elasticity of the bones was neglected for many years. Other micronutrients are magnesium, potassium, vitamin C, and vitamin K. Calcium has been shown to have beneficial effects on bone mineral-mass at all ages.[52]

Osteoarthritis (OA) is the most common form of arthritis affecting synovial joints.[53] It is associated with age and an increasingly important health problem in an older, generally fitter, and more health-conscious society. Except for the spine, three most frequently affected joint areas are the hips, knees, and hands. Although the benefits of exercise in osteoarthritis are clear, today there is not enough evidence on what the optimal exercise dose is and whether exercise can prevent the development of OA in people.[54–56] Table 16.2 summarizes some nutrients to protect the bones and Table 16.3 illustrates some of the active materials to be used to help OA problems.

Table 16.1 Some actives recommended in hair loss and hair greying as antiaging components.

	Active	Function	Suggested daily dose	Ref.
Primary	Vitamin B_7 (biotin)	Catalyses steps in fatty acid, glucose, and amino acid metabolism	30 µg	36, 37
	Vitamin B_2 (riboflavin)	Although rare, its deficiency can cause hair loss	1.4 mg	38
	Folic acid	Synthesis of nucleic acids and in amino acid metabolism	400 µg	37
	Vitamin B_{12} (methylcobalamin)	Synthesis, neurological function, and red blood cell formation	2.4 µg	37
	Methylsulfonyl methane	Hair growth promotion	1–3 g	39
	Vitamin C	Antioxidant and collagen synthesis	>2 g or high concentration	37, 40
	Vitamin D	Synthesized in epidermal keratinocytes and promotes hair growth	As needed	41
	Zinc	Prevents hair loss	50–100 mg	37, 40
Secondary	Copper	Prevents hair loss	3 mg	40
	Iron	Prevents hair loss	100 mg	37
	Selenium	Antioxidant and necessary for normal cellular function	55 µg	37
	Serenoa repens	Antioxidant (β-sitosterol)	100–200 mg	42–44

Table 16.2 Calcium, vitamin D, and other actives for bone health.

	Active	Statement	Suggested daily dose	Ref.
Primary	Calcium	1–3 years	500 mg	52, 54
		4–8 years	800 mg	52, 54
		9 years and above	1300 mg	52, 54
	Vitamin D (D_3)	0–18 years	5 µg	52
		18 years and above	20 µg	52
Secondary	Potassium	Production of bone mass	Low potassium diets increase urinary calcium losses and high potassium diets reduce them	57, 58
	Magnesium	Production of bone mass	1000 mg	40
	Phosphorus	Production of bone mass	1250 mg	40
	Type I collagen (collagen hydrolysate)	Type I collagen represents 30% of the bone mass in a young person	8–12 g	53
	Vitamin C	Production of bone mass	>3 g or high concentration	40
	Vitamin K	Production of bone protein	90 µg (women), 120 µg (man)	59
	Vitamin A (REA)	Important in retarding the aging process	800 µg (women) and 1000 µg (men)	40

Abbreviation: REA, retinol equivalent activity.

Table 16.3 Actives known to have an effect on osteoarthritis symptoms.

	Active	Statement	Suggested daily dosage	Ref.
Primary	Chondroitin sulphate	Supports for joints, ligaments, and tendons	500–1000 mg	40
	Glucosamine sulphate or *N*-acetylglucosamine	Supports for bones, tendons, cartilage, and joint fluid	1500 mg	40
	Methylsulfonylmethane	Supports joint and tissue repair	500–1000 mg	40
Secondary	Bromelain	Reduces inflammation	400 mg	40
	Calcium	Repair of connective tissue	1500–2000 mg	40, 54
	Type II collagen	Undenatured, from chicken sternum	40 mg	60

16.3.5 Vision and Hearing Loss

When people age, the lens and retina in the eye deteriorate. This means that the old eye needs more light than a younger one. The lens of the eye can also grow unevenly and this can cause some glare problems. Older adults may find it difficult to see pastel colours clearly such as pale blue, light green, pale yellow, and pale pink. It becomes difficult to tell the difference between dark brown, navy blue, and black. The aged eye cannot adapt quickly to changes in the brightness of light and sudden changes can be dangerous and may cause accidents and falls. Due to many factors related to the lens or retina, moderate vision loss is common in the elderly.[61] Carotenoids such as lutein and zeaxanthin are the essential components for a healthy vision along with docosahexaenoic acid (DHA) and other antioxidants such as zinc, vitamin C, and vitamin E. Regular use of such a formula may help against vision loss in the elderly.[62,63] On the other hand, hearing loss is a consequence of human aging. It has been argued that dysfunctions in central processing are important factors fostering hearing loss in advanced age. Aging has also well-documented precipitations in the neural structure and function. However, it is unclear how these effects interact with the development of deficiency in hearing ability because normal aging in humans is usually accompanied by hearing loss.[64,65]

16.3.6 Aging Heart and Vascular System

Without any association with any disease, changes occur in the organs of the cardiovascular system with advanced age. The leading cause of death in the elderly is heart diseases. If summarized, diseases such as atherosclerotic coronary heart disease, hypertensive cardiovascular disease, calcific aortic stenosis, pulmonary embolism, and congestive heart failure are some of them.[64] The loss of arterial elasticity is related to aging, the change in the ratio of collagen and elastin in the vascular structure is mainly the cause of artery dilatation, and particularly the vascular stiffening due to loss of flexibility leads to increased pressure in the vessels.[65,66] The beneficial effects of exercise and dietary consumption of omega-3 fatty acids are some of the non-pharmacological interventions on the arterial health.[67,68] Currently, enriched functional foods containing omega-3 fatty acids are available worldwide. Specifically, eicosapentaenoic acid (EPA) and DHA from fatty fish and microalgae and alpha-linolenic acid (α-LnA) from plant-based foods have been shown to slightly lower serum triacylglycerol and raise high-density lipoprotein concentrations. An important source of omega-3 fatty acids is fish such as herring, bluefish, mackerel, sardines, salmon, tuna, trout, and anchovy. The major plant sources of α-LnA are flaxseed, chia seeds, canola and soybean oil, walnuts, and avocados.[69] The Mediterranean diet has been shown to reduce cardiovascular morbidity and mortality in both primary and secondary prevention. Other dietary patterns that have been shown to provide benefits in specific medical conditions include a low-fat diet for individuals at high cardiovascular risk and low-carbohydrate diets for overweight people and metabolic syndrome.[70] Table 16.4 shows some of the recommended actives.

Table 16.4 Actives known to have an effect on the health of the cardiovascular system.

	Active	Statement	Suggested daily dosage	Ref.
Primary	Coenzyme Q10	Antioxidant and increases oxygenation	100–300 mg	40, 70
	Omega-3 (PUFA)	DHA and EPA	>2 g	70
	Dietary fibre	Oat flakes, bran, and whole grain bread	4 g of β-glucan and 3 g of soluble dietary fibre from oats	70
Secondary	Calcium	For cardiac muscle	1500 mg	40
	Magnesium	For cardiac muscle	350 mg	40, 70
	Potassium	Needed for electrolyte balance	100 mg	40
	Type I collagen (collagen hydrolysate)	Type I collagen represents 25% of the total body protein and 80% of connective tissue in humans	8–12 g	53
	L-Carnitine	High efficiency fatty acid processing and lower TAG levels in blood	1–3 g	40

Abbreviations: DHA, docosahexaenoic acid; EPA, eicosapentaenoic acid; PUFA, polyunsaturated fatty acids; TAG, triacylglycerols.

16.4 Nutrition and Antiaging

16.4.1 Flavonoids

Flavonoids are a large family of polyphenolic plant compounds. The physicochemical properties of flavonoids influence their metabolic fate. Flavonoids have been shown to exhibit antiinflammatory, antithrombogenic, antidiabetic, anticancer, antiaging, and neuroprotective activities through different mechanisms.[71] They are classified into twelve major subclasses based on their chemical structures, six of which are anthocyanidins, flavan-3-ols, flavonols, flavones, flavanones, and isoflavones. Some specific examples of flavonoids include isorhamnetin, kaempferol, myricetin, quercetin, apigenin, luteolin, baicalein, chrysin, eriodictyol, hesperetin, naringenin, and proanthocyanidins. Figure 16.1 shows the chemical structures of flavonoid derivatives of the flavan ring. Bioavailability of these polyphenols *in vivo* is a major determinant in their ability to exert biological activities relevant to human health such as antiaging.[72] The biological effects of flavonoids appear to be related to their ability to modulate a number of cell-signalling cascades. Therefore, we can suggest a hypothesis that if there is a high enough concentration of a specific flavonoid on the site of action such as the basal cells in the epidermis or the fibroblasts in the dermis, the aging process can be slowed down and perhaps be reversed.

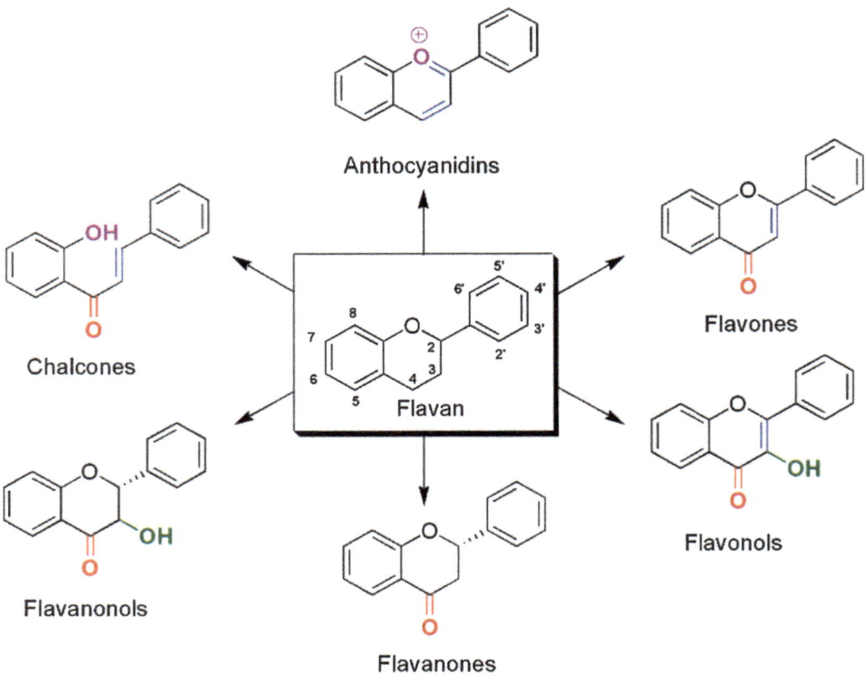

Figure 16.1 Chemical structures of flavonoids found in plant-based products.

16.5 Specific Examples of Food Components With Antiaging Effects

16.5.1 Resveratrol

The major sources of antioxidant phenolic compounds in daily life are mostly coffee, tea, berries, grapes, certain nuts, and unrefined vegetable oils such as extra virgin olive oil. For instance, red grape skin contains a special antioxidant, resveratrol, in the range of 0.5–10 ppm.[73] Resveratrol is one of the activators of deacetylases known as sirtuins (SIRT1-7) with a potential to delay aging and age-related diseases.[74] As a natural molecule, *trans*-resveratrol (Figure 16.2) was claimed to be one of the most potent of these natural activators[75] and of antiaging molecules.

16.5.2 Olive Oil

A daily intake of 0.5–2.0 g of phenolics may come from food sources.[76] With a regular olive oil consumption, significant increases in plasma antioxidant capacity and antioxidant enzyme activity were observed.[77] As for the total phenolic content, virgin olive oils from different sources were reported to contain an average of 232 mg kg^{-1} phenolic compounds.[78] Among the specific chemicals that are found in olive oil are lutein, tyrosol, hydroxytyrosol, squalene, and oleuropein.[79] It was reported that long-term dietary virgin olive oil which formed 10% of the diet with a polyphenol content of 6 mg per kg per day reversed age related brain dysfunctions in mice.[80] Squalene can be found at high concentrations in unrefined olive oils between 0.2 and 7.5 g kg^{-1}

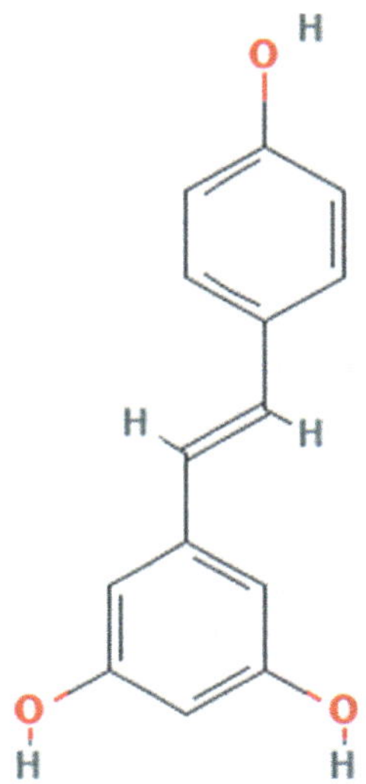

Figure 16.2 Chemical structure of resveratrol found in red grape skin.

which protects the skin from lipid peroxidation caused by ultra-violet (UV) light exposure. Squalene is a good moisturizer for the skin and fights against manifestations of skin aging such as wrinkles.[81]

16.5.3 Tea

Skin aging which is often called "premature aging of the skin" is almost entirely due to UV radiation exposure from the sun (290–400 nm).[82] This is also known as photo-aging. Otherwise, chronic aging of the skin is a much slower process, showing significant differences based on genetic disposition and skin pigmentation level of humans. One can fight against aging by increasing daily antioxidant intake. Brewed tea (*Camellia sinensis*) is one of the most widely available drinks in the world, and its antioxidant polyphenols are extremely important contributors to human health. The antioxidant effect of tea flavonoids protects against skin damage and skin-aging. Therefore, tea extracts either topically applied to the skin[83] or simply drinking tea are both effective antioxidant-based interventions against aging. In a recent paper,[84] in cell cultures, green tea extract caused an upregulation of *hyaluronan synthase-2* gene and a significant downregulation of *matrix metalloproteinase-9*. Hence, green tea extract was found to have a substantial anabolic effect on hyaluronic acid, collagen, and elastin in the skin. During the fermentation, leading to the production of black tea, smaller catechins are influenced by polyphenol oxidase which leads to the formation of orange-red coloured dimeric theaflavins and dark-brown polymeric pigment thearubigins. Partially fermented oolong tea or unfermented green tea contains epicatechin, epigallocatechin, epicatechin gallate, and epigallocatechin gallate as the basic polyphenols of tea leaves. Tea leaves also contain caffeine, L-theanine, myricetin, quercetin, and kaempferol as examples of alkaloids, amino acids, and flavonols, respectively. They all contribute to the antiaging effect and the health benefits of tea.[85] Figure 16.3 shows the molecular structure of epigallocatechin gallate which is one of the most important polyphenols of the tea plant.

16.5.4 Apple

Apple (*Malus domestica*) skin components such as ursolic acid and pectin were reported to protect muscle strength and to prevent wrinkle formation in the human skin. Ursolic acid is a lipophilic pentacyclic triterpenoid that is found in the waxy skin of apples. It inhibits muscle atrophy and sarcopenia due to advanced age and stimulates muscle growth in non-obese mice.[86] Apple pectin oligogalacturonides contribute to wrinkle prevention in the skin.[87] Life extending effects of apple extracts were studied in fruit flies and in yeast cells. When a whole apple extract, containing the flesh, skin, and polyphenolic fractions, was used in mutant *Saccharomyces cerevisiae* which shows premature aging, the life span of the yeast was increased by 40%. In the case of fruit flies, when apple polyphenols were part of the diet, the

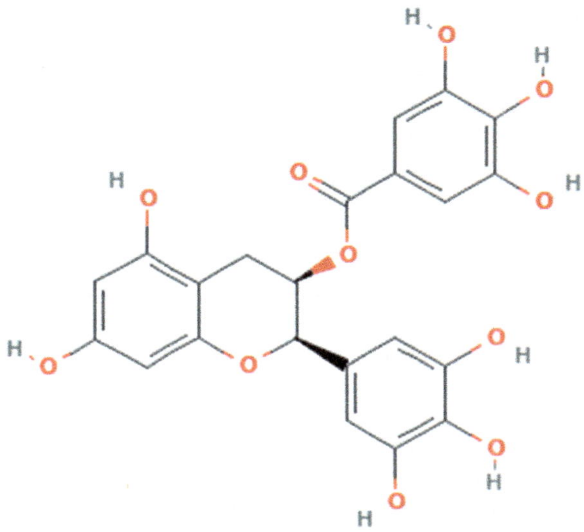

Figure 16.3 Chemical structure of epigallocatechin gallate (EGCG) found in green tea.

mean lifespan was extended by 10%. The antiaging activity of apple polyphenols was mediated by their interaction with the gene expression of superoxide dismutase, catalase, and methuselah.[88–90]

16.5.5 Berries

When it comes to the antiaging effect, berries are a very important group of fruits which mainly grow in temperate climate zones such as northern Europe. Some of the better known berries are strawberries, blueberries, raspberries, lingonberries, cranberries, red currant, and gooseberries. They are rich sources of phenolic substances and anthocyanins which are responsible for their red, purple, and blue colours.[91] In berries, the predominant flavonol is quercetin. In red currant, the phenolic compounds such as *trans*-resveratrol, *p*-coumaric acid, and cinnamic acid are abundant.[92] It seems that the specific antiaging effect of berries mainly targets the brain and memory. Strawberries and blackberries were reported to enhance cognitive and motor behaviour in aged rats. In humans, it was reported that blueberry juice supplementation in elderly adults (600 mL per day, equivalent to 0.6 g anthocyanins) improved memory after 12 weeks of regular use. The results indicated that several aspects of learning and memory, including list recall, paired associate learning, and intrusion errors, were improved in the juice-supplemented subjects. Antiaging effects of three different berry extracts were shown in cell cultures against cataract, alopecia, skin whitening, and weight loss. Particularly, black raspberry extract was found to be superior.[93]

16.6 Conclusion

Antioxidants have a strong effect on antiaging by preventing oxidative stress at the cellular level and downregulating the expression of genes related to inflammatory pathways. Together with chronic inflammation, oxidative stress is related to several chronic diseases, especially sarcopenia and frailty in the elderly population. Biomarkers of oxidative stress and inflammatory factors can be a useful diagnostic tool to determine the extent of damage in aging. Antioxidant supplements, such as resveratrol and other nutritional compounds, together with moderate aerobic exercise and a high quality daily sleep may slow down the detrimental effects of aging. The best way of utilizing antioxidants is regular consumption of a plant-based diet. In addition, adopting a Mediterranean diet is a plus. Coffee, tea, olive oil, berries, and other fruits seem to contribute to the antiaging approach not only for a younger skin, but also for better cognitive skills. Lastly, hydrolysed-collagen intake and light physical activity such as walking would contribute to healthier joints, younger brain, sharper eyesight, and smoother skin which are some of the basic indicators of antiaging.

References

1. S. J. Olshansky, L. Hayflick and B. A. Carnes, No truth to the fountain of youth, *Sci. Am.*, 2002, **286**, 92–95.
2. K. P. Sanjoy, Food-based interventions for cancer management: an ayurvedic perspective, in *Ayurvedic Science of Food and Nutrition*, ed. S. Rastogi, Springer Science & Business Media, New York, NY, 2014, pp. 81–105.
3. A. Sharma, H. K. Verma, S. Joshi, M. S. Panwar and C. C. Mandal, A link between cold environment and cancer, *Tumour Biol.*, 2015, **36**, 5953–5964.
4. E. M. John, D. M. Dreon, J. Koo and G. G. Schwartz, Residential sunlight exposure is associated with a decreased risk of prostate cancer, *J. Steroid Biochem. Mol. Biol.*, 2004, **89**, 549–552.
5. P. Boffetta and M. Hashibe, Alcohol and cancer, *Lancet Oncol.*, 2006, **7**, 149–156.
6. A. Giacosa, R. Barale, L. Bavaresco, P. Gatenby, V. Gerbi and J. Janssens, *et al.*, Cancer prevention in Europe: the Mediterranean diet as a protective choice, *Eur. J. Cancer Prev.*, 2013, **22**, 90–95.
7. V. Dilis and A. Trichopoulou, Mediterranean diet and olive oil consumption-estimations of daily intake of antioxidants from virgin olive oil and olives, in *Olive Oil: Minor Constituents and Health*, ed. D. Boskou, CRC Press, Boca Raton, FL, 2009, pp. 201–210.
8. D. Trichopoulos and A. Trichopoulou, Traditional Mediterranean diet and health, in *Olive Oil: Minor Constituents and Health*, ed. D. Boskou, CRC Press, Boca Raton, FL, 2009, pp. 7–10.

9. S. Kuriyama, T. Shimazu, K. Ohmori, N. Kikuchi, N. Nakaya and Y. Nishino, *et al.*, Green tea consumption and mortality due to cardiovascular disease, cancer, and all causes in Japan: the Ohsaki study, *JAMA*, 2006, **296**, 1255–1265.
10. A. J. Bailey, T. J. Sims, E. N. Ebbesen, J. P. Mansell, J. S. Thomsen and L. Mosekilde, Age-related changes in the biochemical properties of human cancellous bone collagen: relationship to bone strength, *Calcif. Tissue Int.*, 1999, **65**, 203–210.
11. A. Chandrasekaran, M. D. P. S. Idelchik and J. A. Melendez, Redox control of senescence and age-related disease, *Redox Biol.*, 2017, **11**, 91–102.
12. L. Baumann and I. J. Allemann, Antioxidants, in *Cosmetic Dermatology*, ed. L. Baumann, McGraw Hill, New York, NY, 2nd edn, 2009, pp. 292–311.
13. K. Venkataraman, S. Khurana and T. C. Tai, Oxidative stress in aging-matters of the heart and mind, *Int. J. Mol. Sci.*, 2013, **14**, 17897–17925.
14. M. Genestra, Oxyl radicals, redox-sensitive signalling cascades and antioxidants, *Cell. Signalling*, 2007, **19**, 1807–1819.
15. D. Salisbury and U. Bronas, Reactive oxygen and nitrogen species: impact on endothelial dysfunction, *Nurs. Res.*, 2015, **64**, 53–66.
16. A. Phaniendra, D. B. Jestadi and L. Periyasamy, Free radicals: properties, sources, targets, and their implication in various diseases, *Indian J. Clin. Biochem.*, 2015, **30**, 11–26.
17. E. Birben, U. M. Sahiner, C. Sackesen, S. Erzurum and O. Kalayci, Oxidative stress and antioxidant defense, *World Allergy Organ J.*, 2012, **5**, 9–19.
18. M. Christofidou-Solomidou and V. R. Muzykantov, Antioxidant strategies in respiratory medicine, *Treat. Respir. Med.*, 2006, **5**, 47–78.
19. J. Q. Wu, T. R Kosten and X. Y. Zhang, Free radicals, antioxidant defense systems, and schizophrenia, *Prog. Neuro-Psychopharmacol. Biol. Psychiatry*, 2013, **46**, 200–206.
20. L. He, T. He, S. Farrar, L. Ji, T. Liu and X. Ma, Antioxidants and cellular homeostasis, *Cell. Physiol. Biochem.*, 2017, **44**, 532–553.
21. A. M. Pisoschi and A. Pop, The role of antioxidants in the chemistry of oxidative stress: a review, *Eur. J. Med. Chem.*, 2015, **97**, 55–74.
22. T. Flatt, A new definition of aging?, *Front. Genet.*, 2012, **3**, 148.
23. K. B. Beckman and B. N. Ames, The free radical theory of aging matures, *Physiol. Rev.*, 1998, **78**, 547–581.
24. M. de la Fuente and J. Miquel, An update of the oxidation-inflammation theory of aging: the involvement of the immune system in oxi-inflamm-aging, *Curr. Pharm. Des.*, 2009, **15**, 3003–3026.
25. I. Liguori, G. Russo, F. Curcio, G. Bulli, L. Aran and D. Della-Morte, *et al.*, Oxidative stress, aging, and diseases, *Clin. Interventions Aging*, 2018, **13**, 757–772.
26. L. Ferrucci, M. Gonzalez-Freire, E. Fabbri, E. Simonsick, T. Tanaka and Z. Moore, *et al.*, Measuring biological aging in humans: a quest, *Aging Cell*, 2020, **19**, e13080.

27. H. N. Chalise, Aging: basic concept, *Am. J. Biomed. Sci. Res.*, 2019, **1**, 8–10.
28. R. M. Trüeb, Aging of hair, *J. Cosmet. Dermatol.*, 2005, **4**, 60–72.
29. F. Lolli, F. Pallotti, A. Rossi, M. C. Fortuna, G. Caro and A. Lenzi, *et al.*, Androgenetic alopecia: a review, *Endocrine*, 2017, **57**, 9–17.
30. R. M. Trüeb, Molecular mechanisms of androgenetic alopecia, *Exp. Gerontol.*, 2002, **37**, 981–990.
31. A. M. Kligman, The comparative histopathology of male-pattern baldness and senescent baldness, *Clin. Dermatol.*, 1988, **6**, 108–118.
32. A. B. Kumar, H. Shamim and U. Nagaraju, Premature greying of hair: review with updates, *Int. J. Trichol.*, 2018, **10**, 198–203.
33. M. D. Tollenaere, E. Chapuis, P. Auriol, D. Auriol, A. Scandolera and R. Reynaud, Global repigmentation strategy of grey hair follicles by targeting oxidative stress and stem cells protection, *Appl. Sci.*, 2021, **11**, 1533.
34. A. R. Mendelsohn and J. W. Larrick, The danger of being too sympathetic: norepinephrine in Alzheimer's disease and greying of hair, *Rejuvenation Res.*, 2020, **23**, 68–72.
35. A. A. el Faramawy, I. S. Hanna, R. M. Darweesh, A. S. Ismail and H. I. Kandil, The degree of hair greying as an independent risk marker for coronary artery disease, a CT coronary angiography study, *Egypt. Heart J.*, 2018, **70**, 15–19.
36. J. D. O'Sullivan, C. Nicu, M. Picard, J. Chéret, B. Bedogni and D. J. Tobin, *et al.*, The biology of human hair greying, *Biol. Rev. Cambridge Philos. Soc.*, 2021, **96**, 107–128.
37. H. M. Almohanna, A. A. Ahmed, J. P. Tsatalis and A. Tosti, The role of vitamins and minerals in hair loss: a review, *Dermatol. Ther.*, 2019, **9**, 51–70.
38. N. Mahabadi, A. Bhusal and S. W. Banks, *Riboflavin Deficiency*, StatPearls Publishing, Treasure Island, FL, 2021.
39. N. Muizzuddin and R. Benjamin, Beneficial effects of a sulfur-containing supplement on hair and nail condition, *Nat. Med. J.*, 2019, **11**, 11.
40. P. A. Balch, *Prescription for Nutritional Healing*, Avery Trade, New York, NY, 4th edn, 2006.
41. V. Vegesna, J. O'Kelly, M. Uskokovic, J. Said, N. Lemp and T. Saitoh, *et al.*, Vitamin D_3 analogs stimulate hair growth in nude mice, *Endocrinology*, 2002, **143**, 4389–4396.
42. J. I. Kalwat, The use of *Serenoa repens* (saw palmetto) in hair care products, *Biomed. J. Sci. Tech. Res.*, 2019, **13**, 9725–9728.
43. M. Pezza, V. Carlomagno and G. Casucci, Telogen effluvium treated with *Serenoa repens* supplement, *Senses Sci.*, 2014, **1**, 41–43.
44. M. Narda, S. Aladren, E. Cestone and V. Nobile, Efficacy and safety of a food supplement containing L-cystine, Serenoa repens extract and biotin for hair loss in healthy males and females: A prospective, randomized, double-blinded, controlled clinical trial, *J. Cosmo. Trichol.*, 2017, **3**, 127.

45. S. C. Manolagas and A. M. Parfitt, What old means to bone, *Trends Endocrinol. Metab.*, 2010, **21**, 369–374.
46. P. J. Marie, Bone cell senescence: mechanisms and perspectives, *J. Bone Miner. Res.*, 2014, **29**, 1311–1321.
47. P. S. Pohl, C. J. Winstein and B. E. Fisher, The locus of age-related movement slowing: sensory processing in continuous goal-directed aiming, *J. Gerontol*, 1996, **51**, P94–P102.
48. Z. Erim, M. F. Beg, D. T. Burke and C. J. de Luca, Effects of aging on motor-unit control properties, *J. Neurophysiol.*, 1999, **82**, 2081–2091.
49. K. N. Tu, J. D. Lie, C. K. Wan, M. Cameron, A. G. Austel and J. K. Nguyen, *et al.*, Osteoporosis: a review of treatment options, *Pharmacol. Ther.*, 2018, **43**, 92–104.
50. T. Sözen, L. Özışık and N. Ç. Başaran, An overview and management of osteoporosis, *Eur. J. Rheumatol. Inflammation*, 2017, **4**, 46–56.
51. N. C. Wright, A. C. Looker, K. G. Saag, J. R. Curtis, E. S. Delzell and S. Randall, *et al.*, The recent prevalence of osteoporosis and low bone mass in the United States based on bone mineral density at the femoral neck or lumbar spine, *J. Bone Miner. Res.*, 2014, **29**, 2520–2526.
52. J. W. Nieves, Osteoporosis: the role of micronutrients, *Am. J. Clin. Nutr.*, 2005, **81**, S1232–S1239.
53. B. Kwatra, Collagen supplementation: therapy for the prevention and treatment of osteoporosis and osteoarthritis: a review, *World J. Pharm. Pharm. Sci.*, 2020, **9**, 589–604.
54. C. M. Weaver, The growing years and prevention of osteoporosis in later life, *Proc. Nutr. Soc.*, 2000, **59**, 303–306.
55. M. V. Hurley, The role of muscle weakness in the pathogenesis of osteoarthritis, *Rheum. Dis. Clin. North Am.*, 1999, **25**, 283–298.
56. J. Cushnaghan and P. Dieppe, Study of 500 patients with limb joint osteoarthritis. I. analysis by age, sex, and distribution of symptomatic joint sites, *Ann. Rheum. Dis.*, 1991, **50**, 8–13.
57. D. E. Sellmeyer, M. Schloetter and A. Sebastian, Potassium citrate prevents increased urine calcium excretion and bone resorption induced by a high sodium chloride diet, *J. Clin. Endocrinol. Metab.*, 2002, **87**, 2008–2012.
58. S. Jehle, A. Zanetti, J. Muser, H. N. Hulter and R. Krapf, Partial neutralization of the acidogenic Western diet with potassium citrate increases bone mass in postmenopausal women with osteopenia, *J. Am. Soc. Nephrol.*, 2006, **17**, 3213–3222.
59. P. Weber, Vitamin K and bone health, *Nutrition*, 2001, **17**, 880–887.
60. F. Bakilan, O. Armagan, M. Ozgen, F. Tascioglu, O. Bolluk and O. Alatas, Effects of native type II collagen treatment on knee osteoarthritis: a randomized controlled trial, *Eurasian J. Med.*, 2016, **48**, 95–101.
61. E. S. Abdel-Aal, H. Akhtar, K. Zaheer and R. Ali, Dietary sources of lutein and zeaxanthin carotenoids and their role in eye health, *Nutrients*, 2013, **5**, 1169–1185.

62. H. E. Bartlett and F. Eperjesi, Effect of lutein and antioxidant dietary supplementation on contrast sensitivity in age-related macular disease: a randomized controlled trial, *Eur. J. Clin. Nutr.*, 2007, **61**, 1121–1127.
63. V. Cardin, Effects of aging and adult-onset hearing loss on cortical auditory regions, *Front. Neurosci.*, 2016, **10**, 199.
64. L. E. Humes, J. R. Dubno, S. Gordon-Salant, J. J. Lister, A. T. Cacace and K. J. Cruickshanks, *et al.*, Central presbycusis: a review and evaluation of the evidence, *J. Am. Acad. Audiol.*, 2012, **23**, 635–666.
65. R. Stewart and A. Wingfield, Hearing loss and cognitive effort in older adults' report accuracy for verbal materials, *J. Am. Acad. Audiol.*, 2009, **20**, 147–154.
66. C. F. Jackson and N. K. Wenger, Cardiovascular disease in the elderly, *Rev. Esp. Cardiol.*, 2011, **64**, 697–712.
67. A. Avolio, D. Jones and M. Tafazzoli-Shadpour, Quantification of alterations in structure and function of elastin in the arterial media, *Hypertension*, 1998, **32**, 170–175.
68. A. M. Dart and B. A. Kingwell, Pulse pressure-a review of mechanisms and clinical relevance, *J. Am. Coll. Cardiol.*, 2001, **37**, 975–984.
69. A. Jędrusek-Golińska, D. Górecka, M. Buchowski, K. Wieczorowska-Tobis, A. Gramza-Michałowska and K. Szymandera-Buszka, Recent progress in the use of functional foods for older adults: a narrative review, *Compr. Rev. Food Sci. Food Saf.*, 2020, **19**, 835–856.
70. S. Eilat-Adar, T. Sinai, C. Yosefy and Y. Henkin, Nutritional recommendations for cardiovascular disease prevention, *Nutrients*, 2013, **5**, 3646–3683.
71. T. Y. Wang, Q. Li and K. S. Bi, Bioactive flavonoids in medicinal plants: Structure, activity and biological fate, *Asian J. Pharm. Sci.*, 2018, **13**, 12–23.
72. A. Mouradov and G. Spangenberg, Flavonoids: a metabolic network mediating plants adaptation to their real estate, *Front. Plant Sci.*, 2014, **5**, 620.
73. E. Celotti, R. Ferrarini, R. Zironi and L. S. Conte, Resveratrol content of some wines obtained from dried Valpolicella grapes: Recioto and Amarone, *J. Chromatogr. A*, 1996, **730**, 47–52.
74. K. M. Kasiotis, H. Pratsinis, D. Kletsas and S. A. Haroutounian, Resveratrol and related stilbenes: their anti-aging and anti-angiogenic properties, *Food Chem. Toxicol.*, 2013, **61**, 112–120.
75. B. P. Hubbard and D. A. Sinclair, Small molecule SIRT1 activators for the treatment of aging and age-related diseases, *Trends Pharmacol. Sci.*, 2014, **35**, 146–154.
76. V. G. Corrêa, C. Tureck, G. Locateli, R. M. Peralta and E. A. Koehnlein, Estimate of consumption of phenolic compounds by Brazilian population, *Rev. Nutr.*, 2015, **28**, 185–196.
77. R. Zamora-Ros, J. A. Rothwell, A. Scalbert, V. Knaze, I. Romieu and N. Slimani, *et al.*, Dietary intakes and food sources of phenolic acids in

the European Prospective Investigation into Cancer and Nutrition (EPIC) study, *Br. J. Nutr.*, 2013, **110**, 1500–1511.

78. R. W. Owen, W. Mier, A. Giacosa, W. E. Hull, B. Spiegelhalder and H. Bartsch, Phenolic compounds and squalene in olive oils: the concentration and antioxidant potential of total phenols, simple phenols, secoiridoids, lignans and squalene, *Food Chem. Toxicol.*, 2000, **38**, 647–659.
79. S. Cicerale, X. A. Conlan, A. J. Sinclair and R. S. Keast, Chemistry and health of olive oil phenolics, *Crit. Rev. Food Sci. Nutr.*, 2008, **49**, 218–236.
80. V. Pitozzi, M. Jacomelli, D. Catelan, M. Servili, A. Taticchi and A. Biggeri, *et al.*, Long-term dietary extra-virgin olive oil rich in polyphenols reverses age-related dysfunctions in motor coordination and contextual memory in mice: role of oxidative stress, *Rejuvenation Res.*, 2012, **15**, 601–612.
81. H. Masaki, Role of antioxidants in the skin: anti-aging effects, *J. Dermatol. Sci.*, 2010, **58**, 85–90.
82. M. Türkoğlu and N. Cigirgil, Evaluation of black tea gel and its protection potential against UV, *Int. J. Cosmet. Sci.*, 2007, **29**, 437–442.
83. U. Heinrich, C. E. Moore, S. De Spirt, H. Tronnier and W. Stahl, Green tea polyphenols provide photoprotection, increase microcirculation, and modulate skin properties of women, *J. Nutr.*, 2011, **141**, 1202–1208.
84. M. Türkoğlu, E. Pekmezci and S. Kılıç, Green tea extract exerts anabolic effects on extracellular matrix of the skin, *Indian J. Pharm. Sci.*, 2020, **82**, 368–373.
85. M. Türkoğlu, T. Uğurlu, G. Gedik, A. M. Yılmaz and A. S. Yalçin, *In vivo* evaluation of black and green tea dermal products against UV radiation, *Drug Discoveries Ther.*, 2010, **4**, 362–367.
86. S. D. Kunkel, C. J. Elmore, K. S. Bongers, S. M. Ebert, D. K. Fox and M. C. Dyle, *et al.*, Ursolic acid increases skeletal muscle and brown fat and decreases diet-induced obesity, glucose intolerance and fatty liver disease, *PLoS One*, 2012, 7, e39332.
87. C. Lebreton-Decoster, P. Rousselle, C. Laperdrix, C. Lubrano, J. R. Robin and B. Coulomb, Oligogalacturonides improve tissue organization of *in vitro* reconstructed skin, *Int. J. Cosmet. Sci.*, 2011, **33**, 455–461.
88. C. Peng, H. Y. Chan, Y. Huang, H. Yu and Z. Y. Chen, Apple polyphenols extend the mean lifespan of *Drosophila melanogaster*, *J. Agric. Food Chem.*, 2011, **59**, 2097–2106.
89. V. Palermo, F. Mattivi, R. Silvestri, G. La Regina, C. Falcone and C. Mazzoni, Apple can act as anti-aging on yeast cells, *Oxid. Med. Cell. Longevity*, 2012, 491759.
90. J. A. Joseph, B. Shukitt-Hale and L. M. Willis, Grape juice, berries, and walnuts affect brain aging and behaviour, *J. Nutr.*, 2009, **139**, S1813–S1817.
91. S. Ehala, M. Vaher and M. Kaljurand, Characterization of phenolic profiles of Northern European berries by capillary electrophoresis and determination of their antioxidant activity, *J. Agric. Food Chem.*, 2005, **53**, 6484–6490.

92. R. Krikorian, M. D. Shidler, T. A. Nash, W. Kalt, M. R. Vinqvist-Tymchuk and B. Shukitt-Hale, *et al.*, Blueberry supplementation improves memory in older adults, *J. Agric. Food Chem.*, 2010, **58**, 3996–4000.
93. K. Dan, A. Takada, Y. Kanaho, Y. Kusumi and H. Banno, Anti-aging effects of black raspberry extract on cataract, alopecia, skin whitening, and weight loss, *Funct. Foods Health Dis.*, 2018, **8**, 17–34.

Subject Index